# Biology, Management, and Conservation of Lampreys in North America

# Biology, Management, and Conservation of Lampreys in North America

*Edited by*

LARRY R. BROWN
*U.S. Geological Survey, California Water Science Center*
*Placer Hall, 6000 J Street, Sacramento, California 95819, USA*

SHAWN D. CHASE
*Sonoma County Water Agency*
*Post Office Box 11628, Santa Rosa, California 95406, USA*

MATTHEW G. MESA
*U.S. Geological Survey, Western Fisheries Research Center*
*Columbia River Research Laboratory*
*5501 Cook Underwood Road, Cook, Washington 98605, USA*

RICHARD J. BEAMISH
*Pacific Biological Station, Fisheries and Oceans Canada*
*Nanaimo, British Columbia V9R 5K6, Canada*

PETER B. MOYLE
*Department of Wildlife, Fish, and Conservation Biology, University of California*
*One Shields Ave, Davis, California 95616, USA*

**American Fisheries Society Symposium 72**

Proceedings of the Symposium
"Biology and Conservation of Lampreys in North America"
Held in San Francisco, California, USA
6 September 2007

American Fisheries Society
Bethesda, Maryland, USA
2009

Suggested citation formats are

*Entire book*

Brown, L. R., S. D. Chase, M. G. Mesa, R. J. Beamish, and P. B. Moyle, editors. 2009. Biology, management, and conservation of lampreys in North America. American Fisheries Society, Symposium 72, Bethesda, Maryland.

*Chapter in book*

Sower, S. A., M. Freamat, and S. I. Kavanaugh. 2009. Insight from lamprey genomics: brain and pituitary reproductive hormones of lampreys. Pages 57–70 *in* L. R. Brown, S. D. Chase, M. G. Mesa, R. J. Beamish, and P. B. Moyle, editors. Biology, management, and conservation of lampreys in North America. American Fisheries Society, Symposium 72, Bethesda, Maryland.



Printed in the United States of America on acid-free paper.

Library of Congress Control Number 2009913081
ISBN 978-1-934874-13-4
ISSN 0892-2284

American Fisheries Society Web site: *www.fisheries.org*
American Fisheries Society
5410 Grosvenor Lane, Suite 110
Bethesda, Maryland 20814
USA

# Contents

# Preface

Lampreys have the reputation of being ugly, slimy bloodsuckers that wantonly kill fish we like to eat, such as salmon. With the exception of slimy, none of this characterization is true for lampreys. Instead, they are intriguing organisms that filter-feed on benthic algae, microbes, and other organic material as juveniles, with only a few species consuming flesh and blood as adults. All are beautifully adapted for their lifestyles. Like their most distant ancestors, the ostracoderms, they spend most of their lives sucking up organic matter through jawless mouths. The jawed bony fishes eventually drove most of the jawless ostracoderms to extinction; however, the lamprey lineage persisted by adapting to prey on the very fishes that displaced their ancestors. In the process, they developed the same lifestyle that we admire in salmon—freely moving between marine and freshwater environments and being very successful in both. In fact, lampreys are an indicator of freshwater habitat suitable for salmon. Like salmon, adults build nests (redds, if they were salmon), with most dying after spawning. The young (ammocoetes) rear in the backwaters of streams and are important parts of stream ecosystems, providing food for predators from insects to fish to birds and recycling organic waste. The anadromous forms transform into big-eyed silvery "smolts" (macropthalmia), which migrate to sea (or a large lake) and attack fish larger than themselves. A few species always kill their prey, but for other species, many of their prey survive attacks, some multiple times! The large migratory lampreys are themselves important prey for sea lions and other predators, especially when lampreys concentrate in estuaries on their way upstream to spawn. For Native Americans, these fat, migrating lampreys were historically a highly valued and important food source. Lampreys are still valued by Native Americans as food and for their cultural importance. Lampreys are also part of the cuisine of some European countries (e.g., Portugal). Amazingly, most lamprey species have lost the adult predatory stage of the life cycle and metamorphose, spawn, and die in the same stream in which they were spawned. Unfortunately, the bad reputation of large lampreys and the inconspicuous nature of small nonpredaceous lampreys have resulted in a long history of ignoring the importance and special requirements of lampreys in aquatic ecosystems. If sufficient information were available, it is likely that a majority of lamprey species would be listed as threatened species under the U.S. Endangered Species Act (ESA).

The inspiration for this book was a petition to list four species of lamprey under the ESA in 2003. The 2004 decision of the U.S. Fish and Wildlife Service was to not list the species because of insufficient information. This decision energized many researchers and others to gather existing and new information on lampreys. Three of the editors (Brown, Chase, and Moyle) organized a one-day symposium on lampreys at the 2005 annual meeting of the California–Nevada Chapter of the American Fisheries Society. The high attendance and interest expressed for lampreys at the symposium led to the organization of a more ambitious symposium at the 2007 annual meeting of the American Fisheries Society in San Francisco. The American Fisheries Society also agreed to publish the proceedings of the symposium as a book. As with most such book projects, papers have come and gone, as have overly ambitious timelines. The editors express their sincere thanks to all of the authors that have stayed with us to the end of the project. We also thank our many colleagues that participated in the review of papers. This book is intended to present new information on lampreys while demonstrating their fascinating nature. The book starts with a paper on the importance of lampreys in the lives of Native Americans in the Klamath River drainage. We then have a number of articles that provide new insights into the systematics, evolution, and biology of lampreys. We end with several papers on the status and conservation of lampreys, including several that provide very practical advice on identifying and protecting them. We hope the book provides you, the reader, with useful information and inspiration to both study these fascinating organisms and support lamprey conservation efforts.

**Note on scientific names:** Pacific lamprey is designated as *Lampetra tridentata* or *Entosphenus tridentatus* in various chapters of this book. Both designations were accepted because the change from *Lampetra tridentata* to *Entosphenus tridentatus* is in progress. *Lampetra tridentata* appears in the current 2004 issue of *Common and Scientific Names of Fishes from the United States, Canada, and Mexico, 6th edition*, but *Entosphenus tridentatus* will appear in the next edition of this book (J. S. Nelson, Chair, AFS Committee on Names of Fishes, personal communication). In addition, changes in generic names to *Entosphenus, Lethenteron,* and *Tetrapleurodon* of species recognized in *Lampetra* in 2004 and, in some cases, endings of species names are based on H. S. Gill, C. B. Renaud, F. Chapleau, R. L. Mayden, and I. C. Potter, 2003, Copeia 2003(3):687–703.

# Symbols and Abbreviations

The following symbols and abbreviations may be found in this book without definition. Also undefined are standard mathematical and statistical symbols given in most dictionaries.

| | |
|---|---|
| A | ampere |
| AC | alternating current |
| Bq | becquerel |
| C | coulomb |
| °C | degrees Celsius |
| cal | calorie |
| cd | candela |
| cm | centimeter |
| Co. | Company |
| Corp. | Corporation |
| cov | covariance |
| DC | direct current; District of Columbia |
| D | dextro (as a prefix) |
| d | day |
| *d* | dextrorotatory |
| df | degrees of freedom |
| dL | deciliter |
| E | east |
| *E* | expected value |
| *e* | base of natural logarithm (2.71828…) |
| e.g. | (exempli gratia) for example |
| eq | equivalent |
| et al. | (et alii) and others |
| etc. | et cetera |
| eV | electron volt |
| F | filial generation; Farad |
| °F | degrees Fahrenheit |
| fc | footcandle (0.0929 lx) |
| ft | foot (30.5 cm) |
| ft$^3$/s | cubic feet per second (0.0283 m$^3$/s) |
| g | gram |
| G | giga ($10^9$, as a prefix) |
| gal | gallon (3.79 L) |
| Gy | gray |
| h | hour |
| ha | hectare (2.47 acres) |
| hp | horsepower (746 W) |
| Hz | hertz |
| in | inch (2.54 cm) |
| Inc. | Incorporated |
| i.e. | (id est) that is |
| IU | international unit |
| J | joule |
| K | Kelvin (degrees above absolute zero) |
| k | kilo ($10^3$, as a prefix) |
| kg | kilogram |
| km | kilometer |
| *l* | levorotatory |
| L | levo (as a prefix) |
| L | liter (0.264 gal, 1.06 qt) |
| lb | pound (0.454 kg, 454g) |
| lm | lumen |
| log | logarithm |
| Ltd. | Limited |
| M | mega ($10^6$, as a prefix); molar (as a suffix or by itself) |
| m | meter (as a suffix or by itself); milli ($10^{-3}$, as a prefix) |
| mi | mile (1.61 km) |
| min | minute |
| mol | mole |
| N | normal (for chemistry); north (for geography); newton |
| *N* | sample size |
| NS | not significant |
| n | ploidy; nanno ($10^{-9}$, as a prefix) |
| *o* | ortho (as a chemical prefix) |
| oz | ounce (28.4 g) |
| *P* | probability |
| *p* | para (as a chemical prefix) |
| p | pico ($10^{-12}$, as a prefix) |
| Pa | pascal |
| pH | negative log of hydrogen ion activity |
| ppm | parts per million |

| | |
|---|---|
| qt | quart (0.946 L) |
| *R* | multiple correlation or regression coefficient |
| *r* | simple correlation or regression coefficient |
| rad | radian |
| S | siemens (for electrical conductance); south (for geography) |
| SD | standard deviation |
| SE | standard error |
| s | second |
| T | tesla |
| tris | tris(hydroxymethyl)-aminomethane (a buffer) |
| UK | United Kingdom |
| U.S. | United States (adjective) |
| USA | United States of America (noun) |
| V | volt |
| *V*, Var | variance (population) |
| var | variance (sample) |
| W | watt (for power); west (for geography) |
| Wb | weber |
| yd | yard (0.914 m, 91.4 cm) |
| α | probability of type I error (false rejection of null hypothesis) |
| β | probability of type II error (false acceptance of null hypothesis) |
| Ω | ohm |
| μ | micro (1026, as a prefix) |
| ′ | minute (angular) |
| ″ | second (angular) |
| ° | degree (temperature as a prefix, angular as a suffix) |
| % | per cent (per hundred) |
| ‰ | per mille (per thousand) |

American Fisheries Society Symposium 72:1–39, 2009


# Yurok and Karuk Traditional Ecological Knowledge: Insights into Pacific Lamprey Populations of the Lower Klamath Basin

Robin S. Petersen Lewis*
*Portland Oregon*

*Abstract.*—Until recently, Pacific lamprey *Lampetra tridentata* was generally considered insignificant and overlooked in West Coast fisheries management. Substantial population declines in the past two decades have triggered increased interest, yet current studies lack historical context. The Yurok and Karuk tribes of the Klamath River in northern California have long had a relationship with Pacific lamprey, utilizing its subsistence and cultural value for thousands of years. In this research, ethnographic methods were used to gather traditional ecological knowledge of Pacific lamprey from Yurok and Karuk tribal community members. This system of knowledge can provide key insights regarding the life history, decline, and ecological significance of local lamprey populations. According to tribal lamprey fishers (eelers), lamprey populations began declining in the Klamath basin more than 40 years ago, with average harvests dropping from 1,000 to less than 15. Harvest methods utilized by eelers provide insights into stream morphology and flows, as well as lamprey distribution and relative abundance. Eelers observed corresponding shifts in lamprey behavior, with electromagnetic changes in the atmosphere. Tribal community members emphasize the significant role lampreys play in the river system, both as prey and essential contributors of marine-derived nutrients.

## Introduction

Pacific lampreys *Lampetra tridentata* are anadromous lampreys native to most streams of the Pacific coast, from the Aleutian Islands southward to Baja California, Mexico. Pacific lamprey is one of at least six lamprey species that reside in the Klamath River basin (Moyle 2002). The Klamath River flows over 260 mi (418 km) from its source in south-central Oregon (Figure 1), journeying through four major dams and the Klamath Mountains to its mouth at the Pacific Ocean. As a native fish species that has evolved for millions of years with this river system and as an important predator, prey, and nutrient source, a decline in Pacific lamprey populations may have a serious ecological impact.

* Corresponding author: rpl@rsplewis.com

The Klamath River is the geographical and spiritual reference point of Yurok and Karuk life (Huntsinger et al. 1994; King 2004). The river serves as the basis for their spatial orientation, so words such as yurok, which means downriver in Karuk, or won, which means uphill or away from the river in Yurok, are used to indicate direction (Kroeber 1925; Hotelling 1978; Bell 1991). Hundreds of Karuk and Yurok villages were spread along the Klamath River, at the mouths of streams, and up into some of the tributaries (Bright 1978; Pilling 1978; Salter 2003). Yurok ancestral territory includes a narrow strip of land that extends along the coastline from 6 mi (9.7 km) north of the Klamath River south to Trinidad, then up along the Klamath River to just above its confluence with the Trinity River (Curtis 1924; Pilling 1978; Hunt-

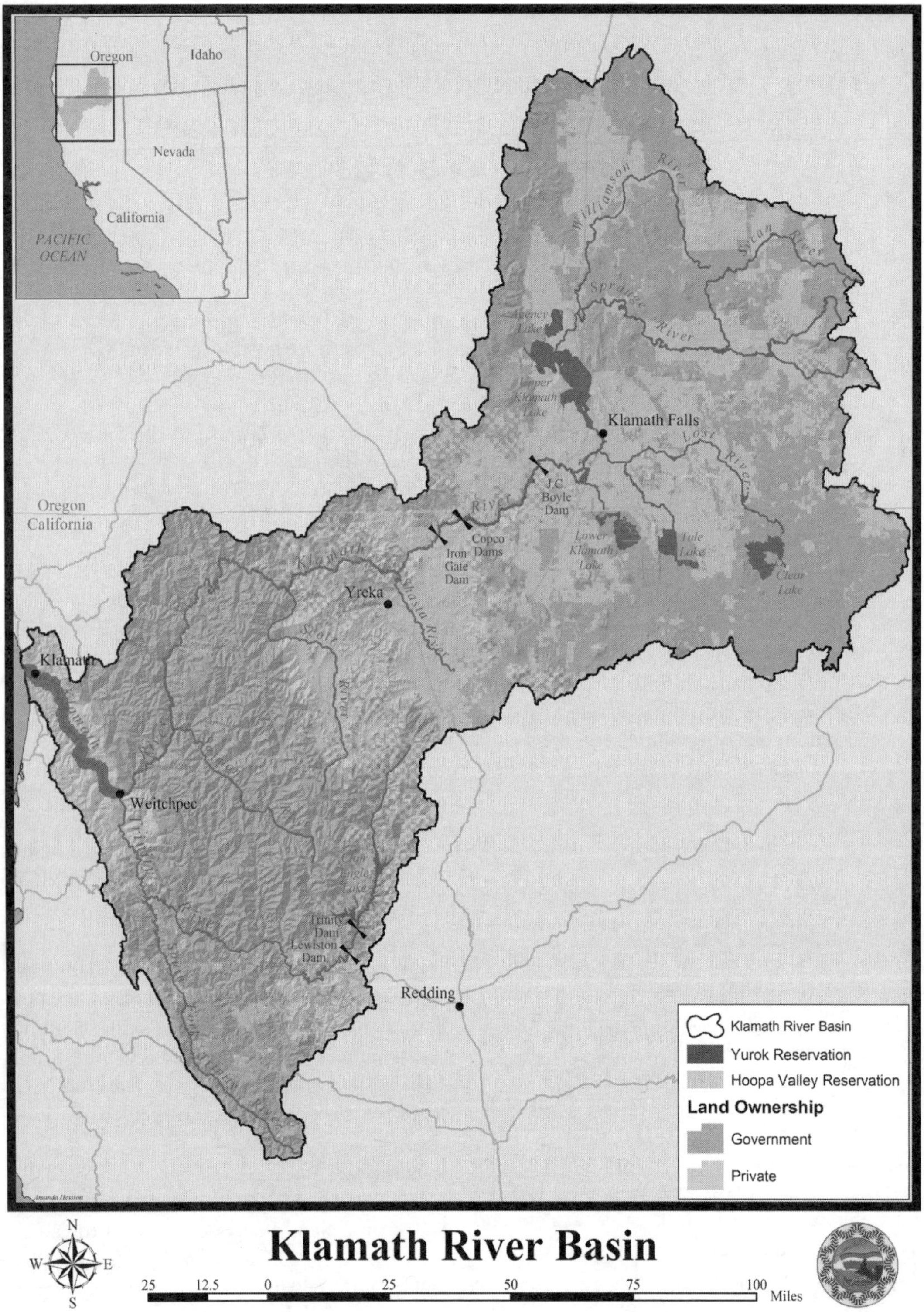


Figure 1. Klamath River basin (map courtesy of the Yurok tribal fisheries program).

singer et al. 1994). Karuk ancestral territory begins at Bluff Creek, extending upstream along the Klamath River to just below the Seiad Valley and including the Salmon River halfway up to the Forks of Salmon (Curtis 1924; Bright 1978; Bell 1991). "For these tribes, the Klamath River was a highway connecting them as a cultural unit" (Salter 2003:5). The Yurok and Karuk, as well as the neighboring Hupa on the Trinity River, speak very different languages but share a similar culture. They have always intermarried, traded, and attended one another's ceremonies (Thompson 1916; Kroeber and Barrett 1960; Bell 1991; Huntsinger et al. 1994; Beckman 1998). While the Hupa are associated with the Klamath River basin and have an intertwined history with the Yurok and Karuk, this research is only focused on those tribes that are historically associated with the middle and lower main stem of the Klamath River (Kroeber and Barrett 1960; Wallace 1983).

The middle and lower Klamath tribes value Pacific lamprey, *key'ween* in Yurok and *akraah* in Karuk, as an important ecologically and culturally significant species (Kroeber and Barrett 1960). Native people have harvested adult Pacific lampreys, known as "eel," for thousands of years as an important subsistence fishery (Pletcher 1963; Hammond 1979; Downey et al. 1993; Close et al. 1995). "Just like everything in life that's sacred to us, they've been here as long as we can remember and I hope they're here long after we're gone" (Karuk eeler). Lamprey has an important place in the spiritual lives of people that live along the river.

> The eel has its place in our religion and in our survival. Not only do we have stories of them in our creation stories, but also physically, he allows us to survive. I really can't articulate further to say that there's really no separation between the Eel people and us because we go together. One without the other... Even myself, I'm an urban Indian…I miss having eel. So sometimes I come home and have an eel, I think, 'oh, the world is good'. It's just what we do. It's hard to articulate the separation because there really isn't any. They're a part of us, a part of our way of life, part of our identity. We're eel eaters, we're salmon eaters, we're deer eaters. It makes our culture who we are (Karuk tribal elder).

The harvesting of lamprey, known as eeling, is an important time for the communication of transgenerational knowledge from elders to younger generations.

> It had a lot of social value. When you're sitting around that campfire down in there, you're hearing your elders tell you stories or tell you about the times when we were catching eels all night long for three straight days and nights. I mean just hearing the stories and just getting that feeling that you were moving into manhood. I think that was very key. I think right now that's what our children are missing. They're not having that experience to spend time with the elders, to be taught. Now they just go down to eel, without being taught the way it's supposed to be and the way it's been since the beginning of time (Karuk tribal community member).

Native knowledge of lamprey is sacred and passed on from generation to generation (Close et al. 2004). This knowledge is based on thousands of years of close observation and integrated in a complex understanding of natural systems. Only in the past few decades have the natural resource management skills of indigenous cultures been recognized by the dominant culture. Hundreds of years of suppression and exploitation limited the abilities of tribes to practice their traditional management.

Personal insights into a species life history are often disregarded by scientists as merely anecdotal so therefore unreliable. However, while such insights may not be understood in the context of Western scientific knowledge, the connection of these insights to the larger picture may not be understood mainly because of a lack of scientific studies. As the fishers of a tribal community, eelers are on the river for most of their lives; they know the place of things and recognize where they should be during a certain time of year or time of day. While these fishers may seem to be combining aspects that should not or could not be together, traditional ecological knowledge (TEK) may reveal relationships that the nonnative observer may not initially see (Deloria and Wildcat 2001; Senos et al. 2006).

TEK is knowledge of a particular geographical location that includes politics and ethics; it is reflexive, accumulated through empirical observation, and passed down transgenerationally and within group membership of the community. F. Berkes (1999:8) defines TEK as

> a cumulative body of knowledge, practice, and belief, evolving by adaptive processes and handed down through generations by cultural transmission, about the relationship of living beings (including humans) with one another and with the environment...is both cumulative and dynamic, building on experience and adapting to changes.

Social, spiritual, and physiological adaptations emerge as a group changes in response to the dynamic environment and accumulates knowledge of this environment over long periods of time (Turner et al. 2003). People learn from their experiences, encode this knowledge, and pass it on to the next generation. Culture-based behavior adapts quickly as conditions and opportunities change (Redman 1999). The larger ethics of connection and relation also emerge based on ancestral ties, ecological community-focused theories of nature, and the causal relationships of all biological organisms.

TEK is often compared to Western scientific knowledge; however, it is important to understand a key distinction. According to Berkes et al. (1998), while both focus on a holistic, long-term understanding of nature, Western scientific knowledge has its roots in the ideology of human dominion over nature, which differs from native beliefs that humans are a part of nature. Additionally, "indigenous, contextuali[z]ed ways of knowing contrast with the linear, compartmentali[z]ed ways of thinking that are integral to societies with written traditions" (Smith et al. 2000:22). This distinction is important to consider when working with both TEK and Western scientific knowledge. TEK is often "encoded both in distinctive paradigms and into everyday practice, and these may get stripped away in translation because they do not travel easily across cultural boundaries" (Cruikshank 1998:52). The tendency is "to shatter the holism of local knowledge by simply eliminating the parts that do not . . . migrate easily" (Cruikshank 1998:62).

Many community-based resource management projects have emerged since the 1980s that utilize ethnographic methods to document TEK and integrate it with Western science (Johnson 1992; Huntington 2000; Senos et al. 2006). While these researchers apply the information they acquire to current scientific studies, these projects are rarely included in the larger body of Western scientific literature, and many scientists have differing opinions on its validity and utility (Bielawski 1996; Senos et al. 2006). Additionally, TEK continues to be categorized into parts for use by science "rather than a system of knowledge that could inform science" (Cruikshank 1998:50).

Three studies that utilized TEK in researching Pacific lamprey in the Pacific Northwest are Chapin et al. (1998), Nadasdy (2003), and Close et al. (2004). All were short-term studies but each emphasized the applied aspect of the information acquired. Researchers for the Confederated Tribes of the Siletz Indians describe the importance of utilizing TEK in their lamprey research projects,

> The interviews with the elders gave us insights to aquatic habitat parameters, local lamprey ecology, and the population decline as it relates specifically to the Pacific lamprey in Rock and Little Rock creeks. Additionally, the interviews substantiated the local concern about the Pacific lamprey and served

as a guide for further research (Chapin et al. 1998:1).

Research was also conducted for the Confederated Tribes of the Umatilla Indian Reservation to gather TEK of Pacific lamprey from tribal members "to gain additional natural history insights and baseline life history information for a poorly understood species" (Close et al. 2004:141). These studies are important stepping-stones to more in-depth studies that continue to refute historical biases and demonstrate the importance of this parallel system of knowledge. However, even these studies demonstrate the pressure that tribes feel to fit their traditional knowledge into Western scientific management terms and systems.

> The imperative of incorporating TEK into the state management system has caused researchers to focus on extracting from communities only that kind of information that can be expressed in a few very specific ways. . .forms that can be utilized within the institutional framework of scientific resource management. . .and then to interpret that information in a manner consistent with the assumptions of scientific wildlife management (Nadasdy 2003:129).

While working with TEK, it is important to keep in mind the social, historical, and political conditions that have allowed Western culture to translate and redefine traditional cultural systems of knowledge into their own categories of understanding.

Current declines of Klamath River Pacific lamprey populations may have both cultural and ecological impacts (Petersen 2006). According to Karuk and Yurok tribal fishers, the lamprey fishery has experienced a dramatic decline in the past decade. In a preliminary study conducted by the Yurok Tribe, local eelers described harvests of 1,500 lamprey in one day during historical runs. During this 1998 study, lamprey catch per unit effort was reported as ranging from zero to 100 lampreys per day, with 20 lampreys found to be extremely good (Larson and Belchik 1998). However, much of the literature regarding lamprey is focused on populations in Oregon and British Columbia. "The limited information that exists concerning... Pacific lamprey populations in the Klamath and other nearby drainages reflects the gap in our knowledge of these species' population trends and man-caused factors which may affect them" (Larson and Belchik 1998:10).

While there are limits to species-specific approaches to resource management, here this approach was used as a process to explore and learn from the holistic foundation of Karuk and Yurok TEK. Pacific lampreys were used as a lens through which to view and understand the Klamath River basin environment as a whole. This paper illustrates how Karuk and Yurok TEK provides a more comprehensive understanding of Pacific lamprey populations in the lower Klamath basin, specifically looking at life history, presence and distribution, relative abundance, population decline factors, and ecological significance. TEK is particularly important in this situation because Western scientific knowledge is largely lacking on Klamath basin lampreys. As is the nature of TEK, there is overlap among these categories and the intention is to provide this knowledge in a manner that is accessible to outside groups.

## Methods

The protection of intellectual property rights is an important issue when working with indigenous groups (Grenier 1998). Outsiders often use aspects of a culture that center on religious and spiritual knowledge in exploitative and inappropriate manners. Technically, tribes can use their inherent right of sovereignty to assert their own values and to control access to and the use of cultural knowledge. Sovereignty is limited, however, by the plenary power of the federal government (Pevar 2002; Wilkins 2002). While it "guarantees" that laws can be passed that protect nonmaterial resources within a reservation, there is little protection beyond the context of the community (Pinel and Evans

1994; Berkes 1999). Many tribes recognize the role of anthropologists and other outsiders in helping to protect native cultural knowledge. However, it can become a two-edged sword when information must be shared in order to protect knowledge that tribes want to keep secret because it is sacred and has specific rules for the process and use of its transmission. The very act of disclosure destroys the sacredness that gives the meaning in the first place (Pinel and Evans 1994; Holmlund 2006).

In each tribal fisheries program, a tribal member who worked in the program became a key informant. A key informant is a person in the position to know the culture and community as a whole, especially the particular area of interest, while also willing to "walk you through their culture and show you the ropes" (Bernard 2002:187). They were fundamental in contacting and connecting with participants. The key informants also provided valuable personal insight into their world as eelers and members of their native community, as well as the greater Klamath River ecological community.

Data collection was triangulated using participant observation, informal interviews, semidirected interviews, focus groups, and archival and historical research. This approach ensured the reliability and validity of data by obtaining "comparable, confirmatory data from multiple sources at different points in time, and through the use of multiple methods" (Bernard 1998:719). According to the 2000 U.S. Census, the Yurok and Karuk tribes currently have populations around 4,029 and 2,702 people, respectively. During this study, a total of 83 Yurok and Karuk tribal and community members participated through informal interviews, focus groups, and semidirected interviews.

### Participant Observation

Participant observation included observing the activities involved in the process of eeling. Both men and women participate at different times in this cyclical process, which is more than just the act of harvesting the lampreys. The process of eeling includes understanding the "right way and time" to gather materials for making the tools for harvest (hooks, gaffs, baskets, and nets) and acting in the proper way while making these tools. It also means being respectful of fellow eelers, the ocean and river that is providing this food source, and the lamprey that are giving themselves to the eelers. The process of eeling includes the cleaning and processing of the harvested lamprey. Additionally, following the proper way of living ensures that the waters are prepared so that the lampreys return again and again.

Observing the process of eeling was an important step in developing rapport with local eelers and learning more about the cultural and social importance of eeling and lampreys (Bernard 1998; Ervin 2000). Often informal interviews and discussion came out of these experiences. It was an ideal time to discuss and see harvest methods firsthand, as well as to place into context the changes in the river and landscape that had been described during interviews. While a person was in the act of eeling, making an eel basket, or cleaning a lamprey, it often triggered thoughts and ideas about the species and the river system that they may not have thought of during a more formalized interview (Ervin 2000).

Table 1 includes each of the eeling activities observed and the number of participants involved in each activity. A participant was counted only once for a specific activity, even if they were involved in more than one observation session. They were also counted for each activity in which they were involved. Because there is overlap in the number of participants, a cell was not included with total participants for all activities.

### Informal Interviews

Many informal interviews came out of participant observation experiences. Additionally, some participants preferred to have an informal conversation to a more structured interview. Not including conversations that occurred during observation sessions, 14 participants were spoken with informally. Informal interviews

Table 1. Participant observation activities by number of sessions, as well as the eeler status and number of participants per activity.

| Activity | Number of observation sessions | Participants: Elder eelers | Participants: Eelers | Participants: Non-eelers | Total participants |
|---|---|---|---|---|---|
| Eeling at the mouth | 4 | 6 | 6 | 1 | 13 |
| Setting basket traps | 2 | 3 | 2 | | 5 |
| Taking out eel baskets | 2 | 3 | | | 3 |
| Processing eel | 1 | 2 | | | 2 |
| Weaving eel baskets | 3 | | 3 | | 3 |
| Eel fishing holes | 2 | 1 | | | 1 |

with the eelers or other local people often filled in missing pieces of data. In talking with them about other subjects, it was an opportunity to learn indirectly about their views and knowledge on lamprey (Bernard 2002; Senos et al. 2006). The makeup of these informal discussions allowed the conversation to move beyond the given topic, and in essence, they became the connectors that helped make the picture more complete.

### *Semidirected Interviews*

Interviews were focused on the men who harvest Pacific lampreys because they are the ones on the water and see the fish while they are moving through the river system. Women were also interviewed because they participate in the cleaning and processing of lampreys. Surprisingly, two of the eelers interviewed were women who had been eeling their entire lives. Women harvesting lampreys is generally seen as taboo. While these two women provided their own personal experiences as women eelers, the cultural mechanisms were not revealed that allowed them to eel throughout their lives. Additionally, both men and women who have not regularly participated in lamprey harvest were interviewed. These participants provided insight into knowledge and experience that is gained through living a life that revolves around the river.

Interviews included both elders and non-elders to explore the transmission of transgenerational knowledge. Status as an elder is not easily defined; respect rather than age is often the determining factor when considering who is an elder in the community. Elders provided knowledge that they were taught from their elders, as well as their experiences. Many have experienced significant changes even within their own lifetimes. Non-elders also shared knowledge that they have been taught from their elders, many of whom have already passed on. They also offered new perspectives that intertwined Western scientific knowledge with traditional knowledge.

In total, 59 semidirected interviews were conducted with 66 participants; ages ranged from 24 to 90 years old. While many participants were Yurok and Karuk tribal members, others had Yurok and Karuk ancestry but were members of another tribe. Participants were included in the tribal community to which they expressed a connection either as a member, by ancestry, or in which they resided. Table 2 breaks down semidirected interview participants by tribal community, gender, and eeler status.

The length of each semidirected interview ranged from 25 min to 2.5 h, with a total of more than 60 interview hours. Participants were offered the option to have a tribal member conduct the interview; however, no one took this option. The purpose of the interview was explained, as well as how the knowledge would be used, and then they were asked if they had any questions. Each Karuk participant was presented with a one-page summary of the research goals and objectives. In response

Table 2. Semi-directed interview participant by tribal community, gender, and eeler status.

| Tribal community | Gender | Participants: Elder eelers | Eelers | Non-eelers | Total participants |
|---|---|---|---|---|---|
| Yurok | Male | 15 | 12 | 3 | 30 |
| | Female | 1 | 1 | 7 | 9 |
| Karuk | Male | 10 | 11 | 1 | 22 |
| | Female | – | – | 5 | 5 |
| Total participants | | 26 | 24 | 16 | 66 |

to questions from participants regarding what fisheries biologists knew about lamprey, a handout was developed to give to eelers during semidirected interviews and focus group meetings. It drew upon the most current research available on the biology and natural history of Pacific lamprey from the perspective of Western science (Pletcher 1963; Kan 1975; Beamish 1980; Kostow 2002). The purpose of the handout was both to share this information and as an act of reciprocity for the time and knowledge that participants were sharing. It also provided a means to initiate conversation.

Participants were interviewed individually or, per situation, in groups and asked questions to start a discussion. As the conversation evolved, more questions were asked that pertained to the new topics and direction that had developed. This type of questioning allowed the information to come directly from the participants (Huntington 2000). While it is impossible to completely eliminate bias in research, by allowing participants to "define the content of the discussion" (Bernard 2002:209), the hope is to limit the researcher's "culturally derived standards of relevance" (Nadasdy 2003:121). To make sure specific topics were covered in each interview, a more detailed interview guide was developed from which to obtain reliable, comparable data (Nadasdy 2003). The interview guide was categorized by general themes that were the basis of the research. These themes helped to maintain focus within the interviews while still allowing participants to direct their own answers and the opportunity to ask additional questions based on these new insights. In this manner, new information arose that was very important in answering the research questions but would not have been revealed if only the questions on the interview guide had been asked.

## *Focus Groups*

Two focus groups with Yurok and Karuk tribal community members were conducted for a total of 4 h. The intent of the focus groups was to initiate conversation in a group setting about topics that might be difficult to discuss on an individual level. The thoughts and ideas that came from one person sparked additional comments from another and uncovered cultural norms and underlying ideologies (Morgan 1997; Bernard 1998). The first focus group was conducted in Klamath and included nine Yurok eelers. Eight men and one woman participated, with ages ranging from 24 to 78 years old. Participants were recruited by the researcher and other eelers. A general question guideline was utilized to initiate conversation and to keep the group focused. The open-ended quality of the approach allowed the content to come from the group.

The second focus group took place on the Quartz Valley Indian Reservation along the Scott River. Thirteen families and original allottees were moved to this area after the violence along the middle Klamath during the mid- to late 1800s. The reservation is now home to Karuk, Shasta, and Klamath Indians. The four people who participated in the focus group are closely related to Karuk people living along the main-stem Klamath River. Due to the nature of the focus group, the structure was much different from the first group. Two men and two women participated, two who had eeled when they were younger and two who had never eeled. The two elders who participated had lived on the Scott River for most of their lives. This meeting facilitated discussion around a map of the Scott River Valley, utilizing questions initially from the focus group guide and then from the conversations that followed.

### *Follow-Up Summaries*

Audio recordings were transcribed for each interview. Transcriptions of the interviews and focus groups totaled more than 400 pages. Summaries were developed from interview transcripts for each person. These summaries were framed under the main themes that had emerged from the interviews. Each person received a summary of their interview to provide them with the opportunity to review, edit, and comment on this first level of analysis. It was important to provide participants with the opportunity to remain actively involved in all steps of the project, including the analysis. Overall, 21% of participants responded to the follow-up summaries. Comments included correcting errors in spelling and place names, as well as elaborating on previous responses to questions.

### *Archival and Historical Research*

Access was granted to the Yurok tribal archives to watch and take notes from videotaped fisheries interviews from the early 1990s. Research was also conducted into Karuk and Yurok cultural, social, and political history in the Humboldt Room at Humboldt State Library and the California Indian Library Collections at Humboldt County Library. Additionally, state and federal agencies provided local fisheries management policy, research, and history.

### *Analysis*

Data analysis was an ongoing process throughout the fieldwork. Field jottings, field notes, a field journal, and a personal journal were maintained (Bernard 1998). These tools allowed the gathered information to be processed while the research was conducted (Bernard 2002). A coded system was developed based on the themes and subthemes that emerged throughout the data collection and transcription process. The six main themes that emerged include eeling experiences, eel biology, population decline, ecological significance, cultural significance, and political ecology. Each main theme included 8 to 11 subthemes. These themes and subthemes were given codes based on words that came from the theme subject and developed naturally from the study (Miles and Huberman 1994). The summaries of each interview were then coded. A spreadsheet was developed detailing the themes and subthemes that were evident in each informal and semidirected interview. Due to the immense amount of data collected, not all themes were included in this paper. It was possible to follow up the analysis by talking directly with the tribal community member that provided the knowledge originally or by checking it with other community members to verify the accuracy of the analysis.

Information presented in the Results is based on the knowledge and opinions shared during my discussions and interviews with Yurok and Karuk eelers and other tribal community members during 2004–2005. While many participants use Western biological terminology and concepts because they are interested in and informed about aspects of Western scientific knowledge, they are still speaking from a traditional perspective. Corresponding ethnographic and biological studies are included as footnotes or citations where applicable. References to Western scientific knowledge are indicated by "WSK" and to Traditional Ecological Knowledge by "TEK."

## Results

### *Life History*

Participants found ammocoetes buried in the sand and fine gravel along the sides of the Klamath River and its tributaries. Most of these larvae were found in groups whose numbers increased as substrate size decreased (e.g., more were found in sand while fewer were found in larger gravel areas). Participants believe ammocoetes come out at night to avoid predators. Those who fish at night recall seeing larvae making tracks on the shore, crawling along the shallows when the tide went out or river levels drop. They are thought to filter feed in the silt (WSK, see Pletcher 1963; Kan 1975; Hammond 1979; Beamish and Northcote 1989), and when all of the nutrients are consumed, they move to another spot. They have also been observed feeding on decomposing adult lamprey.

Ammocoetes were observed at lengths of 1–6 in (2.5–15 cm). The larger ammocoetes measuring around 6 in (15 cm) long were described as being a pinkish color with eyes. Tribal community members remarked that the number of young lampreys they see when they go swimming now versus when they were children has declined dramatically. More than 30 years ago, they could easily pull up hundreds of different-sized ammocoetes from the sand with their hands. As they disturbed the sand while walking in the water, people also remember watching the larvae pop up, swim around, and bury themselves back into the substrate.

Participants noted that during other times of the year, hundreds of ammocoetes were found underneath their boats or under the floats (inner tubes) for their net anchors. The last time participants saw these larvae ranged from last spring to more than 20 years ago, with most people speculating that it had been at least 10 years ago. One person remarked that he still sees spurts of young lamprey towards the end of September. Most of the participants who had seen ammocoetes or juveniles recently work in some capacity for the fisheries programs where they catch them in screw traps. The ones they find are of different age-classes but are within 3–4 years of one another. Not many participants recall seeing juvenile Pacific lampreys migrating out to the ocean. A number of reasons were speculated, including that most people are not on the river if the juveniles are out-migrating during late fall into winter when the waters are higher. They are also less likely to be seen if they travel down the middle of the river where they are not fighting the current as they go back out.

Tribal eelers use harvest methods for Pacific lampreys that capitalize on adult behavior and their own understanding of the complex interactions between this behavior and changes in the river system. Lamprey is well known as resilient, strong survivors (Figure 2). "Well, jeez, you can't hardly kill a damn eel. You take them out of the shittin' water and throw them in your damn kitchen sink and hell, they'll be alive

Figure 2. Pacific lamprey with eel hook wound, Klamath River, 2005.

the next day down crawling underneath your bed" (Karuk eeler). Many people recall seeing lampreys crawl straight up over dams, as long as there is plenty of water coming over. One participant described how they wiggle their tail to go up an inch or two, then suck back on, then go up some more, sometimes falling back off to start all over again. They are also known to make a suctioning or "popping" sound as they latch onto the rocks. In describing the large numbers of lamprey that used to move through the Klamath River system, one elder remarked, "You would be surprised how many damn eels run when they're running. There's no way of keeping track of them. You go along at Ishipishi Falls, these falls up here, you can hear them. Little suction cups on the bottom of the river."

Eelers find that lamprey are drawn to the path of least resistance, so eel baskets (Figures 3 and 4) are placed to create eddies and changes in water current that can be felt downstream. The lampreys follow this slowing of the water current and swim up into the traps where they cannot escape.

> Eel always go upriver; the board at the upper end of a basket makes a wake so that the water goes around it and comes back together at the front end of the trap where the funnel starts. The wake will go way down river and the eel can feel it and follow it because they don't swim very good and it's easier swimming. The eel follow the wake figuring there's a resting place above where the water makes a whirlpool and that's where they'll rest. They funnel right in up against the backboard and then they're trapped in there (Karuk eeler).

When the water is moving fast, Pacific lampreys come in close to the shore and eelers will use dip nets to harvest them (Figure 5). They are seen moving along the top of the water when it is muddy where they are thought to use a "breather hole" at the top of their body. Participants also indicated that lamprey "cruise the top" of the water when there is a lot of sand in the water because the sand clogs their gills.

Figure 3. Traditional hazel stick eel basket, Klamath River, 2005.

Figure 4. Setting bicycle rim eel baskets along the middle Klamath River, 2005.

Figure 5. Half-size dip net, Klamath River, 2005.

When there is a lot of gravel in the water, the lampreys will go to the bottom and hang onto a rock. Eelers utilize the knowledge of these behaviors to choose the most appropriate harvest method. "Eel is not a good swimmer in freshwater, which is why he sucks up the rocks in a swift area, whereas a fish will jump up the stream" (Karuk eeler). When an eeler is not able to see into the water because it is too muddy or when the lampreys are along the sides of the river, they use a dip net to capture them.

> When it gets muddy in the high water with all of the debris going down through the center of the river, the fish and the eels go down along the edges of the banks where the brush knocks all of the debris away coming down through the stream. There are these pockets in the high water when it's all muddy where you can dip into where the fish are swimming upstream. You can catch all different types of trout, eels, steelhead, sometimes salmon (Karuk tribal community member).

The tides also influence where lampreys are swimming in the river. Eelers at the mouth go out 2–3 h before the low tide and eel during the "slack tide" (Figure 6). They catch more during the outgoing tide when the water is swift because the lampreys come in closer to shore (Figure 6). Many participants attributed lamprey behavior, such as moving along the sides of streams or utilizing eddies, as predator evasion strategies and taking the path of least resistance because they are "lazy swimmers."

Lampreys use one another to climb over rock faces and dams. Eelers can use their gloved hands or an eel hook (Figure 7) to harvest at these locations,[1] but they must first understand important aspects of lamprey behavior to be successful at capturing all of the fish in this "wedge of eels."

[1] TEK: "Many lampreys are even now caught by hand, the fisherman either wearing a glove or carrying in his hand a rag or something else to prevent the lamprey from slipping through his grasp...Lampreys were also taken by the Karok with a gaff, anciently an acute-angled point of bone or antler attached to a wooden handle three feet or more in length" (Kroeber and Barrett 1960:71).

Figure 6. Eeling at the mouth of the Klamath River, 2005.

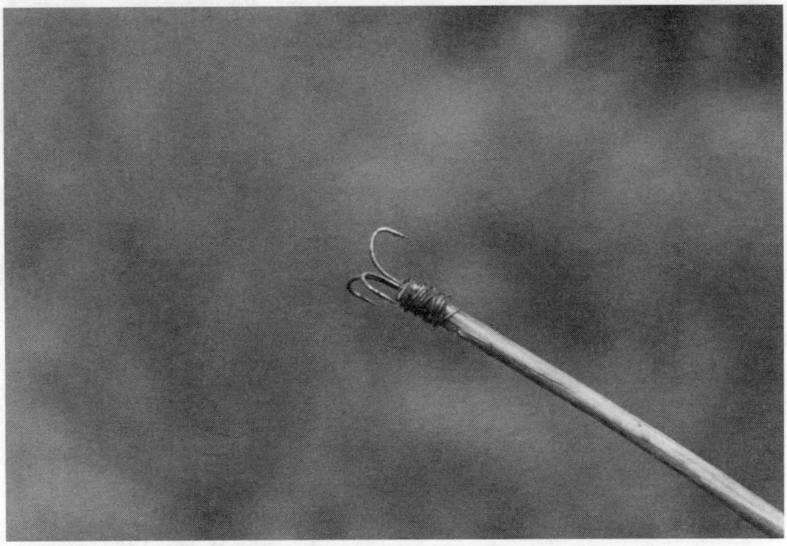

Figure 7. Six- to 12-ft long treble hook used at riffles and falls, Klamath River, 2005.

> The head eel he goes as far up and another eel'll come along and he'll hold the head back like that and he'll catch a hold and he'll keep on doing that until he catches up with this top guy and he'll jump up there. And then he'll get a hold and the next eel comes up over the top of him and he'll come up and grab a hold. And they just keep on going 'til they get to the top of that rock and then they just start going like this…they throw their head…they throw themselves up like that and then they'll just kind of give a flip and they'll flip their head up and catch hold (Yurok eeler).

When an eeler starts to pull off a lamprey from this wedge, he must start at the bottom or they will all drop off. Any kind of scent or blood in the water will also cause them to drop back into the water. Eelers believe lamprey has a strong sense of smell, so at the falls they take 30 min to an hour breaks between harvesting sessions to allow the next group of lampreys to come up. Otherwise, they find that the group will go deeper or move away from where they can be harvested.

Lampreys are harvested during the day and night at the mouth of the Klamath River and only at night upstream. Some participants feel that lampreys become more sensitive to light as they move upstream while others see it as a predator evasion strategy.[2] One of the more interesting observations made by eelers was that any change in electromagnetic activity brings lampreys out of the water onto the rocks.[3] During 1997, an eeler caught 200–300 lampreys in 10 min during a solar burst. That same year, another eeler went out the night after a lunar eclipse and caught more than 300 lampreys.

[2] TEK: "It is this tendency of the lamprey to hide among the rocks that makes it an easy prey to the fisherman's gaff…Into the crevices in these rock piles lampreys crawl to hide during the bright hours of the day, resuming their journey with the coming of night" (Kroeber and Barrett 1960:71).

[3] WSK: Lampreys have been found to have an electrosensory system that is as sensitive as those of other electroreceptive fish (Bodznick and Northcutt 1981). They also possess electroreceptors on their head and trunk regions that may be useful in finding prey (Bodznick and Preston 1983). Scientists speculate that lightning may be a source of "noise" that interferes with the detection of significant electrosignals or limits the resolution. It has been shown to cause millisecond pulses in natural water bodies, up to hundreds of miles in distance (Hopkins 1973, cited in Bullock 2005). During electromagnetic activity, it is speculated that lamprey may not be able to utilize their electrosensory system, so they move out of the water as an evasion strategy or the pulses formed by lightning or other sources may overwhelm their sensory system causing them to try to "escape."

Lightning storms and other phenomenon also seem to bring them out of the water.

> [W]hen I came back, you could see the northern lights because it sent a bunch of negative ions towards the earth and it made to where you could see the northern lights all the way down here and so I ran down to the eeling hole and there was, oh, there was eels everywhere. They were out on dry rock and we just caught a whole bunch of them real quick and left…They were like two and four feet up on dry rocks. That's the first time I've ever seen that (Karuk eeler).

More than one participant remarked that temperature and light are involved in the running of lamprey. They start moving up the river a few hours after it gets dark. Even at night, they move away from the moonlight, so eelers find out where and when the shadow of the moon is going to fall before they go out eeling. When it is cold, they do not move and ball up at the bottom of the river. Warmth can make lampreys very lively. Eelers that run eel baskets mentioned that when it is warm in the air and water, the lampreys move wildly around inside or come shooting out when they open up the basket. For those using other harvesting methods, it has to warm up before the lampreys surface enough to catch them.

At Ishipishi and Ikes Falls (near the Klamath and Salmon Rivers confluence), eelers shine a light out onto the rocks to see the flash of the lamprey's belly and then reach out in the dark with their hook (Figure 7) to grab it. Coon Creek Falls are farther downstream and geomorphically different from Ishipishi and Ikes Falls. Lamprey are able to stay on rocks under the water at Coon Creek, so eelers use a different method of hooking called "scraping the rocks." This method allows an eeler to hook without seeing the lamprey first. Some eelers have switched over to using this method and combined it with the "hook what you see" methods at Ishipishi and Ikes to capture lampreys during colder temperatures and other conditions that force them deeper into the water or into the willows.

Lampreys are also sensitive to water conditions upstream. Eelers at the mouth mentioned seeing lampreys ganged up, waiting for the right current before they begin moving up into the river. They see the lampreys swimming together in groups of at least three or more (WSK, see Moyle 2002). Upstream eelers found that these lamprey groups run a specific route or slot described as looking like an ant trail. They speculate that the lampreys are leaving some kind of scent for the others to follow. The next group finds a different route as the water level changes.

> Eels follow certain paths and stay in those paths unless the water drops. It all depends on the water level. It seems that when the water is real low, the eels go right up the middle and don't have to struggle so hard. When it's higher or when the water is cold and the sides are a little warmer, then the eel stay to the sides more. When it's too cold they ball up and wait at the bottom of a hole (Karuk eeler).

Some people that live upriver made a morphological distinction between two runs of lamprey that come through the system, one bluer and one darker. It was speculated that participants may be describing those lampreys that have just come into the river system (larger, bluer) and those that have wintered over and are ready to spawn (darker, smaller). Some participants only described one type of lamprey. One explanation given to me for this discrepancy was those people that live up in the tributaries or at the mouths of the tributaries are only seeing the lampreys that have been there for a year and are on their way up to spring spawn. Participants who have seen two different-looking lampreys found that the blue eels come in later during June when the other darker lampreys are dying; both are found to have the same teeth structure and are larger than the river lamprey *Lampetra ayresii*. One eeler noticed that a final run comes in during April and spawns in the estuary. Another eeler mentioned that he always goes down to the mouth to eel on his birthday in the middle of April and catches more than 100 lampreys on the same day every year.[4]

Differences between those lampreys caught at the mouth and those harvested upstream are apparent in both meat quality and relative size. Lampreys harvested upstream have a lower fat quantity and are found to take on the taste of the river. Interestingly though, while lampreys downstream by the mouth take on worms during the warmer months, those upstream do not. Many people make mention of "a yard of eels" or "out there chasing the 'yards,'" referring to the length of a lamprey. "Everybody said they could eat a couple yards of eels" (Yurok eeler). Upstream participants noticed that lampreys become half the size that they were at the mouth once they reach the falls, then half the size again once they reach the Scott River Valley. Most people recognize that lampreys throughout the Klamath River basin are not as big as they used to be.

### *Presence and Distribution*

Eelers utilize their understanding of the landscape to identify where fish will be and how they should be harvested.

> Human beings can be as bad as the otters. They want the most amount of fish, the easiest fishing spot, and the quickest way they can get the most fish as possible, which is generally at the mouth of streams where the creeks are coming into the river and bringing cooler water and more oxygen and fish are naturally moving toward those places. In low water years, you can see the fish holding up in the streams, trying to figure out how to solve the primal drive to get where they are supposed to get, spawn, and lay their eggs (Karuk tribal community member).

Historically, families had fishing holes up and down the river and up into the tributaries. These fishing holes were utilized for many of the fish species that come through the system, depending on the season and the water conditions. Participants described spring freshets that came through each year to cleanse the system, removing built-up sediments, debris, and contaminants, while keeping the morphology relatively the same. Fishing holes remained unchanged for generations, allowing the knowledge of these fisheries to remain intact. Many people reason that land management practices and policy within the Klamath basin impacted both these fisheries and the associated family fishing holes.

Some lamprey harvest practices have continued while others cannot, due to low numbers of lampreys coming through the system or changes in stream structure.

> Even though they are from the same tribe, everyone has different eeling spots. There aren't very many fishermen anymore so the competition for the fishing hole isn't like it used to be. There used to be family rights by clans. The head of clans went first, prestige, wealth, elders, lots of different factors. You don't see that anymore, there aren't enough fish to let that happen (Karuk eeler).

All participants remarked on changes they had seen in lamprey population numbers and runs. Historically, the lamprey run always started at the same time every year. Now with fewer lampreys moving through the system, it is especially difficult for eelers to know if and when the fish have started running. When the water is colder at the beginning of the season, lampreys move deeper in the water; if the water is slow, they move out to the middle of the river where it is difficult to see or fish for them. Only when

[4] WSK: Moyle (2002) suggests that the two forms may represent resident versus migratory individuals or separate runs, one that spawns immediately after the spring upstream migration and one that migrates in the fall, holds over, and spawns the following spring. Bayer et al. (2000) reports that tagged lampreys in the John Day River have been observed by fisheries biologists to hide under boulders when they arrived in August and remain there until the following March during which time they moved to their spawning grounds.

conditions force the lampreys into narrow channels are people able to see them come through during the beginning of their run. One of the eelers recalled an experience eeling during 2002, a low water year, when the stream had narrowed so much he was able to see the unusual sight of hundreds and hundreds of lampreys moving through the same area of stream at the same time.

Table 3 summarizes the methods eelers use to harvest lampreys, as determined by specific

Table 3. Eeling harvest technology corresponding to stream morphology and hydrology, as well as Pacific lamprey behavior and relative abundance. Data summarized from interviews.

| Technology | Morphology | Hydrology | Methods | Lamprey behavior | Abundance |
|---|---|---|---|---|---|
| Eel basket | riffles; eddies | swift | Force of the water hits the basket to create an eddy downstream. | Take the path of least resistance; look for resting places | variable |
| | estuary | outgoing tide; swift | Wade out during the high tide when the water is slower. | Use margins for protection during fast water to avoid predation and look for slower-moving water. | variable |
| Dip net | banks; eddies | muddy | Use when the water is muddy. | Move near the top of the water to breathe; cannot see the eeler. | variable |
| | banks; eddies | high water; swift | "High water dipping." | Use margins for protection during fast water to avoid predation and look for slower-moving water. | many |
| Platform fishery | eddies | high water; swift | Use a trigger, A-frame, or dip net. | Take the path of least resistance; look for resting places. | many at once |
| | wide channels | varies | Go out over the mainstream channel where lamprey are moving through the system. | Move through the slower water. | many at once |
| Eel hook | mouth–ocean side | waves | Eel the outgoing tide and the start of the incoming. | Surf the waves into the river. | variable |
| | mouth–river side | swift | Stand along the edge to hook lamprey during the low tide. | Move along the shoreline during swifter water. | variable |
| | banks | varies | Scratch lamprey out of the water. | Large biomass so moving throughout the stream. | many |
| | falls; narrow channels | swift | Hook what can be seen out on the rocks. | Move up the rocks out of the water or at the surface. | variable |
| | falls; narrow channels | slow | Scrape off from under the rocks by feeling for lamprey. | Move up the rocks under the water. | variable |
| Hand | falls; riffles | swift | Use a glove or sock to grab off the rocks; originally used a fern. | Form a wedge as they make their way up a surface. | variable |
| | narrow channel | varies | Create a channel if not naturally there. | Forced to move in close together. | variable |

stream system structures and flows, as well as lamprey behavior and relative abundance. Many of these methods require strong flows of water. Those methods that utilized platform fisheries are no longer in use today as a result of the decline in subsistence fisheries populations in the Klamath basin. Historically, many of the family fishing holes had platform fisheries where fishers utilized trigger nets and dip nets for both salmon and lampreys. As an eeler explained,

> If you're getting everything off of your platform then you're good. All of these traditional platform fisheries, trigger net, dip net, the lamprey fishery was right off of that same platform in the same area so you didn't have to move it…there are probably different reasons why they don't fish there now, but a lot of it is because there's no fish (Karuk eeler).

A trigger net, or A-frame, has a deep pocket with a string tied to the end of the net. Your finger holds the string so when the fish hits the net, you feel the trigger and pull the net up quickly to close the pocket with the fish inside. Tribal fishers commented that both trigger net and dip net fisheries from platforms require the fish to be coming up through the system in large groups so they are moving through the entire channel of the stream.

With fewer lampreys coming through the system, eeling now happens in the rapids where the flow is hard and the lampreys have to use the rocks along the margins. "If you have less flow, the eels have it easier and can slide on through and not be caught because they're in the middle of the river where you can't fish for them. You need the flow to force the eels to the side" (Karuk eeler). In the past, some families constructed ledges for the lampreys to crawl up so they could catch them. As one Karuk eeler pointed out, "If they relied on the eel for subsistence, life or death, they'd be down there building those runways for them." They would throw rocks down to make the lampreys expose themselves when they swam up, filling in the crevices to almost create a staircase for them. One of the families that lives along a tributary of the Klamath used to fill in one of the riffles to create one channel so the lampreys would come in to where they could get to them; otherwise, they never saw them. "It's very important to realize that this is a flow-dependent fishery…low flow means less opportunity, more flow means more opportunity…fish are going to take the easiest migration route up the river and with less flow they can go up the middle of the river" (Karuk eeler).

### *Relative Abundance*

The elders tell stories of how Pacific lampreys used to be so thick, you could hear them. When eelers went out to harvest them at the mouth, at the falls, or up in the tributaries, they could easily pull out one lamprey after another because there were so many; anyone could harvest as much as they needed. "I remember seeing the rocks where we got the eels were just a solid mass, hundreds and hundreds" (Karuk tribal community member). Just over 45 years ago, the lampreys were still so thick that crews were sent in to unclog the creeks because they had no flows. Up and down the river, tales are told of a biomass so great that lampreys were poisoned in those creeks, as well as at the dams where they were caught up in the turbines. For those who were the fishers of their village, eeling "was no mystery back then." The elders have no recollection of ever going eeling and not catching lampreys. They saw millions moving up the river; enough that you could drop a basket in the water and within an hour or two, it was full. The baskets were so heavy with lampreys that they had to position the boat and pull the basket up at an angle. A Yurok eeler recalled that before the floods of 1955 and 1964, they could set three baskets and pull in 1,500 lampreys at a time, enough that they had to tie a chain to the baskets to pull them up with a truck.

Those who process lampreys recall as children helping their mother or grandmother flatten 1,500 lampreys at a time. Smokehouses could hold up to 1,000 lampreys and were always packed solid. "Most eels I've ever caught, [he] and I got 600 in one tide…I had a thousand eels in that tide, in that smokehouse…it was two tides" (Yurok elder eeler). In the past, the Yurok never went eeling at night at the mouth because there were enough lampreys to harvest during the day. The lamprey were described as so plentiful in the past that someone could go down anytime when they were running and see them; there were always plenty. Eeling at night is a very dangerous activity both at the mouth and upstream. Almost every year, someone is washed out into the river or ocean. Important lessons taught to all young or new eelers are to "never turn your back on the ocean" and "always know where you are in relation to the river." When someone drowns in the river, fishing and other activities in the river stop for a few days.[5]

More than 30 years ago, everyone came home with 200–300 lampreys, or a couple of gunny sacks, after a night of eeling. "In one tide, you fish the outgoing tides, sometimes you'd grab a hundred by the time you got down to the bottom" (Yurok elder eeler). One participant mentioned that his grandmother spoke of "picking out the better ones." They never had shortages or bad years for eeling. Even 50 years ago, participants have memories of catching 3,000–5,000 lampreys in one night during a peak run.

It was a lot of hard work carrying the sacks of lamprey off the beach or back up from the falls, so even though they could have caught more, most people stopped after they reached the number they needed. Forty lampreys equals about 80 lbs and is a good load up one of those hills. As the fishers in their village, it was important for the eelers to give away most of their harvest to other community members, especially elders. With a catch of 3,000 during one night, an eeler would only come home with 100–150 lampreys because they had given the rest away. Even out in the Scott Valley, stories of catching 500 lampreys up in some of the tributaries are common. Just as it was told of the salmon, the elders recall the lampreys as another fish that you could almost "walk across the backs of them" as they were coming through the Klamath River system by the millions. One Karuk eeler's uncle told him about his grandfather being down at Ike's Falls one night watching them take out 4,000 lampreys. The same eeler mentioned that with a couple hundred people living in that vicinity during those days, that would amount to around two lampreys per person, which would be gone in a couple of days.

Pacific lamprey populations began to decline rapidly in the 1960s. One participant recalled that the last time he had seen a full smokehouse was more than 45 years ago. Nowadays, most smokehouses are smaller and hold only about 100 lampreys, but even that size is difficult to fill in a whole season. While eeling was "no mystery" in the past, these days an eeler has to go down every day during the run to catch a lot of lampreys. "Now knowing where and when is not enough, you have to be there to get eel" (Yurok eeler). Those eelers that harvest 100 lampreys happened to hit a good run and are the only ones down there.

Participants remember that in the 1980s, an eeler was lucky to catch 50–100 lampreys, which was considered a lot. By the 1990s, they were lucky to harvest any. Some years are good, while other years there are hardly any lamprey. Now anywhere from enough to eat to 30 lampreys is considered a good catch, and it can take all day or night. An eeler has to work at it and spend a couple of nights to harvest 100. "When I was a kid, there was just thousands and thousands of eels. By the time you'd go down to check your basket, you could catch 300 in one basket. Nowadays you're lucky to catch, on a good night, maybe 10 or 15" (Yurok eeler). Anything in a basket these days is considered successful. A good night is if you even catch a lamprey; for many, it is not even worth it anymore to spend the night. Eeling has become a form of recreation rather than a means of subsistence. Providing for your elders is a demonstration of respect and a primary responsibility for a Yurok or Karuk person; yet the older people say that they

[5] TEK, WSK: F. Lake suggests that this could serve as an indirect lamprey population escapement benefit (U.S. Forest Service, personal communication).

hardly receive lampreys anymore because no one has any to bring them. One elder only received six lampreys last year while another had not had any in the past 15 years. "I haven't had any since God knows when" (Karuk elder).

One of the Karuk eelers recalled lamprey populations overlapping into one another so there was always a constant run. "[N]ow you just have remnant populations and so when one population goes through, then it's over, and the next one comes up a couple weeks later. There's no continuous flow like there used to be and the peak season is a lot shorter." Another eeler has found that the peak season is not even there anymore. The lamprey are also not in all of the tributaries like they used to be, even up in the Salmon River.[6] "It's not like the eel are going someplace else, they're just not there." Figure 8 outlines the changes eelers have noted regarding the number of lamprey harvested over time using different gathering methods. Harvest sessions for eel baskets are based on three baskets overnight; all other methods are based on 3–4 h sessions.

[6] TEK: See Kroeber and Gifford (1980) for the Origin story of the rock at Forks of Salmon, known as the place where lamprey are supposed to return to every year.

## *Factors Contributing to Population Decline*

Tribal community members point to a number of key limiting factors that have influenced the decline of Pacific lamprey populations in the Klamath basin, as well as other species that are dependent on river resources. These factors can be grouped into prevailing resource management approaches that include upslope management, natural flow regimes, contamination, and predation (WSK, see Potter et al. 1986; Close et al. 1995; Kostow 2002). According to participants, it has been through the continued combined influence of these management approaches that lamprey populations have not been able to recover (WSK, see Moyle 2002). They believe that because lamprey is a flow-dependent species, the loss of water quantity is the main limiting factor for both its decline and their recovery.

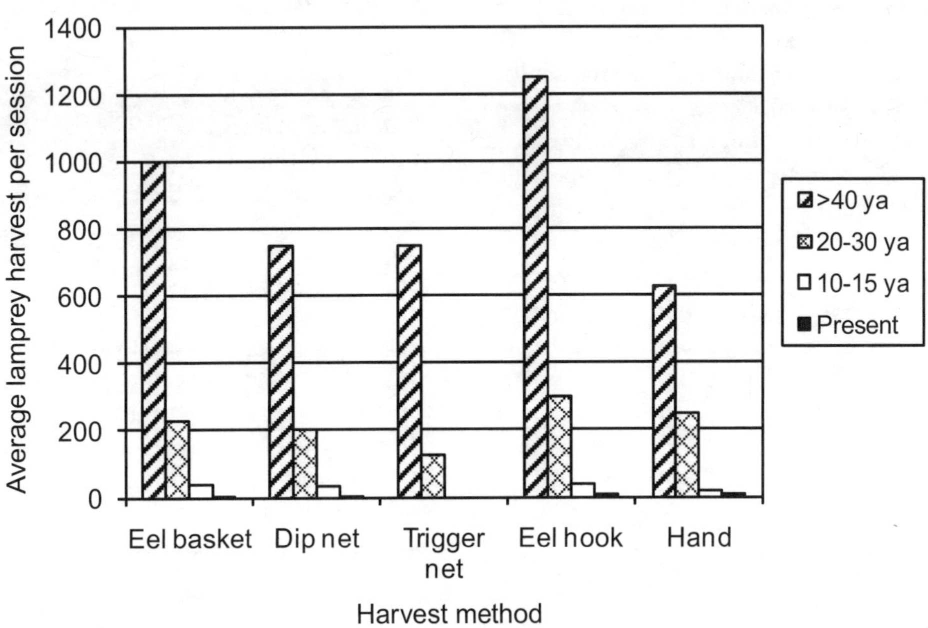


Figure 8. Changes in Pacific lamprey harvest over time (ya = years ago), based on 2004–2005 data. Harvest sessions for eel baskets are based on three baskets overnight; all other methods are based on 3–4 h sessions.

*Upslope management.*—Many participants commented on changes in upslope management, the management of ecosystems that are upslope from aquatic ecosystems, which developed after European settlement. These were described as changes in the fire regime, draining of wetlands, and intensive logging practices. They commented that fire suppression leads to increased brush and an environment of dry fuel that contributes to large, hot fires, which eventually destroy large tracts of forest. Traditional fire management practices that were in place before European contact contributed to the health of a forest by increasing plant diversity and limiting the impacts of large, devastating fires.[7]

> Fire was controlled because there wasn't as much brush and it was done every year so you don't have as much duff and insects that affect the acorns. There are great big huge acorn trees where there hasn't been a burn and out of 20 acorns only 3 will be good. Where there has been a burn, a majority of the 20 acorns will be good (Karuk eeler).

According to participants, today's upslope management practices allow the undergrowth to grow into the canopy, thereby blocking sunlight so that plants, like huckleberry, relied on by both humans and animals, cannot ripen their berries. Additionally, basketry materials need to be young and pliable, but without fire, they become infected by insects and disease and break easily.[8]

Tribal community members have observed that changes in the fire regime not only affect the plants native people rely on for food and basketry materials, it also impacts both the surface and subsurface hydrograph. They explained that whether set through lightning or controlled burns, when fire is suppressed it leads to an increase in forest density and higher rates of evapotranspiration, which uses up the water that goes into the streams, thus changing the hydrograph and the stream flow (TEK, see Stewart 2002). It was stated that when understory burns were set by local people, they happened during the right time of year to contribute to the moisture content for plants and the fish by adding more water into the streams. Increased amounts of brush draw up the water so that it is no longer available to go into the tributaries and the river. One participant remarked that the springs he used to know as a child are no longer there.

> It'll change with every little fluctuation in the water. It used to be a lot easier because the water was always in the same place every season and then lately it's been way lower, I don't understand it. I used to go down there all of the eeling season and then fish right off of these certain, you know, there'd be three or four levels. The water would be to where I knew where they were at that level and now it's like you stand down by the river and you're looking four or five feet up the bank at where you're used to being that time of year, you know, it's that much lower (Karuk community member).

According to many participants, fire suppression has made a direct impact on lamprey populations: "Ammocoetes that are in the fine mouth of these tributaries aren't getting the kind of hydrograph with the quantity and quality of water that they had historically." As another participant pointed out,

> So, if we don't burn, that affects our eels. I know it sounds weird, but it does. Because what happens is that now that we have all this new growth, they all have root structures. And what do roots like

[7] WSK: Investigations into the role of fire in the Klamath Mountains reveal considerable change around the 1940s, at which time fire suppression policies were acted upon. Prior to this time, the landscape revealed that frequent, low-moderate severity fires were a common ecological process (Skinner 1995).

[8] WSK, TEK: Lake (2007) includes extensive discussion on the effects of fire suppression on vegetation, as well as the associated cultural resources. His interview with a tribal elder relates the use of cultural burning by her mother-in-law for managing hazel sticks for eeling: "It was a nice little prairie up there that had a bunch of hazel growing on it that she burnt. And she'd go up and set it on fire. She set it on fire in the fall; she said she was doing it for eel sticks" (Lake 2007:542).

more than anything? Water. So, a lot of our water that is meant for our river is being sucked up by all this new growth (Karuk tribal community member).

Additionally, tribal community members mentioned the historical presence of functioning wetlands that would soak up the water like a sponge and then release it into the river over time. "When you get rid of all those sponges alongside the river all the way up into the Upper Basin, and there were huge amounts of wetlands that got disrupted, then when you have a winter event now, instead of it all soaking up into that sponge, it runs down into the river" (Karuk eeler). They have noticed that with the loss of wetlands, large flood events are becoming more common and are occurring at a greater magnitude.

Many participants believe that logging practices in the past also affected the streams through the loss of gravel bars, increased silt and sedimentation, and the build-up of debris and gravel filling in the creeks. The Klamath basin has a long history of logging and sediment build-up. Management practices historically did not have all of the stringent laws they have today. It was not uncommon for the streams to be used as roads and to transport logs. As one participant put it, "the laws were made after the damage was already done." Other participants spoke of small creeks that used to have salmon and lamprey fisheries in them but are almost dry now, either filled in with sediment or they no longer have the contributing flows from the surrounding upslope environment.

Large redwoods used to grow up to the stream's edge, shading and cooling the river. One participant said that his uncle described it as a cave of trees hanging over as he made his way up the river. Another participant remarked that as the water quantity decreases, the water temperature increases. They continued by saying that since Iron Gate Dam is an earthen dam, the water heats up as it comes out over the top.[9] They believe that less water also results in more fish bodies moving around in the same space, heating up the water and creating an environment susceptible to greater disease transfer and higher mortality. One of the Karuk eelers remarked that the water temperature needs to be a certain way because the Maker made everything in balance. People think they can regulate this way or that way, but it is not the way it was designed to work. As many people I talked with attested to, the lamprey is a hardy species, a primitive creature that has evolved in response to many environmental changes throughout time. The decline of one of the most resilient species is an indicator that the system as a whole is in decline. A Yurok tribal fisher concluded, "If we're losing him and he's the toughest along with the sturgeon, holy mackerel, no wonder we're losing our salmon because of the temperature, the water, and all that."

*Natural flow regimes.*—Almost every tribal community member interviewed remarked on the loss of spring freshets and its impact on the health of the river system, including the loss of fish populations. The disruption of natural flow regimes through the building of dams, logging, mining, and changing weather patterns has contributed to a number of detrimental changes in stream morphology (Yurok and Karuk tribal community members; WSK, see Doremus and Tarlock 2008). Historically, streams were flushed out every year, cleaning out the sand and the gravel and taking out the willows along the sides of the banks. Participants noticed that impacts from management practices are magnified without annual high waters to flush and cleanse the system, contributing to higher rates of sedimentation and woody debris contribution, the loss of spawning and rearing habitat, decreased water quality, and the intensification of flood events.

Tribal community members observed historically that the clean sand areas along the river where the ammocoetes live were washed and cleaned when the river was high. Now when people dig into the sand bars, they run into layers of old leaves. They noticed that hydromin-

[9] Iron Gate Dam is constructed of rock-fill with compacted clay core and concrete cut-off wall (Source: www.pacificorp.com/File/File16143.pdf).

ing took all of the soils from the hills and put them into the streams, affecting the spawning grounds. They also believe that pesticides and other pollutants stay in the system longer because they are not flushed through by spring freshets.

The floods of 1955 and 1964 had extensive impacts on the Klamath basin. Participants remarked that historically, the rains and the snows would come every year and the river would rise in the springtime. Soon the river was back to where it was the year before and nothing changed. They had the same fishing holes. Now family fishing holes that had been used for generations have changed dramatically or no longer exist. One Karuk tribal community member remarked that the river and creeks are not as big as when she was younger. She described pools so enormous, up to 10 m deep, that children refused to swim in them because they could not see the bottom. Another person recalled going up the creeks when he was younger to areas where there was a lot of sand and small holes behind logs where they would catch trout. Now he can go up a mile and not catch anything because the holes are filled in with rock from the 1964 flood.

Floods are not an unusual occurrence on the Klamath River, but participants found that the effects from the 1964 flood, especially, were exacerbated by changes in land topography due to management practices within the basin.

> Now these hundred year events have been hastened to where now you have every ten years a hundred year event and I can look up to this upslope and say that's the reason why, logging, things like that, to where now instead of functioning wetlands which actually soak up the water like a sponge and as it gets in here and the water goes down, then that water releases into the river. When you get rid of those sponges all alongside the river all the way up to the Upper Basin, where there are huge amounts of wetlands that got disrupted, so now you have a winter event, instead of it all soaking up into that sponge, now it runs down into the river and now they're having these catastrophic events happening quicker, but they're having more magnitude; then you have the sediment and everything else behind that (Karuk elder eeler).

One participant remarked that prior to the floods, he remembers seeing black clouds of lampreys coming through the system, something he never saw again afterward. Eelers noticed that the floods affected the spawning grounds of many fish species, including lamprey. They explained that in the 1960s, there was a timber harvest boom, with more roads being built throughout the lower basin. When the 1964 flood occurred, logging roads washed out into landslides and clear cuts contributed massive amounts of sediment into the tributaries. Big rocks piled up at the tributary mouths, and when the large amounts of woody debris came down the streams, they dammed up until enough force blew open the mouths. The tributaries were changed dramatically; holes were filled in with sediment and sandbars were washed out or dirtied up with mud and leaves. One avid tribal fisher remarked, "The leaves stop the water and all the fine silt settles there." He now finds that when he goes fishing, he has to wade out to about thigh deep to get past the leaves and the mud that will sometimes be a foot deep.

Many tribal community members attribute the decline of lamprey populations to factors affecting the larval stage.

> One of the main reasons that the eels are being killed off is that there is no sand for them to incubate in. The kids used to find eels in a big pool at the sandbars below Clear Creek. The sand was fairly clean and they'd find a lot of eels in there, now it's all mud and clay. The sand needs to be open enough that the water can slowly flow through it and bring oxygen to the eels. When the sandbars are all mud, it suffocates them (Karuk tribal community member).

Many participants noticed a decline in water quality soon after Iron Gate Dam was built. The Klamath is one of the few rivers that becomes cleaner as it moves downstream, due to all of the tributaries contributing freshwater into the system. Many people who left the area for a period, upon returning, noticed a dramatic change in the river, noting stagnant, slower flows, strong odors, dirtier water, more moss and algae, and higher temperatures. Those who used to swim in the main-stem river refuse to now because of the decline in water quality. "There were always a few eddies where the water tended to slow down and the mud would grow up, but now there's algae and moss all along the river and you can see the line of leaves up along the river. The mud stays mixed in with the sand" (Karuk tribal community member).

The disruption of a natural flow regime has created what one participant called, "a perpetual state of drought." He remarked that lampreys are a hardy animal, but there is something in the system to which they have been unable to adapt, especially regarding water quality in their larval stage. The lack of spring flows and the fluctuation of flows, in general, are conditions related to the dams (Karuk tribal elder). Especially during the summer or early fall, many ammocoetes are found stranded when there are drastic increases and then decreases in water released from the dams, which dry up the sandbars and take away the edge habitat; some people refer to them as "invisible fish kills."[10] Many participants recalled finding hundreds, even thousands, of stranded ammocoetes left in little side pools where the water had become hot or dried up. A tribal eeler who works for one of the fisheries programs remarked,

> And also, they got a long life span. They got to be in the river for so many years. Well, all these generations are in there through all these years and when they put that Iron Gate in there it...when irrigation time came around, boom, first thing they did they'd be pumping out maybe twenty, twenty-two [hundred] cfs up, down. Overnight they'd shut it down to a thousand. And it'd leave all these, where their beds are, where they spawn, leave them high and dry. And then they couldn't get back to the river fast enough before they died. And then the eight-year cycles, it takes a long time before it can rebuild. And you get all them spawning beds dead. I think that has a lot to do with it (Karuk eeler).

During natural flows, the water would go up and then decline gradually, so the fish would be out of those high water areas once the water was down and they would have minimal stranding.

The diversion of water from the Klamath basin for agriculture in the upper basin, the Shasta and Scott River valleys, and the Central Valley impacts water flows for the lower Klamath. Tribal fishers observe that each species of fish spawns at certain elevations and topography, so if the water drops too much, then the larval and juvenile stages do not have a chance for survival. The water has not been high enough in the tributaries to achieve spawning habitat for lampreys or other fisheries. Participants remember that when the creeks were big, the fish did most of their spawning up there, where there was not as much of a chance of being wiped out when the water did come up. Lampreys are not going into the tributaries as much now because the river is so low. According to one participant's grandmother, the fish never spawned in the river, they all went up into the creeks. Tribal fishers explained that the river did not have the sediment loads that it has now, which come down Ishipishi Falls, settle into, and fill up the deep fishing holes. The fish have no holding room like they used to or the hole

[10] WSK: "Karuk Fishery crews located large ammoc[o]ete kills when rescuing fish from stranding pools. Current ramping practices are causing potentially irreparable damage to future lamprey populations by ramping water too fast from Irongate Dam. When the water level is dropped unnaturally fast, the ammoc[o]etes that occupy the fine sediments along the river margins end up stranded in dry sand bars." (Karuk Tribe 2004).

might be out in the middle of the stream, so fishers have to build platforms to get out there. In the past, the fish were closer to the sides so they only fished in the backwaters. Eelers now risk their lives building bridges or platforms to get out to where the fish are moving.

As children swimming in the Klamath River in the summer, community members remember trying to swim straight across the river and ending up a third of a mile downstream. Their elders would often warn them about the river. Nowadays, most people find it easy to swim straight across. The entire river margin habitat for lamprey is drying up, which is mainly impacting the larval stage and has the greatest impact on the depletion of lamprey runs. Participants in the Scott River Valley witness lampreys going up into the Scott River to spawn in the irrigation ditches; these are later dewatered and the ammocoetes become stranded and die.

> The decline has to do with when they are filtering in the silt. We've found strandings of hundreds and hundreds of eels, all dried up in the sand. The ammocoetes like the silt and sand, so you find them along the margins of the rivers. When they ramp the water down 200 cfs overnight, they don't really come out and wiggle across the land to get back in the water, they pretty much dry up where they are. We're monitoring a few spots where it happens regularly, but I'm sure it happens all up and down the river (Karuk eeler).

*Contamination.*—Nonpoint sources of contamination in the Klamath watershed are attributed to agriculture, road runoff, and spraying. A common comment during interviews was that lamprey populations began their major decline during the 1960s when the timber companies began spraying up and down both sides of the river with chemicals like 2,4-D. Spraying was heaviest during the 1980s when many plant, animal, and insect populations were affected (WSK, see Norris et al. 1991). Community members mentioned not seeing any more bees, birds, squirrels, chipmunks, or porcupines. Stories were common about deer with tumors, deteriorating skin, or blindness, as well as birds with large and mutated eggs. "Wherever they sprayed, there was no life" (Yurok elder). Human impacts ranged from birth defects to high rates of rare cancers. Millions of candlefish, or eulachon *Thaleichthys pacificus*, used to come into the estuary to spawn during the same time that lamprey were migrating into the river. One participant remarked that in 1963–1964, it seemed like the river was poisoned and the candlefish tried to find another habitat. By 1975–1976, candlefish numbers were so small, it was as if they were completely gone (TEK, see Larson and Belchik 1998; Yurok tribal archival records).

Participants have found that pesticides, fertilizers, and other agricultural runoff to the river create an imbalance in chemical and nutrient loads in the system. Toxins are observed accumulating as they move down the river and, with a decrease in water quantity in the system, are not diluted as quickly. One of the frequent observations made by tribal community members is seeing an increase in large amounts of white foam in the river. Even as late as 1987, it was not uncommon to see fire retardant drops into the river after a forest fire had been put out. Only in recent history have agencies been required to keep records of these drops.

Participants told stories of vast numbers of lamprey that once came through the Klamath River system and clogged dam turbines and irrigation diversions. They spoke of eradication programs apparently carried out by California Fish and Game in the 1970s to poison lampreys that were seen as a nuisance to dam operators and farmers. While records of these programs on the Klamath are not readily available and were not mandatory at the time, participants claim that dam operators on the Columbia River have records of their own lamprey eradication programs during the same period of time (WSK, see Close et al. 1995). By analogy, similar Klamath programs may have been in operation at the same time. While fishing for other species, tribal fishers recall hundreds of dead poisoned lampreys getting caught up in their nets in the early 1970s.

One elder eeler summed up the multitude of contamination factors that have contributed to the decline of lamprey populations in the Klamath basin:

> Well, I think they should be looking into why they're allowing all that poison to be put on the earth and they could care less for the animals, birds, or whatever, Indians and their eels or fish, salmon to live. And the fact that there is poison being put all in our Klamath River Basin from...all the way up there and the farmers are putting something on the earth to make their potatoes bigger and better. And then there's Simpson Timber Company and other companies, logging companies are allowed to spray. CalTrans is putting a lot of bad things on their roads and highways and byways. They're not concerned, all they're concerned about is progress and, if anybody's objecting to that then they probably would be considered a terrorist (Yurok elder eeler).

*Predation.*—Historically, native people of the Klamath River took fur seals and sea lions for their teeth, hides, oils, and meat to use for ceremonial and subsistence purposes (TEK, see Kroeber and Barrett 1960). The inception of the Marine Mammal Protection Act of 1972 prohibited the taking of marine mammals in U.S. waters.[11] Many fishers speculate that sea lions and seals are getting more abundant because they are so protected. "Yeah, you know, that's one of our big problems down there, too many seals and sea lions down there." A story was told of cutting open a dead sea lion and watching the lamprey roll out "like candy." Another Yurok elder eeler said, "I really feel terrible because we can't do anything about them anymore and you have to stand there and watch them eat the eels and you don't have any in your sack" (WSK, see Merrell 1959; Bowlby 1981; Roffe and Mate 1984; Williamson and Hillemeier 2001).

## Ecological Significance

Many participants recognize the significant role Pacific lamprey have in the health and balance of the Klamath River system. For most, the loss of the lamprey is another symptom of what is going on with the river. They see its recovery as key to the survival of the river. As a Karuk tribal community member explained,

> When the eels don't come, it affects everything. If people aren't eating them, then they may have to harvest more salmon than they'd like to. The eels are here for a reason and are not here to be desecrated or made extinct by man. They're supposed to be here to help feed the animals. Everything will suffer when they're gone.

*Seasons.*—The people who live along the Klamath River use environmental cues to determine when it is time to go eeling. "The old timers long time ago, boy, they seen that swallow, they knew those eels were on their way" (Karuk tribal community member). Especially upriver from the mouth, these indicators from nature told people what time it was when they did not have clocks or calendars. Nature moved the seasons along and every year it was the same. "[T]hat's when we're inclined to say the temperature and parameters are right to where we should go fishing right now" (Karuk elder eeler). Depending on where they live on the river system, each family has their own indicators that tell them when the lampreys are running. One person would tell me that when the dogwoods are blooming and the crickets are singing, then the lampreys are at Ikes Falls. Another would say that when the big creek fern blooms, the lampreys are in the water. "They call it the eel cleaning fern because they use it to hold onto the eel when they clean it. It still happens now, when the fern blooms, the Great Spirit is telling them that it's the cycle of the eel" (Karuk elder).

[11] U.S. Waters, except for specific Alaskan native activities.

Other indicators include moths flying around, buzzards and osprey returning, hummingbirds, crickets singing, and seals and sea lions feeding. Some people went more by weather conditions, "Water is warm and it rains, it's good eeling" (Yurok elder).

> The eels like to run when it's warm. It's gotta be up to about around mid-50s, 60s, between 50 and 60 degrees and the frogs really start hollerin' and the moths start flying around. Can really tell that's the time that they love to run. Catch 'em in your basket and you pull your basket in and they're just floppin' all over the place 'cause they're runnin' so hard, you know, in the morning. They really get lively because of the warmth (Yurok eeler).

In general, though, most people agree that the calendar is off these days or needs to be recalibrated. The trees are blooming when it is still winter and the fish are coming in later. One tribal eeler remarked that the migration timings may be obscured by the bell-shaped distribution of the lamprey run. Historically, the runs were so large that everyone knew when they started coming up into the system. These days, the numbers are low enough that the few that are coming up will be moving up through the middle of the river where they are not noticed until the run reaches its peak. In general, however, tribal members are noticing that the return of neotropical birds and the flowering of the plants are occurring 1 or 2 weeks earlier than in the past. As some fishers noted, sometimes the lampreys just do not come at all. "You see the dogwoods bloom, the buzzards are here, and the swallows are flying around, but no eels" (Karuk eeler). If you get disconnected with one species, then it disconnects you from season to season.

Lamprey has its place to feed the people and to be a part of the cycle, like salmon, deer meat, and berries.

> That was the importance of eels, eating eels that time of year because different times of year we'd be eating those trout, but then at that time of year that was not only a trigger for the fish but it was also a trigger to say quit fishing for those other species that are spawning" (Karuk eeler).

A Yurok elder told me that the Creator made the lamprey and gave the Indian people the river; the river brings up different kinds of fish each season. Historically there were no dead spots in between runs of lamprey; the runs of different species would overlap. Even in the recent past, lampreys were harvested at the mouth near the end of November in time for Thanksgiving. Now eelers find a few coming through during that time period, but most eeling at the mouth begins in January and goes through April when they turn to other fishing or the lampreys begin to get wormy. Lamprey harvest in the Orleans and Somes Bar area, near the confluence of the Klamath and Salmon rivers, usually begins at the end of March and finishes around mid-June. In the Scott River Valley, harvest takes place during a short window from July to mid-August.

*Interspecies dependency.*—An understanding of the interdependency of all species in the Klamath River system goes back through the stories. The story of the Eel and Sucker gambling game illustrates this deeper knowledge.

> Sturgeon was drumming for Eel during the betting and Bullhead (sculpin) was drumming for Sucker. Eel and Sturgeon lost everything they had, so in the last game they put up their bones and lost. This is why Sucker and Bullhead have so many bones and Eel and Sturgeon have none. Sturgeon is still going up and down sucking on the bottom looking for enough gold to buy back his bones (Yurok elder eeler).

The people who live along the Klamath River and those who harvest lamprey recognize the important connection the populations have with the health and preservation of other species in the system. "As a fisherman, you just know that they're connected, you know you've got to have one for the other. And if you lose one, you're going to lose the next one" (Yurok tribal community member). Lampreys are here for a reason and affect even those things that do not feed on them. They are an integral part of the ecosystem. As a Karuk tribal community member remarked,

> There's a reason the eels evolve, a reason they came into the Klamath River, and there's a reason they need to always be there. We may not know what those reasons are, but because we don't know what those reasons are, it's even more important to protect them (Karuk tribal community member).

Lampreys were described as water quality indicators. If the river is not healthy, then the lampreys and other resources cannot be healthy.

> It's all connected…there's nothing you're going to do, you're not going to just save the salmon, you're not going to do it. If you're going to save the salmon, you're going to have to save all of them because you're going to have to fix the system. I think the more species you look at, the further ahead that you are in the game and fixing the entire system. Because you're just looking at it from one species, what that one species needs and you're going to come up short (Karuk eeler).

Many participants believe, at one time, lampreys probably had the largest biomass of any living species in the river, when considering both the mass of ammocoetes and migrating adults. This mass of lampreys moving through the system helped to feed the predators and acted as a decoy for the salmon coming through during the same time of year. Once lampreys spawn, they die and are recycled back into the system. They become "food for other fish or help prepare the waters" (Yurok eeler) (WSK, see Kostow 2002). Almost all participants remarked on the significant difference they noticed in the number of dead lampreys moving through the system in the past decade. When they were children, they remember seeing many dead ones settling in the swimming holes in the creeks and rivers in the summer. When he was younger, one participant recalled thinking that the leaves in the water were "eel beds." It was difficult to swim without stepping on or seeing lampreys float by in the water.

Fishers would have to pick the dead lampreys out of their nets at the beginning of steelhead *Oncorhynchus mykiss* and spring salmon seasons around July. Most people recall noticing a decline in the "floaters" in the 1970s and early 1980s. Before they started to decline, it was difficult to drive up and down the river without smelling the rotting lampreys.[12] Thousands of dead lampreys used to float down in the late spring or summer. "Old float back eel" or per-ner-keesh, in Yurok, was a derogatory name you call someone. One participant told me how his grandmother and great-grandfather would walk along and collect the dead lampreys that used to be all over the beach. They would boil them down and use the grease to saturate their dugout to keep it soft so it would not chip when it hit a rock. It took a lot of lampreys to obtain the thick amounts of oil they used for one dugout. These days, there is no odor around the river and except for those who work with the fisheries programs, most people have not seen a dead lamprey in years.

In the past, the holes would be so loaded with dead lampreys that people said they would only drink out of the side streams; the dead lampreys turn into a white "juice-like powder." Karuk tribal members recall seeing them on the shores and at the bottom of the pools of the creeks from the end of July to early August. The dead lampreys fertilized the river and replenished the riparian zone. They were also an important food source for salmon fry and other predators. "Well, they aren't only important to the people, they're important to our river. The

[12] TEK: Lake (2007) interview with a tribal elder recalls the role of dead lamprey in the river system: "There is no more eels. In fact at one time in the year the eel were so thick you, didn't want to go close to the water because of the smell of rotten god damn eels…and what that was, was the fishes' food" (Lake 2007:425).

river needs all the nutrients. The dead eel that goes up to spawn and it dies and it floats back. The buzzard eats it. Everything eats off it when they come back" (Yurok elder eeler). In the past, carcasses were seen deposited at the bottom of the deeper holes where the sturgeon came in to spawn, which would then eat the dead lampreys before making their way back to the ocean. "Eel fed a lot of fish, he did after it died, that's why there were so many of them. That's why there was an abundance, that's why the Great Spirit had them in the river as an abundant food" (Karuk elder). Participants recalled seeing a variety of animals feeding on them, including gulls, eagles, ravens, crows, bears, and buzzards.

> The dead eels, when they came back down the river, my god, buzzards and ospreys and all of those birds had ample food to feed their young and they were in abundance around here. But now it's a thrill just to see any bird, let alone know they're around all the time. You know it's affected their food chain, as well (Karuk eeler).

The dead lampreys would fall apart and wash close to the shores, feeding all of the young fish and decomposing into nutrients that the ammocoetes then filtered through the system.

> I think the eel was one of the most valuable assets to all the fish in the river because I think he created all the food for them. Because the salmon in there, there wasn't enough dead salmon to do it, but there were so many dead eels that they all washed up along the shores, all your creeks, and the main river here thick, real thick and heavy and then they would just turn into that milk and foam that would float down the edge of the shore and the bank that the little bitty fish live on. Without the eel, we ain't got that (Karuk elder eeler).

> The eel has its place, you know, they go up, they feed our people. And when they'd go up there and they died, they floated back. And after, there's a cycle thing, you know. They come up and they feed our people, they go up and spawn, and when they died they floated back and they'd land on the beaches. And about that time, the buzzards would come back, they'd come back down South and they'd go along and clean the beaches. Things fell pretty bad there for a while; they'd clean them up real quick. That's the buzzard meal when they first came back in the spring. Come along and clean the beaches up. Plus it fertilizes our rivers and all, because the river it will die if it ain't, if nothing is coming up in it, no life, because all the vegetation along the rivers, all the nutrients will wash out after awhile because there ain't no dead matter" (Yurok eeler).

Both the live adults and the ammocoetes are an important food source for other species (WSK, see Stewart et al. 1983; Whyte et al. 1993; Close et al. 1995). Adults are seen being fed on by osprey, eagles, hawks, river otters, fishers, mink, martin, blue heron, and bear, anything that is fish eating and can get a hold of them. Many participants downriver remarked on not seeing any of the old gray-haired whiskered otters anymore that rely on lamprey in the winter. Everything begins to decline when there is not any food in the river system. Tribal fishers witnessed ammocoetes serving as food for cutthroat trout *O. clarkii*, rainbow trout (nonanadromous *O. mykiss*), mergansers, kingfishers, gulls, terns, and many other species (WSK, see Moyle 2002).

## Summary

The following are the major contributions to our understanding of Pacific lamprey populations in the Klamath basin that came out of this study of Karuk and Yurok TEK, as well as the ecological and cultural characteristics of the Klamath River system.

1. Two morphologically distinct Pacific lampreys were observed along the main-stem

Klamath River. One run was blue and larger while the other was darker and smaller. While some participants have seen both types, those who eel primarily at the mouth only mentioned seeing the larger blue lamprey. Eelers who live and eel up in the tributaries only saw the smaller dark lampreys.

2. Adult lamprey have been observed to display specific behaviors in response to changing water conditions such as temperature, water flows, and particulates in the water. Lampreys are thought to utilize a "breather hole" at the top of their body during muddy conditions. They also use one another to move over barriers, creating a wedge of lampreys that leapfrog their way over rocks and dams.
3. Participants observed groups of lampreys moving together through the system and following a specific "trail" through the water. That trail altered when the next group came through and responded to changes in water conditions.
4. Pacific lampreys were observed to be sensitive to electromagnetic activity such as lightning and solar flares, during which they either moved deeper into the water or came out of the water onto the rocks in abundance.
5. During normal run years, the harvest of Pacific lampreys in the Klamath River basin begins at the mouth in late November, at the Somes Bar area (the confluence of the Salmon and Klamath rivers) near the end of March, and upriver in the Scott River Valley by late July.
6. Historically, Pacific lamprey mostly spawned in the tributaries of the Klamath River. They were found in all of the tributaries, moving higher up in the system to spawn than steelhead populations. While eelers and other fishers typically do not see lampreys spawning, they have witnessed dead lampreys floating down from above where they fish for other species.
7. Eeling is a flow-dependent fishery. Eelers have developed over thousands of years the most appropriate technology to harvest lampreys based on their knowledge of the morphology, flow, and substrate of the stream, as well as the behavior of lampreys in response to specific conditions.
8. Tribal community members noticed a correlation between the extirpation of eulachon populations, another high fat species, from the river system during the same time period that lamprey populations began to decline. Participants expressed concern over the toxins in the river system as a limiting factor for populations, as well as a potentially serious threat to the humans who consume them.
9. Tribal community members noted predation from seals and sea lions as one of the major limiting factors for Pacific lamprey populations.
10. Participants observed that the number of ammocoetes, adults, and dead lampreys seen in the system has declined dramatically in the past 40–50 years. They correlate-these changes with intensified upslope management practices (i.e., logging, herbicide spraying, dams, fire suppression, and wetland destruction) and changes in the natural flow regimes, which have affected spawning and rearing habitats. Tribal community members believe that rapid increases and decreases in water released from Iron Gate Dam, especially during the summer and early fall, create invisible fish kills because the larvae are left stranded in the up per substrate.
11. The combination of intensive logging practices, hydraulic mining, wetland destruction, water diversions, loss of spring freshets, and road building have created a system prone to the magnified effects of natural episodic floods. These flood events have depleted lamprey spawning and rearing habitat, as well as generated a scarce primary food base and limited opportunities for population regeneration. Historically, seasonal spring freshets were on a smaller scale and cleaned out the river sys-

tem by washing out sediments and taking out the willows along the banks.

12. Fire suppression has led to an increase in forest density and higher rates of evapotranspiration, so water is used up that would otherwise be available in controlled or cultural burn areas to contribute to the hydrograph and increased stream flows.
13. Tribal community members recalled a considerable biomass of Pacific lampreys moving through the Klamath River system in the past, serving a significant role in the ecological integrity of the ecosystem. They believe that lampreys help prepare the waters in their own way of being a part of the circle of life. Whether as prey for fish-eat ing species or scavenging matter for salmon fry and sturgeon, the lampreys are a prima ry food source. Pacific lampreys are also essential contributors of marine-derived nutrients and organic matter.
14. Tribal fishers place a high value on environmental indicators to know when lampreys are running in the system. These indicators may include dogwoods blooming, crickets singing, or frogs croaking. This "calendar" is a management strategy for knowing when to harvest a species and when to leave a species alone. If these indicators are affected by changes in the environment, then indigenous people will be the first to observe that change.

## Discussion

Western science and TEK are each limited in their abilities to interpret the complexities of today's natural systems. Human impacts on the environment and inadequate management decisions of the past have created natural systems that require a more holistic approach to resource issues. Modern resource management lacks an adequate framework to link ecological and social components of the environment.

> It is a cultural landscape. Everything in it is related. And that's the way the tribal people have always looked at it. And finally, some biologists, in certain circles, are beginning to use the words, 'ecosystem management', which means everything has to work together. The only thing with ecosystem management is they're still, mostly, leaving the people and their hearts and their feelings and their culture out of it, which is also a part of it. It's a whole (R. Pierce quoted in a documentary by Chamber 2001).

To mitigate these issues, it is important to include TEK in the process formally (Sallenave 1994). TEK can provide a complementary perspective to the standard scientific approach, providing avenues for further research and collaboration. The everyday experiences of the local fisher and their knowledge of natural history can provide important information for managers (Stoffle et al. 1994). In traditional management systems, the core is derived "from the observations and experiences of generations of fishermen and fisherwomen working in environments with which they are intimately familiar" (Ruddle 1994:175). Additionally, TEK can be viewed "as a mechanism to implement co-management and self-government, and as a mechanism to integrate local values into decision making" (Berkes and Henley 1997:31).

While it is ethnocentric to assume, or insist, that all TEK can be or should be translated into Western concepts, the alignment of these knowledge systems is an important step to developing comprehensive, system-based fisheries management.

Since the historical perspective of lamprey as a trash fish created an environment in which it was essentially ignored in Western fisheries management, there is little data available about Pacific lamprey populations in the Klamath basin. Tribal participants were able to provide a wide range of information about both lamprey and the Klamath River system, including knowledge that is currently lacking in Western science. Additionally, participants were able to share a local perspective and understanding that has been developed and fine-tuned over thousands of years about the specific ecology of the lower Klamath basin. They are able to make connections between lampreys and other aspects of the river system that are just beginning to be understood and/or respected by Western science.

Tribal fishers and community members provided a number of key insights into Pacific lamprey populations in the lower Klamath basin. Participants keyed in on critical points in history when lamprey numbers declined and speculated on corresponding factors that may have triggered detrimental changes in spawning and rearing habitat or directly impacted the health of different life stages. They found that the loss of a natural flow regime has had the most devastating impact, reducing water flows and water quality necessary for productive spawning and rearing habitat. More frequent large-scale flooding has had a dramatic impact on family fishing holes and fish habitat by altering stream structure and dynamics significantly. A crucial point is the dewatering of ammocoetes habitat, creating "invisible fish kills" by stranding ammocoetes with rapid drops in water levels.

Participants described biological and behavioral characteristics of lamprey that may provide insights into both its resilience and sensitivity to changes in the stream system. The persistence of toxins in lampreys and in the river system is an important concern to address. The oil rich properties of lamprey may contribute to the bioaccumulation and biomagnification of contaminants being brought into the river system. Lampreys pick up toxins as they filter feed in the backwater areas of freshwater environments and when they feed on fish and whales as adults and juveniles in the ocean. When they return to spawn, there are likely bioaccumulated toxins in their bodies. These toxins may be passed on to the humans who eat them and to the river system once the adults decompose. Tribal fishers also noted previously undocumented behaviors involving light and electromagnetic reactions or sensitivities. Additionally, elders noticed that lampreys have changed their spawning behavior; they used to only spawn in the tributaries of the Klamath River, whereas now they mainly spawn in the main stem.

The function of dead lampreys in the system has not been considered in current scientific literature. The considerable biomass contributes both to freshwater nutrient loads, as well as being as a food source for multiple species, including sturgeon and salmon fry. Along with salmonids species, Pacific lamprey are essential contributors of marine-derived nutrients and organic matter (Beamish 1980). Especially within the nutrient-limited systems of the Pacific Northwest, biologists have found that even small nutrient pulses are found to increase benthic macroinvertebrate abundance and elevate freshwater productivity (Wipfli et al. 1998, 1999; Naiman et al. 2002). Biological studies also emphasize the importance of anadromous species as a significant link between aquatic and terrestrial ecosystems (Willson and Halupka 1995). This research brings to light that a drastic reduction of dead lampreys in the Klamath system has occurred in the past 20 or more years, and they are now only seen in the river in small numbers. The consequence of this loss is not only that Pacific lamprey populations are at risk of extinction; it may be a warning that a comprehensive river system collapse is soon to follow or is already in progress.

Tribal fishers believe eeling is a flow-dependent fishery. The harvest methods utilized by eelers are determined by the morphology, flow, and substrate of the stream, as well as the behavior and relative abundance of the lampreys (Table 3).

> Thus, different environmental conditions ruled the life cycles of these species, controlled the methods by which each might be taken, and gave rise to the different devices: weirs, nets, traps, spears, harpoons, gaffs, and other inventions of the primitive fishermen (Kroeber and Barrett 1960).

Conversely, the chosen methods for a reach of stream are indicators of that stream's morphological characteristics and the lampreys' relative abundance for that area. Similar descriptions of the methods portrayed by tribal participants can be found in older ethnographies. Kroeber and Barrett (1960) draw from ethnographic records dating between the 1860s and the 1950s. For example, these records describe how an eeler chooses where to place his eel basket, and in doing so, also illustrate stream conditions and lamprey behavior:

> The fisherman always chooses a shallow spot in the river where the current is strong, but not too swift, and where the bottom is gravelly, the sort of place favored by the lampreys for spawning. The lamprey builds a kind of nest by carrying pebbles with its suctorial mouth and dropping them where its eggs have been deposited until it has built up a small gravel mound. The lamprey, particularly the male, tends to travel chiefly at night and to spend its days attached or concealed among rocks on the river bottom. The eel pot is, therefore, an ideal place for the lamprey to seek shelter (Kroeber and Barrett 1960:70).

These historical descriptions support evidence that similar harvest methods have been used for multiple generations. Specific harvest techniques used in the past for a given location may be indicators of lamprey presence in those streams, as well as indicators of the historical stream morphology.

## *Working with Traditional Ecological Knowledge*

As the federal government continues to pass legislation requiring state and federal agencies to work closely with local tribal communities in resource management issues (e.g., Secretarial Order 3206; National Environmental Policy Act; Executive Order 13175), creating opportunities for aligning TEK and Western scientific knowledge becomes more imperative. Many agencies are finding this assignment overwhelming or unnecessary, so oftentimes the extent to which tribes and local communities are involved in resource management decisions is minimal.

> Science-based management, however, appears to offer managers a way of restricting the uncertainty inherent in natural-resource management; it is, after all, "scientific," thus promising a degree of rationality and precision that nonscientific approaches are believed to lack. It is therefore to be expected that decision makers, grounded in the belief that science offers a powerful means of exerting control, and largely ignorant of alternative nonwestern cultural traditions, will necessarily remain partial to the advice originating from science-based state-management institutions (Freeman 1989).

From the Western perspective, the idea of incorporating local knowledge into management decisions, let alone supporting local resource management initiatives, may seem daunting at the very least. However, it is increasingly imperative in issues concerning Pacific lamprey, where the "lack of available evidence" may influence decisions that could adversely affect populations.

A number of lessons came out of this research regarding working with TEK. One is that TEK cannot be compartmentalized; it includes not only ecological, but also social, cultural, religious, and political dimensions. This holistic perspective means that working with both Western science and TEK can sometimes create translation issues. When a tribal community member utilizes a word to describe a behavior or physical characteristic, that same word may have a very different meaning to someone coming from a Western perspective. Additionally, some TEK may not have a counterpart in Western science but are insights into a more comprehensive understanding of Pacific lamprey and the river system. Participants are products of their history; so many times, they utilize both Western scientific knowledge and TEK to develop their own understanding of

Pacific lamprey and the Klamath River system. They may use Western concepts because they have a Western science background; however, these same words are incorporated into and understood from both a Western and traditional worldview.

The complexity of TEK means that as outsiders we are not merely looking for common ground or data to support our own hypothesis. The tendency "is to find where the tangent points exist with Western science and to proclaim, quite rightly, that Indians arrived at the same conclusions using a much different epistemology or metaphysics" (Deloria and Wildcat 2001:5). This approach provides the opportunity for communication, but it is not the whole goal, nor is it always culturally appropriate. Additionally, we must be cautious in assuming that what is true for one indigenous group is true for all indigenous groups. As we recognize the diversity among tribes, we need to go beyond superficial appearances and understand that most cultural differences arise from divergent ideas about nature and human connections with nature. As "experts" from the dominant worldview, it is easy to fall into the trap of ignoring these differences by analyzing and redefining TEK into our own categories of understanding.

Rather than try to work with TEK on their own, it is suggested that Western fisheries managers include the people who have the traditional knowledge into each step of the management process. "The consultative integration of local folk knowledge in management decisions can increase the legitimacy of those decisions" (Jentoft and Mikalsen 1994:313). One participant who works in a Western scientific field, explained to me that,

> Managers need to start listening to Indian people and those scientists that think they can't learn anything from the indigenous population. It is very interesting to read what the Indian people said to Kroeber in 1880 or so about how it should be managed and what's wrong. And it's true, those things that they told him are true, those cause and effects, they happened. The Indian people were very observant and they had to know things very intimately to be able to survive. And for thousands of years of observing or given special knowledge from the spirit people, they gained that information and they had a knowledge of this area. A lot of times Western science wants to have these control plots and measurements and if doesn't happen five times in a row, then it's not legitimate. Most of what I have seen is that the Indian people know when things are going wrong (Karuk tribal community member).

The acceptance of traditional ecological knowledge does not mean we are utilizing a knowledge base that is centered on historical conditions. Nor does the utilization of Western scientific knowledge in traditional management mean that we are losing aspects of the traditional. "'Modern' influences do not necessarily make contemporary local knowledge less 'traditional,' as they are incorporated into a framework of existing knowledge" (Ruddle 1994:175). A society's TEK is

> an encyclopedic and complex organized body of information that has evolved through generations and is still evolving. A local knowledge system is "traditional" by virtue of its long and deep roots and its origin in a specific culture and a local ecological system, but it is not static (Ruddle 1994:174).

In utilizing aspects of both Western science and TEK to inform one another, we can develop a more comprehensive understanding of a species within the context of and relationship to its local ecological and sociocultural system.

## Acknowledgments

I would like to express my gratitude to the Yurok and Karuk tribal community members who shared with me their wisdom and experience. Arnie Nova of the Yurok fisheries program and Ron Reed of the Karuk fisheries program provided unfailing guidance and commitment in connecting me with participants, imparting valuable insight into their world as eelers and native people living on the river, and overall being the glue that helped fit all the pieces together.

## References

Bayer, J. M., T. C. Robinson, and J. G. Seelye. 2000. Upstream migration of Pacific lampreys in the John Day River: behavior, timing and habitat use. Annual Report 2000. Bonneville Power Administration Project Number 2000-052, Portland, Oregon.

Beamish, R. J. 1980. Adult biology of river lamprey (*Lampetra ayresi*) and the Pacific lamprey (*Lampetra tridentata*) from the Pacific coast of Canada. Canadian Journal of Fisheries and Aquatic Sciences 37:1906–1923.

Beamish, R. J., and T. G. Northcote. 1989. Extinction of a population of anadromous parasitic lamprey, *Lampetra tridentata*, upstream of an impassable dam. Canadian Journal of Fisheries and Aquatic Sciences 46:420–425.

Beckman, T. 1998. Chapter 10: the Yurok and Hupa of the northern coast. Harvey Mudd College. Available: www4.hmc.edu:8001/humanities/Beckman/Indian/CA/Ch10.htm (November 2004).

Bell, M. 1991. Karuk: the upriver people. Naturegraph Publishers, Inc, Happy Camp, California.

Berkes, F. and T. Henley. 1997. Co-management and traditional knowledge: threat or opportunity? Policy Options (March):29–31.

Berkes, F., M. Kislalioglu, C. Folke, and M. Gadgil. 1998. Exploring the basic ecological unit: ecosystem-like concepts in traditional societies. Ecosystems 1:409–415.

Berkes, F. 1999. Sacred ecology. Taylor and Francis, Philadelphia.

Bernard, H. R., editor. 1998. Handbook of methods in cultural anthropology. AltaMira Press, New York.

Bernard, H. R. 2002. Research methods in anthropology: qualitative and quantitative approaches, 3rd edition. AltaMira Press, New York.

Bielawski, E. 1996. Inuit indigenous knowledge and science in the artic. Pages 216–227 *in* L. Nadar, editor. Naked science: anthropological inquiry into boundaries, power, and knowledge. Routledge, New York.

Bodznick, D., and R. G. Northcutt. 1981. Electroreception in lampreys: evidence that the earliest vertebrates were electroreceptive. Science 212(4493):465–467.

Bodznick, D., and D. G. Preston. 1983. Physiological characterization of electroreceptors in the lampreys *Ichthyomyzon unicuspis* and *Petromyzon marinus*. Journal of Comparative Physiology A Neuroethology, Sensory, Neural, and Behavioral Physiology 152(2):209–217.

Bowlby, C. E. 1981. Feeding behavior of pinnipeds in the Klamath River, northern California. Master's thesis. Humboldt State University, Arcata, California.

Bright, W. 1978. Karok. Pages 180–189 *in* W. C. Sturtevant, editor. Handbook of North American Indians, volume 8. California. Smithsonian Institution, Washington, D.C.

Bullock, T. H. 2005. Electric fish, electric organ discharges, and electroreception [Online]. Electrochemistry encyclopedia. Available: http://electrochem.cwru.edu/encycl/art-f01-fish.htm (July 2009).

Chamber, A. 2001. Uknii: "And so it was." Karuk Department of Natural Resources and Tribal Council, Orleans, California.

Chapin, K., J. Vergun, and D. Wright. 1998. Development of an aquatic habitat geographical information system for Pacific lamprey habitat in Rock Creek (Siletz), Oregon. Confederated Tribes of Siletz. Available: http://dusk.geo.orst.edu/siletz.html (September 2004).

Close, D. A., M. Fitzpatrick, H. Li, B. Parker, D. Hatch, and G. James. 1995. Status report of the Pacific lamprey (*Lampetra tridentata*) in the Columbia River basin. Bonneville Power Administration, Contract No. 95BI39067, Portland, Oregon.

Close, D. A., A. D. Jackson, B. P. Conner, and H. W. Li. 2004. Traditional ecological knowledge of Pacific lamprey (*Entosphenus tridentatus*) in northeastern Oregon and southeastern Washington from indigenous peoples of the Confederated Tribes of the Umatilla Indian Reservation. Journal of Northwest Anthropology 38(2):141–162.

Cruikshank, J. 1998. The social life of stories: narrative and knowledge in the Yukon Territory. University of Nebraska Press, Lincoln.

Curtis, E. S. 1924/1976. The North American Indian, volume 13. In E. W. Hodge, editor. Report of E.

Curtis. The North American Indian, 20 volumes. Johnson Reprint Corporation, New York.

Deloria, V, Jr., and D. Wildcat. 2001. Power and place: Indian education in America. Fulcrum Resources, Golden, Colorado.

Doremus, H. D., and A. D. Tarlock. 2008. Water war in the Klamath basin: macho law, combat biology, and dirty politics. Island Press, Washington, D.C.

Downey, T., D. Rilatos, A. Sondenaa, and B. Zybach. 1993. The Siletz eels: oral history interviews with Siletz tribal elders and neighboring residents regarding the decline in Siletz River lamprey populations. Oregon State University, Corvallis.

Ervin, A. M. 2000. Applied anthropology: tools and perspectives for contemporary practice. Allyn and Bacon, Needham, Massachusetts.

Freeman, M. M. R. 1989. Graphs and gaffs: a cautionary tale in the common-property resources debate. Pages 92–109 *in* F. Berkes, editor. Common property resources: ecology and community-based sustainable development. Belhaven Press, London.

Grenier, L. 1998. Working with indigenous knowledge: a guide for researchers. International Development Research Centre, Ottawa, Ontario.

Hammond, R. J. 1979. Larval biology of the Pacific lamprey *Lampetra tridentata* (Gairdner), of the Potlatch River, Idaho. Master's thesis. University of Idaho, Moscow.

Holmlund, R. C. 2006. Fish, forests, fire, and freedom: infringements of Karuk religious freedom through federal natural resource management. Master's thesis. Humboldt State University, Arcata, California.

Hopkins, C. D. 1973. Lightning as background noise for communication among electric fish. Nature 242:268–270.

Hotelling, W. E. 1978. My life with the Kar-ooks, miners and forestry. Willow Creek, California.

Huntington, H. P. 2000. Using traditional ecological knowledge in science: methods and applications. Ecological Applications 10:1270–1274.

Huntsinger, L., S. McCaffrey, L. Watt, and M. Lee. 1994. A Yurok forest history. University of California, Berkeley.

Jentoft, S., and K. H. Mikalsen. 1994. Regulating Fjord fisheries: folk management or interest group politics? Pages 287–316 *in* C. L. Dyer and J. R. McGoodwin, editors. Folk management in the world's fisheries: lessons for modern fisheries management. University Press of Colorado, Boulder.

Johnson, M., editor. 1992. Lore: capturing traditional environmental knowledge. Dene Cultural Institute, International Development Research Centre, Ottawa, Ontario.

Kan, T. T. 1975. Systematics, variation, distribution and biology of lampreys of the genus *Lampetra* in Oregon. Doctoral dissertation. Oregon State University, Corvallis.

Karuk Tribe. 2004. Karuk Tribe comments to the scoping of the Klamath hydroelectric project. Federal Energy Regulatory Commission, FERC No. P-2082–027, Washington, D.C.

King, T. F. 2004. First salmon: the Klamath cultural riverscape and the Klamath River hydroelectric project. Klamath River Intertribal Fish and Water Commission, Hoopa, California.

Kostow, K. 2002. Oregon lampreys: natural history status and problem analysis. Oregon Department of Fish and Wildlife. Available: http://nrimp.dfw.state.or.us/nrimp/information/docs/fishreports/OregonLampreys.pdf (2009).

Kroeber, A. L. 1925. Handbook of the Indians of California. Bureau of American Ethnology, Bulletin 78, Washington, D.C.

Kroeber, A. L., and S. A. Barrett. 1960. Fishing among the Indians of northwestern California. Anthropological records, volume 21, no. 1. University of California Press, Berkeley.

Kroeber, A. L., and E. W. Gifford. 1980. Karok myths. University of California Press, Berkeley.

Lake, F. K. 2007. Traditional ecological knowledge to develop and maintain fire regimes in northwestern California, Klamath-Siskiyou bioregion: management and restoration of culturally significant habitats. Doctoral dissertation. Oregon State University, Corvallis.

Larson, Z. S., and M. R. Belchik. 1998. A preliminary status review of eulachon and Pacific lamprey in the Klamath River basin. Yurok Tribal Fisheries Program, Klamath, California.

Miles, M. B., and A. M. Huberman. 1994. Qualitative data analysis, 2nd editio. Sage Publications, Thousand Oaks, California.

Merrell, T. R. 1959. Gull food habits on the Columbia River. Fish Commission of Oregon Research Briefs 7(1):82.

Morgan, D. L. 1997. Focus groups as qualitative research, 2nd edition. Sage Publications, Thousand Oaks, California.

Moyle, P. B. 2002. Inland fishes of California. University of California Press, Berkeley.

Nadasdy, P. 2003. Hunters and bureaucrats: power, knowledge, and aboriginal-state relations in the southwest Yukon. UBC Press, Vancouver, British Columbia.

Naiman, R. J., R. E. Bilby, D. E. Schindler, and J. M. Helfield. 2002. Pacific salmon, nutrients, and the dynamics of freshwater and riparian ecosystems. Ecosystems 5:399–417.

Norris, L. A., H. W. Lorz, and S. V. Gregory. 1991. Forest chemicals. Pages 207–296 *in* W. R. Meehan, editor. Influences of forest and rangeland management on salmonid fishes and their habitats. American Fisheries Society, Special Publication 19, Bethesda, Maryland.

Petersen, R. S. 2006. The role of traditional ecological knowledge in understanding a species and river system at risk: Pacific lamprey in the lower Klamath basin. Master's thesis. Oregon State University, Corvallis.

Pevar, S. L. 2002. The rights of Indians and tribes, 3rd edition. Southern Illinois University Press, Carbondale/Edwardsville.

Pilling, A. R. 1978 Yurok. Pages 137–154 *in* W. C. Sturtevant, editor. Handbook of North American Indians, volume 8. California. Smithsonian Institution, Washington, D.C.

Pinel, S. L., and M. J. Evans. 1994. Tribal sovereignty and the control of knowledge. In T. Greaves, editor. Intellectual property rights for indigenous peoples, a sourcebook. Society for Applied Anthropology, Oklahoma City, Oklahoma.

Pletcher, T. F. 1963. The life history and distribution of lampreys in the Salmon and certain other rivers in British Columbia, Canada. Master's thesis. University of British Columbia, Vancouver.

Potter, I. C., R. W. Hilliard, J. S. Bradley, and R. J. McKay. 1986. The influence of environmental variables on the density of larval lampreys in different seasons. Oecologia 70:433–440.

Redman, C. L. 1999. Human impact on ancient environments. The University of Arizona Press, Tucson.

Roffe, T. J., and B. R. Mate. 1984. Abundances and feeding habits of pinnipeds in the Rogue River, Oregon. Journal of Wildlife Management 48(4):1262–1274.

Ruddle, K. 1994. Local knowledge in the folk management of fisheries and coastal marine environments. Pages 161–206 *in* C. L. Dyer and J. R. McGoodwin, editors. Folk management in the world's fisheries: lessons for modern fisheries management. University Press of Colorado, Boulder.

Sallenave, J. 1994. Giving traditional ecological knowledge its rightful place in environmental impact assessment. Northern Perspectives (Canadian Arctic Resources Committee) 22(1):16–19.

Salter, J. F. 2003. White paper on behalf of the Karuk Tribe of California: a context statement concerning the effect of the Klamath hydroelectric project on traditional resource uses and cultural patterns of the Karuk people within the Klamath River corridor. PacifiCorp, Contract no. 3000020357, San Francisco, California.

Senos, R., F. K. Lake, N. J. Turner, and D. Martinez. 2006. Traditional ecological knowledge and restoration practice. Pages 393–426 *in* D. Apostol and M. Sinclair, editors. Restoring the Pacific Northwest: the art and science of ecological restoration in Cascadia. Island Press, Washington, D.C.

Skinner, C. N. 1995. Change in spatial characteristics of forest openings in the Klamath Mountains of northwestern California, USA. Landscape Ecology 10:219–228.

Smith, C., H. Burke, and G. K. Ward. 2000. Globalisation and indigenous peoples: threat or empowerment? Pages 1–24 *in* C. Smith and G. K. Ward, editors. Indigenous cultures in an interconnected world. UBC Press, Vancouver, British Columbia.

Stewart, D. J., D. Weininger, D. V. Rottiers, and T. A. Edsall. 1983. An energetics model for lake trout, *Salvelinus namaycush*: application to the Lake Michigan population. Canadian Journal of Fisheries and Aquatic Sciences 40:681–698.

Stewart, O. C., 2002. Forgotten fires: Native Americans and the transient wilderness. H. T. Lewis and M. K. Anderson, editors. University of Oklahoma Press, Norman.

Stoffle, B. W., D. B. Halmo, R. W. Stoffle, and C. G. Burpee. 1994. Folk management and conservation ethics among small-scale fishers of Buen Hombre, Dominican Republic. Pages 115–138 *in* C. L. Dyer and J. R. McGoodwin, editors. Folk management in the world's fisheries: lessons for modern fisheries management. University Press of Colorado, Boulder.

Thompson, L. 1916/1991. To the American Indian: reminiscences of a Yurok woman. Heyday Books, Berkeley, California.

Turner, N. J., I. J. Davidson-Hunt, and M. O'Flaherty.

2003. Living on the edge: ecological and cultural edges as sources of diversity for social-ecological resilience. Human Ecology 31(3):439–461.

U.S. Census. 2000. American FactFinder. U.S. Census Bureau. Available: http://factfinder.census.gov/home/aian/sf_aian.html (2004).

Wallace, D.R. 1983. The Klamath knot. Sierra Club Books, San Francisco.

Whyte, J. N. C., R. J. Beamish, N. G. Ginther, and C. E. Neville. 1993. Nutritional condition of the Pacific lamprey (*Lampetra tridentata*) deprived of food for periods of up to two years. Canadian Journal of Fisheries and Aquatic Sciences 50(3):591–599.

Wilkins, D. E. 2002. American Indian politics: and the American political system. Rowman & Littlefield Publishers, Inc., New York

Williamson, K. and D. Hillemeier. 2001. An assessment of pinniped predation upon fall-chinook salmon in the Klamath River estuary, CA, 1999. Yurok Tribal Fisheries Program, Klamath, California.

Willson, M. F., and K. C. Halupka. 1995. Anadromous fish as keystone species in vertebrate communities. Conservation Biology 9(3):489–497.

Wipfli, M. S., J. P. Hudson, and J. P. Caouette. 1998. Influence of salmon carcasses on stream productivity: response of biofilm and benthic macroinvertebrates in southeastern Alaska, U.S.A. Canadian Journal of Fisheries and Aquatic Sciences 55:1503–1511.

Wipfli, M. S., J. P. Hudson, D. T. Chaloner, and J. P. Caouette. 1999. Influence of salmon spawner densities on stream productivity in southeast Alaska. Canadian Journal of Fisheries and Aquatic Sciences 56:1600–1611.

American Fisheries Society Symposium 72:41–55, 2009


# Novel Relationships among Lampreys (Petromyzontiformes) Revealed by a Taxonomically Comprehensive Molecular Data Set

Nicholas J. Lang*
*Division of Fishes, Department of Zoology, Field Museum of Natural History*
*1400 South Lake Shore Drive, Chicago, Illinois 60605, USA*

Kevin J. Roe
*Natural Resource Ecology and Management, 339 Science II, Iowa State University*
*Ames, Iowa 50011, USA*

Claude B. Renaud
*Research Services Division, Canadian Museum of Nature*
*Post Office Box 3443, Station D, Ottawa, Ontario, K1P 6P4, Canada*

Howard S. Gill
*School of Biological Sciences and Biotechnology, Murdoch University*
*Perth, Western Australia 6150, Australia*

Ian C. Potter
*School of Biological Sciences and Biotechnology, Murdoch University*
*Perth, Western Australia 6150, Australia*

Jörg Freyhof
*Leibniz Institute of Freshwater Ecology and Inland Fisheries*
*Müggelseedamm 310, 12587 Berlin, Germany*

Alexander M. Naseka
*Zoological Institute of the Russian Academy of Sciences*
*Universitetskaya nab. 1, St. Petersburg, 199034, Russia*

Philip Cochran
*Saint Mary's University of Minnesota*
*700 Terrace Heights #10, Winona, Minnesota 55987, USA*

Hector Espinosa Pérez
*Instituto de Biología, Universidad Nacional Autónoma de México, México, D.F., México*

Evelyn M. Habit
*Unidad de Sistemas Acuáticos, Centro de Ciencias Ambientales, EULA-Chile*
*Universidad de Concepción, Casilla 160-C, Concepción, Chile*

* Corresponding author: oligocephalus@gmail.com

Bernard R. Kuhajda
*Department of Biological Sciences, Box 870345, University of Alabama Tuscaloosa, Alabama 35487, USA*

David A. Neely
*California Academy of Sciences, 875 Howard Street, San Francisco, California 94103, USA*

Yuri S. Reshetnikov
*Severtsov Institute of Ecology and Evolution, Russian Academy of Sciences Leninskii prospect, 33, 119071, Moscow, Russia*

Vladimir B. Salnikov
*National Institute of Deserts, Flora and Fauna, Ministry of Nature Protection of Turkmenistan Bitarap Turkmenistan Street, 15, 744000, Ashgabat, Turkmenistan*

Maria Th. Stoumboudi
*Hellenic Centre for Marine Research, Institute of Inland Waters Post Office Box 712, 190 13, Anavyssos, Greece*

Richard L. Mayden
*Department of Biology, 3507 Laclede Ave, Saint Louis University St. Louis, Missouri 63103, USA*

*Abstract.*—The systematics of lampreys was investigated using complete mitochondrial cytochrome *b* sequences from all genera and nearly all recognized species. The families Geotriidae and Petromyzontidae are monophyletic, but the family Mordaciidae was resolved as two divergent lineages at the base of the tree. Within Petromyzontidae, the nonparasitic *Lethenteron* sp. S and *Okkelbergia aepyptera* were recognized as distinct lineages, *Lethenteron morii* and *Lampetra zanandreai* were moved to new genera, a sister species relationship was recovered between *Caspiomyzon wagneri* and *Eudontomyzon hellenicus*, and a clade was recovered inclusive of *Entosphenus hubbsi* and western North American *Lampetra* (*L. ayresii* and *L. richardsoni*). The placement of *E. hellenicus* as the sister species to *C. wagneri* reduces the number of genera comprised entirely of parasitic species to two, *Geotria* and *Petromyzon*. The recognition of distinct lineages for *O. aepyptera* and *Lethenteron* sp. S recognizes, for the first time, lineages comprised entirely of nonparasitic species. Apart from the results mentioned above, monophyly was supported for the multispecific genera *Entosphenus*, *Eudontomyzon*, *Ichthyomyzon*, *Lampetra* (restricted to European species), and *Lethenteron*. Intergeneric relationships within Petromyzontidae were poorly resolved, but separate clades inclusive of *Entosphenus* and *Tetrapleurodon* (subfamily Entospheninae) and one comprised of *Eudontomyzon*, *Lampetra*, and *Okkelbergia* were recovered.

## Introduction

Lampreys (Order Petromyzontiformes) are not fishes in a taxonomic sense. They are, possibly along with the nonvertebrate hagfishes, surviving members of a lineage that is the sister clade to all jawed vertebrates (sharks and rays, bony fishes, and tetrapods; Meyer and Zardoya 2003; Nelson 2006). Although there are relatively few species (Table 1) in the group, several factors, including a two-stage life cycle, a relative lack of measurable features, and the wide geographic

Table 1. Recognized species of lamprey based on Kottelat and Freyhof (2007), Potter and Gill (2003), and Yamazaki and Goto (1996). Species are listed by family, genus (Potter and Gill 2003), and stem-satellite species groups (Potter 1968; Vladykov and Kott 1979). Parasitic species are labeled with an asterisk.

Geotriidae

*Geotria australis**

Mordaciidae

*Mordacia lapicida**

*Mordacia mordax**
*Mordacia praecox*

Petromyzontidae

*Caspiomyzon wagneri**

*Entosphenus folletti*

*Entosphenus hubbsi*

*Entosphenus lethophagus*

*Entosphenus macrostomus**

*Entosphenus minimus**

*Entosphenus similis**

*Entosphenus tridentatus**

*Eudontomyzon danfordi**
*Eudontomyzon mariae*
*Eudontomyzon vladykovi*

*Eudontomyzon hellenicus*

*Eudontomyzon morii**

*Eudontomyzon stankokaramani*

*Ichthyomyzon bdellium**
*Ichthyomyzon greeleyi*

*Ichthyomyzon castaneus**
*Ichthyomyzon gagei*

*Ichthyomyzon unicuspis**
*Ichthyomyzon fossor*

*Lampetra aepyptera*
*Lampetra ayresii**
*Lampetra pacifica*
*Lampetra richardsoni*

Table 1. Continued.

Petromyzontidae

*Lampetra fluviatilis**
*Lampetra planeri*

*Lampetra lanceolata*

*Lethenteron alaskense*
*Lethenteron appendix*
*Lethenteron camtschaticum**
*Lethenteron reissneri*

*Lethenteron kessleri*

*Lethenteron* sp. N

*Lethenteron* sp. S

*Lethenteron zanandreai*

*Petromyzon marinus**

*Tetrapleurodon geminis*
*Tetrapleurodon spadiceus**

distribution of the group, make the study of lamprey systematics problematic. The life cycle of all lampreys includes both an extended larval stage and a relatively brief adult stage. During the larval stage, which lasts several years, the lamprey is referred to as an ammocoete and lacks the characters found in adult specimens that distinguish species and genera of lampreys from one another. During a transformative period of 1 or 2 months, the lamprey develops eyes, fins, and a tooth-bearing oral disk and then usually lives 1 or 2 years before spawning and dying shortly after (Hardisty and Potter 1971a, 1971b). This life cycle results in a relatively short period of time in which specimens exist with fully formed diagnostic adult characteristics, time often spent migrating to and through rivers, lakes, and oceans, making capture potentially difficult.

Once a lamprey has metamorphosed into its adult form, morphological characters are largely limited to three character sets: the number of trunk myomeres; the position of the dorsal fins; and, historically most valuable, the pattern and morphology of teeth in the oral disk. The value of tooth characters in global lamprey systematics is, however, hampered by another aspect of lamprey natural history, the presence of both parasitic and nonparasitic species. Following metamorphosis, parasitic lampreys feed on host fishes for up to 2 years, growing and maturing before spawning and death (Hardisty and Potter 1971b). Nonparasitic lampreys, however, transform with already maturing gonads and within several months spawn and die, having never fed as adults. In general, nonparasitic species do not fully form the tooth patterns that are primarily used to classify parasitic lampreys, and the taxonomic placement of nonparasitic species is based on the hypothesis that such species evolved from parasitic ancestors (Hardisty and Potter 1971c). This hypothesis is supported by the fact that the distributional ranges of nonparasitic species often overlap those of parasitic species with similar numbers of myomeres and/or tooth patterning. Potter (1968) and Vladykov and Kott (1979) identified groupings of lampreys composed of a parasitic (stem) species and one or more nonparasitic (satellite) species (Table 1), putting forth the hypothesis that single parasitic species gave rise to multiple nonparasitic species at different times and/or in different areas.

When faced with a group, such as the lampreys, that possesses so few morphological characters, molecular data represent an incredibly valuable source of information. Herein, we present the first taxonomically comprehensive phylogenetic analysis of lampreys to include both parasitic and nonparasitic species.

## Systematic and Taxonomic Review

The three major lamprey lineages are currently recognized as distinct families (Hubbs and Potter 1971; Gill et al. 2003). As a whole, lampreys exhibit an antitropical distribution, with two families endemic to the Southern Hemisphere and the third restricted to the Northern Hemisphere. The family Geotriidae contains a single species, *Geotria australis*, which occurs throughout New Zealand and the southern regions of Australia (including Tasmania) and South America. The other Southern Hemisphere family, Mordaciidae, is represented by two species in Australia, *Mordacia mordax* and *M. praecox*, and one in Chile, *M. lapicida* (Potter and Gill 2003).

All lampreys in the Northern Hemisphere belong to the family Petromyzontidae, which contains 38 of the 42 recognized species in the order (Table 1). As currently recognized, the genera *Entosphenus* (*En.*, western North America and eastern Asia), *Eudontomyzon* (*Eu.*, central/eastern Europe and eastern Asia), *Lampetra* (*La.*, western Eurasia and western North America), *Lethenteron* (*Le.*, North America, Asia, and southern Europe), and *Petromyzon* (eastern North America and Europe) are widespread while *Caspiomyzon* (Caspian Sea basin), *Ichthyomyzon* (eastern North America), *Okkelbergia* (eastern North America), and *Tetrapleurodon* (central Mexico) have more restricted distributions (Potter and Gill 2003). Although these nine lineages of petromyzontid lamprey have been recognized since Creaser and Hubbs (1922), relative taxonomic ranks suggested by subsequent studies were highly variable (Figure 1). A 10th lineage, the subgenus *Reighardina* (genus *Ichthyomyzon*), was named by Creaser and Hubbs (1922) but subsequently synonymized by Hubbs and Trautman (1937).

Only *Okkelbergia*, erected as a subgenus for *Lampetra aepyptera* by Creaser and Hubbs (1922), has had its validity challenged over time. *Okkelbergia* is the only lineage of lamprey that does not include a parasitic species, and interpretation of the highly variable dentition of *La. aepyptera* has driven the shifting rank of the group from generic recognition (Hubbs and Potter 1971) to not being recognized at all (Vladykov and Kott 1976). The taxonomic placement of another nonparasitic species, *Lethenteron zanandreai*, has also been controversial. Vladykov (1955) originally described this species in the genus *Lampetra* based on the species' perceived lack of posterior circumoral teeth and possession of tricuspid middle lateral circumoral teeth. Zanandrea (1958) showed, however, that both of these characters are variable, and most subsequent workers have placed this species in *Lethenteron* (Hubbs and Potter

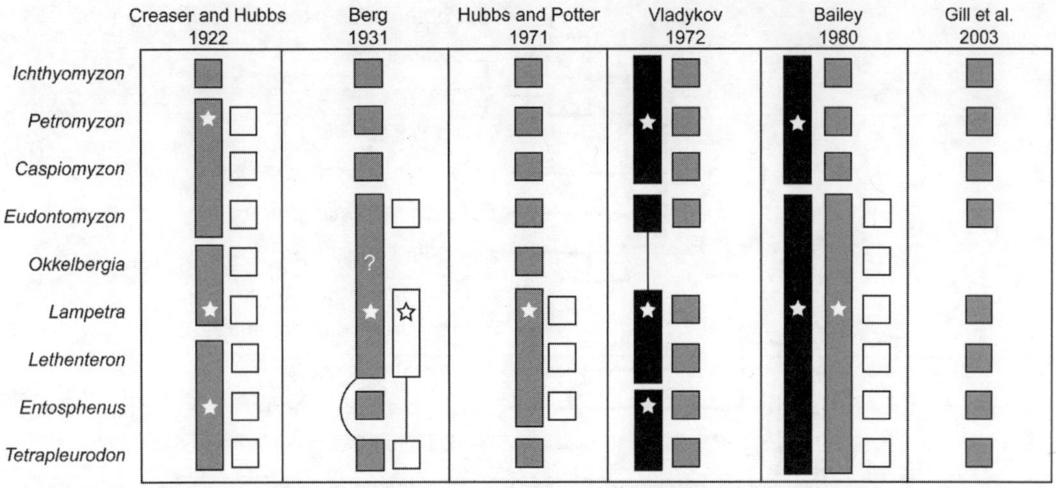


Figure 1. Visual representation of the taxonomic history of petromyzontid lampreys. The *x*-axis indicates major papers in lamprey systematics arranged chronologically while the *y*-axis indicates major lineages that have been proposed within Petromyzontidae. Black boxes indicate the taxonomic breadth of subfamilies (Vladykov 1972) or tribes (Bailey 1980), gray boxes indicate the taxonomic breadth of genera, and white boxes indicate the taxonomic breadth of subgenera. Stars indicate the name-bearing lineage for a given taxonomic group. Lines connect members of a group that could not be situated next to each other. The question mark indicates that Berg (1931) was unsure of his placement of *Okkelbergia* but provisionally synonymized the lineage with *Lampetra*, a move followed by Vladykov (1972). Gill et al. (2003) did not comment on *Okkelbergia* because it was not included in their data set.

1971; Potter and Gill 2003; but see Kottelat and Freyhof 2007).

There have been two interspecific phylogenetic studies on lampreys. The taxonomically most comprehensive study utilized morphological and chromosomal characters but included only parasitic species (Gill et al. 2003). Gill et al. (2003) recognized three families of lamprey and supported eight genera within Petromyzontidae but were unable to comment on the status of *Okkelbergia* (Figure 2). Although several multi-generic clades were well supported by the analysis of Gill et al. (2003), only two, the subfamily Lampetrinae sensu Vladykov (1972) and the tribe Lampetrini sensu Bailey (1980), had been previously hypothesized. The subfamilies Entospheninae and Petromyzontinae (Petromyzontini sensu Bailey [1980]) sensu Vladykov (1972) were both resolved as grades at the base of Petromyzontidae. The second study, Docker et al. (1999), utilized mitochondrial DNA (mtDNA) sequences and focused on the subfamilies Lampetrinae and Entospheninae sensu Vladykov (1972). Docker et al. (1999), expanded upon by Docker (2006), hypothesized that both *Lampetra* and *Entosphenus* are nonmonophyletic due to recovery of a clade inclusive of *La. ayresii*, *La. richardsoni*, and *En. hubbsi* that was not closely related to clades comprising the remaining species in those genera. Docker et al. (1999) and Docker (2006) also hypothesized that *Okkelbergia* should not be recognized at any taxonomic level, that *Lethenteron zanandreai* should be moved to *Lampetra*, and that *Le.* sp. S of Yamazaki and Goto (1996) may be more closely related to species of *Entosphenus* than species of *Lethenteron*. Species of *Eudontomyzon* and *Tetrapleurodon* were not available for inclusion in the mtDNA studies, however, and several species, such as *Le. zanandreai* and *Le.* sp. S, were represented by very short, partial sequences (267 base pairs [bp] of cytochrome *b* [cyt *b*]). Finally, although not specifically a phylogenetic study, Silver et al. (2004) presented a tree based on a nuclear coding gene, in which

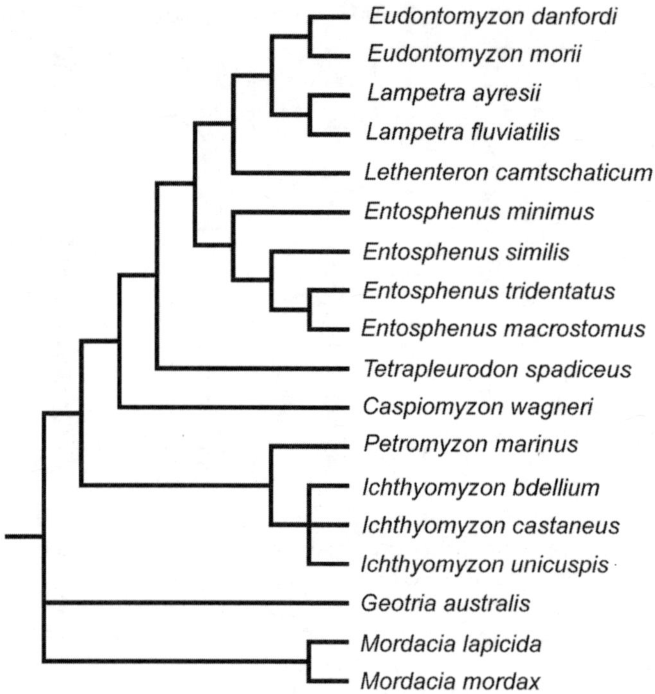


Figure 2. Phylogenetic relationships of lampreys recovered by Gill et al. (2003) based on morphology of parasitic species.

they recovered a clade uniting Petromyzontidae and Geotriidae, which was in turn sister to Mordaciidae.

## Methods

Specimens were collected using a variety of methods (hand, nets, traps, and/or electroshocking equipment) and were either frozen or fixed in 95% ethanol. When only a portion of the specimen was fixed in ethanol, generally a fin or a small muscle plug from the dorsolateral surface, the remaining specimen was fixed in formalin to serve as voucher material. Whenever possible, fully metamorphosed adult specimens, which were identified by ichthyologists expert in both the local fauna and the global diversity of lampreys, were utilized during this study. The only species represented solely by ammocoetes are *Entosphenus hubbsi*, which was represented by adult specimens in Docker et al. (1999), and *Eudontomyzon morii*. The data set includes all species recognized by Potter and Gill (2003) except for *En. folletti*, *Lampetra pacifica*, and *Tetrapleurodon spadiceus*, from which tissues were unavailable. Several species that were either identified (*Lethenteron* sp. N and *Le.* sp. S of Yamazaki and Goto [1996]) or elevated (*Eudontomyzon stankokaramani* [Holčík and Šorić 2004] and *Eu. vladykovi* [Kottelat and Freyhof 2007]) recently, but were not included in Potter and Gill (2003), were also included. Although most species are represented by a single specimen, two specimens each were included of *Eu. hellenicus* and species in monotypic genera, except for *Petromyzon marinus*, which has been shown to possess extremely little divergence across its range (Rodríguez-Muñoz et al. 2004). The only species represented exclusively by data from GenBank is *Le.* sp. S (Yamazaki et al. 2006). Materials are listed in Appendix 1.

Whole DNA was extracted using either a standard phenol-chloroform method or the DNEasy Kit (QIAGEN, Valencia, California). The use of cyt *b* in lamprey systematics presents a series of problems due to the facts that not only is the gene flanked on the 3' end by a large noncoding region of DNA (Lee and Kocher 1995; Delarbre et al. 2000), but the three petromyzontiform families are also sufficiently divergent to preclude the use of a single set of conserved primers. Preliminary cyt *b* amplifications utilized the conserved primers "cytbL" and "H15149," and lineage-specific primers were developed from the preliminary sequence data and available mitogenomic information (Table 2). Amplification products of the expected size were purified using a gel extraction kit (QIAGEN, Valencia, California) and sequenced utilizing the same primers as the amplifications, except for "cytbL," which fails to sequence due to the noncoding region to the 3' end of cyt *b*, resulting in single stranded data for the first ~300 bp. Sequence reaction products were visualized using the CEQ 8000 Genetic Analysis System (Saint Louis University, St. Louis, Missouri) or an ABI 3700 (Auburn University Genomics Laboratory, Auburn, Alabama or Macrogen Inc., Seoul, Republic of Korea).

Sequence files were edited, contigs were assembled, and sequences were aligned by eye using BioEdit. Maximum parsimony analyses were implemented in PAUP*4.0b10 (Swofford 2002), with 100 repetitions of random stepwise addition and TBR branch swapping. *Branchiostoma belcheri* (AJ404477), the Japanese lancelet, and *Myxine glutinosa* (AJ278504), a hagfish, were designated as outgroups for all parsimony analyses, and gaps were treated as a fifth base. Support for nodes was estimated using 1,000 "fast-stepwise" addition replicates of bootstrapping.

MrModelTest 2.2.1 for Classic (Nylander 2004) was used to hypothesize the best-fit model for position-defined partitions of the data set. These models were used in mixed model Bayesian analyses in MrBayes3.1 (Ronquist and Huelsenbeck 2003) that ran for 5,000,000 generations. Priors for these analyses were flat, four chains were utilized, and trees were sampled every 5,000 generations. *Branchiostoma belcheri* was designated as the outgroup in Bayesian analyses. Stationarity of negative log-likelihood values was evaluated by plotting these values against generation, and all trees before the value stabilized were discarded.

Table 2. Primer combinations used to generate sequence data.

| |
|---|
| Petromyzontiformes |
| CytbL: GTGACTTGAAAAACCACCGTTG |
| H15149: AAACTGCAGCCCCTCAGAATGATATTTGTCCT |
| Geotriidae |
| Geotria496L: CTTTTTCCTCTGTAATCCACATCTGCCG |
| Phe1612H: CTTCAGTGCTCTGCTTTAATG |
| *Mordacia lapicida* |
| Mlap348L: ATGAAACGTAGGAGTAATCT |
| 12SLamp: TCACGGGAGTGCGGAAACTTGC |
| *Mordacia mordax* and *M. praecox* |
| Mord540L: TCGATTCTTTACCTTTCATTTTATTCTTCC |
| 12SLamp: TCACGGGAGTGCGGAAACTTGC |
| Petromyzontidae |
| Cytb494L: AGCCTTCTCTTCAGTTATACACATTTGTCG |
| Phe1612H: CTTCAGTGCTCTGCTTTAATG |

## Results

The data matrix was truncated to 1,133 characters due to alignment ambiguity past that point. Although minor length variation exists among the outgroups, *Geotria australis*, and all remaining sampled taxa, the sequences are easily aligned by eye and gaps were placed that do not interrupt the reading frame of cyt *b* (positions 7–15 in the outgroups and 13–15 in *G. australis*). All included sequences are complete except that of *Lethenteron* sp. S from GenBank (missing bp 385–1,133). There are 718 variable characters of which 610 are parsimony informative. Parsimony analysis recovered a set of eight trees with a length of 2,393 steps.

MrModelTest suggested a mixture of models for each codon position. The third position approached complete agreement, with all hierarchical likelihood ratio tests (hLRTs) and the Akaike information criterion (AIC), suggesting the general time-reversible model with the gamma shape parameter (GTR+Γ) and all but one LRT adding the invariant sites parameter (GTR+I+Γ), which was used in all Bayesian analyses. The models suggested for the first and second positions, however, were a mixture of the GTR+Γ and the Hasegawa-Kishino-Yano (HKY+G) model. For the first position, GTR+I+Γ was used based on it being suggested by two of four LRTs and the AIC. Although the AIC suggested GTR+Γ for the second position, only a single LRT agreed while the other three suggested HKY+Γ. Initial analyses were run using both GTR+Γ and HKY+Γ as the model for the second position, and because no substantive differences were found between the approaches, HKY+Γ was used in the analysis reported herein. Bayesian analyses resulted in two sets of 1,001 trees, both of which conservatively reached stationarity at 100,000 generations. Fifty-percent majority rule consensus trees were generated for each set using all but the first 21 trees. Support values of at least 0.95/1.0 for Bayesian posterior probability (BPP; the average of two runs, rounded to two decimal points) and 90/100 for parsimony bootstrap (PB) are considered well supported in the following discussion.

## *Interfamilial Relationships*

All analyses recovered the Petromyzontiformes (PB: 95, BPP: 1.0), the Geotriidae (PB: 100, BPP: 1.0), and the Petromyzontidae (PB: 100, BPP: 1.0) as monophyletic clades (Figures 3 and 4). Mordaciidae was recovered in all analyses as two distinct lineages at the base of the tree. Although the node uniting the Australian Mordacia (*M. mordax* and *M. praecox*) with Geotriidae

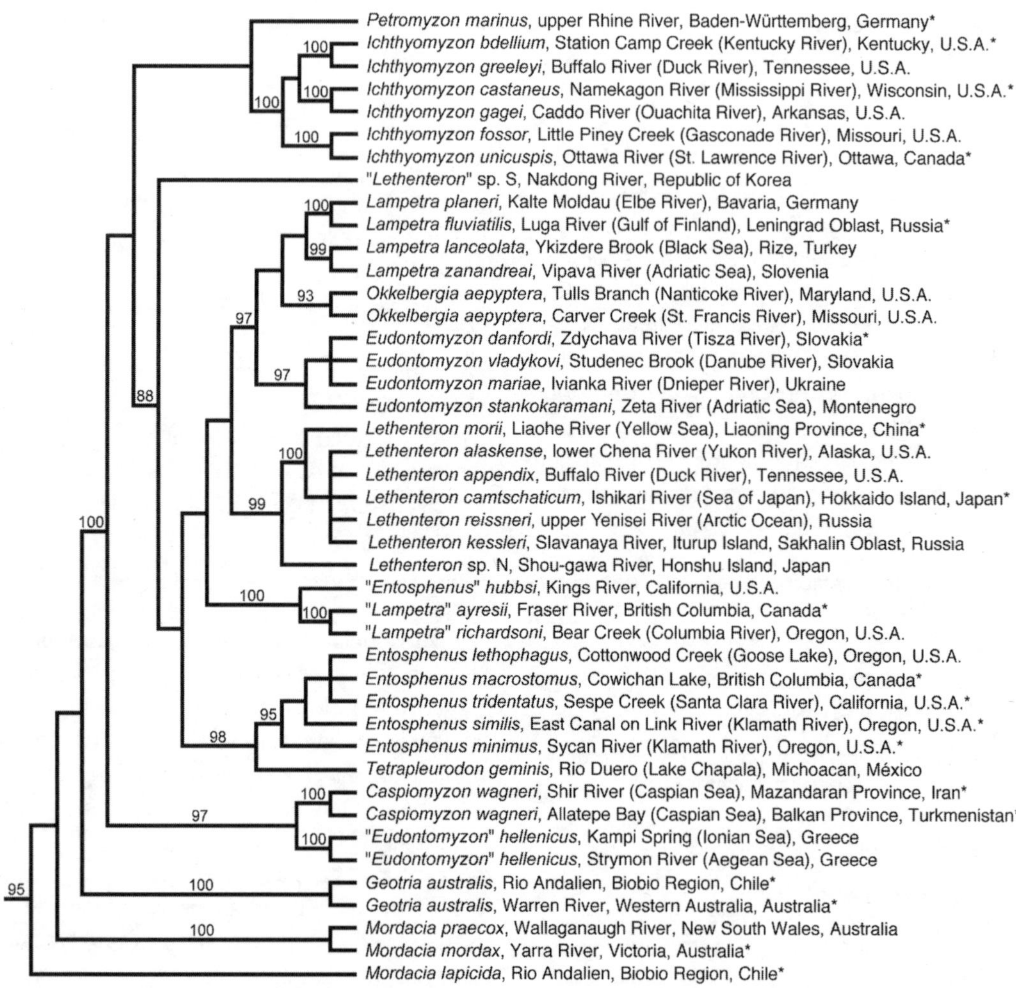


Figure 3. Strict consensus of all equally parsimonious trees recovered in PAUP*. Nodes are labeled with bootstrap values greater than 85. Parasitic species are labeled with an asterisk.

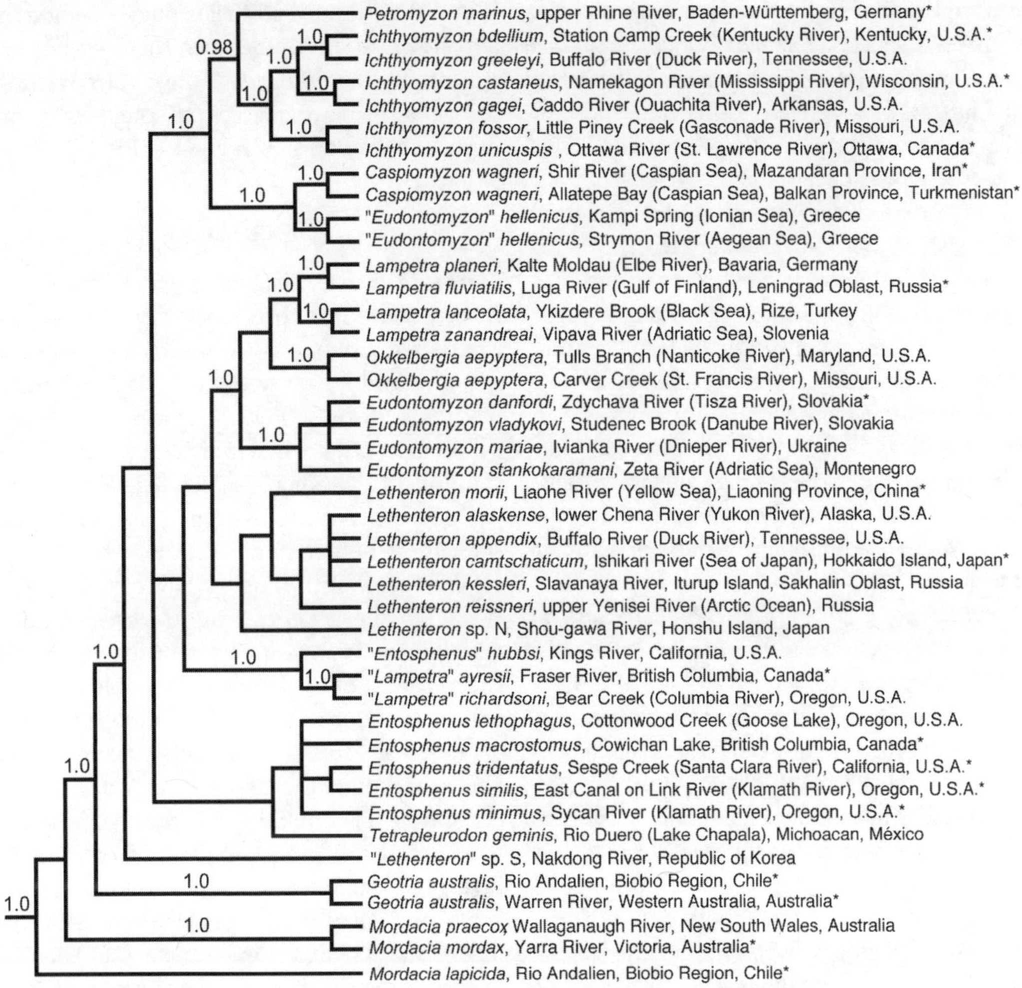


Figure 4. Majority-rule consensus of postburnin trees resulting from one of the Bayesian analyses. Nodes are labeled with Bayesian posterior probabilities greater than 0.95. Parasitic species are labeled with an asterisk.

and Petromyzontidae is present in all consensus trees, it receives no bootstrap and very low (0.76) posterior probability support, effectively creating an unresolved trichotomy among the Chilean M. lapicida, the Australian Mordacia (PB: 100, BPP: 1.0), and Geotriidae plus Petromyzontidae (no bootstrap support; BPP: 1.0).

## *Relationships within Petromyzontidae*

Several hypothesized relationships require early mention in order to facilitate further discussion. First, we recover the previously recognized *Eudontomyzon hellenicus* (PB: 100. BPP: 1.0) as the sister species to *Caspiomyzon wagneri* (PB: 100, BPP: 1.0) in all analyses (PB: 97, BPP: 1.0). Second, all analyses recover species of *Lampetra* from western North America, *La. ayresii* and *La. richardsoni*, as a clade (PB: 100, BPP: 1.0) that is not only distinct from species of *Lampetra* from both eastern North America and Eurasia, but is also sister to *Entosphenus hubbsi* (PB: 100, BPP: 1.0). Also, *Lethenteron* sp. S of Yamazaki and Goto (1996), although clearly placed within the Petromyzontidae in our analyses, is not well supported as a member of any hypothesized clade. Finally, two species, *Eu. morii* and *Le. zanandreai*, were recovered in the genera *Lethenteron* and *Lampetra*, respectively, rather than those in which they were placed by Potter and Gill (2003). In cases where a species was recovered within a well-supported clade with an available generic name (e.g., *Eu. morii* in *Lethenteron*), the generic designation is simply changed. For those clades that have no available name (e.g., western North American *Lampetra* plus *En. hubbsi*), generic designation are retained in quotation marks.

The relationships within Petromyzontidae differ significantly between the Bayesian and parsimony analyses due to the collapse of several parsimony-supported nodes in the Bayesian analysis. Genera represented by multiple samples for which monophyly is supported by all analyses are *Ichthyomyzon* (PB: 100, BPP: 1.0), *Caspiomyzon* (PB: 100, BPP: 1.0), *Eudontomyzon* (PB: 97, BPP: 1.0), and *Okkelbergia* (PB:93, BPP: 1.0). Clades comprised of *Lampetra fluviatilis* plus *La. planeri* (PB: 100, BPP: 1.0) and *La. lanceolata* plus *La. zanandreai* (PB: 99, BPP: 1.0) are recovered as sister clades in all analyses, but only supported as such by the Bayesian analyses (BPP: 1.0). The genera *Entosphenus* (PB: 95) and *Lethenteron* (PB: 99) both fail to find support in the Bayesian analysis. Support for multi-generic clades also differs between analyses, with only a clade inclusive of *Eudontomyzon*, *Lampetra*, and *Okkelbergia* receiving both parsimony and Bayesian support (PB: 97, BPP: 1.0). The parsimony analysis recovers a well-supported clade uniting *Entosphenus* and *Tetrapleurodon* (Entospheninae sensu Vladykov (1972), PB: 98) while the Bayesian analysis supports not only a clade uniting *Ichthyomyzon* and *Petromyzon* (BPP: 0.98), but also a sister relationship between this clade and one composed of *Caspiomyzon* and "*Eu.*" *hellenicus* (Petromyzontinae sensu Vladykov (1972), BPP: 1.0). The only well-supported intrageneric relationships were among species of *Ichthyomyzon*. Not only were the satellite species pairs of Vladykov and Kott (1979) (*I. bdellium* plus *I. greeleyi* [PB: 100, BPP: 1.0], *I. castaneus* plus *I. gagei* [PB: 100, BPP: 1.0], and *I. unicuspis* plus *I. fossor* [PB: 100, BPP: 1.0]) all well supported, but a sister clade relationship between the *I. bdellium–I. greeleyi* and *I. castaneus–I. gagei* pairs was also recovered in both analyses, although only supported by the Bayesian analysis (BPP: 1.0).

## Discussion

Our study recovered a high number of well-supported clades that correspond to previously hypothesized groupings. All nine of the previously hypothesized petromyzontid lineages were recovered as distinct clades by at least one analytical method and are herein treated as genera. Except for the placement of *Mordacia lapicida*, for which molecular data have not been previously presented, the interfamilial relationships we recovered agree with those of Silver et al. (2004). Although the generic name *Caragola* Gray is available for *M. lapicida*, we choose to retain the species in *Mordacia* based on available morphological data (Gill et al. 2003) and await additional data, perhaps from more slowly evolving genes, before rejecting the monophyly of Mordaciidae.

Given the similarities in their general distribution, the difference in the level of sequence divergence exhibited across the Southern Hemisphere in *Geotria* (<2%) and *Mordacia* (~21%) is intriguing. Differences in life history and habitat, or that of their host species, may have restricted *Mordacia* to a nearshore lifestyle, leaving them intimately tied to the drifting continents while *Geotria* was able to traverse long distances at sea. Indeed, *Geotria* breeds in widespread areas, including Tasmania, the entire southern tip of South America (Chile and Argentina), New Zealand, the Falkland Islands, and both southeastern and southwestern Australia, while breeding *Mordacia* are restricted to relatively small regions of the southern continents (southeastern Australia, Tasmania, and Chile). *Geotria australis* is also known to make up a large proportion of the diet of nesting albatrosses on South Georgia Island, ~1,400 km ESE of the nearest known breeding grounds in the Falkland Islands, although it is unclear how much of that distance is covered by either prey or predator (Potter et al. 1979).

We cannot explain the failure of the Bayesian analyses to recover *Lethenteron*, *Entosphenus*, and *Entosphenus* plus *Tetrapleurodon* as monophyletic groups at this time. In order to test the effect of outgroup choice on relationships within Petromyzontidae, we removed *Branchiostoma belcheri*, *Myxine glutinosa*, and the species of *Mordacia* from the data set and performed additional analyses using *Geotria australis* as the sole outgroup. These analyses also failed to recover the problematic clades mentioned above. Although there is reason to expect the results of a parsimony analysis to be problematic when using a data set composed of highly divergent lineages (i.e., Long Branch Attraction, Bergsten [2005]), we are aware of no reason to expect a Bayesian analysis to fail to support clades composed of sequences that are less than 0.9% divergent (e.g., *Entosphenus*). While this is an issue that requires further investigation with a larger data set, the number of clades on which the methods agree and the fact that those clades recovered by parsimony analysis alone correspond to previously hypothesized clades make us confident in our results.

Given that all three studies rely on mtDNA, it is not surprising that our results largely agree with those of Docker et al. (1999) and Docker (2006). We also recover a clade comprised of western North American "*Lampetra*" plus "*Entosphenus*" *hubbsi*, remove *Lampetra zanandreai* from *Lethenteron*, and hypothesize that "*Le.*" sp. S is a distinct lineage within Petromyzontidae. These hypotheses are strengthened in our analysis by the inclusion of not only species of *Eudontomyzon* and *Tetrapleurodon*, but also a more complete sequence for *La. zanandreai*. The only major change facilitated by the inclusion of additional species is our recognition of *Okkelbergia* as a distinct genus. Inclusion of species of *Eudontomyzon* in our data set revealed that the well-supported node uniting *O. aepyptera* and *La. fluviatilis* in Docker et al. (1999) actually defines a trichotomy among *Eudontomyzon*, *Lampetra*, and *Okkelbergia*. Although additional data may resolve this node by placing *O. aepyptera* in the genus *Lampetra*, we refute the hypothesis of Vladykov and Kott (1979) that the closest relatives of *O. aepyptera* are found in western North America rather than Europe.

Apart from the trichotomous clade composed of *Eudontomyzon*, *Lampetra*, and *Okkelbergia*, there are no intergeneric relationships that are well supported by all of our analyses. Petromyzontinae (Bayesian) and Entospheninae (parsimony) (both sensu Vladykov [1972]) are both well supported by one analytical method but not the other. Hypotheses regarding relationships among genera and subfamilial designation will require additional data, and current hypotheses should be further tested using data from the nuclear genome.

The changes to the generic designation of *Lampetra zanandreai* and *Lethenteron morii* place both species into genera that seem more plausible than previous designations, given their geographic distribution. For instance, *La. zanandreai* is distributed in drainages of the northern Mediterranean Sea. These and neighboring drainages are home to other species of *Lampetra*, but the nearest species of *Lethenteron* occupy the Arctic Ocean drainage. Similarly, when *Le. morii* was placed in the genus *Eudontomyzon*, it was the only species in the genus east of the Caspian Sea drainage. Other species of *Lethenteron*, however, are widespread across northeast Asia, in drainages surrounding the distribution of *Le. morii*. In the original description of *Lethenteron morii*, Berg (1931) commented on the remarkable nature of such a disjunct distribution for the subgenus *Eudontomyzon* and raised the possibility that the subgenus, as he constructed it, was not monophyletic. The specimen of *Le. morii* in our data set is a metamorphosing individual with developing diagnostic teeth in the lateral fields of the oral disk (exolaterals). It is also from the Liaohe River drainage, immediately west of the Yalu River drainage, which contains the type locality. Although the haplotype sampled from our specimen is distinct within *Lethenteron*, and Berg (1931) commented that ammocoetes he examined from the Liaohe drainage most likely represented *Le. morii*, exolateral teeth are not unknown in other species of *Lethenteron*, the distribution of which in the northern Korean Peninsula and neighboring regions of the People's Republic of China is poorly known. It is therefore possible that our specimen is an individual of a *Le. reissneri*-like species, and the collection of adult *Le. morii* must be a primary goal of future efforts.

It is clear from the results of our study that molecular characters will play an important role in shaping our future views of lamprey systematics. This source of data not only allowed us to include nonparasitic species in our analyses, leading to several novel hypotheses, but also revealed a deep divergence between parasitic species that are indistinguishable using tooth characteristics. Until Vladykov and Follett (1958) separated them based on pigmentation and morphometrics, the western North American "*Lampetra*" *ayresii* and the European *Lampetra fluviatilis* were considered conspecific (Creaser and Hubbs 1922). In general, the satellite species groups of Potter (1968) and Vladykov and Kott (1979) are herein supported. Our analyses indicate, however, that two lineages, *Okkelbergia* and "*Lethenteron*" sp. S, are composed solely of nonparasitic species, and it is possible that better sampled studies of other groups will reveal additional cryptic lineages within nonparasitic species complexes. Although "*Le.*" sp. S is morphologically indistinguishable from *Lethenteron* sp. N, it has been shown to be genetically distinct using both allozyme and mitochondrial sequence data (Yamazaki and Goto 1998; Yamazaki et al. 2006). Unfortunately, we were unable to obtain tissue of this species, and its inclusion in our data set is based on a partial sequence. The placement of this species within Petromyzontidae must be explored using additional sources of data in the future. Finally, the well-supported sister species relationship hypothesized between *Caspiomyzon wagneri* and "*Eudontomyzon*" *hellenicus* is a very exciting result of our analyses. This hypothesis not only aligns a nonparasitic species with *C. wagneri* for the first time, leaving *Geotria* and *Petromyzon* as the only strictly parasitic lineages, but also moves "*Eu.*" *hellenicus* across the subfamilial/tribal divide proposed by Vladykov (1972) and Bailey (1980). Although it is possible to collect samples of *C. wagneri* from additional portions of its range (e.g., the Volga River drainage), there are no additional known localities for "*Eu.*" *hellenicus*. Testing of this hypothesis will have to rely on characters from the nuclear genome and/or a reappraisal of morphological characters in these two species.

While our study suggests that more morphological characters may be homoplasious within lampreys than previously hypothesized, our failure to resolve well-supported intergeneric clades precludes us from truly testing the hypothesis of Gill et al. (2003). It is our intention for this, the most comprehensive study on lamprey systematics to date, to be a starting point from which future questions of lamprey systematics can be effectively pursued.

## Acknowledgments

This study was funded in part by NSF 89507 and 296162 awarded to RLM and KJR (CO-PI's) and ICP, HSG, and CBR (International Collaborators). NAM was supported by RFBR grant #09-04-01584. Specimens, tissues and/or field assistance were provided by the following people: A. Abdoli, J. A. Anderson, S. A. Barndt, R. E. Blanton, H. T. Boschung, J. S. Boyce, B. M. Burr, B. W. Coad, A. L. George, J. E. Gersmehl, F. B. Gutermuth, S. He, A. J. Henry, D. B. Henry, J. Holčík, A. P. Kinziger, T. Lambert, T. J. Landewe, B. A. Levin, J. Lyons, E. L. Macdonald, D. F. Markle, D. L. Morgan, L. Pekárik, J. S. Peyton, T. W. Pietsch, S. L. Powers, J. M. Ray, S. R. Reid, R. Šanda, J. M. Serb, A. M. Simons, D. E. Stevenson, J. F. Switzer, L. E. Thompson, J. R. Tomelleri, D. Turan, R. M. Wood, X-C. Wu, Y. Yamazaki, and M. Zontág.

## References

Bailey, R. M. 1980. Comments on the classification and nomenclature of lampreys: an alternative view. Canadian Journal of Fisheries and Aquatic Sciences 37:1626–1629.

Berg, L. S. 1931. A review of the lampreys of the northern hemisphere. Annuaire du Musée

Zoologique de l'Académie des Sciences de l'URSS 32:87–116.

Bergsten, J. 2005. A review of long-branch attraction. Cladistics 21:163–193.

Creaser, C. W. and C. L. Hubbs. 1922. A revision of the holarctic lampreys. Occasional Papers of the Museum of Zoology University of Michigan 120:1–14.

Delarbre, C., H. Escriva, C. Gallut, V. Barriel, P. Kourilsky, P. Janvier, V. Laudet, and G. Gachelin. 2000. The complete nucleotide sequence of the mitochondrial DNA of the agnathan *Lampetra fluviatilis*: bearings on the phylogeny of cyclostomes. Molecular Biology and Evolution 17:519–529.

Docker, M. F. 2006. Bill Beamish's contributions to lamprey research and recent advances in the field. Guelph Ichthyology Reviews 7:1–52.

Docker, M. F., J. H. Youson, R. J. Beamish, and R. H. Devlin. 1999. Phylogeny of the lamprey genus *Lampetra* inferred from mitochondrial cytochrome *b* and ND3 gene sequences. Canadian Journal of Fisheries and Aquatic Sciences 56:2340–2349.

Gill, H. S., C. B. Renaud, F. Chapleau, R. L. Mayden, and I. C. Potter. 2003. Phylogeny of living parasitic lampreys (Petromyzontiformes) based on morphological data. Copeia 2003:687–703.

Hardisty, M. W., and I. C. Potter. 1971a. The behaviour, ecology and growth of larval lampreys. Pages 85–125 *in* M. W. Hardisty and I. C. Potter, editors. The biology of lampreys, volume 1. Academic Press, New York.

Hardisty, M. W., and I. C. Potter. 1971b. The general biology of adult lampreys. Pages 127–206 *in* M. W. Hardisty and I. C. Potter, editors. The biology of lampreys, volume 1. Academic Press, New York.

Hardisty, M. W., and I. C. Potter. 1971c. Paired species. Pages 249–277 *in* M. W. Hardisty and I. C. Potter, editors. The biology of lampreys, volume 1. Academic Press, New York.

Holčík, J., and V. Šorić. 2004. Redescription of *Eudontomyzon stankokaramani* (Petromyzontes, Petromyzontidae): a little known lamprey from the Drin River drainage, Adriatic Sea basin. Folia Zoologica 53:399–410.

Hubbs, C. L., and I. C. Potter. 1971. Distribution, phylogeny and taxonomy. Pages 1–65 *in* M. W. Hardisty and I. C. Potter, editors. The biology of lampreys, volume 1. Academic Press, New York.

Hubbs, C. L., and M. B. Trautman. 1937. A revision of the lamprey genus *Ichthyomyzon*. Miscellaneous Publications of the Museum of Zoology University of Michigan 35:7–109.

Kottelat, M., and J. Freyhof. 2007. Handbook of European freshwater fishes. Kottelat, Cornol, Switzerland and Freyhof, Berlin, Germany.

Lee, W.-J., and T. D. Kocher. 1995. Complete sequence of a sea lamprey (*Petromyzon marinus*) mitochondrial genome: early establishment of the vertebrate genome organization. Genetics 139:873–887.

Meyer, A., and R. Zardoya. 2003. Recent advances in the (molecular) phylogeny of vertebrates. Annual Review of Ecology Evolution and Systematics 34:311–338.

Nelson, J. S. 2006. Fishes of the world, 4th edition. Wiley, Hoboken, New Jersey.

Nylander, J. A. A. 2004. MrModeltest v2. Program distributed by the author. Evolutionary Biology Centre, Uppsala University, Norbyvägen 18, SE-752 36 Uppsala, Sweden.

Potter, I. C. 1968. *Mordacia praecox* n. sp., a non-parasitic lamprey (Petromyzonidae), from New South Wales, Australia. Proceedings of the Linnean Society of New South Wales 92:254–261.

Potter, I. C., and H. S. Gill. 2003. Adaptive radiation of lampreys. Journal of Great Lakes Research 29 (Supplement 1):95–112.

Potter, I. C., P. A. Prince, and J. P. Croxall. 1979. Data on the adult marine and migratory phase in the life of the Southern Hemisphere lamprey, *Geotria australis* Gray. Environmental Biology of Fishes 4:65–69.

Rodríguez-Muñoz, R., J. R. Waldman, C. Grunwald, N. K. Roy, and I. Wirgin. 2004. Absence of shared mitochondrial DNA haplotypes between sea lamprey from North American and Spanish rivers. Journal of Fish Biology 64:783–787.

Ronquist, F., and J. P. Huelsenbeck. 2003. MrBayes 3: Bayesian phylogenetic inference under mixed models. Bioinformatics 19:1572–1574.

Silver, M. R., H. Kawauchi, M. Nozaki, and S. A. Sower. 2004. Cloning and analysis of the lamprey GnRH-III cDNA from eight species of lamprey representing the three families of Petromyzontiformes. General and Comparative Endocrinology 139:85–94.

Swofford, D. L. 2002. PAUP*. phylogenetic analysis using parsimony (* and other methods) 4.0. Sinauer Associates, Sunderland, Massachusetts.

Vladykov, V. D. 1955. *Lampetra zanandreai*, a new species of lamprey from northern Italy. Copeia 1955:215–223.

Vladykov, V. D. 1972. Sous-division en trois sous-familles des lamproies de l`hémisphère nord de la famille Petromyzonidae. Association canadienne-française pour l`avancement des sciences 39:148.

Vladykov, V. D., and W. I. Follett. 1958. Redescription of *Lampetra ayresii* (Günther) of western North America, a species of lamprey (Petromyzontidae) distinct from *Lampetra fluviatilis* (Linnaeus) of Europe. The Journal of the Fisheries Research Board of Canada 15:47–77.

Vladykov, V. D., and E. Kott. 1976. Is *Okkelbergia* Creaser and Hubbs, 1922 (Petromyzonidae) a distinct taxon? Canadian Journal of Zoology 54:421–425.

Vladykov, V. D., and E. Kott. 1979. Satellite species among the holarctic lampreys (Petromyzonidae). Canadian Journal of Zoology 57:860–867.

Yamazaki, Y., and A. Goto. 1996. Genetic differentiation of *Lethenteron reissneri* populations, with reference to the existence of discrete taxonomic entities. Ichthyological Research 43:283–299.

Yamazaki, Y., and A. Goto. 1998. Genetic structure and differentiation of four *Lethenteron* taxa from the Far East, deduced from allozyme analysis. Environmental Biology of Fishes 52:149–161.

Yamazaki, Y., R. Yokoyama, M. Nishida, and A. Goto. 2006. Taxonomy and molecular phylogeny of *Lethenteron* lampreys in eastern Eurasia. Journal of Fish Biology 68 (Supplement B):251–269.

Zanandrea, G. 1958. Posizione sistematica e distribuzione geografica di *Lampetra zanandreai* Vladykov. Memorie del Museo Civico di Storia Naturale di Verona 6:207–237.

## Appendix 1

Materials examined. Institutional abbreviations are as follows: CMNFI – Canadian Museum of Nature Fish Collection, Ottawa, Ontario, Canada; OSU – Oregon State University Fish Collection, Corvallis, Oregon, USA; STL – Tissue Collection of Saint Louis University, St. Louis, Missouri, USA; UAIC – University of Alabama Ichthyological Collection, Tuscaloosa, Alabama, USA. Genbank accession numbers are listed in parentheses.

Geotriidae

*Geotria australis*: STL 1260.01, Rio Andalien, Biobio Region, Chile (GQ206164); UAIC 11977.01, Warren River, Western Australia, Australia (GQ206165).

Mordaciidae

*Mordacia lapicida*: STL 1260.01, Rio Andalien, Biobio Region, Chile (GQ206185)

*Mordacia mordax*: UAIC 15569.01, Yarra River, Victoria, Australia (GQ206188).

*Mordacia praecox*: UAIC 15570.01, Wallaganaugh River, New South Wales, Australia (GQ206186).

Petromyzontidae

*Caspiomyzon wagneri*: STL 801.01, Allatepe Bay (Caspian Sea), Balkan Province, Turkmenistan (GQ206152); UAIC 12933.01, Shir River, Mazandaran Province, Iran (GQ206151).

"*Entosphenus*" *hubbsi*: UAIC 11547.01, Kings River, California, USA (GQ206150).

*Entosphenus lethophagus*: UAIC 15571.01, Cottonwood Creek, Oregon, USA (GQ206153).

*Entosphenus macrostomus*: UAIC 15572.01, Cowichan Lake, Vancouver Island, British Columbia, Canada (GQ206154).

*Enstosphenus minimus*: UAIC 12934.01, Sycan River, Oregon, USA (GQ206155).

*Entosphenus similis*: UAIC 12935.01, East Canal on Link River, Oregon, USA (GQ206156).

*Entosphenus tridentatus*: UAIC 11044.01, Sespe Creek, California, USA (GQ206157).

"*Eudontomyzon*" *hellenicus*: STL 1253.01, Kampi Spring, Greece (GQ206159); UAIC 15573.01, Strymon River, Greece (GQ206160).

*Eudontomyzon danfordi*: CMNFI 2008-0002, Zdychava River, Slovakia (GQ206158).

*Eudontomyzon mariae*: STL 957.01, Ivianka River, Ukraine (GQ206162).

*Eudontomyzon stankokaramani*: STL 1331.01, Zeta River, Montenegro (GQ206189).

*Eudontomyzon vladykovi*: CMNFI 2002-0009, Studenec Brook, Slovakia (GQ206161).

*Ichthyomyzon bdellium*: UAIC 12459.01, Station Camp Creek, Kentucky, USA (GQ206166).

*Ichthyomyzon castaneus*: UAIC 11978.01, Namekagon River, Wisconsin, USA (GQ206168).

*Ichthyomyzon fossor*: UAIC 11156.01, Little Piney Creek, Missouri, USA (GQ206170).

*Ichthyomyzon gagei*: UAIC 10065.01, Caddo River, Arkansas, USA (GQ206169).

*Ichthyomyzon greeleyi*: UAIC 10601.06, Buffalo River, Tennessee, USA (GQ206167).

*Ichthyomyzon unicuspis*: CMNFI 2008-0004, Ottawa River, Ontario, Canada (GQ206171).

"*Lampetra*" *ayresii*: UAIC 15574.01, Fraser River, British Columbia, Canada (GQ206174).

"*Lampetra*" *richardsoni*: OSU 15388, Bear Creek, Oregon, USA (GQ206177).

*Lampetra fluviatilis*: STL 982.01, Luga River, Leningrad Oblast, Russia (GQ206175).

*Lampetra lanceolata*: STL 954.01, Ykizdere Brook, Turkey (GQ206176).

*Lampetra planeri*: STL 1118.01, Kalte Moldau, Bavaria, Germany (GQ206149).

*Lampetra zanandreai*: STL 1252.01, Vipava River, Slovenia (GQ206184).

"*Lethenteron*" sp. S: Naktong River, Republic of Korea (AB220183).

*Lethenteron alaskense*: CMNFI 1999-0046, lower Chena River, Alaska, USA (GQ206178).

*Lethenteron appendix*: UAIC 10655.01, Buffalo River, Tennessee, USA (GQ206179).

*Lethenteron camtschaticum*: UAIC 11987.01, Ishikari River, Hokkaido Island, Japan (GQ206180).

*Lethenteron kessleri*: STL 965.01, upper Yenisei River, Russia (GQ206183).

*Lethenteron morii*: CMNFI 2001-0042, Liaohe River, Liaoning Province, China (GQ206163).

*Lethenteron reissneri*: UAIC 12326.01, Slavanaya River, Iturup Island, Russia (GQ206181).

*Lethenteron* sp. N: STL 1333.01, Shou-gawa River, Honshu Island, Japan (GQ206182).

*Okkelbergia aepyptera*: STL 613.01, Tulls Branch, Maryland, USA (GQ206172); uncataloged, Carver Creek, Missouri, USA (GQ206173).

*Petromyzon marinus*: STL 1116.01, upper Rhine River, Baden-Württemberg, Germany (GQ206148).

*Tetrapleurodon geminis*: UAIC 15575.01, Rio Duero, Michoacan, México (GQ206187).

American Fisheries Society Symposium 72:57–70, 2009


# Insight from Lamprey Genomics: Brain and Pituitary Reproductive Hormones of Lampreys

Stacia A. Sower*, Mihael Freamat, and Scott I. Kavanaugh
*Center for Molecular and Comparative Endocrinology*
*Department of Molecular, Cellular and Biomedical Sciences*
*University of New Hampshire, 46 College Road, Durham, New Hampshire 03824, USA*

*Abstract.*—This chapter summarizes reproduction and the latest findings on reproductive endocrinology in one of the only two living representatives of the most ancient lineage of vertebrates, agnathans. Modern vertebrates are classified into two major groups, the gnathostomes (jawed vertebrates) and the agnathans (jawless vertebrates). The agnathans are classified into two groups, myxinoids (hagfishes) and petromyzonids (lampreys), while the gnathostomes constitute all the other living vertebrates, including the bony and cartilaginous fishes and the tetrapods. During the past two decades, there have been rapid advances in our knowledge of the structure and function of reproductive hormones in lamprey. Lampreys are the earliest evolved vertebrates for which there are demonstrated functional roles for two (possibly three) gonadotropin-releasing hormones (GnRHs) that act via the hypothalamic-pituitary-gonadal axis controlling reproductive processes. From our structural and functional studies, we have determined the primary amino acid and cDNA sequences of two forms of GnRH, lamprey GnRH-I and -III, one GnRH receptor, and one gonadotropin-beta (GTH-β) hormone. Since 2006, with the availability of the lamprey genome, we have identified an additional GnRH isoform (lamprey GnRH-II) and two glycoprotein hormone receptors (one gonadotropic-like and one thyrotropic-like). The high conservation of these hormones and their receptors throughout vertebrate species makes the lamprey model highly appropriate for examining the neuroendocrine system. Here, we present a summary on our current knowledge of reproductive endocrinology in these basal vertebrates.

## Lamprey Phylogeny

One of the major goals of our laboratory has been to understand and delineate the complex neuroendocrine axis of reproduction in one of the few agnathan species remnant in the fauna—the sea lamprey *Petromyzon marinus*. The agnathans probably arose as the first vertebrates about 530 million years ago, immediately after the evolutionary explosion of multicellular organisms in the Cambrian period. Modern extant agnathans are represented by two groups, the lampreys and the hagfishes. Paleontological analysis of extinct agnathans had suggested that lampreys were more closely related to gnathostomes than either group is to the hagfishes (Forey and Janvier 1993, 1994). However, Janvier and his collaborators suggested from later analysis of the complete mitochondrial DNA that lampreys and hagfishes form a clade (Delarbre et al. 2002), which is also supported by mitochondrial DNA analysis (Suga et al. 1999). Ota et al. (2007) showed putative neural-crest cells in hagfish embryos and that these cells delaminate and migrate like those of all other living vertebrates, supporting the more recent claims of one clade. Further information on the evo-

* Corresponding author: sasower@cisunix.unh.edu

lution of vertebrate brain/pituitary hormones and their genes in lampreys and hagfishes that can be obtained from the lamprey genome can contribute to the ongoing phylogenetic analysis that may help in resolving the phylogenetic relationships between hagfishes, lampreys, and jawed vertebrates.

## Summary of Neuroendocrine System

The hypothalamic-pituitary system is considered to be a vertebrate innovation and seminal event that emerged prior to or during the differentiation of the ancestral agnathans (Sower et al. 2009; Figure 1). Reproduction in vertebrates is controlled by a hierarchically organized endocrine system. The typical hypothalamic-pituitary axis consists of gonadotropin-releasing hormones and gonadotropins. In spite of the very diverse patterns of life cycles and reproductive strategies and behaviors, this endocrine system is remarkably conserved throughout the gnathostome lineages. To date, our previous biochemical, molecular, immunocytochemical, and functional studies during the past 28 years on the structure and function of the gonadotropin-releasing hormones (GnRH) and GTH in lampreys have established that similar to all other vertebrates, the lampreys have a hypothalamic-pituitary-gonadal axis and that there is a high conservation of the mechanisms of GnRH action, although with unique structural elements (Sower et al. 2009). Further comparative studies on these hormone systems will help provide clues on the evolution of reproductive mechanisms and insights into our understanding of gene duplication, structure-activity relations, and the molecular evolution and functional diversity of reproductive hypothalamic and pituitary hormones. With increasing knowledge from genomics and proteomics, it will be necessary for us to understand hormone-induced changes in gene expression levels, interactions between signaling pathways and structural and functional changes in cells under a variety of circumstances.

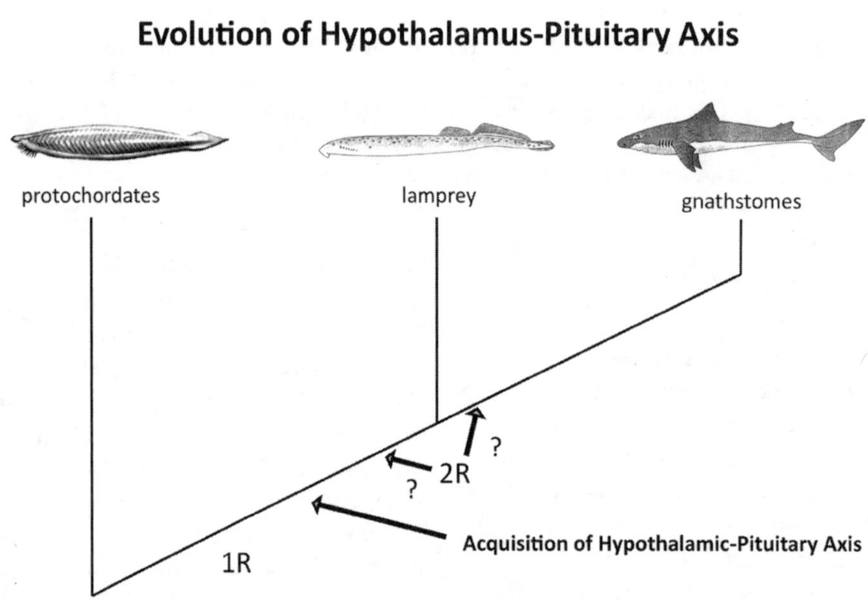


Figure 1. A schematic diagram depicting the hypothalamic-pituitary axis that emerged prior to or during the differentiation of the ancestral agnathans. 1R or 2R: one round or two rounds of whole genome duplication.

## Lamprey Genome Project

As an agnathan, the oldest extant lineage of vertebrates, the sea lamprey has become a model system for analysis of the evolution of many genes and systems, including the evolution of the neuroendocrine regulation of reproduction (Sower 2003; Freamat et al. 2006; Kawauchi et al. 2006; Silver and Sower 2006) and the evolution of development (EvoDev; Kuratani et al. 2002). Lampreys as basal vertebrates were identified in a key position. Nine nonmammalian organisms were chosen by National Institutes of Health (NIH) for mapping the genome, each of which represents a position on the evolutionary timeline marked by important changes in animal anatomy, physiology, development, or behavior (NIH News Release August 4, 2004). The mapping of the lamprey genome started in January 2005 (www.genome.gov/12511858; Figure 2). It is now estimated that the current coverage of the lamprey genome is about 5.9×. The lamprey genome is available for download and on site for analysis using the National Center for Biotechnology Information trace archives and the partially assembled genome by Ensembl (http://genome.wustl.edu/genomes/view/petromyzon_marinus). The phylogenetic position of lampreys as a basal vertebrate allows lampreys to be a basis for understanding the molecular evolution of the genes encoding receptors and hormones that arose in the vertebrates.

## Summary of Known Lamprey Hypothalamic-Pituitary Hormones and Receptors

Lampreys are the earliest evolved vertebrates for which there are demonstrated functional roles for two or possibly three GnRHs that act

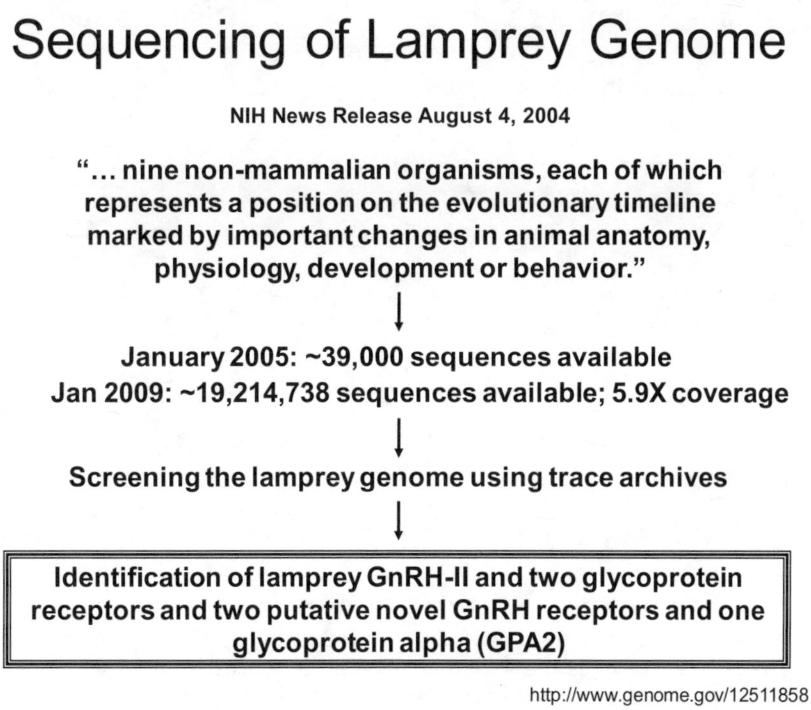


Figure 2. The mapping of the lamprey genome started in January 2005 (www.genome.gov/12511858) that led to the screening and identification of lamprey GnRH-II and two glycoprotein receptors and two putative GnRH receptors and one glycoprotein alpha (GPA2).

as neurohormones controlling the reproductive processes. To date, biochemical, molecular, immunocytochemical, and functional studies on the structure and function of the GnRHs in lampreys have established that similar to all other vertebrates, the lamprey has a hypothalamic-pituitary-gonadal axis and that there is a high conservation of the mechanisms of GnRH action (Sower 2003; Kavanaugh et al. 2008; Figure 3). Generally, gnathostomes have one or two GnRHs that act as hypothalamic hormones, two pituitary gonadotropins, luteinizing hormone (LH) and follicle stimulating hormone (FSH), and one gonadal FSH receptor and one LH receptor. In comparison, the lampreys have two, possibly three, hypothalamic GnRHs, only one known pituitary gonadotropin and one gonadal glycoprotein receptor (Sower et al. 2009). To date, the identification of the primary amino acid and cDNA sequences of two forms of GnRH, lamprey GnRH-I and -III, a novel GnRH (lamprey GnRH-II), the cDNA of one GnRH receptor, the identification of one pituitary gonadotropin-beta, as well as several other brain/pituitary hormones and receptors have been completed (Sherwood et al. 1986; Sower et al. 1993, 2006; Suzuki et al. 2000; Silver et al. 2004, 2005; Sower and Kawauchi 2006; Kavanaugh et al. 2008; Table 1 and 2). The lamprey GnRH receptor reported so far shares several characteristics of both (lGnRHR-1) type-I and type-II vertebrate GnRH receptors (Silver et al. 2005; Silver and Sower 2006). The high conservation of the GnRH and its receptor throughout vertebrate species makes the

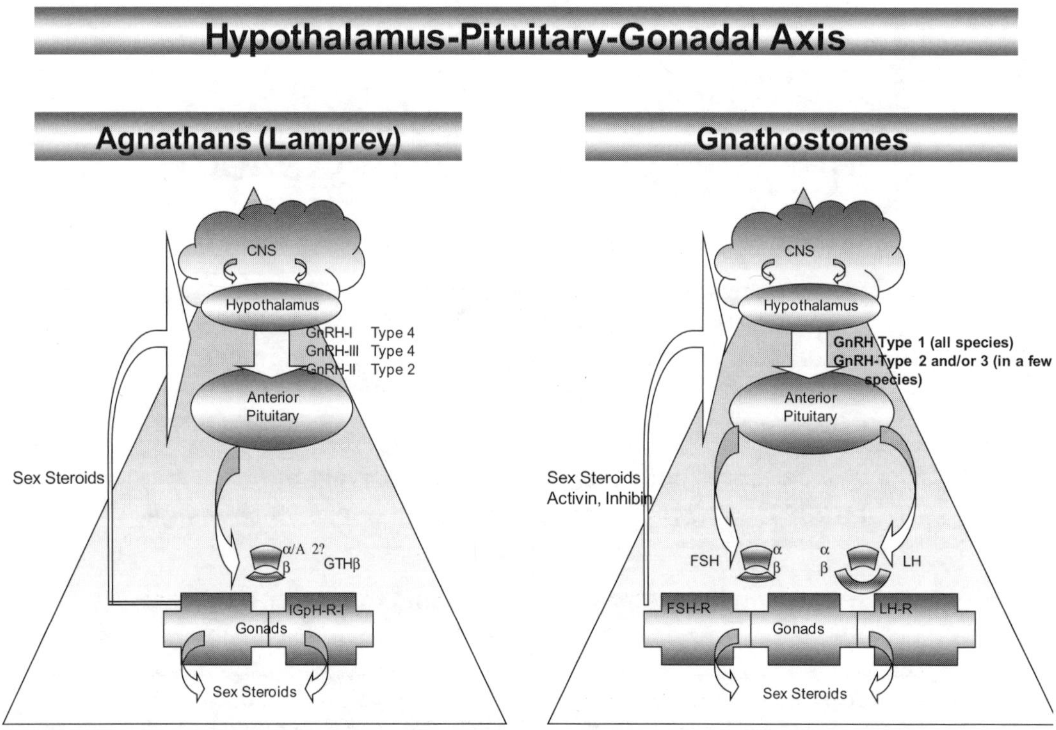


Figure 3. Two schematic diagrams are shown of the hypothalamus-pituitary-gonadal axis in lampreys versus gnathostomes. Lampreys as basal vertebrates are the earliest evolved vertebrates for which there is a clearly defined conserved hypothalamic-pituitary axis. In gnathostomes, there is generally only one major hypothalamic gonadotropin-releasing hormone and two pituitary gonadotropins compared to two, or possibly three, hypothalamic GnRHs and one pituitary gonadotropin in lampreys. Gnathostomes have two identified glycoprotein gonadal receptors and lampreys have one glycoprotein gonadal receptor. (Modified from Sower et al. 2009.)

Table 1. Summary of identified hormones and genes from the brain and pituitary in lampreys. GnRH, gonadotropin releasing hormone; PMY, NPY, neuropeptide Y-related peptide; RFa-A and RFa-B, RFamide peptides; ACTH, adrenocorticotropin; MSH-A, MSH-B, melanotropins; NHF, nasohypophysial factor; AVT, arginine vasotocin; POM, proopiomelanotropin; POC, proopiocortin; GTH, gonadotropin; GH, growth hormone; IGF, insulin like growth factor.

| Tissue | Hormone | Gene/cDNA | Function | References |
|---|---|---|---|---|
| *Brain* | lGnRH-I | | Hypothalamic hormone | Sherwood et al. 1986 |
| | lGnRH-III | | Hypothalamic hormone | Sower et al. 1993 |
| | Somatostatin-14 | | Lipid metabolism | Sower et al. 1994 |
| | NPY (PMY-like) | | Reproduction | Conlon et al. 1994 |
| | Lamprey tachykinin | Unknown | | Waugh et al. 1994 |
| | RFa-A and RFa-B | | Inhibition of GH/ melanotropin | Moriyama et al. 2007 |
| | | NPY | | Soderberg et al. 1994 |
| | | PYY | | Soderberg et al. 1994 |
| | | lGnRH-I | | Gamble et al. 1997; Suzuki et al. 2000 |
| | | lGnRH-II | | Kavanaugh et al. 2008 |
| | | lGnRH-III | | Silver et al. 2004 |
| | | Rfa-A and Rfa-B | | Moriyama et al. 2007 |
| *Pituitary* | ACTH | | Steroidogenesis | Takahashi et al. 1995a |
| | MSH-A | | Melanotropic | Takahashi et al. 1995a |
| | MSH-B | | Melanotropic | Takahashi et al. 1995a |
| | NHF | | Unknown | Sower et al. 1995 |
| | AVT | | Unknown | Lane et al. 1988 |
| | GTHβ | | Gonadotropic | Sower et al. 2006 |
| | | POM | | Takahashi et al. 1995b, 2006 |
| | | POC | | Heinig et al. 1995; Takahashi et al. 2006 |
| | | AVT | | Suzuki et al. 1995 |
| | | GH | IGF expression | Kawauchi et al. 2002; Moriyama et al. 2006 |

Table 2. Summary of identified brain and pituitary hormone receptors and cDNA/genes in lampreys. GnRH, gonadotropin-releasing hormone; PMY, NPY, neuropeptide Y-related peptide.

| Target tissue | Receptor-gene/cDNA of brain-pituitary hormones | References |
|---|---|---|
| *Brain* | NPY receptor | Salaneck et al. 2001 |
| *Pituitary* | lGnRHR-1 | Silver et al. 2005 |
| | lGnRHR-2 and -3 | A. Aquilina-Beck and S. A. Sower, Center for Molecular and Comparative Endocrinology, unpublished |
| *Skin* | Melanocortin receptors (MCa and MCb) | Haitina et al. 2007 |
| *Liver, heart, skeletal muscle* | Melanocortin receptor (MCb) | Haitina et al. 2007 |
| *Testis* | Lamprey glycoprotein receptor 1 | Freamat et al. 2006 |
| *Thyroid tissue* | Lamprey glycoprotein receptor 2 | Freamat et al. 2008 |

lamprey model highly appropriate for examining the GnRH system in terms of its ligands and novel receptors. Similarly, the recent identification of a novel GnRH (Kavanaugh et al. 2008), lamprey GTHβ (Sower et al. 2006), and two glycoprotein receptors (gonadotropic and a thyrotropic-like; Freamat et al. 2006, 2008), provides an opportunity for comparative and evolutionary analysis of the neuroendocrine system in vertebrates.

## Reproductive Cycle of the Lampreys

There are approximately 40 species of lamprey that are classified as parasitic or nonparasitic. Lampreys spawn only once in their lifetime, after which they die. Sexual maturation is thus a synchronized process coordinated with the life stages of the lamprey. The sea lampreys begin their lives in freshwater as blind, filter-feeding larvae. After approximately 5 to 7 years in freshwater streams, metamorphosis occurs and the ammocoetes become free-swimming, sexually immature lampreys. Lamprey metamorphosis is a highly synchronized and programmed process that involves major physiological and morphological changes, reviewed in Youson and Sower (2001). In the parasitic sea lamprey, sexual maturation is a seasonal, synchronized process. During the approximately 15-month-long parasitic sea phase, gametogenesis progresses. In males, spermatogonia proliferate and develop into primary and secondary spermatocytes; in females, vitellogenesis occurs. After approximately 15 months at sea, lampreys return to freshwater streams and undergo the final maturational processes resulting in mature eggs and sperm, spawning, and then death. In nonparasitic species, gonadal differentiation and metamorphosis occur together. Shortly after the end of the maturation of nonparasitic lamprey species, spawning occurs and the lampreys die.

## Gonadotropin-Releasing Hormone

The synthesis and secretion of GnRH is the key neuroendocrine function in the hypothalamic regulation of the hypothalamic-pituitary-gonadal axis. In response to GnRH, the gonadotropins are released from the pituitary gland, which in turn regulate gametogenesis and steroidogenesis (Figure 3). GnRH was first isolated from porcine (Matsuo et al. 1971) and ovine (Burgus et al. 1972) hypothalamic extracts. The GnRH family currently now includes 28 isoforms, 15 and 13 from representative vertebrate and invertebrate species (Kavanaugh et al. 2008; Tsai et al. 2008; Zhang et al. 2008). To date, two to three isoforms have been identified in representative species of all classes of vertebrates. The number of GnRH molecular variants is surprising, particularly in view of the relative constancy of their physiological role in vertebrates (Gorbman and Sower 2003; Silver et al. 2004). Growing evidence reveals that almost all vertebrates synthesize at least two isoforms of GnRH in the brain (Dubois et al. 2002; Silver et al. 2004; Guilgur et al. 2006; Kah et al. 2007; Kavanaugh et al. 2008; Okubo and Nagahama 2008). During the past few years, as more GnRH primary sequences and their respective cDNAs have been identified, more phylogenetic and functional studies have been done leading to proposed phylogenetic relationships of the GnRHs (Grober et al. 1995; Fernald et al. 1999; Parhar 2002; Morgan and Miller 2004; Silver et al. 2004; Guilgur et al. 2006; Kah et al. 2007; Zhang et al. 2008). Early analyses by Grober et al. (1995) and Fernald and White (1999) suggested that there are three paralogous groups of GnRH in gnathostome brains. In recent reviews of GnRH and respective receptors, various scenarios on the phylogenetic relationships among just the gnathostome GnRHs and receptors have been suggested (Guilgur et al. 2006; Kah et al. 2007; Okubo and Nagahama 2008). We had shown from an extensive analysis all available cDNA GnRH sequences, including the eight cloned lamprey GnRH-III cDNAs, that the vertebrate GnRHs are grouped into four paralogous lineages: GnRH 1, 2, 3, and 4 (Silver et al. 2004). We based our analysis on phylogenetic analysis, function, neural distribution, and developmental origin (Silver et al. 2004). This phylogenetic analysis confirmed the earlier model of three GnRH lineages and, in addition, showed that lamprey GnRH-I and -III form a fourth group. The identification of two more invertebrate mollusk GnRHs and one in *Annelid* has suggested a fifth grouping of GnRHs (*Aplysia*, octopus, limpet, annelid GnRH) in invertebrates and supports the fourth group of lamprey GnRH-I and –III (Tsai et al. 2008; Zhang et al. 2008). In addition, the primary structure of a type 1 GnRH (mammalian GnRH) was determined by protein purification and mass spectrometry in the cephalochordate Amphioxus (Chambery et al. 2009); however, the cDNA has not been identified from the Amphioxus genome or protein coupled receptor (PCR) despite numerous attempts (Tello and Sherwood 2009). The phylogenetic relationship of the lamprey type 4 (lamprey GnRH I and III) to the gnathostome GnRHs was unclear from these studies. From the information to date on the identification of the GnRHs, there is likely an ancestral GnRH that gave rise to the GnRH lineages in the deuterostome and proteomes (Tsai and Zhang 2008). Much further information and identification of GnRH will be required across the metazoans to elucidate the evolution of GnRH.

Recently, a cDNA encoding a preproGnRH was cloned from the sea lamprey (Kavanaugh et al. 2008). As with all other vertebrate GnRHs, the lamprey GnRH-II is highly conserved with 10 amino acids and conserved C- and N-termini. Similar to GnRH-II expression in gnathostomes, lamprey GnRH-II was shown to be widely distributed and expressed in a number of tissues. In the brain, lamprey GnRH-II was shown to be located in the preoptic-anterior hypothalamus by in situ hybridization and immunocytochemistry and stimulated the pituitary-gonadal axis. Lamprey GnRH-II was shown to stimulate the hypothalamic-pituitary axis using in vivo and in vitro studies. Lamprey GnRH-II was also shown to activate the inositol phosphate signaling system in COS-7 cells transiently transfected with the lamprey GnRH receptor. These studies provide evidence for a novel lamprey GnRH that has a role as a third hypothalamic GnRH. In summary, the newly discovered lamprey GnRH-II offers a new paradigm of the origin of the vertebrate GnRH family. Likely due to a genome/gene duplication event, an ancestral gene gave rise to two lineages of GnRHs—the gnathostome GnRH and lamprey GnRH-II (Kavanaugh et al. 2008). The gene duplication events that generated the different fish and tetrapod paralogous groups likely took place within the gnathostome lineage, after its divergence from the ancestral agnathans. Lamprey GnRH-I and III (type 4) can be identified now as paralogous homologs of gnathostome GnRH and lamprey GnRH II group (type 2), resulting from a duplication within the lamprey lineage. This implies that there probably was a genome/gene duplication event that gave rise to all forms of vertebrate GnRH affecting the common ancestor of lamprey and gnathostome isoforms. The gnathostome branch of the lamprey GnRH-I and -III orthologous group was probably lost during evolution. Future studies on the identification of GnRH and its cDNA as well as the functions and distribution in hagfishes, chondrichthyes, and amphioxus will be necessary to have a fuller understanding of the evolution of chordate GnRH.

## Lamprey GnRH Receptor

At the anterior pituitary, GnRH action is mediated through high-affinity binding with the GnRH receptor, a class A or rhodopsin-like seven transmembrane G-protein coupled receptor (GPCR). GnRH receptors have been classified typically as type-I (without a C-terminal tail) and type-II (with a C-terminal tail; Silver et al. 2005; Guilgur et al. 2006). The type-I GnRH receptor is unique among all GPCRs in that these receptors lack the highly conserved intracellular carboxy-terminal (C-terminal) tail. In the sea lamprey, a full-length transcript was isolated and cloned, encoding a functional type-II GnRH receptor from the pituitary of the sea lamprey (Silver et al. 2005). The cloned receptor retains the conserved structural features and amino acid motifs of other known GnRH receptors and notably includes a C-terminal intracellular tail of ~120 amino acids. The lamprey GnRH receptor was initially shown to activate the inositol phosphate (IP) signaling system; stimulation with either lamprey GnRH-I or lamprey GnRH-III led to dose dependent responses in transiently transfected COS7 cells. Expression of the receptor transcript was demonstrated in the pituitary and testes using RT-PCR while in situ hybridization showed expression and localization of the transcript in the proximal pars distalis of the pituitary (Silver et al. 2005). The phylogenetic placement, structural, and functional features of this GnRH receptor suggested that it is representative of an ancestral GnRH receptor. The extensive C-terminal tail of this lamprey GnRH receptor may have great significance for understanding the evolutionary change of this vital structural feature within the GnRH receptor family.

In recent functional studies, the lamprey GnRH receptor was also shown to activate the cAMP signaling system, which required the first 40 amino acids of the C-terminal tail (Silver and Sower 2006). Pharmacological profiling provided evidence that the lamprey GnRH receptor is lamprey GnRH-III selective, which supports the hypothesis that lampreys express two more receptors selective for GnRH-I and -II. Truncations of the lamprey GnRH receptor's C-terminal tail were shown to reduce binding affinity. Two additional partial/complete clones of two more GnRH receptors have been identified in lampreys from the lamprey genome (accession numbers: DQ915103, DQ915102/EF166083, EF166082). Once these GnRH receptors have been cloned, structure-function activities of the lamprey GnRHs can be performed with the transfected GnRH receptors and distribution studies can be done to better understand the role of each of the receptors in relation to the GnRH in the lamprey neuroendocrine axis. Comparisons and phylogenetic analyses can then be done on the molecular evolution of GnRH receptors and their respective ligands.

## Thyrotropin Releasing Hormone

In mammals, thyrotropin releasing hormone (TRH) is considered a major hypothalamic hormone that acts on the pituitary to stimulate the synthesis and release of thyrotropin hormone (a member of the pituitary glycoprotein family) that in turn acts on the thyroid gland to stimulate the synthesis and/or release of the thyroid hormones, thyroxine and triiodothyronine. The thyrotropin-releasing hormone has also been shown to release pituitary growth hormone (GH; Guillemin 1978; Schally 1978) and prolactin (PRL; Jackson and Reichlin 1977). The role of TRH in nonmammalian vertebrates is less established, although it has been shown to activate PRL, TSH, GH, and α-melanocyte-stimulating hormone from the pituitary in amphibians and frogs (reviewed in Del Carmen De Andres et al. 2002). In addition, corticotropin-releasing hormone (CRH) has also been shown to activate pituitary TSH release in nonmammalian vertebrates, and it is suggested that CRH might be a common regulator of both the thyroidal and adrenal/interrenal axes (De Groef et al. 2006). A potential role of CRH as a pituitary GpH regulator in lampreys has yet to be investigated.

To date, there are only two reports on the presence of TRH in lampreys. Youngs et al. (1985) showed that TRH content was present in the pituitary, brain, and spinal cord of larval and adult sea lampreys and adult river lampreys *Lampetra ayresii*, as determined by radioimmunoassay. A more extensive immunocytochemistry study was done in which the distribution of TRH mainly occurred in the preoptic region and the hypothalamus in large larvae and adult upstream migrating sea lamprey (Del Carmen De Andres et al. 2002). There are no reports that have examined the biological activity of TRH in lampreys. Sower et al. (1985) reported that adult lampreys treated with a partly purified salmon gonadotropin or a GnRH analog significantly elevated plasma thyroxine. It was hypothesized from these studies that one hypothalamic GnRH stimulated both the pituitary-thyroid and pituitary-gonadal axes. In later studies, lamprey GnRH-I and –III were shown to be significantly correlated with the seven stages of lamprey metamorphosis (Youson et al. 2001). Unlike the induction of frog metamorphosis and even amphioxus metamorphosis by thyroid hormones (Denver 2008), thyroid hormones do not stimulate the process of metamorphosis in lamprey. Metamorphosis in sea lampreys is characterized by a significant decline in thyroid hormones, changes in lipid metabolism, and elevated GnRH (Youson and Sower 2001). Subsequently, we have determined that lampreys likely have only one pituitary glycoprotein (GTH) and one gonadal glycoprotein receptor and one thyroid glycoprotein receptor. Therefore, our working hypothesis is that the glycoprotein hormone/glycoprotein hormone receptor systems emerged as a link between the neuro-hormonal and peripheral control levels during the early stages of Gnathostome divergence.

## Gonadotropin

The traditional glycoprotein hormone (GpH) family consists of two GTHs, LH and FSH, and one thyroid-stimulating hormone (TSH). They are part of the cystine-knot family of growth factors with two noncovalently bound subunits, alpha and beta. The alpha subunit is common within a single species (Kawauchi et al. 1989). The beta subunits are homologous and convey hormone specificity (Themmen and Huhtaniemi 2000; Szkudlinski et al. 2002). Therefore, these glycoprotein hormones are believed to have evolved from a common ancestral molecule through duplication of beta-subunit genes and subsequent divergence (Li and Ford 1998; Moyle et al. 1994). Two GTHs have been identified in all taxonomic groups of gnathostomes, including actinopterygians (Suzuki et al. 1988; Kawauchi et al. 1989; Querat et al. 2000), sarcopterygians (Querat et al. 2004), chondrichthyans (Querat et al. 2000), and only recently in jawless vertebrates (Sower et al. 2006).

## Lamprey GTH

To date, only one GTH has been identified in jawless vertebrates (Sower et al. 2006). We have identified the first and perhaps only gonadotropin beta-like protein by cDNA cloning in sea lamprey (Sower et al. 2006). Because the sea lamprey GTHβ protein is a clear outgroup compared to those of the LH and FSH family based on phylogenic analysis, we had proposed that an ancestral GpH gave rise to only one GTH in lampreys and to the GpH family that gave rise to LH, FSH, and TSH during the early evolution of gnathostomes. A fifth heterodimeric pituitary GpH (after FSH, LH, and TSH) was discovered in 2002 and termed "thyrostimulin" due to its thyroid-stimulating activity (Nakabayashi et al. 2002). However, the thyrostimulin alpha subunit, called GPA2, is homologous but not identical to the common alpha subunit (GPA1 or a) in the other GpHs. With the discovery of GPA2 (Thyrostimulin alpha) and GPB5 (Thyrostimulin beta) homologs in invertebrates (including *Drosophila melanogaster*), it is proposed by Sudo et al. (2005) that an ancestral heterodimeric GpH hormone existed before the divergence of vertebrates and invertebrates and that a later gene duplication event in vertebrates produced the thyrostimulin (GPA2 and GPB5) and GTH/TSH (GPA1 and LHβ/FSHβ/TSHβ). Preliminary evidence to date shows only one glycoprotein alpha subunit in lampreys that has higher similarity with mammalian GPA2 compared to the GPA1 (a) subunits (Sower et al. 2009). Despite extensive research by both molecular cloning and blast analysis of the lamprey genome, no evidence of a GPAα or GPA1 was found. In addition, there is preliminary data for a GPB5 (Sower et al. 2009). Thus, it appears that lampreys may have an ancestral GPA2, GTHβ, and GPB5, but this will have to be confirmed by cloning and functional studies. Our current working hypothesis is that the GPA2 identified is the ancestral alpha subunit; after the gnathostome-agnathan divergence, gene duplications produced the two alpha (GPA1 and GPA2) subunits and four beta subunits (FSHβ, LHβ, TSHβ, and GPBb), and potentially, lampreys may have two GpHs: GPA2/b and GPA2/GPB5 (Sower et al. 2009).

## Glycoprotein Hormone Receptors

Gonadotropin and thyrotropin hormone actions are mediated through a subfamily of GPCRs, namely the GpH receptors (Combarnous 1992). Known GpH-Rs share a number of unique features. They are composed of two functionally distinct modules of similar size, an extracellular N-terminal domain followed by a prototypical GPCR segment. The extracellular N-terminal domain is primarily responsible for high-affinity hormone binding and contains a central portion of nine Leu-rich repeat motifs, flanked by N- and C-terminal Cys-rich clusters. The C-terminal half of the receptor contains a transmembrane region with seven hydrophobic transmembrane α-helices, connected by intra- and extracellular loops and an intracellular C-terminal domain (Grossmann et al. 1997; Dufau 1998; Ascoli et al. 2002; Moyle et al. 2005). To date, approximately 79 GpH-Rs have been identified and described in 36 different species, mostly in mammals and also in three species of birds, two species of reptiles, one amphibian, and 10 species of fish (Hovergen database, http://pbil.univ-lyon1.fr).

Until recently, there had been no GpH-Rs described in any species of agnathans. One functional GpH receptor (lGpH-R I; Freamat et al. 2006) was identified from lamprey testis and a second functional GpH-R receptor (lGpH-R II) was identified and shown to be expressed mainly in the thyroid tissue (Freamat and Sower 2008). We hypothesize that lGpH-R I and lGpH-R II are the only members of the GpH receptor subfamily in lamprey (Freamat and Sower 2008). They are descendants of the TSH receptor-like molecular ancestors of the GpH-Rs in Gnathostomes and are likely the result of the genome duplication event hypothesized to have taken place before the divergence of the lamprey lineage (Sidow 1996; Kuratani et al. 2002).

The 719-amino acid full-length cDNA encoding lGpH-R I is highly similar and likely a homolog of the vertebrate GpH-Rs (including LH, FSH, and TSH receptors; Freamat et al. 2006). The key motifs, sequence comparisons, and characteristics of the identified GpH-R reveal a mosaic of features common to all other classes of GpH-Rs in vertebrates. The lGpH-R I was shown to activate the cAMP signaling system using human chorionic gonadotropin in transiently transfected COS-7 cells. The highest expression of the receptor transcript was demonstrated in the testes using reverse transcriptase-PCR. The high expression of lGpH-R I in the testis and the high similarity with gnathostome gonadotropin hormone receptors suggest that lGpH-R I functions as a receptor for lamprey gonadotropin hormone(s).

The second GpH receptor (lGpH-R II) in the agnathan sea lamprey has 781 protein residues and was approximately 43% identical with mammalian TSH-R and FSH-R representative sequences (Freamat and Sower 2008). Similar to these two classes of mammalian receptors, lGpH-R II is assembled from 10 exons. A synthetic ligand containing the lamprey GpH β chain tethered upstream of a mammalian α chain activated the lGpH-R II expressed in COS-7 cells but to a lesser extent than lGpH-R I. The most obvious feature of the coding sequence of lGpH-R II is the presence of a long linker fragment called signaling specificity domain (SSD) or "hinge" located in between the Leu-rich domain of the extracellular segment and the transmembrane domain. This is one of the longest linker fragments described so far in all vertebrate GpH receptors. This is in contrast with the similar region of the lGpH-R I, which is the shortest SSD/hinge segment among all vertebrate GpH-Rs (Freamat et al. 2006). Molecular phylogenetic analysis of vertebrate GpH-R protein sequences suggests a closer relationship between lGpH-R II and Gnathostome thyrotropin receptors.

Therefore, at this point, a comparative perspective on this endocrine compartment in lampreys relative to the gnathostome well-established paradigm suggests the involvement of one pituitary GpH (only β subunit was found) and two GpH-Rs as opposed to three or four dimeric hormones and three receptors in gnathostomes. This role of the GpH/GpH-R system in lampreys has yet to be confirmed experimentally. This requires identification and characterization of a GpH α chain homolog. The existence of a second GpH ligand in lampreys with a distinct binding specificity to GpH-R I and II cannot be excluded, but if it exists, it is likely even less similar to gnathostome sequences than the one already described (Freamat and Sower 2008). From the studies completed to date, we hypothesize that there is lower specificity of GTH and its receptor in agnathans and that during coevolution of the ligand and its receptor in gnathostomes, there were increased specificities of interactions between each GpH (TSH, LH, and FSH) and its receptor.

## Summary

Reproduction in vertebrates is controlled by a hierarchically organized endocrine system. In spite of the very diverse patterns of life cycles and reproductive strategies and behaviors, this endocrine system is remarkably conserved throughout the Gnathostome lineages. Understanding the hormonal control of the reproduction in Agnathans, and in particular in the sea lamprey, is of foremost importance for knowing the earliest events in the evolution of the vertebrates (Kuratani et al. 2002; Kawauchi and Sower 2006). Further comparative studies on the GnRH-GTH proteins and respective receptors will provide clues on the evolution of reproductive mechanisms and insights into our understanding of gene duplication, structure-activity relations, and the molecular evolution and functional diversity of these hypothalamic-pituitary systems.

## Acknowledgments

Funded by NSF Grant # 0421923 and 0849569. This is Scientific Contribution Number 2350 from the New Hampshire Agricultural Experiment Station.

## References

Ascoli, M., F. Fanelli, and D. L. Segaloff. 2002. The lutropin/choriogonadotropin receptor, a 2002 perspective. Endocrine Reviews 23:141–174.

Burgus, R., M. Butcher, M. Amoss, N. Ling, M. Monahan, J. Rivier, R. Fellows, R. Blackwell, W. Vale, and R. Guillemin. 1972. Primary structure of the ovine hypothalamic luteinizing hormone-releasing factor (LRF) (LH-hypothalamus-LRF-gas chromatography-mass spectrometry-decapeptide-Edman degradation). Proceedings of the National Academy of Sciences of the United States of America 69:278–282.

Chambery, A., A. Parente, E. Topo, J. Garcia-Fernàndez, and S. D'Aniello. 2009. Characterization and putative role of a type I gonadotropin-releasing hormone in the cephalochordate amphioxus. Endocrinology 99:346–353.

Combarnous, Y. 1992. Molecular basis of the specificity of binding of glycoprotein hormones to their receptors. Endocrine Reviews 13:670–691.

De Groef, B., S. Van der Geyten, V. Darras, and E. Kuhn. 2006. Role of corticotropin-releasing hormone as a thyrotropin-releasing factor in non-mammalian vertebrates. General and Comparative Endocrinology 146:62–68.

Del Carmen De Andres, M., R. Anadon, M. J. Manso, and M. J. Gonzalez. 2002. Distribution of thyrotropin-releasing hormone immunoreactivity in the brain of larval and adult sea lampreys, *Petromyzon marinus* L. Journal of Comparative Neurology 453:323–335.

Delarbre, C., C. Gallut, V. Barriel, P. Janvier, and G. Gachelin. 2002. Complete mitochondrial DNA of the hagfish, *Eptatretus burgeri*: the comparative analysis of mitochondrial DNA sequences strongly supports the cyclostome monophyly. Molecular phylogenetics and evolution 22:184–192.

Denver, R. J. 2008. Chordate metamorphosis: ancient control by iodothyronines. Current Biology 18:R567–R569.

Dubois, E. A., M. A. Zandbergen, J. Peute, and H. J. Goos. 2002. Evolutionary development of three gonadotropin-releasing hormone (GnRH) systems in vertebrates. Brain Research Bulletin 57:413–418.

Dufau, M. L. 1998. The luteinizing hormone receptor. Annual Review of Physiology 60:461–496.

Fernald, R. D., and R. B. White. 1999. Gonadotropin-releasing hormone genes: phylogeny, structure, and functions. Frontiers in Neuroendocrinol 20:224–40.

Forey, P., and P. Janvier. 1993. Agnathans and the origins of jawed vertebrates. Nature 361:129–134.

Forey, P., and P. Janvier. 1994. Evolution of the early vertebrates. American Scientist 82:554–566.

Freamat, M., H. Kawauchi, M. Nozaki, and S. A. Sower. 2006. Identification and cloning of a glycoprotein hormone receptor from sea lamprey, *Petromyzon marinus*. Journal of Molecular Endocrinology 37:135–146.

Freamat, M., and S. A. Sower. 2008. A sea lamprey glycoprotein hormone receptor similar with gnathostome thyrotropin hormone receptor. Journal of Molecular Endocrinology 41:219–228.

Gorbman, A., and S. A. Sower. 2003. Evolution of the role of GnRH in animal (Metazoan) biology. General and Comparative Endocrinology 134:207–213.

Grober, M. S., T. R. Myers, M. A. Marchaterre, A. H. Bass, and D. A. Myers. 1995. Structure, localization, and molecular phylogeny of a GnRH cDNA from a paracanthopterygian fish, the plainfin midshipman (*Porichthys notatus*). General and Comparative Endocrinology 99:85–99.

Grossmann, M., B. D. Weintraub, and M. W. Szkudlinski. 1997. Novel insights into the molecular mechanisms of human thyrotropin action: structural, physiological, and therapeutic implications for the glycoprotein hormone family. Endocrine Reviews 18:476–501.

Guilgur, L. G., N. Moncaut, A. Canario, and G. Somoza. 2006. Evolution of GnRH ligands and receptors in gnathostomata. Comparative Biochemistry and Physiology Part A Molecular and Integrative Physiology 144:272–283.

Guillemin, R. 1978. Peptides in the brain: the new endocrinology of the neuron. Science 202:390–402.

Jackson, I. M., and S. Reichlin. 1977. Thyrotropin-releasing hormone: abundance in the skin of the frog, *Rana pipiens*. Science 198:414–415.

Kah, O., C. Lethimonier, G. Somoza, L. G. Guilgur, C. Vaillant, and J. J. Lareyre. 2007. GnRH and GnRH receptors in metazoa: a historical, comparative, and evolutive perspective. General and Comparative Endocrinology 153:346–364.

Kavanaugh, S. I., M. Nozaki, and S. A. Sower. 2008. Origins of gonadotropin-releasing hormone (GnRH) in vertebrates: identification of a novel GnRH in a basal vertebrate, the sea lamprey. Endocrinology 149:3860–3869.

Kawauchi, H., and S. A. Sower. 2006. The dawn and evolution of hormones in the adenohypophysis. General and Comparative Endocrinology 148:3–14.

Kawauchi, H., K. Suzuki, H. Itoh, P. Swanson, N. Naito, Y. Nagahama, M. Nozaki, Y. Nakai, and S. Itoh. 1989. The duality of teleost gonadotropins. Fish Physiology and Biochemistry 7:29–38.

Kuratani, S., S. Kuraku, and Y. Murakami. 2002. Lamprey as an evo-devo model: lessons from comparative embryology and molecular phylogenetics. Genesis 34:175–183.

Li, M. D., and J. J. Ford. 1998. A comprehensive evolutionary analysis based on nucleotide and amino acid sequences of the alpha- and beta-subunits of glycoprotein hormone gene family. Journal of Endocrinology 156:529–542.

Matsuo, H., Y. Baba, R. M. Nair, A. Arimura, and A. Schally. 1971. Structure of the porcine LH- and FSH-releasing hormone. I. The proposed amino acid sequence. Biochemical and Biophysical Research Communications 43:1334–1339.

Morgan, K., and R. P. Millar. 2004. Evolution of GnRH ligand precursors and GnRH receptors in protochordate and vertebrate species. General and Comparative Endocrinology 139:191–197.

Moyle, W. R., R. K. Campbell, R. V. Myers, M. P. Bernard, Y. Han, and X. Wang. 1994. Co-evolution of ligand-receptor pairs. Nature 368:251–255.

Moyle, W. R., W. Lin, R. V. Myers, D. Cao, J. E. Kerrigan, and M. P. Bernard. Models of glycoprotein hormone receptor interaction. Endocrine 26:189–205.

Nakabayashi, K., H. Matsumi, A. Bhalla, J. Bae, S. Mosselman, S. Y. Hsu, A. J. Hsueh. 2002. Thyrostimulin, a heterodimer of two new human glycoprotein hormone subunits, activates the thyroid-stimulating hormone receptor. Journal of Clinical Investigation 109:1445–1452.

Okubo, K., and Y. Nagahama. 2008. Structural and functional evolution of gonadotropin-releasing hormone in vertebrates. Acta Physiologica (Oxford) 193:3–15.

Ota, K. G., S. Kuraku, and S. Kuratani. 2007. Hagfish embryology with reference to the evolution of the neural crest. Nature 446:672–675.

Parhar, I. S. 2002. Cell migration and evolutionary significance of GnRH subtypes. Progress in Brain Research 141:3–17.

Querat, B., Y. Arai, A. Henry, Y. Akama, T. J. Longhurst, J. M. Joss. 2004. Pituitary glycoprotein hormone beta subunits in the Australian lungfish and estimation of the relative evolution rate of these subunits within vertebrates. Biology of Reproduction 70:356–363.

Querat, B., A. Sellouk, and C. Salmon. 2000. Phylogenetic analysis of the vertebrate glycoprotein hormone family including new sequences of sturgeon (*Acipenser baeri*) beta subunits of the two gonadotropins and the thyroid-stimulating hormone. Biology of Reproduction 63:222–228.

Schally, A. V. 1978. Aspects of hypothalamic regulation of the pituitary gland. Science 202:18–28.

Sherwood, N. M., S. A. Sower, D. R. Marshak, B. A. Fraser, M. J. Brownstein. 1986. Primary structure of gonadotropin-releasing hormone from lamprey brain. Journal of Biological Chemistry 261:4812–4819.

Sidow, A. 1996. Gen(om)e duplications in the evolution of early vertebrates. Current Opinion in Genetics and Development 6:715–722.

Silver, M. R., H. Kawauchi, M. Nozaki, and S. A. Sower. 2004. Cloning and analysis of the lamprey GnRH-III cDNA from eight species of lamprey representing the three families of Petromyzoniformes. General and Comparative Endocrinology 139:85–94.

Silver, M. R., N.V. Nucci, A.R. Root, K. L. Reed, and S. A. Sower. 2005. Cloning and characterization of a functional type II gonadotropin-releasing hormone receptor with a lengthy carboxy-terminal tail from an ancestral vertebrate, the sea lamprey. Endocrinology 146:3351–3361.

Silver, M. R., and S. A. Sower. 2006. Functional characterization and kinetic studies of an ancestral lamprey GnRH-III selective type II GnRH receptor from the sea lamprey, *Petromyzon marinus*. Journal of Molecular Endocrinology 36:601–610.

Sower, S. A. 2003. The endocrinology of reproduction in lampreys and applications for male lamprey sterilization. Journal of Great Lakes Research 29:50–65.

Sower, S. A., Y. C. Chiang, S. Lovas, and J. M. Conlon. 1993. Primary structure and biological activity of a third gonadotropin-releasing hormone from lamprey brain. Endocrinology 132:1125–31.

Sower, S. A., M. Freamat, and S. I. Kavanaugh 2009. The origins of the vertebrate hypothalamic-pituitary-gonadal (HPG) & hypothalamic-pituitary thyroid (HPT) endocrine systems: new insights from lampreys. General and Comparative Endocrinology 161:16120–16129.

Sower, S. A., S. Moriyama, M. Kasahara, A. Takahashi, M. Nozaki, K. Uchida, J. M. Dahlstome, and H. Kawauchi. 2006. Identification of sea lamprey GTHbeta-like cDNA and its evolutionary implications. General and Comparative Endocrinology 148:22–32.

Sower, S. A., E. Plisetskaya, and A. Gorbman. 1985. Steroid and thyroid hormone profiles following a single injection of partly purified salmon gonadotropin or GnRH analogues in male and female sea lamprey. Journal of Experimental Zoology 235:403–408.

Sudo, S., Y. Kuwabara, J.-I. Park, S. Y. Hsu, and A. J. Hsueh. 2005. Heterodimeric fly glycoprotein hormone-alpha2 (GPA2) and glycoprotein hormone-beta5 (GPB5) activate fly leucine-rich repeat-containing G protein-coupled receptor-1 (DLGR1) and stimulation of human thyrotropin receptors by chimeric fly GPA2 and human GPB5. Endocrinology 146:3596–3604.

Suga, H., D. Hoshiyama, S. Kuraku, K. Katoh, K. Kubokawa, and T. Miyata. 1999. Protein tyrosine kinase cDNAs from amphioxus, hagfish, and lamprey: isoform duplications around the divergence of cyclostomes and gnathostomes. Journal of Molecular Evolution 49:601–608.

Suzuki, K., R. L. Gamble, and S. A. Sower. 2000. Multiple transcripts encoding lamprey gonadotropin-releasing hormone-I precursors. Journal of Molecular Endocrinology 24:365–376.

Suzuki, K., H. Kawauchi, and Y. Nagahama. 1988. Isolation and characterization of subunits from two distinct salmon gonadotropins. General and Comparative Endocrinology 71:302–306.

Szkudlinski, M. W., V. Fremont, C. Ronin, and B. D. Weintraub. 2002. Thyroid-stimulating hormone and thyroid-stimulating hormone receptor structure-function relationships. Physiological Reviews 82:473–502.

Tello, J. A., and N. M. Sherwood. 2009. Amphioxus: beginning of vertebrate and end of invertebrate type GnRH receptor lineage. Endocrinology 150:2847–2856.

Themmen, A. P. N., and I. T. Huhtaniemi. 2000. Mutations of gonadotropins and gonadotropin receptors: elucidating the physiology and pathophysiology of pituitary-gonadal function. Endocrine Reviews 21:551–583.

Tsai, P. S., and L. Zhang. 2008. The emergence and loss of gonadotropin-releasing hormone in protostomes: orthology, phylogeny, structure, and function. Biology of Reproduction 79:798–805.

Youngs, L. J., A. Winokur, and M. E. Selzer. 1985. Thyrotropin-releasing hormone in lamprey central nervous system. Brain Research 338:177–180.

Youson, J. H., and S. A. Sower. 2001. Theory on the evolutionary history of lamprey metamorphosis: role of reproductive and thyroid axes. Comparative Biochemistry and Physiology B 129:337–345.

Zhang, L., J. A. Tello, W. Zhang, and P.-S. Tsai. 2008. Molecular cloning, expression pattern, and immunocytochemical localization of a gonadotropin-releasing hormone-like molecule in the gastropod mollusk, *Aplysia californica*. General and Comparative Endocrinology 156:201–209.

American Fisheries Society Symposium 72:71–114, 2009


# A Review of the Evolution of Nonparasitism in Lampreys and an Update of the Paired Species Concept

Margaret F. Docker*
*Department of Biological Sciences, University of Manitoba*
*Winnipeg, Manitoba R3T 2N2, Canada*

*Abstract.*—In most lamprey genera, "paired" species exist in which the larvae (which are microphagous filter feeders) are morphologically similar but the adults differ dramatically, becoming parasitic on teleost fishes or nonparasitic (i.e., do not feed at all) following metamorphosis. Parasitic lampreys feed for several months to several years (either in their natal stream or after migrating to larger fresh or marine water bodies) before embarking on a nontrophic upstream migration, sexual maturation, and spawning (followed by death); nonparasitic lampreys eliminate the parasitic phase, begin sexual maturation toward the end of metamorphosis, and spawn and die within 6–10 months of metamorphosis. In each species pair, the reduction in the length of postlarval life in nonparasitic lampreys is generally accompanied by an increase in the length of the larval period (and size at metamorphosis) so that the evolution of nonparasitism appears to have occurred without a change in the overall life span. Rather, nonparasitism appears to have evolved as a result of a change in the timing of metamorphosis relative to the timing of sexual maturation. Conspicuous morphological (e.g., adult body size, relative eye and oral disk size) and histological (e.g., lack of a functional digestive tract) differences distinguish nonparasitic adults from parasitic forms, and most lamprey taxonomists recognize life history type as a species-specific characteristic. However, plasticity of feeding type (e.g., facultative parasitism) has been observed in some lamprey populations, and molecular data on a number of paired species show no genetic differentiation between sympatric species pairs and suggest a polyphyletic origin for several nonparasitic species. This paper reviews the paired species concept, the repeated and independent evolution of nonparasitism in different genera and even within species, the evidence for facultative parasitism or facultative nonparasitism in some lamprey species, and the potential for hybridization between paired species and attempts to answer the question, are brook lampreys "real" species? The tentative answer is that there likely is not a single answer for all lamprey species pairs; different species pairs represent speciation at different stages. Some pairs appear to be distinct species according to both the biological and phylogenetic species concepts (i.e., they are reproductively isolated and show reciprocal monophyly), although each is not necessarily fixed for feeding type. In contrast, other pairs may represent incipient speciation and others yet may be experiencing ongoing gene flow. Parallels are therefore drawn between different lamprey species pairs and the divergent life history types found in other animal taxa (e.g., echinoderms and amphibians) and other temperate fish species (e.g., anadromous and freshwater-resident salmonids).

* Corresponding author: dockerm@cc.umanitoba.ca

## Introduction

In most lamprey genera, "paired" species exist in which the larvae (ammocoetes) are morphologically and ecologically similar but the juveniles and adults differ dramatically (Zanandrea 1959; Vladykov and Kott 1979; Potter 1980). All lampreys spend the first years of life as blind larvae in the silt bottoms of rivers and streams and feed on microscopic organisms and detritus, but following metamorphosis, lampreys may be nonparasitic (i.e., nontrophic) in their natal stream (spawning and dying within 6–10 months) or they may become parasitic on teleost fishes (either in their natal stream or after migrating to larger fresh or marine water bodies). Following metamorphosis, the digestive tract is not functional in nonparasitic (brook) lampreys, and other conspicuous morphological differences (most notably differences in adult body size) distinguish nonparasitic adults from parasitic forms. Since it is generally thought that size-assortative mating would result in reproductive isolation between parasitic and nonparasitic species (e.g., Beamish and Neville 1992), most lamprey taxonomists recognize life history type as a species-specific characteristic (see below).

There has long been an interest in these closely related pairs. Loman (1912) recognized that the European river lamprey *Lampetra fluviatilis* and European brook lamprey *L. planeri* were morphologically similar but that European brook lamprey exhibited delayed metamorphosis and accelerated sexual maturation relative to European river lamprey (see Heterochrony, below). The morphological similarity between several other nonparasitic and parasitic lampreys was likewise recognized by Hubbs (1925), who suggested several cases in which a particular brook lamprey species had apparently evolved from a form similar to that of an extant parasitic lamprey. The geographic distributions of the two forms often overlap, although the nonparasitic forms may be allopatric to the presumed ancestor (e.g., American brook lamprey *Lethenteron appendix* (formerly *Lampetra appendix*) and Arctic lamprey *Lethenteron camtschaticum* (formerly *Lampetra camtschatica*; see Renaud 1997), suggesting local extinction of the ancestor or range expansion of the nonparasitic form. The term "paired species" was later coined by Zanandrea (1959) and used for the first time to refer to the European river and brook lampreys. Vladykov and Kott (1979) introduced the more general term "satellite species" because there are several cases in which more than one brook lamprey (satellite) species has apparently been derived from a single parasitic (stem) species[1]. For example, it appears that the anadromous Arctic lamprey has given rise to at least four brook lampreys (Table 1). Satellite species may also include parasitic freshwater lampreys (e.g., the freshwater-resident *Entosphenus* species; Table 1) that are presumed to be derived from an anadromous stem species. In this review, however, I focus on the relationship between parasitic and nonparasitic species (rather than between anadromous and freshwater lampreys), and I use the term "paired species" rather than "satellite species" because the emphasis here is on the parasitic-to-nonparasitic relationship or transformation, regardless of the number of times it has occurred in each group. However, freshwater parasitic or "praecox" (i.e., lampreys that feed at sea but reduce their marine feeding phase by at least 1 year)[2] forms may sometimes be intermediaries in the evolution of nonparasitism

[1] Note, however, that not all authors are consistent in describing whether satellite species are derived from an ancestor similar to an extant parasitic lamprey (e.g., Beamish and Thomas 1983) or from the extant parasitic lamprey itself (e.g., Zanandrea 1959; Vladykov and Kott 1979). Recent derivatives may indeed be derivatives of extant species (see below), although having extant ancestors of living species is inconsistent with cladistics.

[2] Some authors (e.g., Hardisty 1986a) use the term "praecox" to refer to smaller forms of parasitic species without explicitly stating whether they are freshwater-resident or anadromous but with a shortened marine phase, often because adult habitats were not definitively known. In this review, where possible, I try to distinguish between freshwater-resident parasitic lampreys and those with a reduced marine phase (e.g., where Adams et al. (2008) used stable isotope analysis to show that the small and large size-classes of the European river lamprey corresponded to those feeding in fresh- and saltwater, respectively).

Table 1. List of "paired" parasitic and nonparasitic lamprey species; for "satellite" species (see text), two or more nonparasitic derivatives are listed for a given parasitic species. Geographic distribution and details of the adult phase (e.g., whether they are anadromous) are given in Potter (1980) and Renaud (1997).

| Parasitic | Nonparasitic |
|---|---|
| Pacific lamprey *Entosphenus tridentatus* (formerly *Lampetra tridentata*) (Gairdner in Richardson 1836) | Pit-Klamath brook lamprey *Entosphenus lethophagus* (formerly *Lampetra lethophaga*) Hubbs 1971[a,b] |
| Carpathian lamprey *Eudontomyzon danfordi* Regan 1911 | Ukrainian brook lamprey *Eudontomyzon mariae* (Berg 1931)[c] |
| | Drin brook lamprey *Eudontomyzon stankokaramani* Holčík and Šorić 2004 |
| Ohio lamprey *Ichthyomyzon bdellium* Jordan 1885 | Mountain brook lamprey *Ichthyomyzon greeleyi* Hubbs and Trautman 1937[d] |
| Chestnut lamprey *Ichthyomyzon castaneus* Girard 1858 | Southern brook lamprey *Ichthyomyzon gagei* Hubbs and Trautman 1937 |
| Silver lamprey *Ichthyomyzon unicuspis* Hubbs and Trautman 1937 | Northern brook lamprey *Ichthyomyzon fossor* Reighard and Cummins 1916 |
| River lamprey (also known as western river lamprey) *Lampetra ayresii* (Günther 1870) | Western brook lamprey *Lampetra richardsoni* Vladykov and Follett 1965[e] |
| European river lamprey *Lampetra fluviatilis* (Linnaeus 1758) | European brook lamprey *Lampetra planeri* (Bloch 1784) |
| Arctic lamprey *Lethenteron camtschaticum* (formerly *Lampetra camtschatica*) (Tilesius 1811)[f] | American brook lamprey *Lethenteron appendix* (formerly *Lampetra appendix*) (DeKay 1842) |
| | Alaskan brook lamprey *Lethenteron alaskense* Vladykov and Kott 1978b[g] |
| | Siberian brook lamprey *Lethenteron kessleri* (Anikin 1905)[h] |
| | Far Eastern brook lamprey *Lethenteron reissneri* (Dybowski 1869)[i] |
| Short-headed lamprey *Mordacia mordax* (Richardson 1846) | Precocious lamprey *Mordacia praecox* Potter 1968 |
| Mexican lamprey *Tetrapleurodon spadiceus* (formerly *Lampetra spadicea*) (Bean 1887) | Mexican brook lamprey *Tetrapleurodon geminis* (formerly *Lampetra geminis*) Alvarez 1964 |

[a] A second nonparasitic derivative of Pacific lamprey, *Entosphenus folletti* Vladykov and Kott 1976b, was described with fewer myomeres, more robust dentition, and larger adult size than Pit-Klamath brook lamprey; it has since been synonymized with Pit-Klamath brook lamprey (Robins et al. 1980).

[b] Parasitic freshwater satellite species of Pacific lamprey include Vancouver lamprey *Entosphenus macrostomus* (formerly *Lampetra macrostoma*) (R. J. Beamish 1982), Klamath lamprey *E. similis* (formerly *L. similis*) Vladykov and Kott 1979, and Miller Lake lamprey *E. minimus* (formerly *L. minima*) (Bond and Kan 1973).

[c] Another nonparasitic derivative of Carpathian lamprey, Vladykov's lamprey *Eudontomyzon vladykovi* Oliva and Zanandrea 1959, was synonymized with Ukranian brook lamprey (Renaud 1982), although some authors consider it a valid species (e.g., Kottelat 1997; Lang et al. 2009); a third reputably nonparasitic species, *E. gracilis* (Kux 1965), is now thought to be conspecific with Carpathian lamprey (Renaud and Holčík 1988; see Evidence for Plasticity of Feeding Type, in text).

[d] A second nonparasitic derivative of Ohio lamprey, Tennessee brook lamprey *Ichthyomyzon hubbsi* Raney 1952, was synonymized with mountain brook lamprey (Robins et al. 1980).

[e] A second nonparasitic derivative of western river lamprey, Pacific brook lamprey *Lampetra pacifica* Vladykov 1973, was synonymized with western brook lamprey (Robins et al. 1991), although it (and some other populations of *L. richardsoni* sensu lato) are genetically quite distinct from *L. richardsoni* sensu stricto (Boguski 2009); Pacific brook lamprey was described as having fewer myomeres and less well-developed teeth than western brook lamprey.

Table 1. Continued.

[f] Also known as *Lethenteron japonicum* (which is a junior synonym of *L. camtschaticum*; see Renaud et al. 2009a) and the Japanese lamprey (e.g., Fukayama and Takahashi 1982).
[g] Considered a synonym of *Lethenteron appendix* by Nelson et al. (2004), but see Renaud et al. (2009a).
[h] Another nonparasitic derivative of Arctic lamprey (with low myomere counts), *Lethenteron matsubarai* Vladykov and Kott 1978a, was synonymized with Siberian brook lamprey (Iwata et al. 1985).
[i] Far Eastern brook lamprey (s.l.) appears to represent at least three distinct species: *Lethenteron reissneri* (s.s.) from its type locality, and *Lethenteron* sp. N and *Lethenteron* sp. S from northern Japan and southern Japan and southern Korea, respectively (Yamazaki and Goto 1998; Yamazaki et al. 2006); the southern form (*Lethenteron* sp. S) was highly divergent and is not likely a derivative of Arctic lamprey, and *Lethenteron* sp. N, although a derivative of Arctic lamprey, is not considered a closely related paired species in this review; see Figure 3). Both *Lethenteron* sp. N and *Lethenteron* sp. S have a low number of myomeres (51–66 and 50–62, respectively; Yamazaki and Goto 1997) relative to *L. reissneri* (57–63; Vladykov and Kott 1979). Hubbs and Potter (1971) provisionally recognized *L. mitsukurii* (Hatta 1901), a highly dwarfed and degenerate form having a low number of myomeres from southern Japan, as a distinct species; its relationship to *Lethenteron* sp. S is unknown.

(see Conditions under which Brook Lampreys Might Evolve).

Closely related paired species are currently recognized in 7 of the 10 lamprey genera (Table 1). Nonparasitism therefore appears to have arisen repeatedly in lamprey evolution, with different nonparasitic derivatives arising independently at different times or in different locations (Hubbs and Potter 1971; Vladykov and Kott 1979). However, not all lamprey species are paired. Three monotypic genera (*Petromyzon*, *Caspiomyzon*, and *Geotria*) without nonparasitic species have been identified (but see Lang et al. 2009, this volume, regarding *Caspiomyzon*). Furthermore, several so-called relict species (nonparasitic lampreys that occur at or near the extreme southern limits of distribution of the Northern Hemisphere lampreys) cannot be unambiguously paired with an extant parasitic species (Hubbs and Potter 1971; Docker et al. 1999; Lang et al. 2009). These include the Po brook lamprey *Lampetra zanandreai*, least brook lamprey *L. aepyptera*, and Kern brook lamprey *Entosphenus hubbsi* (formerly *L. hubbsi*)[3] and presumably represent more ancient nonparasitic derivatives. In this review, the term "paired species" refers only to parasitic lampreys and their closely related nonparasitic derivatives and not to the relict nonparasitic lampreys and their presumed ancestor.

The paired species concept and the evolution of nonparasitism in lampreys have been reviewed previously, notably by Hubbs and Potter (1971), Vladykov and Kott (1979), Potter (1980), and Hardisty (2006) and has also been discussed recently by Salewski (2003). The fundamentals of these reviews will be reiterated here (e.g., in terms of morphological differences between members of a pair, size-assortative mating leading to reproductive isolation, and conditions under which nonparasitic lampreys might evolve), but the object of the current review is to provide an update on these topics, particularly with the inclusion of recent molecular data, evidence for plasticity of feeding type, and comparisons to species pairs and divergent morphotypes in other vertebrates and invertebrates.

## Morphological and Life History Differences between Paired Species

### *Juveniles and Adults*

There are pronounced morphological differences in the juveniles (i.e., postmetamorphosis but prior to full sexual maturity) and adults (sexually mature individuals) of most paired lamprey species (Figure 1). Morphological differences

[3] The Kern brook lamprey was placed in the genus *Entosphenus* by Vladykov and Kott (1976a), but DNA sequence data suggest that it is a relict species in the *Lampetra ayresii* lineage (see Figure 3).

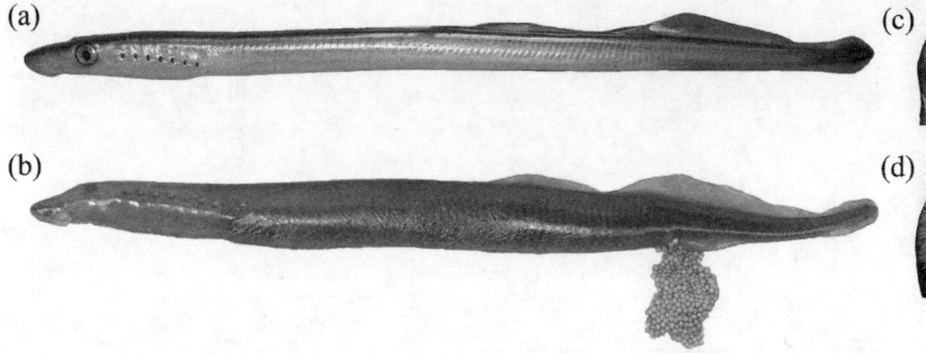


Figure 1. Differences in morphology and dentition between the parasitic North American river lamprey (a, c) and western brook lamprey (b, d). (a) Downstream-migrating river lamprey collected from U.S. Bureau of Reclamation's Tracy Fish Collection Facility (Old River) in Byron, California, on March 4, 2003; total length was 163 mm (Photo: René Reyes). (b) Sexually mature female western brook lamprey collected in Ziegler Creek, Washington, on July 16, 2004; total length was 166 mm (Photo: Damon Goodman and Stewart Reid). (c) Dentition on downstream-migrating river lamprey collected from Siletz River, Oregon, on April 17, 2007; total length was 140 mm (Photo: Steve Johnson). (d) Dentition on sexually mature western brook lamprey collected from Smith Creek, British Columbia, on May 14, 2008; total length was 159 mm (Photo: David Boguski).

between European river and brook lampreys were apparent by approximately 2 to 4 months after the onset of metamorphosis (Hardisty 1970; Bird and Potter 1979a, 1979b); a relatively large eye, oral disk, and silver coloration were apparent in the river lamprey, and marked enlargement and development of the gonad, starting at stage 5 of metamorphosis, was observed in the brook lamprey (Bird and Potter 1979a). As noted above, most of the differences between paired species are related to differences in feeding and the timing of sexual maturation relative to metamorphosis. In lampreys, the adult esophagus or foregut (which originates as a solid cord of cells from the dorsal fold of the pharynx) forms by gradual fusion of vacuoles at metamorphosis (Youson 1980). The lumen of the foregut becomes patent in parasitic species (measuring 5–25 mm in diameter) but measures less than 1 mm in diameter in nonparasitic species (Vladykov and Kott 1979; but see Evidence for Plasticity of Feeding Type, below). Consequently, nonparasitic lampreys do not feed at all following metamorphosis, whereas the feeding phase in parasitic lampreys generally lasts from a few months to 2–3 years (e.g., Beamish 1980; Bergstedt and Swink 1995). In nonparasitic lampreys, the gonad is relatively well developed by the completion of metamorphosis and maturation is rapid; spawning generally occurs within 6–10 months in their natal streams (hence the name "brook lampreys" is universally applied to all nonparasitic lampreys). In parasitic lampreys, the gonad remains immature during the feeding phase, which occurs in the natal stream or river, a freshwater lake, or the ocean. Cessation of feeding (and loss of the ability to osmoregulate in seawater) eventually occurs when increased production of sex steroids in the maturing gonad causes intestinal atrophy (Larsen 1980).

Adult body sizes therefore differ considerably between paired species; in parasitic lampreys, size at spawning is two to three times greater than at metamorphosis, whereas nonparasitic species are shorter than at metamorphosis (Table 2). Given their smaller adult size, the fecundity of nonparasitic lampreys is considerably reduced relative to their parasitic counterparts (Vladykov 1951). Eye and oral disk size also tend to be larger in the parasitic (predatory) members of each pair (Vladykov and Kott 1979; Figure 1a, 1b). Although dentition is generally degenerate in the relict nonparasitic species (in

Table 2. Morphological characteristics of paired and satellite lamprey species from the family Petromyzontidae; p and np indicate parasitic and non-parasitic feeding types (see Table 1 for each species' common name). Myomeres are for ammocoetes unless otherwise indicated (with an asterisk). Size of transformers is length of recently metamorphosed specimens; adult size is at time of spawning or during upstream migration; in some cases, size of males (m) and females (f) is given separately. See text for explanation where multiple adult forms are reported (e.g., praecox or freshwater (fw)-resident Arctic lamprey or giant American brook lamprey). TL = total length.

| | Myomeres | | Size at metamorphosis (TL, mm) | | Adult size (TL, mm) | |
|---|---|---|---|---|---|---|
| Species | Range | Mean | Maximum larval size | Size of transformers—range (mean) | Maximum or range | Mean |
| *Entosphenus:* | | | | | | |
| *tridentatus* (p) | 61–77[a] | 68.1[a] | 142[b,d] | 112–135 (122.6)[d] | 682[f] | 537[f] |
| | 63–69[b] | 66.1[b] | | (78, 120)[e] | 760[g] | |
| | 64–71[c] | 66.6[c] | | | | |
| *lethophagus* (np) | 63–68[c] | 65.5[c] | 155[h] | 134–199 (165)[h] | 116–142[h] | |
| | | | | | 220[g] | |
| *Eudontomyzon*: | | | | | | |
| *danfordi* (p) | 61–67[i] | 63.9[i] | 200[j] | 158–205 (183.3)[k] | 155–207[k] | 182.8[k] |
| | 63–66[j,*] | 64.3[j,*] | 208[i] | | 300[j] | |
| *mariae* (np) | 58–68[l] | 60.8[l] | 213[l] | 191–215[l] | 148–171[l] | |
| | 64–73[j,*] | 69.5[j,*] | | | | |
| *stankokaramani* (np) | 59–61[m] | 59.6[m] | 203[n] | | 125–181[m] | 158.4[m] |
| | 61–65[n] | 63.5[n] | | | | |
| *Ichthyomyzon*: | | | | | | |
| *bdellium* (p) | 53–62[o,*] | 56.9[o,*] | | | 200?–259[o] | 189[o] |
| | | | | | 300[g] | |
| *greeleyi* (np) | 55–61[o,*] | 57.9[o,*] | | 111–182 (138)[p] | 105–161[o] | 128[o] |
| | | | | | 200[g] | 128, 136 (m)[p] |
| | | | | | | 135, 142 (f)[p] |
| *castaneus* (p) | 49–56[o,*] | 52.6[o,*] | | 90–100[f] | 200?–300[o] | 216[o] |
| | | | | 94–165 (122)[q] | 178–250[q] | 207[q] |
| | | | | | 380[g] | |
| *gagei* (np) | 51–54[o,*] | 52.3[o,*] | | 103–146 (124)[r] | 99–126[o] | 107[o] |
| | | | | | 116–143[r] | 126.9[r] |
| | | | | | 170[g] | |

Table 2. Continued.

| Species | Myomeres | | Size at metamorphosis (TL, mm) | | Adult size (TL, mm) | |
|---|---|---|---|---|---|---|
| | Range | Mean | Maximum larval size | Size of transformers–range (mean) | Maximum or range | Mean |
| *unicuspis* (p) | 47–55[o,*] | 50.5[o,*] | | 89–110[f] | 205–328[o] | 224[o] |
| | | | | 91–155 (114)[q] | 157–308[q] | 248[q] |
| | | | | | 390[g] | |
| | | | | | 392[s] | |
| | | | | | 415[t] | |
| *fossor* (np) | 47–56[o,*] | 50.9[o,*] | 135 (m)[u] | 120–150[f] | 94–146[o] | 119[o] |
| | | | 147 (f)[u] | 84–182 (126)[q] | 86–166[q] | 115[q] |
| | | | | 97–127 (114)[u] | 170[g] | |
| *Lampetra*: | | | | | | |
| *ayresii* (p) | 63–67[b] | 65.1[b] | 122[b] | 40–190 (110)[w] | 168–236[v] | 197[v] |
| | 65–71[v] | 66.8[v] | 172[v] | | 173–210[x] | 196[x] |
| | | | | | 310[g] | 169[y] |
| *richardsoni* (np) | 58–67[b] | 62.9[b] | 213[b] | 130– 200[f] (134)[y] | 120–172[x] | 140[x] |
| | 57–65[z] | 60.7[z] | 175[z] | | 90–147[y] | 112, 130[y] |
| | | | | | 101–154[z] | 136[z] |
| *fluviatilis* (p) | 58–61[v] | 59.0[v] | | 78–133 (96–105)[cc] | 185–492[cc] | 290–310[cc] |
| | 58–66[aa] | 62.5[aa] | | (102m, 105f)[dd] | praecox 180–250[cc] | |
| | 61–67[bb] | 64.1[bb] | | | | |
| *planeri* (np) | 57–65[z] | 61.2[z] | 170[ee] | 120–175 (150)[ff] | 95–180[gg] | 137 (m), 140 (f)[gg] |
| | 58–64[ee] | 60.7[ee] | | (116m, 123f)[dd] | | 111 (m), 114 (f)[dd] |
| | 56–69[ff] | 62.3[ff] | | | | |
| *Lethenteron*: | | | | | | |
| *camtschaticum* (p) | 60–74[f] | | | 150–210[f] | 174–350[ii] | 293 (m), 279 (f)[ii] |
| | 67–77[hh] | 71.1[hh] | | | 227–625[jj] | |
| | | | | | praecox 145–220[ii] | praecox 170[ii] |
| | | | | | praecox 224–247[jj] | |
| | | | | | fw 100–165[ii] | fw 155[kk] |
| | | | | | fw 117–188[kk] | fw 150[ll] |
| | | | | | fw 138–161[ll] | |

Table 2. Continued.

| | Myomeres | | Size at metamorphosis (TL, mm) | | Adult size (TL, mm) | |
|---|---|---|---|---|---|---|
| Species | Range | Mean | Maximum larval size | Size of transformers—range (mean) | Maximum or range | Mean |
| *appendix* (np) | 63–70[f] | | 218[nn] | 112–187[f] | 100–109[pp] | 104[pp] |
| | 66–72[mm,*] | 68.9[mm,*] | 231[oo] | | 151–215[qq] | |
| | | | | | 154–218[nn] | 185[nn] |
| | | | | | 159–205[rr] | 179[rr] |
| | | | | | 161–217[ss] | 188[ss] |
| | | | | | "giant" 260–290[tt] | "giant" 286[tt] |
| | | | | | "giant" 272–354[uu] | "giant" 313[uu] |
| *alaskense* (np) | 66–72[w] | | | 150–210[ww] | | |
| *kessleri* (np) | 66–74[xx] | | 230[xx] | 227[xx] | 125–260[xx] | |
| | 63–76[hh] | 68.9[hh] | | | | |
| *reissneri* (np) | 57–72[xx] | | | 140–200[yy] | | |
| | 64–73[hh] | 68.2[hh] | | | | |

[a] Meeuwig et al. (2006); [b] Richards et al. (1982); [c] Bond and Kan (1973); [d] Beamish and Levings (1991); [e] Beamish (1980; mean for Qualicum and Oyster rivers was 120 mm; mean for Morrison Creek was 78 mm, with smallest metamorphosed lamprey 47 mm); [f] Scott and Crossman (1973); [g] Page and Burr (1991); [h] Hubbs (1971); [i] Renaud and Holčík (1986); [j] Vladykov et al. (1982; myomeres for ammocoetes and adults combined); [k] Renaud and Holčík (1988); [l] Holčík and Renaud (1986; *E. mariae* from Vistula River system only); [m] Holčík and Šorić (2004); [n] Karaman (1974); [o] Hubbs and Trautman (1937; myomeres for ammocoetes and adults combined); [p] Medland and Beamish (1987; mean size at spawning calculated separately for males and females in two streams); [q] Morman (1979); [r] F. W. H. Beamish (1982); [s] Cochran and Marks (1995); [t] Cochran (2004); [u] Purvis (1970); [v] Vladykov and Follett (1958); [w] Beamish (1980; one downstream migrating *L. ayresii* was 40 mm, all others > 90 mm); [x] Beamish and Neville (1992); [y] Beamish and Withler (1986; for *L. richardsoni*, mean of 112 mm at spawning in Morrison Creek; mean of 130 mm for seven individuals from other populations); [z] Vladykov and Follett (1965); [aa] Potter and Osborne (1975); [bb] Malmqvist (1978); [cc] Hardisty (1986a; mean length at metamorphosis varied among populations); [dd] Bird and Potter (1979b; mean sizes for males and females, respectively); [ee] Vladykov (1955); [ff] Hardisty (1986b; 150 mm is modal length at metamorphosis); [gg] Malmqvist (1980); [hh] Yamazaki et al. (2006); [ii] Kucheryavyi et al. (2007b); [jj] Holčík (1986a); [kk] Heard (1966); [ll] Yamazaki et al. (1998); [mm] Rohde and Arndt (1975; myomere counts from adults); [nn] Seagel and Nagel (1982); [oo] Mundahl et al. (2006); [pp] Hoff (1988); [qq] Mundahl and Sagan (2005); [rr] Kott (1971); [ss] Kott (1974); [tt] Manion and Purvis (1971); [uu] Cochran (2008); [vv] Vladykov and Kott (1978b); [ww] Houston (1991); [xx] Holčík (1986b); [yy] Fukayama and Takahashi (1983).

terms of number of teeth and cusps), tooth patterns are similar in paired species (which has historically provided evidence of their close relationship; e.g., Hubbs 1925). However, it appears that dentition may be more variable in the nonparasitic member of the pair (e.g., Hubbs 1971), presumably due to a reduction in the selection pressures that stabilize the functional arrangement of teeth in parasitic species (Hardisty and Potter 1971; Potter 1980). As a result, there are sometimes more teeth and cusps in the nonparasitic member of a pair rather than fewer, as might be predicted (Hardisty 2006). There are also differences between paired species in the sharpness of the teeth: in parasitic species, the teeth remain sharp following metamorphosis and tend to be orange-brown in color, are well developed and strongly keratinized, and are replaced frequently (two or more times); in nonparasitic species, they are yellowish-white, are weakly developed, and are replaced only once or not at all (Vladykov and Kott 1979; Figure 1c, 1d). Other differences may include the number and shape of the velar tentacles (e.g., with wings on the velar apparatus of some parasitic species; Vladykov and Kott 1976c; Renaud et al. 2009b) and sometimes the number of trunk myomeres (Table 2 and below).

## *Larvae*

Paired species differ considerably in the extent to which they can be differentiated as larvae, due to varying degrees of morphological differentiation between them (Vladykov and Kott 1979). Hardisty (2006) suggested that when nonparasitic lampreys first appeared within an existing parasitic population, adults would be recognizable only by their smaller size at spawning (and other differences related to feeding) and larvae would not be distinguishable. With reproductive isolation, morphological differences would presumably appear in the larvae as well—either through natural selection or genetic drift—and the rate at which differences accumulate would be related to the degree of reproductive isolation, time since divergence, and such factors as strength of selection and population size (see below). Consequently, as different species pairs likely represent speciation at different stages (Figure 2), there is no universal rule regarding the ability to differentiate paired species in the larval stage[4]. Some pairs are "difficult or impossible" to distinguish in the larval stages (Hardisty 2006); some may be distinguishable on a statistical basis (based on the traits described below), but there is often considerable overlap (Potter 1980; Hardisty 2006). Diagnostic differences are generally evident between the so-called relict brook lampreys (see above) and their presumed ancestor (Vladykov and Kott 1979), but in closely related pairs, individuals can rarely be identified unambiguously to species. Therefore, although significant morphological differences—even if not diagnostic—are indicative of a barrier to gene flow, it is difficult to determine whether such differences exist in mixed populations (i.e., where the paired species occur sympatrically). Virtually all comparisons have been made on allopatric populations.

In some cases, nonparasitic lampreys can be recognized in the larval stage by the presence of fewer trunk myomeres than in the parasitic species. This reduction in myomeres in brook lampreys may be associated with the overall decrease in body length attained at maturity in the nonparasitic species. Lower myomere numbers are particularly the case with the relict brook lampreys (Vladykov and Kott 1979), but the ability to distinguish among paired species based on myomere counts is variable. For example, with means of 66.8 and 60.7 myomeres, respectively, river lampreys can generally be differentiated from western brook lampreys (Vladykov and Follett 1958, 1965), although overlap in observed counts precludes diagnostic identification (Table 2). In contrast, differences

[4] Also, see Polyphyly and the Repeated Evolution of Nonparasitism, below, regarding evidence that some species may be polyphyletic (i.e., that different nonparasitic populations within a species may have arisen independently such that different populations of the same species pair may vary in the degree to which they differ morphologically).

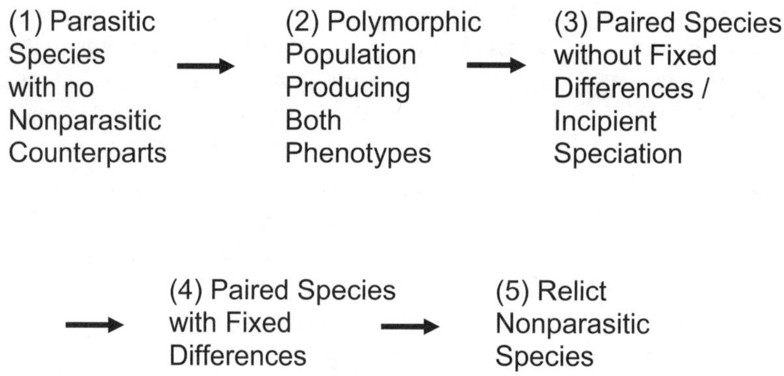


Figure 2. Diagram illustrating that different lamprey species pairs likely represent speciation at different stages, from (1) parasitic species with no nonparasitic counterparts to (5) relict nonparasitic species that have long diverged from a parasitic ancestor. In (3), reproductive isolation would be acting, but there has not been sufficient time for morphological or genetic differences to become fixed (see text). The involvement of a freshwater (or praecox) parasitic stage as an intermediate step in the evolution of nonparasitism is not represented but may be important in many cases (see text).

between parasitic and nonparasitic members of other paired species appear to be more subtle or nonexistent (Table 2). For example, differences in the average number of myomeres in allopatric European river and brook lampreys are statistically significant, but the differences are fractional (e.g., 61.2 for four European brook lamprey populations versus 62.4 for two European river lamprey populations; Potter and Osborne 1975) and are not diagnostic; differences among European brook lamprey populations are as great as the differences between pooled species averages (see Hardisty 1986a, 1986b, 2006). There appear to be no differences in myomere counts in other pairs, such as pairs in the *Ichthyomyzon* genus (Hubbs and Trautman 1937; Table 2) and the Southern Hemisphere species short-headed lamprey and precocious lamprey (Potter 1968).

Likewise, body pigmentation can be used to distinguish between members of some lamprey species pairs, but differences may not be entirely diagnostic or consistent. North American river lampreys in the Fraser River can generally be differentiated from allopatric western brook lampreys by the reduction of pigmentation in the former, and larvae identified by this method were reared through metamorphosis to confirm identification (Richards et al. 1982). In Carpathian lamprey and Ukrainian brook lamprey, live and freshly preserved Ukrainian brook lamprey ammocoetes have mottled pigmentation on their body and generally strong caudal fin pigmentation, whereas there is no mottling on the body in Carpathian lamprey and usually weak caudal fin pigmentation (Holčík and Renaud 1986). Pigmentation differences have been reported in *Ichthyomyzon* species, but there is overlap in these characters and inconsistency among authors. For example, Vladykov and Kott (1980) reported that silver lampreys could be distinguished from northern brook lampreys based on a more heavily pigmented caudal fin and head, whereas Lanteigne (1988) distinguished northern brook lampreys by the nonpigmented band above the branchial pores. Neave et al. (2007) compared these pigmentation characters in silver and northern brook lampreys over a broad geographic range and were unable to distinguish the two species. Differences in the incidence of pigmentation of the tongue precursor have also been examined in some species pairs. Potter and Osborne (1975) found pigmentation in the tongue precursor of 74.2–83.7% of European river lampreys compared to only 5.1–35.6% of European brook lampreys. Conversely, Richards et al. (1982) reported an absence of pigmentation in the tongue precursor of American river lamprey river lamprey larvae, but the tongue precursor was generally pigmented in western brook lampreys.

Differences in ovarian histology have also been suggested as a means of distinguishing between members of a species pair in the larval stage. Given the differences in adult body size, nonparasitic species produce far fewer eggs at spawning than their parasitic counterpart (Vladykov 1951), and it appears that these differences may already be apparent in larvae; in at least some species, ammocoetes of nonparasitic species have a lower potential fecundity. For example, Hardisty (1961) found a range of 65–100 oocytes per cross section in European river lamprey larvae and 25–50 oocytes per cross section in European brook lamprey larvae. A similar relationship was observed in chestnut lamprey and southern brook lamprey (Beamish and Thomas 1983) and short-headed lamprey and precocious lamprey (Hughes and Potter 1969), but there was some overlap so that differences are likely not diagnostic. The above comparisons were made among allopatric populations, and the ability to identify individuals unambiguously to species in mixed populations has not been tested (given the lethality of the method). A population presumed to consist of both silver and northern brook lampreys showed a bimodal distribution in oocytes per cross section, with a range of 15–93 oocytes per section; a population from which only northern brook lamprey adults have ever been collected showed a range of 13–42 (Neave et al. 2007). In the mixed population, 3 individuals out of a total of 42 had high oocyte counts (80, 83, 93) and could be silver lampreys, but another subsample of this population showed equal numbers of silver and northern brook lampreys at metamorphosis. Consequently, this potential method of identification requires better validation and some resolution to the debate regarding the degree of atresia that occurs in nonparasitic species. Histological studies have shown extensive atresia of oocytes during metamorphosis in European brook lampreys (Hardisty 1963, 1964, 1971), Far Eastern brook lampreys (Fukayama and Takahashi 1983) and precocious lampreys (Hughes and Potter 1969). In contrast, no significant differences between the number of oocytes in larvae and the number of mature eggs were found in the southern brook lamprey (Beamish and Thomas 1983) or in the relict least brook lamprey *Lampetra aepyptera* (Docker and Beamish 1991). Beamish and Thomas (1983) and Docker and Beamish (1991) hypothesized that the lack of atresia in these species suggests that they diverged from their parasitic ancestors some time ago, thus giving natural selection sufficient time in which to reduce the oocyte number elaborated in the larvae. The distant origin of the least brook lamprey is supported by all available evidence (e.g., Docker et al. 1999), but—based on morphological (Table 2) and genetic (see below) similarities—it is not likely that southern brook lampreys have been long separated from the chestnut lamprey. Thus, the degree of atresia experienced in nonparasitic lampreys may vary among species pairs or among individuals within a species. If at least some brook lampreys elaborate a large number of oocytes as larvae and then experience significant loss through atresia at metamorphosis, nonparasitic species may not always be diagnosed by low oocyte numbers.

Differences in the timing of gonadal development have also been reported between members of paired species. Differentiation generally appears to occur earlier in nonparasitic species (and, in fact, it has been suggested that the lower number of oocytes elaborated in nonparasitic lampreys is the result of the earlier onset of gametogenesis; Hardisty 1965). Significant events in gonadogenesis, such as the initial onset of mitotic division in the germ cells and initiation of oogenesis, take place a year later in European river lampreys relative to brook lampreys (Hardisty 1970). Likewise, initiation of oogenesis occurs 1 year later in the chestnut lamprey relative to the southern brook lamprey (Beamish and Thomas 1983). It has been postulated that a high larval fecundity can probably be achieved only when gonadogenesis is delayed (Hardisty 2006). There are also significant differences in the timing of gonadal maturation during and after metamorphosis (see Heterochrony, below), but at the onset of metamorphosis, the gonads of nonparasitic and parasitic species in *Mordacia* and *Lampetra* are of a very similar size and stage of development (Hughes and Potter 1969; Hardisty and Potter 1971).

Differences between parasitic and nonparasitic lampreys in terms of the age and size at which metamorphosis occurs have also been suggested. The parasitic species of a pair is generally thought to metamorphose at a younger age and a smaller size. For example, average duration of the larval period in the European river lamprey in the River Teme has been estimated to be 4¼ years, whereas average length of larval life in the European brook lamprey is approximately 6¼ years (Potter 1980; Hardisty 2006). Mean length of metamorphosing and downstream-migrating European river lampreys is approximately 96–105 mm (Bird and Potter 1979b; Hardisty 1986a); length of European brook lampreys at metamorphosis is more variable but modal length is approximately 150 mm (Hardisty 1986b).

It is difficult to say whether the extension of larval life observed in European brook lamprey relative to European river lamprey also applies to other brook lampreys, since age data are rarely available, but size at metamorphosis does appear to be smaller on average in most parasitic members of a pair (Table 2). For example, mean size at metamorphosis has been estimated at approximately 110 and 134 mm for North American river and western brook lampreys, respectively (Beamish 1980 and Beamish and Withler 1986, respectively), 89–110 and 120–150 mm for silver and northern brook lampreys (Scott and Crossman 1973), 150 mm and 140–200 mm for Arctic and Far Eastern brook lampreys (Fukayama and Takahashi 1982, 1983), and 123 and 165 mm for Pacific lampreys and Pit-Klamath brook lampreys (Beamish 1980 and Hubbs 1971, respectively). However, there is considerable intraspecific variation among individuals (e.g., females often appear to metamorphose at larger sizes and an older age; Purvis 1970; Medland and Beamish 1987; Docker and Beamish 1991) and populations (e.g., Manion and Smith 1978; Purvis 1980; Vladykov and Kott 1980; Morkert et al. 1998). Consequently, although the larger larvae in a mixed-species population may often represent the nonparasitic member of a pair, large larvae of parasitic species also have been observed (Table 2).

## Genetic Differences between Paired Species

With very few exceptions, significant genetic differences have not been found between paired lamprey species, despite the use of molecular markers that are expected to provide increasingly greater resolution. In the 1960s and 1970s, chromosome number was compared between short-headed and precocious lampreys (Potter et al. 1968; Robinson and Potter 1969) and the European river and brook lampreys (Robinson et al. 1974) and there were no differences within each pair; however, chromosome number among three genera of Northern Hemisphere lampreys showed little variation (Robinson et al. 1974). Subsequently, hemoglobin electropherograms were also found to be indistinguishable in larvae and adults of these same two lamprey species pairs, although there were differences between the larval and adult stages (Potter and Nicol 1968; Potter and Brown 1975). Comparing allelic frequencies at 21 and 24 allozyme loci, respectively, Beamish and Withler (1986) and Schreiber and Engelhorn (1998) likewise failed to find significant genetic differences between sympatric North American river and western brook lampreys and European river and brook lampreys; genetic differentiation between members of a pair was no greater than variation within each species (i.e., among populations; see Polyphyly and the Repeated Evolution of Nonparasitism, below). Schreiber and Engelhorn (1998) also found that genetic variation among populations of the European brook lamprey was greater where it was outside the range of the European river lamprey than where they were sympatric, and suggested that the migratory river lamprey may be mediating gene flow among the otherwise isolated brook lamprey populations (i.e., that there was gene flow between species where they occurred sympatrically). They suggested that the two are not fully differentiated species.

However, significant—and in fact, diagnostic (i.e., fixed)—allozyme differences have been reported among some of the *Lethenteron* species. Yamazaki and Goto (1998) found that the Arctic and Siberian brook lampreys (as well as *Lethenteron* sp. N and *Lethenteron* sp. S; see Table 1) could be distinguished from each other with diagnostic alleles at 1–11 loci (although Yamazaki et al. (2006) later found the "diagnostic" Siberian brook lamprey allele in sympatric Far Eastern brook lamprey). These fixed differences were observed at three sites where the Arctic and Siberian brook lampreys occurred sympatrically (with no evidence of hybrids), which is the only reported case to date where paired species from the same geographic location can be identified unambiguously using molecular markers.

Fixed differences in mitochondrial DNA sequence have been reported between Arctic lampreys and American brook lampreys from Russia and the Laurentian Great Lakes, respectively (Docker et al. 1999), but other paired species have been shown to share one or more haplotypes (Figure 3). For example, it has not been possible to use mitochondrial DNA sequence to distinguish between North American river and western brook lampreys (Docker et al. 1999; Boguski 2009), Pacific lampreys and Pit-Klamath brook lampreys (Docker et al. 1999; Lorion et al. 2000), silver and northern brook lampreys (Docker 2006; Hubert et al. 2008), Ohio and mountain brook lampreys (Docker 2006), or European river and brook lampreys (Tagliavini et al. 1994; Espanhol et al. 2007; Blank et al. 2008). Yamazaki et al. (2006) likewise failed to find fixed differences in mitochondrial DNA sequence between Arctic and Siberian brook lampreys, even though diagnostic differences were evident with allozymes (Yamazaki and Goto 1998; see above). In some cases, a large portion of the mitochondrial genome was compared; Blank et al. (2008) examined a total of 3,800 base pairs (bp) in the European river and brook lampreys and more than 10,000 bp were compared between silver and northern brook lampreys (see Docker 2006).

This lack of fixed differences in mitochondrial DNA sequence is considered rare for "true" species. Indeed, such fixed differences are the basis of DNA "barcoding" (e.g., Hebert et al. 2004). Using approximately 650 bp at the 5' end of the COI (cytochrome *c* oxidase subunit I) gene (a region included in some of the above studies, e.g., Blank et al. 2008), Ward et al. (2009) surveyed COI sequences in more than 5,000 fish species (approximately 18% of the worldwide total) and found that COI barcodes could separate approximately 98% and 93% of marine and freshwater fish species, respectively. Ward (2009) estimated that if two fish samples are barcode-identical, the probability that they are from the same species is 98–99%. However, some taxa—particularly Canadian freshwater fishes—exhibited lower levels of within-species COI divergence relative to all other fishes, and these also had the highest proportions of barcode indistinguishable species (Ward 2009). Hubert et al. (2008) "barcoded" 190 of the 203 recognized Canadian freshwater fish species (including five lamprey species) and found that the average genetic distance among congeneric species (8.3%) was 27 times higher than differences within species (which averaged 0.3%). However, among these 190 species, 13 (7%) shared barcode haplotypes (Hubert et al. 2008). These 13 species comprised five species pairs or complexes (including silver and northern brook lampreys; Hubert et al. 2008). Species that cannot be distinguished based on DNA barcodes might reflect inadequate taxonomy (where a pair of currently recognized species is in fact a single species) but could also be indicative of recent speciation or past or present hybridization (Hubert et al. 2008; Ward 2009). Therefore, the lack of a single fixed difference between European river and brook lampreys (Blank et al. 2008) and between silver and northern brook lampreys (see Docker 2006) in 3,800 and 10,000

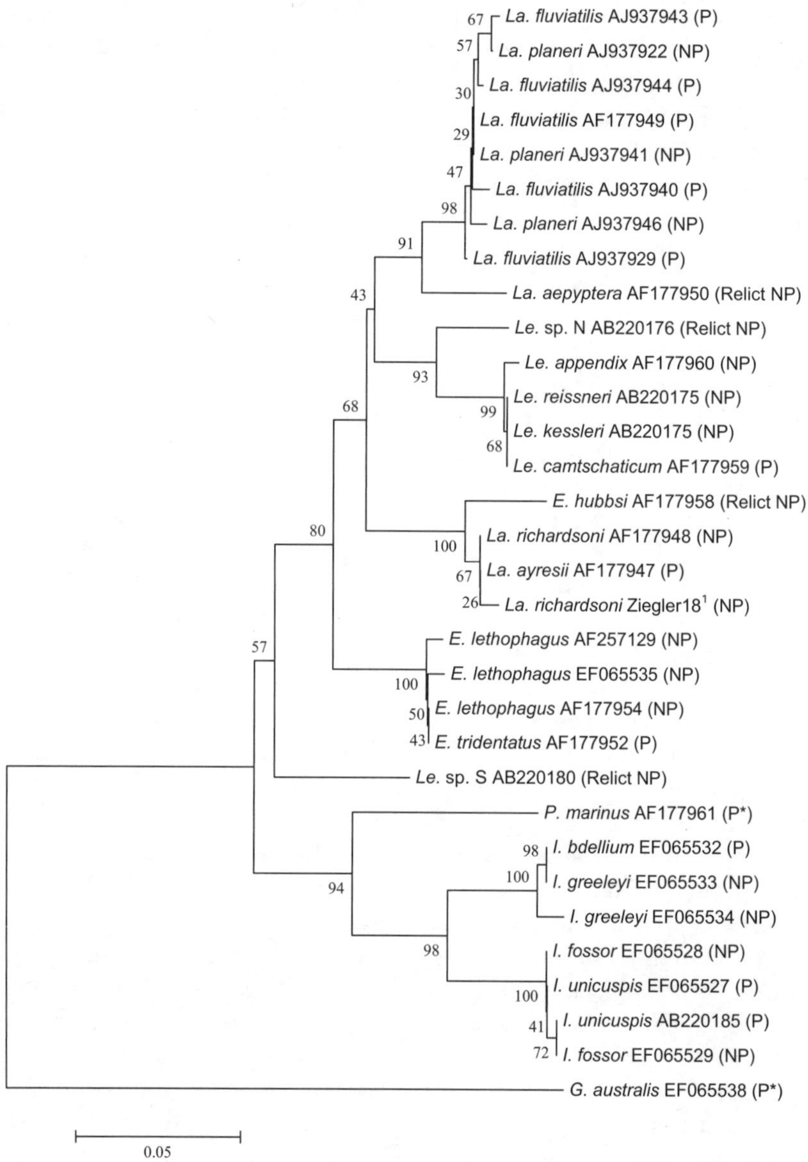


Figure 3. Phylogeny of 22 lamprey species inferred from 384 base pairs of mitochondrial cytochrome b sequence using Kimura's (1980) two-parameter distance model and the neighbor-joining algorithm; numbers at nodes are bootstrap values for 1,000 replications. Abbreviations are as follows: E. = Entosphenus, *G.* = *Geotria*, *I.* = *Ichthyomyzon*, *La.* = *Lampetra*, *Le.* = *Lethenteron*, *P.* = *Petromyzon*, P = parasitic, NP = nonparasitic, * = parasitic lamprey with no nonparasitic counterpart. Sequence data were obtained from GenBank where accession numbers are provided or [1]from Boguski (2009). Tree demonstrates the close genetic relationship between paired species, lack of reciprocal monophyly in a number of pairs (e.g., *La. fluviatilis* and *La. planeri*, *I. unicuspis*, and *I. fossor*), and the genetic distinctiveness of the relict nonparasitic species. Tree topology was similar with maximum parsimony (MP) and Bayesian analysis (not shown), with the exception that the MP tree placed the *E. hubbsi*, *La. richardsoni*, and *La. ayresii* clade as the sister group to the *Le. camtschaticum* clade and Bayesian analysis grouped *La. richardsoni* Ziegler18 with *E. hubbsi*. Note that *E. hubbsi* appears to be a relict species in the *La. richardsoni*–*La. ayresii* clade rather than in the *Entosphenus* clade; this is supported by complete cytochrome *b* sequence data (Boguski 2009; Lang et al. 2009) and nuclear DNA sequence data (Boguski 2009).

bp, respectively, of the mitochondrial genome (i.e., less than 0.03% and 0.01%, respectively) is unusual and should give us pause, but alone cannot lead us to the conclusion that they are a single species. It could be argued that there is a low rate of molecular divergence in lampreys relative to other fishes (see Britten 1986), resulting in an atypical lack of fixed differences between closely related species in this taxon. However, in most of the above lamprey studies, significant intraspecific variation (within or among populations) was detected (e.g., Espanhol et al. 2007; Blank et al. 2008). For example, Pit-Klamath brook lampreys in the Pit basin showed fixed sequence differences relative to those in the Klamath (even though each was genetically indistinguishable from sympatric parasitic lampreys; see Docker 2006). Hebert et al. (2004) proposed a COI sequence threshold of 10 times the mean intraspecific variation in order to define new species, but in lampreys, intraspecific variation is often as great as interspecific variation (between pairs). Consequently, Espanhol et al. (2007), Blank et al. (2008), and others have concluded that paired species have undergone a very recent divergence event or that there is ongoing gene flow between paired species.

Therefore, the similarity or dissimilarity of COI (or other mitochondrial DNA) sequences alone cannot determine whether organisms represent bona fide species where divergence has been recent and rapid. Ward (2009) omitted African cichlids from his analysis since many pairs of species may be identical at the COI barcode and a faster-evolving gene is likely needed to distinguish species. Likewise, studies conducted to date on paired lamprey species do not provide sufficient resolution to decide between very recent divergence and a lack of genetic divergence. Assuming an overall rate of mitochondrial DNA sequence divergence of 2% per million years calculated for several mammalian species (Brown et al. 1979) and used to estimate divergence times in a variety of fish taxa (see Billington and Hebert 1991), it could be estimated that less than 0.03% and 0.01% sequence difference between European river and brook lampreys (Blank et al. 2008) and silver and northern brook lampreys (see Docker 2006), respectively, would suggest that divergences within the past 5,000–13,160 years would not be detectable. Furthermore, with large population sizes and incomplete lineage sorting (whereby shared ancestral polymorphisms in populations have not progressed to reciprocal monophyly during speciation), it is possible fixation might not be achieved even over greater periods of time. Diagnostic differences will not be immediately evident between recently evolved species. Species may have achieved reproductive isolation (i.e., be recognized as biological species; Mayr 1963) but may not yet have had time for fixed differences to arise (e.g., Avise et al. 1990; Pestano et al. 2003).

Consequently, recent molecular analyses are starting to examine differences in allele frequencies that—while not diagnostic—would nonetheless be indicative of barriers to gene flow. In several temperate freshwater fish species with divergent morphotypes, many sympatric species pairs (e.g., sockeye salmon *Oncorhynchus nerka* and kokanee [lacustrine sockeye salmon], benthic and limnetic sticklebacks) lack fixed sequence differences but demonstrate significant differences in allele frequencies that suggest reproductive isolation between the types (e.g., Foote et al. 1989; Taylor 1999; see Divergent Morphotypes in Other Animals, below). This is particularly the case with microsatellite loci, lengths of DNA containing highly variable tandem repeats of two to six nucleotides. They have a high mutation rate, providing information about the recent past of the species, and the use of multiple microsatellite loci from each individual provides a large amount of information about the genome of the study organism (Selkoe and Toonen 2006). Microsatellites also have the advantage of being inherited from both parents; mitochondrial DNA is maternally inherited. Differences in microsatellite allele frequencies between silver and northern brook lampreys were reported by Filcek et al. (2005), who were able to distinguish silver lampreys from tributaries to Lake Michigan from northern brook lampreys from the Lake Superior basin with a high degree of accuracy (using microsatellite loci developed from the sea lamprey *Petromyzon marinus*). However, since the two species were collected from different basins, species differences may have been confounded by geographic differences. Application of these same microsatellite loci to silver and northern brook lampreys from Lake Huron tributaries showed that the two feeding types could not be genetically differentiated when sympatric (M. F. Docker, unpublished data). Geographic differences were apparent, but not differences between the species. More recently, a study using 14 *Ichthyomyzon*-specific microsatellite loci likewise found that silver and northern brook lampreys from the same Lake Huron tributaries were not genetically differentiated (McFarlane 2009) and that intraspecific variation was as great as interspecific variation.

It appears that different species pairs are at different stages of speciation (Figure 2). Some pairs show fixed genetic differences (e.g., Arctic and Siberian brook lampreys and Arctic and American brook lampreys; Yamazaki and Goto 1998; Docker et al. 1999; respectively) and an absence of hybrids in sympatric populations (Yamazaki and Goto 1998), which is indicative of fully differentiated species. In most other pairs, however, the absence of fixed or allele frequency differences suggests either recent or ongoing gene flow (Figure 2). To decide between these possibilities, however, more experimental and field studies will be required to quantify the incidence of interbreeding (see below).

It is interesting to note, however, that Silver et al. (2004) reported large differences in the complementary DNA (cDNA) sequence encoding the prepro-lamprey gonadotropin-releasing hormone (GnRH)-III in silver and northern brook lampreys. Northern brook lamprey cDNA showed only 88.8% sequence identity with that of silver lamprey. However, silver lamprey showed greater than 90% identity with sea lamprey, American brook lamprey, western brook lamprey, and Pacific lamprey, and phylogenetic groupings within the Petromyzontidae did not agree with phylogenies based on dentition or other molecular markers (e.g., Docker et al. 1999; Gill et al. 2003). It is difficult to understand how such sequence differences could arise between silver and northern brook lampreys if the latter is a recent derivative of the former, but sequencing of prepro-lamprey GnRH-III cDNA in silver and northern brook lampreys should be repeated—and conducted in other paired lamprey species—to test whether genetic differences do exist at such functional loci (despite an apparent lack of differences at predominantly neutral markers).

## Polyphyly and the Repeated Evolution of Nonparasitism

It is clear that nonparasitism has evolved independently in different lamprey genera (Hubbs and Potter 1971; Potter 1980). As discussed above, differences in the degree of morphological and genetic differentiation between pairs probably reflects differences in the times when each nonparasitic species evolved (as well as potential differences in the degree of reproductive isolation, population size, and strength of selection pressures—e.g., when the selection pressures on the morphological or genetic traits of a new nonparasitic population are strong, the rate of change is expected to be more rapid). Certain characters in these species undergo quantitative changes with time so that those nonparasitic species with the greatest reduction in number of myomeres or in dentition, for example, are expected to be of the greatest antiquity (Hubbs and Potter 1971). In some cases, the independent origin of nonparasitism from the same presumed ancestor at different times or in different locations may result in morphological differences among the nonparasitic derivatives such that each is recognized as a distinct species (e.g., at least three nonparasitic derivatives of the Arctic lamprey are generally given species status; see Table 1).

However, there have been suggestions that nonparasitism has also evolved independently within named brook lamprey species, that is, different populations of the same species may have originated independently such that a given brook lamprey species is polyphyletic (i.e., a taxon that is derived from two or more ancestral sources). Hubbs (1971) stated that the high frequency of paired lamprey species strongly suggests feeding type plasticity within a species and begets the idea that nonparasitic populations may be polyphyletic even within a species complex. Hubbs and Trautman (1937) found no reason to assume that each parasitic species in the *Ichthyomyzon* genus would give rise to only one nonparasitic form; they speculated that more than one independently evolved nonparasitic derivative of the chestnut lamprey might exist such that the southern brook lamprey, as currently recognized, might have had a polyphyletic origin. This suggestion was reiterated by Cochran (1987) after mature brook lampreys that were morphologically indistinguishable from southern brook lampreys were found in the Namekagon River in northern Wisconsin, 900 km from the nearest previously reported southern brook lamprey population in southern Missouri. Cochran (1987) suggested that this population may represent an independently evolved satellite species of chestnut lamprey (that would have had to occur since glaciation) and referred to it as *Ichthyomyzon* cf. *gagei* to reflect the uncertainty regarding its identity. Lyons et al. (1997), however, subsequently found six significant morphological differences that distinguish *Ichthyomyzon* cf. *gagei* from *Ichthyomyzon gagei*, although the differences observed were not all consistent with one being an older offshoot than the other. According to the predictions of Hubbs and Trautman (1937), older nonparasitic offshoots should have relatively shorter oral disks and snouts and more degenerate teeth; Lyons et al. (1997) found that, on average, *I.* cf. *gagei* had a shorter snout than *I. gagei* but had less degenerate teeth (i.e., a higher number of bicuspid teeth). Lyons et al. (1997) argued that the greatly disjunct geographic range between *I.* cf. *gagei* and *I. gagei* makes gene flow unlikely, thus supporting specific status for *I.* cf. *gagei*, whereas extensive overlap in morphological characteristics would support considering them a single species. Until the issue of polyphyly is resolved (e.g., with molecular genetic analysis, see below), they recommend continuing to use *I.* cf. *gagei*.

There is similar debate regarding whether geographically distinct subsets of other nonparasitic species likewise represent independently derived brook lamprey species. Raney (1952) suggested that brook lampreys from mountain streams and rivers of the upper Tennessee River basin, in western North Carolina and northwestern Georgia, were a second nonparasitic derivative of the Ohio lamprey; he considered this a valid species, distinct from the mountain brook lamprey, and named it the Tennessee brook lamprey. It was assumed to have arisen from Ohio lampreys in the Tennessee River, whereas mountain brook lamprey stemmed from a stock of Ohio lamprey in the upper Ohio River system (Hubbs and Potter 1971). However, Vladykov and Kott (1979) considered Tennessee brook lamprey to be morphologically indistinguishable from mountain brook lamprey and the two species have since been synonymized (Robins et al. 1980). Likewise, the Pacific brook lamprey *Lampetra pacifica* was described by Vladykov (1973) as a second nonparasitic derivative of the North American river lamprey. Vladykov (1973) suggested that the Pacific brook lamprey had fewer myomeres and different body proportions and patterns of pigmentation compared to the western brook lamprey and that the Pacific brook lamprey was limited to streams of California and Oregon (whereas the western brook lamprey was distributed further north in Washington, British Columbia, and possibly Alaska). Robins et al. (1991) considered this species to be synonymous with western brook lamprey, but molecular analysis by Boguski (2009) suggests that Pacific brook lamprey from its type locale is sufficiently distinct from western brook lamprey sensu stricto to merit species designation. Iwata et al. (1985) suggested that *Lethenteron matsubarai*, despite lower myomere counts (Vladykov and Kott 1978a), be synonymized with Siberian brook lamprey, and some authors (e.g., Nelson et al. 2004) consider the Alaskan brook lamprey to be a synonym of American brook lamprey (see Table 1).

It appears, then, that the independent evolution of brook lampreys from a single parasitic species at different times and locations has likely produced morphologically differentiated derivatives. In some species, they are formally recognized as distinct species; in other cases, however, the morphological differences may be too subtle (e.g., where the divergence is more recent or where selection pressures are similar) or the criteria for species status have not been uniformly applied among different taxa (i.e., "splitters" have accorded some of these nonparasitic lampreys distinct species status, whereas "lumpers" have considered others to be all conspecific). Molecular data are thus being used to test the issue of polyphyly, and in virtually all cases where a recognized nonparasitic lamprey species has a wide geographical distribution, the molecular data are consistent with polyphyly. Using allozyme data, Beamish and Withler (1986) proposed a polyphyletic origin for different western brook lamprey populations from the anadromous North American river lamprey along the British Columbia coastline in postglacial times; cytochrome *b* sequence data also indicate that western brook lampreys are polyphyletic (Boguski 2009). Schreiber and Engelhorn (1998) and Espanhol et al. (2007) likewise suggested that the European brook lamprey may have originated from its stem species, the European river lamprey, at least twice. Polyphyly has also been suggested in the Pit-Klamath brook lamprey and northern brook lamprey, with the loss of the parasitic feeding phase occurring independently at least twice in each case (see Docker 2006). Polyphyly has been observed in several other temperate fish species with divergent migratory or feeding types (e.g., anadromous and freshwater-resident sockeye salmon and kokanee or steelhead *Oncorhynchus mykiss* and rainbow trout (nonanadromous *O. mykiss*), dwarf and normal lake whitefish, benthic and limnetic sticklebacks; see Taylor 1999); alternative migratory or feeding types have evolved multiple times independently within each species (see Divergent Morphotypes in Other Animals, below). Since polyphyly is contrary to the phylogenetic species concept (see below), such morphotypes are identified by a single Latin binomial (for example, anadromous steelhead and freshwater-resident rainbow trout are considered a single species, *Oncorhynchus mykiss*; see Docker and Heath 2003).

Thus, in lampreys, further study and greater consistency is needed to clarify the systematics of paired species. Although there are at least three versions of the phylogenetic species concept, they all define species as "the smallest biological entities that are diagnosable and/or monophyletic" (Mayden 1997). In paired lamprey species, for taxa to be monophyletic (i.e., comprising a single common ancestor and all the descendants of that ancestor), each independently derived brook lamprey population would either need to be recognized as a distinct species (if they all possess diagnosable differences) or the parasitic ancestor and all its descendants would have to be grouped together to be considered a single species. As discussed above, there is likely not a single universal rule for all paired species (depending on the presence or absence of diagnosable differences), but a more rigorous and consistent application of these criteria is needed.

## Evidence for Plasticity of Feeding Type

Although adult feeding type has generally been used as a species-specific character in lampreys (see Reproductive Isolation in Paired Species, below), it appears that at least some species are polymorphic for feeding type, that is, some normally nonparasitic lampreys appear capable of facultative parasitism while some normally parasitic lampreys may be capable of maturing without feeding as adults.

### *Facultative Parasitism*

Although the lumen of the gut is generally never patent following metamorphosis in nonparasitic lamprey species (Youson 1980), there appear to be exceptions. For example, Gage (1928) reported that the foregut of a small proportion of American brook lampreys was open following metamorphosis. Likewise, Morris (1972) used X-rays to show that some European brook lamprey individuals were able to swallow water and thus osmoregulate in 30–70% seawater, demonstrating that the gut would have been open from approximately November until January. Hardisty (2006) reiterated 19th-century accounts of feeding by European brook lampreys. Hubbs (1971) concluded that, since it appears that nonparasitism has evolved relatively recently, some individuals within a nonparasitic species may still retain the ability to feed and may show this atavistic reversal to the parasitic feeding type.

There is considerable indirect and direct evidence to support this notion that some nonparasitic lamprey species can produce individuals capable of feeding parasitically after metamorphosis. For example, many thousands of American brook lampreys have been examined during sampling related to the control of sea lampreys in the Great Lakes, and at least six giant adults have been found which were approximately twice as long (260–317 mm total length) and about six times as heavy as normal adult American brook lampreys (Manion and Purvis 1971; Vladykov and Kott 1980; see Table 2). Its parasitic counterpart, the Arctic lamprey, does not occur in the Great Lakes basin (see Renaud et al. 2009a, this volume). Two additional giant American brook lamprey adults (272 and 354 mm total length) were also reported recently by Cochran (2008). Although there is no direct evidence of feeding, it is unlikely that these individuals reached their large size as filter-feeding larvae since no extremely large ammocoetes were found in more than 1 million larvae examined (Manion and Purvis 1971). Manion and Purvis (1971) also argued that American brook lampreys possess morphological and histological adaptations that would permit parasitism. Unlike most other nonparasitic lampreys, this species still has many of the structures required for parasitic life, including a large sucking disk, sharp teeth, and glands that produce anticoagulating secretions. As discussed above, the gut may remain open in some American brook lamprey individuals, and Battle and Hayashida (1965) reported that the gut of this species histologically resembles that of silver and sea lampreys (with the absorptive surface increased three- or fourfold by the development of 20–40 longitudinal mucosal folds during metamorphosis; 40–85 mucosal folds were observed in silver and sea lamprey and none were reported in the northern brook lamprey). Manion and Purvis (1971) suggest that, since it is assumed that brook lampreys evolved from parasitic ancestors, these giant American brook lampreys were genetic reversions to an ancestral type. However, Vladykov and Kott (1980) argued against feeding after metamorphosis. They speculated that giant American brook lampreys could originate from giant larvae that live in deep waters and have not been seen by researchers and sea lamprey control personnel. They further contended that the histological study by Battle and Hayashida (1965) does not confirm that the digestive tract in postmetamorphic American brook lampreys is functional and that, although its dentition is stronger than those of other nonparasitic species (e.g., the Alaskan brook lamprey), their teeth are not as sharp as those of Arctic lampreys; rather, they are more comparable to the teeth found in the Caspian lamprey *Caspiomyzon wagneri*. However, although the Caspian lamprey likely is not parasitic on fish or other vertebrates (Vladykov and Kott 1980), it nevertheless feeds following metamorphosis—either on demersal fish eggs or organic detritus (Vladykov and Kott 1980) or as a scavenger or carrion feeder (Renaud 1997; Renaud et al. 2009b)—supporting the hypothesis that giant American brook lampreys likewise feed as adults. The fact that both giant specimens described by Cochran (2008) were found to have acanthocephalan parasites in the hindgut, which are known only from lampreys that have fed on teleost fish, further supports postmetamorphic feeding in these American brook lamprey specimens.

A population of typically nonparasitic western brook lampreys in Morrison Creek, a small creek on Vancouver Island, British Columbia, also appears to produce a parasitic variety (Beamish 1985, 1987; Beamish and Withler 1986). All larvae in the creek appear similar, but following metamorphosis, some of the lampreys have the well-developed teeth normally found in parasitic species, take on the silver color characteristic of downstream migrating river lampreys, and are capable of feeding in the laboratory. Under these conditions, they spawn and die 1 year later than nonparasitic western brook lampreys (Beamish 1985). However, these parasitic lampreys are not typical river lampreys because they could not osmoregulate in seawater and river lampreys have never been reported in Morrison Creek; the only parasitic species reported is the morphologically distinct Pacific lamprey. Given the absence of river lampreys in Morrison Creek and the observation that none of the larvae found in the creek resembles artificially created hybrids between river and western brook lampreys (Beamish and Withler 1986), it is unlikely that these feeding individuals are *Lampetra ayresii* × *L. richardsoni* hybrids. Since the two varieties could not be separated electrophoretically (Beamish and Withler 1986), they were not considered genetically distinct species. Rather, the parasitic variety was named *L. richardsoni* var. *marifuga* (Beamish 1985), referring to its inability to survive in seawater. Since a "variety" published after 1960 is an excluded taxonomic rank in zoological nomenclature (see Renaud et al. 2009a), this parasitic form in Morrison Creek has not been given formal taxonomic status. It is more common than apparently parasitic American brook lampreys, being captured in the creek each year from 1978 to 1987 (Youson and Beamish 1991), but there is no evidence that the parasitic variety forms a self-sustaining population in the creek. Although capable of feeding in the laboratory, there is no evidence that it feeds in the stream: no scars have been reported on fishes, no increase in lamprey body length after metamorphosis has been noted, and the digestive tracts of wild-caught lampreys were either empty or contained only small amounts of mucus-like material (Beamish 1985). Similarly, although the parasitic variety has spawned in the laboratory and produced viable eggs (Beamish 1985), there is no evidence that it spawns in the wild: development of the gonads and secondary sex characters are delayed and the sex ratios are highly male-biased (Beamish 1985; Youson and Beamish 1991).

It appears, then, that the normally nonparasitic western brook lamprey is capable of producing viable parasitic lampreys (i.e., feeding, growing, and spawning) if proper conditions (such as those in the laboratory) are met, but that they are not presently doing so in Morrison Creek. This population, however, is of great importance for understanding the relationship between members of species pairs and the evolution of divergent feeding types. In particular, it appears to be useful for identifying potential physiological differences between closely related parasitic and nonparasitic forms. Youson and Beamish (1991) compared the internal morphology of adults of the "marifuga" variety (identified by its silver coloration) with those from typical western brook lampreys. As mentioned above, they found highly male-biased sex ratios in marifuga (approximately 80% were male) and that the five females examined were far less developed than males; two of the females showed limited completion of metamorphosis. The marifuga and western brook lamprey adults all showed complete transformation of the endostyle and the thyroid glands were of similar structure. However, development of the gonads and secondary sex characters were delayed in marifuga. The degree of development of the testes was variable in marifuga males, and none was observed with free spermatozoa in the coelomic cavity. In females, egg development was asynchronous (with egg size within a female being highly variable). In terms of the digestive tract, the esophagus of most marifuga specimens had 10–25 longitudinal mucosal folds (with the smallest number observed in females) compared to only a few folds in western brook lamprey adults, where the intestine was atrophied in most cases. The anterior intestine of marifuga had a patent lumen (which was highly dilated in specimens that had fed in the laboratory) and 18–40 longitudinal folds. Interestingly, the gonads in marifuga were more developed than expected in an adult with a functional gut, given that production of sex steroids at maturation normally leads to intestinal atrophy (Larsen 1980). Consequently, differences between marifuga and nonparasitic western brook lampreys suggest that—not surprisingly—the hypothalamic-pituitary-gonadal axis is involved in the transition between feeding types. Hypophysectomy (i.e., removal of the pituitary) interferes with metamorphosis (Joss 1985) and, while not preventing gonadal development (which, unlike all other vertebrates known, occurs in lampreys relatively independently of the pituitary), prevents gonads from producing the hormones responsible for secondary sex characters, intestinal atrophy, and death following spawning (Larsen 1980). Hypophysectomy particularly prolongs survival in males and has differential effects on females and males; oogenesis appears to be more disrupted by hypophysectomy than spermatogenesis in adult European river lampreys (Larsen 1980), likely because vitellogenesis is very dependent on estradiol (Pickering 1976). Therefore, changes in the function of the pituitary in marifuga could lead to disruption—but not prevention—of vitellogenesis or spermatogenesis. More work is required to help identify the endocrinological differences responsible for divergent feeding types in closely related species (see Heterochrony, below).

## *Facultative Nonparasitism*

Evidence suggests that individuals from normally parasitic species are sometimes able to mature and reproduce without feeding following metamorphosis. For example, Heard (1966) reported that anadromous parasitic and freshwater-resident nonparasitic Arctic lampreys occurred sympatrically in the Naknek River system in southwestern Alaska and that the nonparasitic form was more common. The freshwater form rarely exceeded 180 mm (see Table 2). In addition to the smaller size at maturity, lack of postmetamorphic feeding was inferred from the fact that many metamorphosing specimens were observed with maturing gonads. Nonparasitic (dwarf) Arctic lampreys were likewise reported in Japan by Yamazaki et al. (1998). Most recently, Kucheryavyi et al. (2007b) reported that three adult life history types were produced by a single Arctic lamprey population in the Utkholok River basin in Russia: a nonparasitic form, a praecox parasitic form (which fed at sea from several months to a year and was 92% male), and a large anadromous parasitic form (see Table 2). The majority of the lampreys (83% of mature specimens) were the nonparasitic resident form; the anadromous and praecox forms made up 14% and 3% of the specimens, respectively. According to these authors, there was no reproductive isolation among the forms; they propose that the different adult feeding and migratory habits may be linked to different diets at the ammocoete stage (see Reproductive Isolation in Paired Species, and Conditions under which Brook Lampreys May Arise, below).

Facultative nonparasitism has also been suggested in the Carpathian lamprey. Renaud and Holčík (1988) have suggested that the nonparasitic lamprey described as *Eudontomyzon gracilis* by Kux (1965) were nonfeeding individuals of the normally parasitic Carpathian lamprey. According to Kux (1965), the two species could be distinguished only from the second or third to ninth month following metamorphosis, based on differences in intestinal and gonadal development and in their trophic ecology and postmetamorphic life span (21 versus 8–9 months in *E. danfordi* and *E. gracilis*, respectively). Once the adult trophic phase was completed in *E. danfordi*, the intestine and gonads resembled those of *E. gracilis*, and the two species were again indistinguishable. The size range of prespawning adults was similar in both species (160–245 mm in nature and in the laboratory; Kux 1965, 1967), although differences in size at spawning or immediately after death were observed (131–180 mm in *E. gracilis* versus 155–207 mm in *E. danfordi*; Kux 1967). Renaud and Holčík (1988) argued that *E. gracilis* would be the only nonparasitic species of Petromyzontidae with a winged velar apparatus in the adult period; velar wings have been found only in parasitic species (Vladykov and Kott 1976c), including *E. danfordi*, that have begun their adult trophic phase (Renaud et al. 2009b). Thus, they contended that specimens identified as *E. gracilis* were actually facultatively nonparasitic *E. danfordi*, induced by inadequate feeding conditions in the laboratory. According to aquarium experiments by Kux (1967), inadequate feeding conditions during the ammocoete period led to an extension of larval life by 1–2 years and a concomitant reduction in the postmetamorphic phase; following metamorphosis, sexually mature nonfeeding individuals (57.1%) and immature feeding individuals (42.9%) were both observed (although the majority of larvae died).

Although more potential examples of feeding type plasticity or polymorphism in lampreys are thus coming to light, more work is required to determine whether this phenomenon is relatively rare or whether it occurs more commonly than previously believed but has not been frequently observed. Although the evidence for postmetamorphic feeding in the American brook lamprey is very convincing, it is appears to be a rare event. In more than 40 years, fewer than 10 giant American brook lampreys have been reported despite extensive sampling. It is possible that parasitic feeding occurs in other American brook lampreys but to a lesser degree so that the size differences are not apparent; there is considerable range in the size of adult American brook lampreys (Table 2). The gut may be patent and some individuals may possess the potential to feed (as appears to be the case in the marifuga variety of western brook lamprey), but laboratory feeding, histological or X-ray studies (see above), or perhaps acoustic microscopy (see Maeva et al. 2004) would be required to demonstrate this. Furthermore, large American brook lamprey adults may have incorrectly been assumed to be silver or sea lampreys. Where paired species occur sympatrically, feeding individuals of a nonparasitic species would likely be identified as the parasitic species. Given our general inability to clearly distinguish between members of a species pair using characters other than feeding type, it could be argued that feeding brook lampreys are only really noticeable when they do not occur sympatrically with their parasitic counterpart or where they are noticeable due to developmental abnormalities (e.g., marifuga, see above). It might be expected that recently derived nonparasitic populations would still be polymorphic for feeding type, but considerably more research is required to test this (including common garden experiments where offspring from known-species parents are reared through metamorphosis to test heritability of feeding type). It may be that some individuals are capable of showing this reversion to the ancestral parasitic type and are not uncommon—but may be unstable or transient, either dying out or becoming fully parasitic (but see Conditions under Which Brook Lampreys Might Evolve, below, for a discussion of the direction of feeding type evolution).

## Reproductive Isolation in Paired Species

### *Size-Assortative Mating*

Despite what are, at best, subtle morphological or genetic differences between parasitic and nonparasitic members of each pair (see above), lampreys generally have been assigned species status based on adult feeding type because size differences between the types are generally thought to result in immediate reproductive isolation. The spawning act involves the male squeezing the female's body with his tail (generally while attached to the back of her head with his oral disk), so differences in body size affect the precise positioning of the genital regions of the two sexes (Hardisty and Potter 1971). It has been assumed, then, that the origin of dwarf brook lampreys from large parasitic ancestors would have been an abrupt and discontinuous event (Hardisty 2006), resulting in immediate reproductive isolation due to this size-assortative mating (Salewski 2003). As a result, paired species generally are considered "good" biological species.

Size-assortative mating was experimentally demonstrated by Malmqvist (1983) within a species, the European brook lamprey. The importance of size as an isolating mechanism in paired lamprey species has likewise been tested in North American river and western brook lampreys by Beamish and Neville (1992). Three to seven lampreys of each sex were combined in four groups in which the female-to-male size ratios were 0.61 (i.e., size differences exceeded 33%), 0.83 (differences of approximately 25%), 1.19 (less than 22%), and 1.51 (exceeding 29%). Judging from visual observations of spawning activity, the presence of squeeze marks on females, and egg and embryo counts after 25 d, these authors determined that there was a low fertilization success when males and females differed in size by more than 20%. However, there was some successful fertilization even when size differences exceeded 30%. Beamish and Neville (1992) therefore concluded that there are probably additional barriers to hybridization, for example, habitat differences that would also lead to reproductive isolation (see below). As with the North American river and western brook lampreys, many paired species overlap in size within the 20% "acceptable" range, particularly some of the freshwater pairs in which the parasitic species is generally smaller than those that are anadromous (Table 2). This would also apply to paired species that include "praecox" forms (e.g., in the Arctic and European river lampreys; Table 2), in which a reduction in size is thought to be due to a shortening of the parasitic cycle. Kucheryavyi et al. (2007b) found no evidence of assortative mating among three life history types of Arctic lamprey in the Utkholok River and considerable overlap in size. For example, the male praecox form ranged in length from 145 to 220 mm, overlapping in size with female anadromous lampreys that ranged from 174 to 330 mm. Likewise, the male praecox form overlapped in size with females of the freshwater-resident nonparasitic form, which ranged in length from 100 to 165 mm. Even comparing the two extreme size-groups, although there was no overlap in size between male nonparasitic lampreys (100–160 mm) and female anadromous lampreys (174–330 mm), some pairings would still have been within the 20% size range.

Furthermore, some spawning behaviors have been reported that may make fertilization possible even when size differences between the sexes exceed 20%. For example, many species appear to engage in group spawning. Case (1970) noted up to 50 chestnut lampreys in one nest; Morman (1979) reported an average of seven northern brook lampreys per nest, and group spawning was likewise reported for European and American brook lampreys by Malmqvist (1983) and Cochran et al. (2008), respectively. In lampreys, sperm remains motile in water for approximately 1 min (Hardisty 1986), allowing perhaps for accidental fertilization between lampreys of dissimilar sizes (see also Potential Interbreeding, below, for a discussion of mixed-species group spawning). In addition, "satellite" males have been observed in some lamprey species. In the European brook lamprey, Malmqvist (1983) observed one or several males that rapidly circled around the tails of copulating pairs in what he interpreted to be an attempt to fertilize eggs as they were extruded by another male. He also noted polygamous mating, with two males copulating simultaneously with the same female and additional males that attached themselves further back on the head. Cochran et al. (2008) likewise observed satellite males in the American brook lamprey at two separate spawning sites in Minnesota and eastern Wisconsin; at both sites, satellite males were observed in at least half of all matings in nests with three or more lampreys. Satellite male behavior is observed in some other fish species (e.g., in salmonids) where small males "sneak" fertilizations, thus parasitizing the efforts of larger, dominant males (see Taborsky 1998).

## *Other Potential Means of Reproductive Isolation*

Both temporal and spatial separations have been suggested as prezygotic reproductive isolating mechanisms in some paired lamprey species. According to Manion and Hanson (1980), the smaller brook lampreys generally prefer smaller gravel, lower water velocity, and shallower depths than the larger parasitic species. Beamish and Neville (1992) likewise suggested ecological separation in North American river and western brook lampreys; river lampreys appear to spawn in the faster-flowing currents of the Fraser River, whereas western brook lampreys spawn in slower currents at the headwaters of smaller streams. They also suggest that spawning time may be a barrier to hybridization; in the laboratory, at least, western brook lampreys spawned later than the river lampreys. Zanandrea (1959) reported that, although nonparasitic species are found in the same river basin together with their parasitic counterpart, the nonparasitic forms are generally confined to the upper reaches. However, he stated that sometimes both forms can be caught spawning at the same time and place. Furthermore, parasitic and nonparasitic forms may aggregate together in crevices beneath boulders or other cover in areas where current speeds are high (Cochran and Gripentrog 1992), and smaller brook lampreys may take advantage of the nest-building activities of parasitic lampreys in areas of coarse substrate (Brumo 2006; Cochran et al. 2008). Morman (1979) likewise reported that different species sometimes share nests, and different forms of Arctic lamprey have also been found spawning at the same time and location (Kucheryavyi et al. 2007a, 2007b; see Potential Interbreeding, below). Kucheryavyi et al. (2007a, 2007b) also report, however, that in some locations, one form dominates or is found exclusively (e.g., the anadromous form predominated in principal main-stem areas and the resident nonparasitic form was found spawning alone in upper reaches of the Kolkavayam River).

Pheromones have also been suggested as a possible reproductive isolating mechanism in lampreys. Sex pheromones released by spermiating male sea lampreys attract ovulating female sea lampreys (Li et al. 2003), but it is not known whether sex pheromones are species-specific. Although sex pheromones have not been studied in paired species, Yamazaki and Goto (2000) suggested that they may prevent interbreeding in two undescribed *Lethenteron* species (*Lethenteron* sp. N and *Lethenteron* sp. S; see Table 1). These two species are morphologically and genetically distinct (e.g., with fixed allozyme differences), and there is no evidence of hybridization where they occur sympatrically. Spawning seasons and sizes at maturity overlapped (*Lethenteron* sp. S spawned later on average and was significantly larger), but nesting assemblages nevertheless comprised only males and females of the same form.

## *Potential Interbreeding*

Viable offspring have been produced through artificial hybridization between members of species pairs such as European river and brook lampreys (e.g., Enequist 1937), silver and northern brook lampreys (Piavis et al. 1970), and North American river and western brook lampreys (Beamish and Neville 1992). Development was synchronous in the hybrids and controls (e.g., Piavis et al. 1970). However, hybrid offspring were reared only for a few weeks (Piavis et al. 1970) up to 2½ years (Beamish and Neville 1992). Therefore, it is not known what would happen in hybrids at metamorphosis; it is unknown whether metamorphosis would proceed normally in interspecific hybrids and whether they would be fertile. Likewise, it is unknown whether strong selection against hybrids (e.g., in terms of feeding behavior) would result in postzygotic reproductive isolation.

Natural hybrids have not been reported. In the above study by Beamish and Neville (1992), the artificially created hybrids between North American river and western brook lampreys showed pigmentation patterns different from either parental species, and no such pigmentation patterns have been found in nature. There was also no evidence of hybridization between Arctic and Siberian brook lampreys in three river systems where they occur sympatrically; each possessed diagnostic allozyme alleles and no intermediate hybrids were found (Yamazaki and Goto 1998). In most other paired species, however, hybrids would generally not be detectable as larvae since paired species themselves are often indistinguishable or difficult to differentiate in the larval stage (see above). Intermediate adult phenotypes have not been reported in nature (with the possible exception of *L. richardsoni* var. *marifuga*, see above, but in this case, river lamprey does not co-occur with western brook lamprey). Furthermore, as mentioned above, without a better understanding of the heritability of feeding type, we do not know what hybrids would look like at or after metamorphosis.

Therefore, it is believed that prezygotic reproductive isolation—largely size-assortative mating (see above)—limits the production of hybrids in nature, but some studies suggest the potential at least for interbreeding. For example, Huggins and Thompson (1970) observed communal spawning in European river and brook lampreys; they found, however, that size-assortative mating occurred within nests (which presumably would limit interspecific hybridization) but assumed that hybrids would occasionally be produced by accident. Morman (1979) likewise observed mixed-species spawning aggregations in the Great Lakes basin (in at least one nest, both silver and northern brook lampreys were found), and mixed-species or mixed-morph groups have also been reported in *Lethenteron*. Savvaitova and Maksimov (1979) observed large (23–32 cm) and small (10–14 cm) Arctic lampreys and possibly the Far Eastern brook lamprey jointly building nests; as many as 7 large (presumably parasitic) and 15 small (presumably nonparasitic) Arctic lampreys were found in a single nest. Spawning groups consisted of 6–44 adults. Kucheryavyi et al. (2007a) observed both pair and group spawning in Arctic lampreys; pair spawning was most often between the same life history variants (i.e., between anadromous parasitic males and females or freshwater-resident nonparasitic males and females; see above), but mixed pairs of different forms were sometimes noted. In spawning groups, these authors reported the simultaneous presence of all three forms (the above-mentioned two forms as well as an anadromous, but early-maturing praecox form). Cochran et al. (2008) observed a female chestnut lamprey sharing a nest with 25–26 southern brook lampreys (of which approximately 15 were males). During a 5-min period, a southern brook lamprey attached to the back of the female chestnut lamprey's head four times and to the middle of her body once. However, no quivering or release of gametes was observed. To test whether such interspecific fertilizations occur, it should be possible to perform parentage analysis using microsatellite markers. Although species-specific diagnostic markers have not been found (see above), genotyping each potential parent and a subsample of embryos could be used to determine which parent contributed to the offspring.

## Evolution of Nonparasitism

### *Heterochrony*

Heterochrony is the change, over evolutionary time, in the relative rates or timing of developmental events. In particular, it can involve changes in the relative timing of reproductive and somatic development. Such heterochronies can lead to very different-looking organisms, even among closely related species (e.g., Denoël et al. 2004; Gibson and Gibson 2004; see Divergent Morphotypes in Other Animals, below). Heterochrony, as it relates to the evolution of nonparasitism in lampreys, has been reviewed recently—and well—by Hardisty (2006). It therefore will be discussed only briefly here. As discussed above, parasitic species do not become sexually mature until the end of the feeding phase, whereas the gonads begin to mature during metamorphosis in nonparasitic species and the adults bypass the feeding and migration phases (although, to compensate for passive downstream movement during the larval phase, nonparasitic species typically do undergo a short upstream migration to find suitable spawning areas). The gonads are at roughly the same state of development in both parasitic and nonparasitic lampreys at the initiation of metamorphosis, but maturation proceeds more rapidly during and after metamorphosis in nonparasitic species. Since nonparasitic species tend to metamorphose at older ages and larger sizes (see above), it has therefore been suggested that nonparasitism has evolved as a result of a change in the timing of metamorphosis relative to the timing of sexual maturation (Figure 4). This was recognized as early as 1912 by the Dutch zoologist Jan Cornelis Christiaan Loman; he proposed that metamorphosis was delayed and full sexual maturity began prematurely in the European brook lamprey relative to the European river lamprey (see Hardisty 2006). As a result of this developmental shift, fecundity is substantially reduced in the nonparasitic species due to their smaller adult size (e.g., mean fecundity in northern brook and silver lampreys is approximately 1,500 and 15,000 respectively; Vladykov 1951), but presumably there is a fitness tradeoff between fecundity and mortality in the nonparasitic brook lampreys relative to parasitic lampreys. Mortality through predation and other sources associated with the longer and more exposed feeding and migratory phase in parasitic lampreys—although not quantified—is presumably high.

Relatively little is known, however, about the possible mechanism for this change in the relative timing of sexual maturation and metamorphosis. In the past 30 years, there has been considerable progress in understanding the cascade of endocrine events that controls reproduction in lampreys (see Sower and Kawauchi 2001; Sower 2003; Sower et al. 2006; Bryan et al. 2008) and the mechanisms and triggers responsible for metamorphosis in lampreys (see Youson 1980, 2003), but few studies compare parasitic and nonparasitic lampreys (e.g., Youson et al. 1995, 2006; Montpetit et al. 2005) and none examine paired species. For example, Montpetit et al. (2005) compared expression patterns of neuropeptide Y family peptides—important hormones in the control of feeding behavior and reproductive development in mammals—in the brain and gut of sea and southern brook lampreys. Youson et al. (2006) used immunohistochemistry to examine gonadotropin-releasing hormones I and III in the brain of American brook lampreys and compared these data with those available for sea lampreys. Significant differences in the profiles of these hormones have been noted between parasitic and nonparasitic lampreys following metamorphosis, but endo-

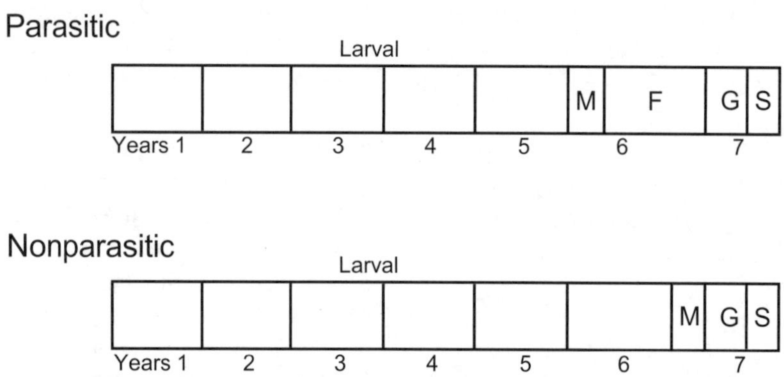


Figure 4. Schematic diagram representing the proposed heterochronic shift that would produce nonparasitic lampreys by altering the timing of metamorphosis relative to the timing of sexual maturation without changing the overall life span. Note that the larval phase in the nonparasitic lampreys will not necessarily be extended by just 1 year. Abbreviations are as follows: M = metamorphosis, F = juvenile (parasitic) feeding phase, G = gonadal maturation, and S = spawning (after Hardisty 2006).

crinological differences have not been detected prior to the point where morphological differences are evident. These previous studies, however, have identified a number of promising candidate genes that may be involved in sexual maturation and metamorphosis. More research comparing hormone profiles and gene expression patterns in paired parasitic and nonparasitic larvae prior to and during metamorphosis is required. Given our inability to reliably distinguish paired species as larvae, however, such paired species comparisons would require using larvae from known allopatric populations or laboratory-reared specimens.

### *Conditions under Which Brook Lampreys Might Evolve*

As with the previous section, this topic has been reviewed recently by Salewski (2003) and Hardisty (2006). Conventional wisdom has suggested that parasitic forms become nonparasitic when they come to live in a habitat where they could not find suitable hosts or become nonmigratory (and concomitantly nonparasitic) in small streams where there is a barrier to migration (e.g., Zanandrea 1959). In terms of becoming nonmigratory as a result of barriers to migration, recent glaciation and deglaciation events during the past 10,000–15,000 years might have promoted the evolution of nonparasitic species by blocking migratory routes with ice or by altering the relative benefits and costs of migration (Espanhol et al. 2007). In terms of developing a nonparasitic life history type in response to a lack of suitable hosts, it may be relevant to note that Cochran and Kitchell (1986) showed that a parasitic lamprey could theoretically optimize its rate of net energy intake in situations where hosts are rare by reducing its rate of blood consumption. This would reduce the lamprey's rate of growth but would presumably allow its host to survive longer (producing more blood during that additional time). As hosts become progressively rarer and lampreys achieve progressively less growth, it is possible to imagine an endpoint in which the lamprey is facultatively nonparasitic. Under extreme conditions, a fully nonparasitic life history type (with a longer larval phase, larger size at metamorphosis, and shortening of the adult phase) would presumably be favored (in terms of optimizing the fecundity-mortality trade-off) over prolonged inefficient feeding. This implies that nonparasitism evolves as a means of "making the best of a bad situation" under conditions where postmetamorphic feeding is poor. Conversely, however, one could argue that nonparasitism might be advantageous where larval conditions are good, that is, that lampreys that can grow large as larvae (and delay metamorphosis to further increase larval size) can more than offset the reduction in fecundity that would accompany a decrease in adult body size with greatly improved survival rates (as the result of remaining burrowed for the vast majority of its life). Under conditions where larval growth is reduced or larval mortality high, one would expect early metamorphosis followed by postmetamorphic feeding to be more advantageous (but see Renaud and Holčík 1988). A similar suggestion has been made by Kucheryavyi et al. (2007b), although in this case, the authors argue that good larval growth conditions promote nonparasitism within a single population of Arctic lamprey producing both feeding types (see above). These authors proposed that individuals with a sufficient quantity of energy resources accumulated during the larval stage, acquired perhaps by feeding on the decomposing carcasses of Pacific salmon, become nonparasitic. Comparable arguments have been proposed regarding jacking in some salmon. In Chinook salmon *Oncorhynchus tshawytscha*, for example, jacks are males that mature sexually after only one summer in seawater; they mature earlier than any of the females and earlier than the large "hooknose" males (see Heath et al. 1991) and gain fertilizations using sneaker or satellite behavior (see above). Rather than making the best of a bad situation, studies suggest that jacks represent the fastest-growing individuals in a cohort and balance a reduction in fertilization success with reduced mortality rates during their shorter time at sea (Heath et al. 1991).

In the transition from parasitism to nonparasitism in lampreys, some authors have suggested that freshwater (or anadromous but praecox) parasitic forms may have been possible intermediaries in the transition from large anadromous to small nonparasitic brook lampreys (e.g., Beamish 1985; Salewski 2003; Hardisty 2006). Beamish (1985) argued that the many changes that must occur in this transition are too major to occur in a single step (i.e., all occurring simultaneously during an extension of the larval period). Instead, he proposed that the first step in this evolutionary pathway was the ability to osmoregulate in freshwater during feeding, possibly acquired as the result of gradual changes in salinity (e.g., during periods of glaciation). This would have been followed by a decrease in the duration of the feeding period and in body size and eventually by the elimination of the feeding period altogether (Beamish 1985). It is arguable, however, whether nonparasitic lampreys need to acquire the ability to osmoregulate in freshwater following metamorphosis. Parasitic species re-enter freshwater as they approach sexual maturity; nonparasitic species begin sexual maturation during the later stages of metamorphosis, thus beginning their freshwater spawning phase prior to the end of metamorphosis. The proposed shift in the timing of sexual maturation relative to metamorphosis would concomitantly eliminate the feeding phase and the marine phase.

More recently, Salewski (2003) and Hardisty (2006) have also discussed whether freshwater parasitic or praecox forms are "jumping off" stages in the evolution of brook lampreys. They noted that the vast majority of brook lampreys are derived from species with praecox or freshwater-resident parasitic forms (e.g., the Arctic lamprey, European river lamprey, Pacific lamprey, and lampreys in the *Ichthyomyzon* and *Eudontomyzon* genera). However, this may demonstrate only that life history type in these lamprey species shows considerable plasticity (in terms of both migratory and feeding habits), without necessarily implying that the praecox or freshwater parasitic forms were intermediaries in the transition to nonparasitism. Furthermore, the North American river lamprey has given rise—likely independently multiple times (see Polyphyly and the Repeated Evolution of Nonparasitism, above)—to nonparasitic populations, despite its inability to osmoregulate in freshwater during its feeding phase (Beamish 1985). Hardisty (2006) argued that the three species that lack nonparasitic derivatives—the sea lamprey, Caspian lamprey, and pouched lamprey *Geotria australis*—are all without praecox forms, but freshwater-resident sea lampreys are extremely successful in the Great Lakes and have likely been present in some North American fresh waters for approximately 10,000 years (Waldman et al. 2004; Docker 2006). Hardisty (2006) suggested that freshwater residency in sea lampreys may represent the first step towards nonparasitism but indicated that the absence of nonparasitic derivatives in these genera could also be explained by a lack of some genetic factor involved in the evolution of nonparasitic species. Alternatively, Hardisty (2006) proposed that it may be selective constraints—not physiological or genetic constraints—that have prevented the evolution of nonparasitism in these species. He suggested that for large, ocean-going lampreys, the decrease in mortality rates that would accompany the shift to nonparasitism could not possibly offset the very drastic reduction in fecundity. In smaller parasitic species with lower levels of fecundity, Hardisty (2006) predicted that the balance would be "more delicately poised" such that unstable environments that reduced feeding opportunities or increased mortality during the adult phase could tip the balance towards nonparasitism.

Although it is debatable, then, whether an intermediate freshwater or praecox stage is necessary for the evolution of brook lampreys, a greater understanding of these forms is sure to shed light on the evolution of nonparasitism. For example, in terms of understanding the heterochronic shift that is thought to be responsible for the elimination of the feeding phase, one would predict that the length of the larval stage (and size at metamorphosis) in the praecox or freshwater-resident forms would be intermediate to the anadromous parasitic and nonparasitic forms. There are some suggestions that this might be the case (e.g., Hardisty 2006), but it has not been studied widely. In fact, where both anadromous and freshwater or praecox forms are known from a single location (e.g., Adams et al. 2008), it is not known whether individuals commit to one life history type or another prior to metamorphosis. Information about potential differences between the sexes in the development of praecox forms would also be informative in understanding sex differences in the evolution of nonparasitism. It appears that both praecox (Beamish 1985; Kucheryavyi et al. 2007b) and nonparasitic (Hubbs 1971) forms develop more readily in males. This would imply that males show a lengthening of the larval period first, although it appears that female brook lampreys ultimately have a longer larval period—and life span in total—relative to conspecific males (e.g., Purvis 1970; Docker and Beamish 1991). Although this review has focused largely on feeding type, with little explicit regard for adult habitat of the parasitic species, understanding the process by which freshwater and praecox forms arise within anadromous parasitic species appears related to the evolution of nonparasitism (i.e., is part of a continuum) and is an area deserving of further study.

### *Direction of Evolution*

As discussed throughout the above sections, the currently held view is that nonparasitic lampreys evolved from parasitic lampreys (Hubbs 1925; Zanandrea 1959; Vladykov and Kott 1979). This was based on the distribution of paired species and on differences in structures related to feeding. For example, rudimentary development of intestinal mucosal folds has been observed in nonparasitic lampreys during metamorphosis (Hilliard et al. 1983; Youson and Beamish 1991), and buccal glands, which produce an anticoagulant necessary for parasitic feeding, develop during metamorphosis in nonparasitic as well as parasitic species (Baxter 1956).

It is not known, however, whether feeding type evolution is always unidirectional in lampreys or whether it is possible to have an evolutionary reversal whereby a nonparasitic species regains the parasitic feeding phase. Facultative parasitism is discussed above (e.g., in the American brook lamprey and *L. richardsoni* var. *marifuga*), but these cases may represent polymorphism in a population that has not yet become fixed for the nonparasitic feeding type rather than a reversal per se. In other organisms that show independent loss of the feeding phase in different lineages (e.g., sea urchins; see Divergent Morphotypes in Other Animals, below), phylogenetic analysis indicates that feeding stages are lost much more often than they are gained and that losses are usually irreversible (Strathman et al. 1992). This may not apply to lampreys, however, if there is no loss of feeding structures in closely related pairs. More research is required, however, to resolve this question. Since there is no a priori reason to believe that fully viable parasitic species that were derived from a nonparasitic lamprey would be recognizable in their morphology or physiology, this will likely require in-depth phylogenetic analysis with molecular loci capable of providing greater resolution between paired species (i.e., to determine if given parasitic species are polyphyletic).

## Divergent Morphotypes in Other Animals

Divergent morphotypes that differ in terms of their phenotypic, behavioral, life history, and physiological characters occur in many species of invertebrates, north temperate fishes, and other vertebrates. These different morphotypes often correspond with the use of different trophic resources. This section will not present a comprehensive review of this topic but is intended to draw some parallels between lamprey species pairs and divergent morphs within species in other taxa, which in some cases, appear to have led to reproductive isolation and speciation. A continuum appears to exist ranging from polyphenism (where divergent phenotypes are produced from a single gene pool in response to different environmental stimuli), through divergent morphotypes with a large genetic component (but which are still considered a single species due to incomplete reproductive isolation or lack of reciprocal monophyly) and incipient speciation, to the existence of two distinct gene pools and reciprocal monophyly (i.e., full species).

Examples of polyphenism in animals include winged or nonwinged morphs in pea aphids, where winged morphs are produced in response to increased predation risk (Podjasek et al. 2005); spadefoot toads that—in addition to omnivorous tadpoles—can also produce a carnivorous phenotype (with an enlarged head, enlarged jaw musculature, shortened intestines, and increased keratinization of the mouth to form a beak) in response to dietary changes or exogenous thyroxine treatment (Pfennig 1992); and tiger salamanders, which are capable of producing a large-headed cannibal morph under conditions of crowding (Hoffman and Pfennig 1999). Similarly, there are several examples of frog and toad species in which the tadpoles can complete development with or without feeding (e.g., Crump 1989; Inger 1992). The nontrophic tadpoles metamorphose in less time than the ones that feed (Crump 1989), and there are pronounced morphological differences between the types in terms of gut and mouth morphology (Inger 1992), but a single adult morphotype has been observed (Inger 1992).

Another example of polyphenism that merits a brief mention here (given its potential parallel to a heterochronic shift leading to nonparasitism in lampreys) is the repeated evolution of paedomorphosis in salamanders. Paedomorphosis—whereby larvae becomes sexually mature without attaining the adult body form, as a result of "metamorphic failure" (Shaffer and Voss 1996)—has evolved independently in different populations and species of salamanders. For example, three families (Proteidae, Cryptobranchidae, and Sirenidae) are obligate paedomorphs, but paedomorphosis is variable in the families Salamandridae (true salamanders and newts) and Ambystomatidae (the mole or tiger salamanders): some species are obligate paedomorphs while other species are facultative paedomorphs (e.g., Shaffer and Voss 1996; Denoël 2006; Takahashi and Parris 2008). Among the facultative paedomorphs, there can be variation in the frequency of paedomorphosis among populations and families (suggesting that there is a genetic basis), but there is also some type of interaction between environmental and genetic control mechanisms (Voss 1995). For example, in newts of the genus *Triturus*, the relative proportion of alternative morphs varies among habitat types (with paedomorphs being predominant in permanent waters and metamorphs being predominant in temporary waters) but also over time (when removed from water, paedomorphs are able to initiate metamorphosis; e.g., Denoël 2006). There has been considerable progress made on understanding the mechanistic basis of metamorphosis and paedomorphosis in these salamanders (e.g., Voss 1995; Voss and Smith 2005), and it has been discovered that different physiological mechanisms are involved in different taxa, that is, that the same phenotype (a paedomorph) has been achieved independently via different routes. For example, some obligate paedomorphic species have a functional metamorphic pathway in the presence of sufficient thyroxine (T4), whereas others do not respond to T4 treatment (demonstrating that paedomorphosis is the result of changes further "down" the hormonal cascade; e.g., at the thyroid hormone receptors; see Shaffer and Voss 1996). As a result, complementation tests, in which obligate paedomorphs with different mechanisms were reciprocally crossed, produced some larvae capable of metamorphosis (Shaffer and Voss 1996). This begs the question of whether the same genetic mechanism is involved in the evolution of nonparasitism in each lamprey species pair and thus whether independently created brook lamprey phenotypes are necessarily equivalent (despite similar morphologies). It would be exciting if similar research could be conducted into the mechanistic basis of nonparasitism in lampreys.

Divergent morphotypes are also common in many north temperate fish species. It appears that there is a large genetic component to morphotype in many of these species, but there is a limited degree of genetic differentiation between them and evidence that environmental triggers can produce the alternative morphotype and that divergence between the morphotypes has occurred independently multiple times, making each morphotype polyphyletic. Many salmonid species, for example, exhibit polymorphisms with respect to migratory life history type (e.g., Taylor et al. 1996; Docker and Heath 2003). The different migratory forms also exhibit morphological differences, and taxonomic distinctions have often been made between them (see Taylor et al. 1996; Docker and Heath 2003). However, genetic evidence clearly suggests that they are polyphyletic, with freshwater-residency evolving independently multiple times (Docker and Heath 2003; Heath et al. 2008). Furthermore, in sympatric populations, mitochondrial and microsatellite markers showed that most sympatric steelhead and rainbow trout pairs did not show significant differences in allele frequencies (Docker and Heath 2003; Olsen et al. 2006; Heath et al. 2008)—although there have been some observations of genetic differentiation (e.g., Narum et al. 2004). There appear to be parallels, therefore, between divergent migratory types in salmonids and the divergent feeding types observed in many paired lamprey species—except that anadromous steelhead and freshwater-resident rainbow trout are now considered a single species, *Oncorhynchus mykiss*. An additional similarity is the observation that—in at least some cases—each of the migratory types can give rise to the other (e.g., Pascual et al. 2001; Zimmerman and Reeves 2002; Thrower et al. 2004; Riva-Rossi et al. 2007).

Perhaps further along the continuum, divergent morphotypes and signs of incipient speciation have been observed in species pairs in the threespine stickleback *Gasterosteus aculeatus*, which occur sympatrically in several lakes in southwestern British Columbia. One form is morphologically and behaviorally specialized for benthivory while the other is adapted for planktivory. Although currently identified by a common Latin binomial, the two morphs have been shown—genetically and experimentally—to act as distinct biological species (Hatfield and Schluter 1996, 1999; Hatfield 1997; Taylor 1999). A low level of natural hybridization has been observed in two lakes, suggesting incipient or recent speciation (Gow et al. 2006), but a breakdown of the species pair into a hybrid swarm was demonstrated in a third lake (see Taylor et al. 2006), illustrating that incipient speciation does not always progress to full speciation.

Examples of divergent feeding types leading to full speciation and reciprocal monophyly—but with occasional plasticity within a species—are found among the echinoderms. In echinoderms (e.g., starfish, sand dollars, and sea urchins), larval development typically proceeds by one of two modes: planktotrophy, which involves obligate larval feeding during an extended period in the plankton; and lecithotrophy, obligate nonfeeding where the larvae (which lack elaborate feeding structures) settle from the plankton relatively quickly. Planktotrophs develop from eggs with only a portion of the energy necessary to reach metamorphic competence, whereas lecithotrophs develop from comparatively large eggs (Strathman et al. 1992; Reitzel et al. 2005; Byrne 2006). Among closely related species with different larval types, adults are morphologically and ecologically similar, but they have diverse egg sizes and modes of larval development (Byrne 2006). Planktotrophy is assumed to be the ancestral mode, and lecithotrophy has evolved multiple times independently (i.e., loss of the larval feeding phase has occurred repeatedly in different lineages); lecithotrophs are found in all five extant echinoderm taxa (Strathman et al. 1992). Taxonomic distribution of planktotrophs and lecithotrophs indicates that feeding larval stages are lost much more often than they are gained and that losses are usually irreversible (Strathman et al. 1992). Presumably, because mortality in the plankton can be high, the benefits of a decrease in development time in lecithotrophs are thought to offset the accompanying reduction in fecundity (Allen et al. 2006). There are clear similarities, therefore, between the repeated and independent loss of the larval feeding phase in echinoderms and the recurrent loss of the adult feeding phase in different lamprey lineages. Furthermore, some echinoderm species show facultative larval feeding and intraspecific variation in feeding type. For example, in addition to the obligate planktotrophs and obligate lecithotrophs found within the brittlestar genus *Macrophiothrix*, two species have been identified with rare intermediate forms (Hart and Podolsky 2005); *M. rhabdota* is a facultative planktotroph that does not require food to complete development, whereas *M. caenosa* retains the external planktotroph morphology but has lost the capacity to feed. Using molecular phylogenetic analysis, these authors determined that the two species with intermediate forms are the most recently derived (*M. rhabdota* even more so than *M. caenosa*) and indicated that loss of larval feeding would occur in sequential steps, development of facultative feeding followed by a loss in the ability to feed and then, finally, a loss of feeding structures. Hart and Podolsky (2005) suggest that the rarity of such intermediate forms likely indicates that they are evolutionarily unstable and short-lived.

In each parasitic-nonparasitic lamprey species pairs, it is not fully known to what extent the morphological and ecological differences between members of each pair are associated with distinct gene pools or whether environmental triggers play a significant role in their divergent feeding types. A brief discussion of where on the above-described continuum lamprey species pairs might fall is presented in the final section below.

## Conclusions

### *Are Brook Lampreys "Real" Species?*

This frequently asked question has been discussed most recently by Salewski (2003) and Hardisty (2006), although it has not yet been resolved. Zanandrea (1959) stated that "the most common opinion to-day is that each of the two paired forms is a 'bona species' and that the nonparasitic species originated from the parasitic one," and most authors continue to share this view (e.g., Salewski 2003); most taxonomists consider lamprey feeding type to be a species-specific characteristic (e.g., Hardisty and Potter 1971; Potter 1980). Other authors, however, have held the view that paired species are morphs or ecological races of a single species (e.g., Enequist 1937; McPhail and Lindsey 1970). Enequist (1937) interpreted the occurrence of both parasitic and nonparasitic European river and brook lampreys to be an adaptation to hydrographical characters (e.g., habitat differences during the larval stage) and considered them alternative life forms of a single species. Considering the two life history types apparently produced by Arctic lampreys in Alaska, McPhail and Lindsey (1970) suggested that the presence or absence of the parasitic phase is no indication of specific differentiation. Beamish and Withler (1986) concurred, stating that "if it can be shown that nonparasitic species produce stable parasitic varieties and vice versa then a species separation based on trophic habits is invalid." Savvaitova and Maksimov (1979) similarly suggested that brook lampreys are most likely freshwater forms of anadromous lampreys, similar to alternative migratory types in salmonids.

As discussed throughout this review, the issue is complex—dealing with morphological, biological, and phylogenetic species concepts—and there likely is not a single answer for all lamprey species pairs. In some species pairs (e.g., Arctic and American brook lampreys; Arctic and Siberian brook lampreys), there is evidence for significant (and diagnostic) genetic differentiation, a lack of hybridization where sympatric, and reciprocal monophyly, which would support considering each of these as distinct species. However, plasticity of feeding type has been reported in the American brook lamprey and, especially, in the Arctic lamprey (see above). However, this does not necessarily preclude considering each of these as distinct species; it only suggests that feeding type alone is not sufficient grounds for species separation. The species status of other lamprey pairs is not as clear-cut, given the variable degree to which they can be differentiated from one another (morphologically and genetically) and whether or not reciprocal monophyly has been demonstrated. North American river and western brook lampreys, for example, show significant differences in myomere counts and pigmentation, as well as evidence for reproductive isolation through both size-assortative mating and ecological separation, but there is strong evidence that the western brook lamprey is polyphyletic (see above). In yet other pairs (presumably those that are very recently derived), there is evidence for polyphyly and little or no morphological and genetic differentiation between the parasitic and nonparasitic member of the pair. Consequently, if comparing the continuum representing the evolution of lamprey species pairs at different stages of the speciation process (Figure 2) with the spectrum observed in other animal species with divergent morphotypes (see above), parasitic-nonparasitic lamprey pairs might be placed within the range of the second to fourth steps described above, that is, corresponding to divergent morphotypes with a large genetic component (but that are still considered a single species due to incomplete reproductive isolation or lack of reciprocal monophyly), incipient speciation, or the existence of two distinct gene pools and reciprocal monophyly. That is, some pairs are on par with anadromous steelhead and freshwater-resident rainbow trout in that they are polyphyletic, with little or no genetic differentiation and evidence for life history type plasticity or polymorphism. Other lamprey pairs appear to be in the process of incipient speciation comparable to that seen in benthic-limnetic sticklebacks; they are polyphyletic but appear to be acting as distinct biological species. In other species pairs (as described above with the Arctic lamprey and some of its derivatives), two distinct gene pools and reciprocal monophyly are evident (as in many echinoderm species where the larval feeding phase has been lost independently and repeatedly). For a clearer understanding of the relationship between parasitic and nonparasitic lampreys and the repeated evolution of nonparasitism, however, further research is required (e.g., higher-resolution phylogenies using both mitochondrial and nuclear markers, quantification of gene flow between sympatric species pairs, and common garden and reciprocal transplant experiments to determine if feeding type is strictly heritable or whether there is an ecological or environmental component). Given the repeated evolution of nonparasitism in most genera worldwide, lamprey species pairs represent an understudied model system in which to investigate life history polymorphism and the speciation process.

## Conservation Implications

The general inability to confidently distinguish paired species during the larval phase—and the larger question of whether paired lamprey species constitute single or separate gene pools—has significant conservation implications. Many lamprey species have been given special conservation status in several jurisdictions (e.g., Renaud 1997; Igoe et al. 2004; Renaud et al. 2009a), but the inability to distinguish them during the prolonged larval stage—morphologically or with diagnostic genetic markers (see above)—presents a major barrier to truly understanding the current distribution and population trends in these species. Paired species presumably have very similar habitat requirements and similar or identical vulnerabilities in the larval stage, but these differ considerably following metamorphosis. In particular, migratory and parasitic adults will be impacted by barriers to migration and by depletion of their prey base, factors that presumably would have little or no effect on nonparasitic brook lampreys (but see below). For example, efforts to control the exotic sea lamprey in the Great Lakes has relied largely on applications of a lampricide that is toxic to lamprey larvae and on barrier dams (Christie and Goddard 2003), and it appears that the silver lamprey has been more affected by these measures than its nonparasitic counterpart, the northern brook lamprey (Schuldt and Goold 1980). This is probably because the silver lamprey has a life history (i.e., parasitic and migratory) and instream distribution (i.e., limited to areas below barrier dams) very similar to that of the sea lamprey. Likewise, the single biggest factor limiting the distribution of the European river lamprey in Ireland is the presence of barriers to upstream migration; brook lampreys are more widely distributed (Igoe et al. 2004). Barriers to migration have also led to local extinctions in the anadromous Pacific lamprey (e.g., Beamish and Northcote 1989).

However, if gene flow is extensive between paired "species," we cannot assume that brook lampreys would be immune to changes in the migration routes or prey base of their parasitic counterpart. If migratory lampreys mediate gene flow among otherwise isolated brook lamprey populations (as suggested by Schreiber and Engelhorn 1998), loss of the parasitic species could also lead to a loss of genetic diversity in small, isolated brook lamprey populations and to greater levels of local extinction. Therefore, measures intended to protect brook lampreys would also need to consider barriers to migration and lake conditions that would affect parasitic lampreys. On the other hand, if lamprey populations are polymorphic for feeding type or if feeding type is determined or can be mediated by environmental factors, it could be possible to rehabilitate endangered populations of one feeding type with the other. For example, it could be possible to "recreate" brook lamprey populations from their parasitic counterpart (or vice versa if a change in feeding type can be bidirectional; see Direction of Evolution). Many unknowns remain, however (e.g., whether the same genetic mechanism is involved in each case and whether independently created brook lamprey phenotypes are necessarily equivalent; see Divergent Morphotypes in Other Animals), and it would be imprudent to assume that such rehabilitation could be achieved easily.

Furthermore, it is important to recognize genetic diversity within currently recognized brook lamprey species. Genetic analysis is increasingly showing that many brook lamprey species, as currently recognized, are polyphyletic (e.g., Espanhol et al. 2007; see Polyphyly and the Repeated Evolution of Nonparasitism), with some small, isolated populations being genetically very distinct (e.g., Boguski 2009). Several of these populations may represent cryptic species, despite what appears to be their highly conserved brook lamprey morphology. Considering these genetically unique brook lampreys as a single, common, well-distributed species would minimize their need for protection. Consequently, until the many knowledge gaps highlighted in this review are bridged, it is important to conserve both phenotypic diversity (e.g., protecting both parasitic and brook lampreys, even if they eventually are recognized under a single Latin binomial) and genetic diversity (e.g., among morphologically similar brook lampreys) in lamprey species pairs.

## Acknowledgments

I thank David Boguski (University of Manitoba, Winnipeg, MB), Damon Goodman (U.S. Fish and Wildlife Service, Arcata, CA), Steve Johnson (Oregon Department of Fish and Wildlife, Corvallis, OR), Stewart Reid (Western Fishes, Ashland, OR), and René Reyes (U.S. Bureau of Reclamation, Byron, CA) for permission to use the photographs in Figure 1, and David Boguski for his help assembling them into a single figure. I also thank three anonymous reviewers for their very helpful comments and insightful suggestions and Larry Brown for his patience as editor.

## References

Adams, C. E., N. Bissett, J. Newton, and P. S. Maitland. 2008. Alternative migration and host parasitism strategies and their long-term stability in river lampreys from the River Endrick, Scotland. Journal of Fish Biology 72:2456–2466.

Allen, J. D., C. Zakas, and R. D. Podolsky. 2006. Effects of egg size reduction and larval feeding on juvenile quality for a species with facultative-feeding development. Journal of Experimental Marine Biology and Ecology 33:186–197.

Alvarez, J. 1964. Ictiología michoacana, IV. Contribución al conocimiento biológico y sistemático de las lampreas de Jacona, Mich., México. Anales de la Escuela Nacional de Ciencias Biologicas México 13:1–4.

Anikin, V. P. 1905. Description of new fish species from Asia. Izvestiya Tomskogo Universiteta. 27:1–17. [In Russian]

Avise, J. C., C. D. Ankney, and W. S. Nelson. 1990. Mitochondrial gene trees and the evolutionary relationship of mallard and black ducks. Evolution 44:1109–1119.

Battle, H. I., and K. Hayashida. 1965. Comparative study of the intraperitoneal alimentary tract of parasitic and nonparasitic lampreys from the Great Lakes region. Journal of the Fisheries Research Board of Canada 2:276–289.

Baxter, E. W. 1956. Observations on the buccal glands of lampreys (Petromyzonidae). Proceedings of the Zoological Society of London 127:95–118.

Beamish, F. W. H. 1982. Biology of the southern brook lamprey, *Ichthyomyzon gagei*. Environmental Biology of Fishes 7:305–320.

Beamish, F. W. H., and E. J. Thomas. 1983. Potential and actual fecundity of the "paired" lampreys, *Ichthyomyzon gagei* and *I. castaneus*. Copeia 1983:367–374.

Beamish, R. J. 1980. Adult biology of the river lamprey (*Lampetra ayresi*) and the Pacific lamprey (*Lampetra tridentata*) from the Pacific coast of Canada. Canadian Journal of Fisheries and Aquatic Sciences 37:1906–1923.

Beamish, R. J. 1982. *Lampetra macrostoma*, a new species of freshwater parasitic lamprey from the west coast of Canada. Canadian Journal of Fisheries and Aquatic Sciences 39:736–747.

Beamish, R. J. 1985. Freshwater parasitic lamprey on Vancouver Island and a theory of the evolution of the freshwater parasitic and nonparasitic life history types. Pages 123–140 *in* R. E. Foreman, A. Gorbman, J. M. Dodd, and R. Olsson, editors. Evolutionary biology of primitive fishes. Plenum, New York.

Beamish, R. J. 1987. Evidence that parasitic and nonparasitic life history types are produced by one population of lamprey. Canadian Journal of Fisheries and Aquatic Sciences 44:1779–1782.

Beamish, R. J., and C. D. Levings. 1991. Abundance and freshwater migrations of the anadromous parasitic lamprey, *Lampetra tridentata*, in a tributary of the Fraser River, British Columbia. Canadian Journal of Fisheries and Aquatic Sciences 48:1250–1263.

Beamish, R. J., and C. M. Neville. 1992. The importance of size as an isolating mechanism in lampreys. Copeia 1992:191–196.

Beamish, R. J., and T. G. Northcote. 1989. Extinction of a population of anadromous parasitic lamprey, *Lampetra tridentata*, upstream of an impassable dam. Canadian Journal of Fisheries and Aquatic Sciences 46:420–425.

Beamish, R. J., and R. E. Withler. 1986. A polymorphic population of lampreys that may produce parasitic and nonparasitic varieties. Pages 31–49 *in* T. Uyeno, R. Arai, T. Taniuchi, and K. Matsuura, editors. Indo-Pacific fish biology: proceedings of the Second International Conference on Indo-Pacific Fishes. Ichthyological Society of Japan, Tokyo.

Bean, T. H. 1887. Descriptions of five new species of fishes sent by Prof. A. Dugès from the Province of Guanajuato, Mexico. Proceedings of the United States National Museum 10:370–375.

Berg, L. S. 1931. A review of the lampreys of the Northern Hemisphere. Ezhegodnik Zoologicheskogo Muzeya Imperatorskogo Akademii Nauk SSSR 32:87–116.

Bergstedt, R. A., and Swink, W. D. 1995. Seasonal growth and duration of the parasitic life stage of the landlocked sea lamprey (*Petromyzon marinus*). Canadian Journal of Fisheries and Aquatic Sciences 52:1257–1264.

Billington, N., and P. D. N. Hebert. 1991. Mitochondrial DNA diversity in fishes and its implications for introductions. Canadian Journal of Fisheries and Aquatic Sciences 48 (Supplement 1):80–94.

Bird, D. J., and I. C. Potter. 1979a. Metamorphosis in the paired species of lampreys, *Lampetra fluviatilis* (L) and *Lampetra planeri* (Bloch). 1. Description of the timing and stages. Zoological Journal of the Linnean Society 65:127–143.

Bird, D. J., and I. C. Potter. 1979b. Metamorphosis in the paired species of lampreys, *Lampetra fluviatilis* (L.) and *Lampetra planeri* (Bloch). 2. Quantitative data for body proportions, weights, lengths and sex ratios. Zoological Journal of the Linnean Society 65:145–160.

Blank, M., K. Jurss, and R. Bastrop. 2008. A mitochondrial multigene approach contributing to the systematics of the brook and river lampreys and the phylogenetic position of *Eudontomyzon mariae*. Canadian Journal of Fisheries and Aquatic Sciences 65:2780–2790.

Bloch, M. E. 1784. Oeconomische Naturgeschichte der Fische Deutschlands. III. Berlin.

Boguski, D. A. 2009. The genetic diversity of brook lampreys, genus *Lampetra* (Petromyzontidae) along the Pacific coast of North America. Master's thesis. University of Manitoba, Winnipeg, Manitoba, Canada.

Bond, C. E. and T. T. Kan. 1973. *Lampetra* (*Entosphenus*) *minima* n. sp., a dwarfed parasitic lamprey from Oregon. Copeia 1973:568–574.

Britten, R. J. 1986. Rates of DNA sequence evolution differ between taxonomic groups. Science 231:1393–1398.

Brown, W. M., M. George, Jr., and A. C. Wilson. 1979. Rapid evolution of animal mitochondrial DNA. Proceedings of the National Academy of Sciences of the United States of America 76:1967–1971.

Brumo, A. F. 2006. Spawning, larval recruitment, and early life survival of Pacific lamprey in the south fork Coquille River, Oregon. Master's thesis. Oregon State University, Corvallis.

Bryan, M. B., A. P. Scott, and W. Li. 2008. Sex steroids and their receptors in lampreys. Steroids 73:1–12.

Byrne, M. 2006. Life history diversity and evolution in the Asterinidae. Integrative and Comparative Biology 46:243–254.

Case, B. 1970. Spawning behaviour of the chestnut lamprey (*Ichthyomyzon castaneus*). Journal of the Fisheries Research Board of Canada 27:1872–1874.

Christie, G. C., and C. I. Goddard. 2003. Sea Lamprey International Symposium (SLIS II): advances in the integrated management of sea lamprey in the Great Lakes. Journal of Great Lakes Research 29 (Supplement 1):1–14.

Cochran, P. A. 1987. The southern brook lamprey (*Ichthyomyzon gagei*) in the St. Croix River drainage of Wisconsin and Minnesota. Copeia 1987:443–446.

Cochran, P. A. 2004. Historical notes on lampreys in Wisconsin. American Currents 30:4–80.

Cochran, P. A. 2008. Observations on giant American brook lampreys (*Lampetra appendix*). Journal of Freshwater Ecology 23:161–164.

Cochran, P. A., D. D. Bloom, and R. J. Wagner. 2008. Alternative reproductive behaviors in lampreys and their significance. Journal of Freshwater Ecology 23:437–444.

Cochran, P. A., and A. P. Gripentrog. 1992. Aggregation and spawning by lampreys (genus *Ichthyomyzon*) beneath cover. Environmental Biology of Fishes 33:381–387.

Cochran, P. A., and J. F. Kitchell. 1986. Use of modeling to investigate potential feeding strategies of parasitic lampreys. Environmental Biology of Fishes 16:219–223.

Cochran, P. A., and J. E. Marks. 1995. Biology of the silver lamprey, *Ichthyomyzon unicuspis*, in Green Bay and the lower Fox River, with a comparison to the sea lamprey, *Petromyzon marinus*. Copeia 1995:409–421.

Crump, M. L. 1989. Life history consequences of feeding versus non-feeding in a facultatively non-feeding toad larva. Oecologia 78:486–489.

DeKay, J. E. 1842. Zoology of New York or the New-York fauna. Part IV, Fishes. W. and A. White and J. Visscher, Albany, New York.

Denoël, M. 2006. Seasonal variation of morph ratio in facultatively paedomorphic populations of the palmate newt *Triturus helveticus*. Acta Oecologica 29:165–170.

Denoël, M., R. Schabetsberger, and P. Joly. 2004. Trophic specialisations in alternative heterochronic morphs. Naturwissenschaften 91:81–84.

Docker, M. F. 2006. Bill Beamish's contributions to lamprey research and recent advances in the field. Guelph Ichthyology Reviews 7:1–52.

Docker, M. F., and F. W. H. Beamish. 1991. Growth, fecundity, and egg size of least brook lamprey, *Lampetra aepyptera*. Environmental Biology of Fishes 31:219–227.

Docker, M. F., and D. D. Heath. 2003. Genetic comparison between sympatric anadromous steelhead and freshwater resident rainbow trout in British Columbia, Canada. Conservation Genetics 4:227–231.

Docker, M. F., J. H. Youson, R. J. Beamish, and R. H. Devlin. 1999. Phylogeny of the lamprey genus *Lampetra* inferred from mitochondrial cytochrome *b* and ND3 gene sequences. Canadian Journal of Fisheries and Aquatic Sciences 56:2340–2349.

Dybowski, B. I. 1869. Vorläufige Mittheilungen über die Fischfauna des Ononflusses und des Ingoda in Transbaikalien. Verhandlungen der K.-K. Zoologisch-Botanischen Gesellschaft in Wien 19:945–958.

Enequist, P. 1937. Das Bachneunauge als Ökologische Modifikation des Flussneunauges—Über die

Fluss—und Bachneunaugen Schwedens Vorläufige Mitteilung. Arkiv för Zoologi 29:1–22.

Espanhol, R., P. R. Almeida, and M. J. Alves. 2007. Evolutionary history of lamprey paired species *Lampetra fluviatilis* (L.) and *Lampetra planeri* (Bloch) as inferred from mitochondrial DNA variation. Molecular Ecology 16:1909–1924.

Filcek, K. B., S. A. Gilmore, K. T. Scribner, and M. L. Jones. 2005. Discriminating lamprey species using multilocus microsatellite genotypes. North American Journal of Fisheries Management 25:502–509.

Foote, C. J., C. C. Wood, and R. E. Withler. 1989. Biochemical genetic comparison of sockeye salmon and kokanee, the anadromous and nonanadromous forms of *Oncorhynchus nerka*. Canadian Journal of Fisheries and Aquatic Sciences 46:149–158.

Fukayama, S., and H. Takahashi. 1982. Sex differentiation and development of the gonad in the Japanese river lamprey, *Lampetra japonica*. Bulletin of the Faculty of Fisheries, Hokkaido University 33:206–216.

Fukayama, S., and H. Takahashi. 1983. Sex differentiation and development of the gonad in the sand lamprey, *Lampetra reissneri*. Bulletin of the Faculty of Fisheries Hokkaido University 34:279–290.

Gage, S. H. 1928. The lampreys of New York State—life history and economics. Biological survey of the Oswego River System. Report of New York State Conservation Department (Supplement) 17:158–191.

Gibson, G. D., and A. J. F. Gibson. 2004. Heterochrony and the evolution of poecilogony: generating larval diversity. Evolution 58:2704–2717.

Gill, H. S., C. B. Renaud, F. Chapleau, R. L. Mayden, and I. C. Potter. 2003. Phylogeny of living parasitic lampreys (Petromyzontiformes) based on morphological data. Copeia 2003:687–703.

Girard, C. F. 1858. Fishes. Pages 1–400 *in* General report on the zoology of the several Pacific railroad routes. United States Pacific railroad route explorations and surveys, volume 10, part 4. U.S. War Department, Washington, D.C.

Gow, J. L., C. L. Peichel, and E. B. Taylor. 2006. Contrasting hybridization rates between sympatric three-spined sticklebacks highlight the fragility of reproductive barriers between evolutionarily young species. Molecular Ecology 15:739–752.

Günther, A. 1870. Catalogue of the fishes in the British Museum. Catalogue of the Physostomi and of the Lophobranchii, Plectognathi, Dipnoi, Ganoidei, Chondropterygii, Cyclostomata, Leptocardii in the British Museum. Taylor and Francis, London.

Hardisty, M. W. 1961. Oocyte numbers as a diagnostic character for the identification of ammocoete species. Nature (London) 191:1215–1216.

Hardisty, M. W. 1963. Fecundity and speciation in lampreys. Evolution 17:17–22.

Hardisty, M. W. 1964. The fecundity of lampreys. Archiv für Hydrobiologie 60:340–357.

Hardisty, M. W. 1965. Sex differentiation and gonadogenesis in lampreys. Part I. The ammocoete gonads of the brook lamprey, *Lampetra planeri*. Journal of Zoology (London) 146:305–345.

Hardisty, M. W. 1970. The relationship of gonadal development to the life cycles of the paired species of lamprey, *Lampetra fluviatilis* (L.) and *Lampetra planeri* (Bloch). Journal of Fish Biology 2:173–181.

Hardisty, M. W. 1971. Gonadogenesis, sex differentiation and gametogenesis. Pages 295–360 *in* M. W. Hardisty and I. C. Potter, editors. Biology of lampreys, volume 1. Academic Press, New York.

Hardisty, M. W. 1986a. *Lampetra fluviatilis* (Linnaeus, 1758). Pages 249–278 *in* J. Holčík, editor. The freshwater fishes of Europe, volume 1, part 1: Petromyzontiformes. AULA-Verlag Wiesbaden.

Hardisty, M. W. 1986b. *Lampetra planeri* (Bloch, 1784). Pages 279–304 *in* J. Holčík, editor. The freshwater fishes of Europe, volume 1, part 1: Petromyzontiformes. AULA-Verlag, Wiesbaden, Germany.

Hardisty, M. W. 2006. Lampreys: life without jaws. Forrest Text, Ceredigion, UK.

Hardisty, M. W., and I. C. Potter. 1971. Paired species. Pages 249–277 *in* M. W. Hardisty and I. C. Potter, editors. The biology of lampreys, volume 1. Academic Press, New York.

Hart, M. W., and R. D. Podolsky. 2005. Mitochondrial DNA phylogeny and rates of larval evolution in *Macrophiothrix* brittlestars. Molecular Phylogenetics and Evolution 34:438–447.

Hatfield, T. 1997. Genetic divergence in adaptive characters between sympatric species of stickleback. American Naturalist 149:1009–1029.

Hatfield, T., and D. Schluter. 1996. A test for sexual selection on hybrids of two sympatric sticklebacks. Evolution 50:2429–2434.

Hatfield, T., and D. Schluter. 1999. Ecological speciation in sticklebacks: environment-dependent hybrid fitness. Evolution 53:866–873.

Hatta, S. 1901. On the lampreys of Japan together with

notes on a specimen of lamprey from Siberia. Annotationes Zoologicae Japonenses 4:21–29.

Heard, W. R. 1966. Observations on lampreys in the Naknek River system of southwest Alaska. Copeia 1966:332–339.

Heath, D. D., C. M. Bettles, S. Jamieson, I. Stasiak, and M. F. Docker. 2008. Genetic differentiation among sympatric migratory and resident life history forms of rainbow trout in British Columbia. Transactions of the American Fisheries Society 137:1268–1278.

Heath, D. D., J. W. Heath, and G. K. Iwama. 1991. Maturation in chinook salmon, *Oncorhynchus tshawytscha* (Walbaum): early identification based on the development of a bimodal weight-frequency distribution. Journal of Fish Biology 39:565–575.

Hebert, P. D. N., M. Y. Stoeckle, T. S. Zemlak, and C. M. Francis. 2004. Identification of birds through DNA barcodes. Public Library of Science Biology 2:1657–1663.

Hilliard, R. W., D. J. Bird, and I. C. Potter. 1983. Metamorphic changes in the intestine of three species of lampreys. Journal of Morphology 176:181–196.

Hoff, J. G. 1988. Some aspects of the ecology of the American brook lamprey, *Lampetra appendix*, in the Mashpee River, Cape Cod, Massachusetts. Canadian Field-Naturalist 102:735–737.

Hoffman, E. A., and D. W. Pfennig. 1999. Proximate causes of cannibalistic polyphenism in larval tiger salamanders. Ecology 80:1076–1080.

Holčík, J. 1986a. *Lethenteron japonicum* (Martens, 1868). Pages 198–219 *in* J. Holčík, editor. The freshwater fishes of Europe, volume 1, part 1: Petromyzontiformes. AULA-Verlag, Wiesbaden, Germany.

Holčík, J. 1986b. *Lethenteron kessleri* (Anikin, 1905). Pages 200–236 *in* J. Holčík, editor. The freshwater fishes of Europe, volume 1, part 1: Petromyzontiformes. AULA-Verlag, Wiesbaden, Germany.

Holčík, J., and C. B. Renaud. 1986. *Eudontomyzon mariae* (Berg, 1931). Pages 165–185 *in* J. Holčík, editor. The freshwater fishes of Europe, volume 1, part 1: Petromyzontiformes. AULA-Verlag, Wiesbaden, Germany.

Holčík, J., and V. Šorić. 2004. Redescription of *Eudontomyzon stankokaramani* (Petromyzontes, Petromyzontidae)—a little known lamprey from the Drin River drainage, Adriatic Sea basin. Folia Zoologica 53:399–410.

Houston, J. 1991. Status of the darktail lamprey, *Lethenteron alaskense*, in Canada. Canadian Field-Naturalist 105:157–160.

Hubbs, C. L. 1925. The life cycle and growth of lampreys. Papers of the Michigan Academy of Science 4:587–603.

Hubbs, C. L. 1971. *Lampetra* (*Entosphenus*) *lethophaga, a* new species, the nonparasitic derivative of the Pacific lamprey. Transactions of the San Diego Society of Natural History 16:125–164.

Hubbs, C. L., and I. C. Potter. 1971. Distribution, phylogeny and taxonomy. Pages 1–65 *in* M. W. Hardisty and I. C. Potter, editors. The biology of lampreys, volume 1. Academic Press, New York.

Hubbs, C. L., and M. B. Trautman. 1937. A revision of the lamprey genus *Ichthyomyzon*. Miscellaneous Publications Museum of Zoology University of Michigan 35:7–109.

Hubert, N., R. Hanner, E. Holm, N. E. Mandrak, E. Taylor, M. Burridge, D. Watkinson, P. Dumont, A. Curry, P. Bentzen, J. Zhang, J. April, and L. Bernatchez. 2008. Identifying Canadian freshwater fishes through DNA barcodes. PLoS ONE 3:e2490. doi:10.1371/journal.pone.0002490.

Huggins, R. J., and A. Thompson. 1970. Communal spawning of brook and river lampreys, *Lampetra planeri* Bloch and *Lampetra fluviatilis*. Journal of Fish Biology 2:53–54.

Hughes, R. L., and I. C. Potter. 1969. Studies on gametogenesis and fecundity in the lampreys *Mordacia praecox* and *M. mordax* (Petromyzonidae). Australian Journal of Zoology 17:447–464.

Igoe, F., D. T. G. Quigley, F. Marnell, E. Meskell, W. O'Connor, and C. Byrne. 2004. The sea lamprey *Petromyzon marinus* (L.), river lamprey *Lampetra fluviatilis* (L.) and brook lamprey *Lampetra planeri* (Bloch) in Ireland: general biology, ecology, distribution and status with recommendations for conservation. Biology and Environment: Proceedings of the Royal Irish Academy 104B:43–56.

Inger, R. F. 1992. A bimodal feeding system in a stream-dwelling larva of *Rhacophorus* from Borneo. Copeia 1992:887–890.

Iwata, A., A. Gotto, and K. Hamada. 1985. A review of the Siberian lamprey, *Lethenteron kessleri* in Hokkaido. Bulletin of the Faculty of Fisheries Hokkaido University 36:182–190.

Jordan, D. S. 1885. A catalogue of the fishes known to inhabit the waters of North America, north of the Tropic of Cancer, with notes on species discovered in 1883 and 1884. Report of the United States Fish Commission 13:789–973, Washington, D.C.

Joss, J. M. P. 1985. Pituitary control of metamorphosis in the Southern Hemisphere lamprey, *Geotria australis*. General and Comparative Endocrinology 60:58–62.

Karaman, M. S. 1974. *Eudontomyson vladykovi stankokaramani* n. ssp. a new subspecies of lamprey from tributaries of the Ohrid-Drim-Skadar system in west Balkan peninsula. Folia Balcanica 3:1–10.

Kimura, M. 1980. A simple method for estimating evolutionary rate of base substitutions through comparative studies of nucleotide sequences. Journal of Molecular Evolution 16:111–120.

Kott, E. 1971. Characteristics of pre-spawning American brook lampreys from Big Creek, Ontario. Canadian Field-Naturalist 85:235–240.

Kott, E. 1974. A morphometric and meristic study of a population of the American brook lamprey, *Lethenteron lamottei* (LeSueur), from Ontario. Canadian Journal of Zoology 52:1047–1055.

Kottelat, M. 1997. European freshwater fishes. Biologia 52 (Supplement 5):1–271.

Kucheryavyi, A. V., K. A. Savvaitova, M. A. Gruzdeva, and D. S. Pavlov. 2007a. Sexual dimorphism and some special traits of spawning behavior of the Arctic lamprey *Lethenteron camtschaticum*. Journal of Ichthyology 47:481–485.

Kucheryavyi, A. V., K. A. Savvaitova, D. S. Pavlov, M. A. Gruzdeva, K. V. Kuzishchin, and J. A. Stanford. 2007b. Variations of life history strategy of the Arctic lamprey *Lethenteron camtschaticum* from the Utkholok River (Western Kamchatka). Journal of Ichthyology 47:37–52.

Kux, Z. 1965. *Lampetra gracilis*, a new nonparasitic species of lamprey from eastern Slovakia. Gasopis Moravského Musea 50:293–302. [In Czech]

Kux, Z. 1967. A contribution to the bionomy of lampreys (Petromyzonidae) in the Tisza region of eastern Slovakia. Gasopis Moravského Musea 52:209–216. [In Czech]

Lang, N. J., K. J. Roe, C. B. Renaud, H. S. Gill, I. C. Potter, J. Freyhof, A. M. Naseka, P. Cochran, H. Espinosa Pérez, E. M. Habit, B. R. Kuhajda, D. A. Neely, Y. S. Reshetnikov, V. B. Salnikov, M. Th. Stoumboudi, and R. L. Mayden. 2009. Novel relationships among lampreys (Petromyzontiformes) revealed by a taxonomically comprehensive molecular data set. Pages 41–55 *in* L. R. Brown, S. D. Chase, M. G. Mesa, R. J. Beamish, and P. B. Moyle, editors. Biology, management, and conservation of lampreys in North America. American Fisheries Society, Symposium 72, Bethesda, Maryland.

Lanteigne, J. 1988. Identification of lamprey larvae of the genus *Ichthyomyzon* (Petromyzontidae). Environmental Biology of Fishes 23:55–63.

Larsen, L. O. 1980. Physiology of adult lampreys, with special regard to natural starvation, reproduction, and death after spawning. Canadian Journal of Fisheries and Aquatic Sciences 37:1762–1779.

Li, W., M. J. Siefkes, A. P. Scott, and J. H. Teeter. 2003. Sex pheromone communication in the sea lamprey: implications for integrated management. Journal of Great Lakes Research 29 (Supplement 1):85–94.

Linnaeus, C. 1758. Systema naturae, 10th edition, volume 1. Laurentii Salvii, Stockholm, Sweden.

Loman, J. C. C. 1912. Über die Naturgeschichte des Bachneunauges *Lampetra planeri*. Zoologisches Jahrbuch (Supplement) 15:243–270.

Lorion, C. M., D. F. Markle, S. B. Reid, and M. F. Docker. 2000. Redescription of the presumed-extinct Miller Lake lamprey, *Lampetra minima*. Copeia 2000:1019–1028.

Lyons, J., P. A. Cochran, and M. E. Sneen. 1997. Taxonomic status and distribution of the lamprey *Ichthyomyzon* cf. *gagei*. American Midland Naturalist 138:69–76.

Maeva, E., I. Bruno, B. Zielinski, M. F. Docker, F. M. Severin, and R. Maev. 2004. The use of pulse-echo acoustic microscopy to non-invasively determine the sex of living larval sea lampreys. Journal of Fish Biology 65:148–156.

Malmqvist, B. 1978. Population structure and biometry of *Lampetra planeri* (Bloch) from three different watersheds in South Sweden. Archiv für Hydrobiologie 84:65–86.

Malmqvist, B. 1980. The spawning migration of the brook lamprey, *Lampetra planeri* Bloch, in a south Swedish stream. Journal of Fish Biology 16:105–114.

Malmqvist, B. 1983. Breeding behaviour of brook lampreys *Lampetra planeri*: experiments on mate choice. Oikos 41:43–48.

Manion, P. J., and L. H. Hanson. 1980. Spawning behavior and fecundity of lampreys from the upper three Great Lakes. Canadian Journal of Fisheries and Aquatic Sciences 37:1635–1640.

Manion, P. J., and H. A. Purvis. 1971. Giant American brook lampreys, *Lampetra lamottei*, in the upper Great Lakes. Journal of the Fisheries Research Board of Canada 28:616–620.

Manion, P. J., and B. R. Smith. 1978. Biology of larval and metamorphosing sea lampreys, *Petromyzon marinus*, of the 1960 year class in the Big Garlic River, Michigan, part II, 1966–72. Great Lakes Fishery Commission Technical Report 30:1–35.

Mayden, R. L. 1997. A hierarchy of species concepts: the denouement in the saga of the species problem. Pages 381–424 *in* H. A. Claridge, H. A. Dawah, and M. R. Wilson, editors. Species: the units of biodiversity. Chapman and Hall, New York.

Mayr, E. 1963. Animal species and evolution. Harvard University Press, Cambridge, Massachusetts.

McFarlane, C. T. 2009. Testing the phylogenetic and biological species concepts in the paired lamprey species, *Ichthyomyzon unicuspis* and *I. fossor*. Honors thesis. University of Manitoba, Winnipeg, Manitoba, Canada.

McPhail, J. D., and C. C. Lindsey. 1970. Freshwater fishes of northwestern Canada and Alaska. Fisheries Research Board of Canada Bulletin 173.

Medland, T. E., and F. W. H. Beamish. 1987. Age validation for the mountain brook lamprey, *Ichthyomyzon greeleyi*. Canadian Journal of Fisheries and Aquatic Sciences 44:901–904.

Meeuwig, M. H., J. M. Bayer, and R. A. Reiche. 2006. Morphometric discrimination of early life stage *Lampetra tridentata* and *L. richardsoni* (Petromyzonidae) from the Columbia River basin. Journal of Morphology 267:623–633.

Montpetit, C. J., V. Chatalov, J. Yuk, I. Rasaratnam, and J. H. Youson. 2005. Expression of neuropeptide Y family peptides in the brain and gut during stages of the life cycle of a parasitic lamprey (*Petromyzon marinus*) and a nonparasitic lamprey (*Ichthyomyzon gagei*). Annals of the New York Academy of Sciences 1040:140–149.

Morkert, S. B., W. D. Swink, and J. G. Seelye. 1998. Evidence for early metamorphosis of sea lampreys in the Chippewa River, Michigan. North American Journal of Fisheries Management 18:966–971.

Morman, R. H. 1979. Distribution and ecology of lampreys in the lower peninsula of Michigan, 1957–75. Great Lakes Fishery Commission Technical Report 33:1–59.

Morris, R. 1972. Osmoregulation. Pages 193–239 *in* M. W. Hardisty and I. C. Potter, editors. The biology of lampreys, volume 2. Academic Press, New York.

Mundahl, N. D., and R. A. Sagan. 2005. Spawning ecology of the American brook lamprey, *Lampetra appendix*. Environmental Biology of Fishes 73:283–292.

Mundahl, N. D., G. Sayeed, S. Taubel, C. Erickson, A. Zalatel, and J. Cousins. 2006. Densities and habitat of American brook lamprey (*Lampetra appendix*) larvae in Minnesota. American Midland Naturalist 156:11–22.

Narum, S. R., C. Contor, A. Talbot, and M. S. Powell. 2004. Genetic divergence of sympatric resident and anadromous forms of *Oncorhynchus mykiss* in the Walla Walla River, USA. Journal of Fish Biology 65:471–488.

Neave, F. B., N. E. Mandrak, M. F. Docker, and D. L. Noakes. 2007. An attempt to differentiate sympatric *Ichthyomyzon* ammocoetes using meristic, morphological, pigmentation, and gonad analyses. Canadian Journal of Zoology 85:549–560.

Nelson, J. S., E. J. Crossman, H. Espinosa-Pérez, L. T. Findley, C. R. Gilbert, R. N. Lea, and J. D. Williams. 2004. Common and scientific names of fishes from the United States, Canada, and Mexico, 6th edition. American Fisheries Society, Special Publication 29, Bethesda, Maryland.

Oliva, O., and C. Zanandrea. 1959. Posizione sistematica di una lampreda de Cilistovo (Cecoslavacchia). Doriana, Annali del Museo Civico di Storia Naturale "G. Doria" 98 (Supplement 2):1–5.

Olsen, J. B., K. Wuttig, D. Fleming, E. J. Kretschmer, and J. K. Wenburg. 2006. Evidence of partial anadromy and resident-form dispersal bias on a fine scale in populations of *Oncorhynchus mykiss*. Conservation Genetics 7:613–619.

Page, L. M., and B. M. Burr. 1991. A field guide to freshwater fishes of North America north of Mexico. The Peterson Field Guide Series, Houghton-Mifflin, Boston.

Pascual, M., P. Bentzen, C. Riva Rossi, G. Mackey, M. T. Kinnison, and R. Walker. 2001. First documented case of anadromy in a population of introduced rainbow trout in Patagonia, Argentina. Transactions of the American Fisheries Society 130:53–67.

Pestano, J., R. P. Brown, N. M. Suárez, and M. Báez. 2003. Diversification of sympatric *Sapromyza* (Diptera: Lauxaniidae) from Madeira: six morphological species but only four mtDNA lineages. Molecular Phylogenetics and Evolution 27:422–428.

Pfennig, D. W. 1992. Proximate and functional

causes of polyphenism in an anuran tadpole. Functional Ecology 6:167–174.

Piavis, G. W., J. H. Howell, and A. J. Smith. 1970. Experimental hybridization among five species of lampreys from the Great Lakes. Copeia 1970:29–37.

Pickering, A. D. 1976. Stimulation of intestinal degeneration by oestradiol and testosterone implantation in the migrating river lamprey, *Lampetra fluviatilis* L. General and Comparative Endocrinology 30:340–346.

Podjasek, J. O., L. M. Bosnjak, D. J. Brooker, and E. B. Mondor. 2005. Alarm pheromone induces a transgenerational wing polyphenism in the pea aphid, *Acyrthosiphon pisum*. Canadian Journal of Zoology 83:1138–1141.

Potter, I. C. 1968. *Mordacia praecox*, n. sp., a nonparasitic lamprey (Petromyzonidae), from New South Wales, Australia. Proceedings of the Linnean Society of New South Wales 92:254–261.

Potter, I. C. 1980. The Petromyzoniformes with particular reference to paired species. Canadian Journal of Fisheries and Aquatic Sciences 37:1595–1615.

Potter, I. C., and I. D. Brown. 1975. Changes in haemoglobin electropherograms during the life cycle of two closely related lampreys. Comparative Biochemistry and Physiology 51B:517–519.

Potter, I. C., and P. I. Nicol. 1968. Electrophoretic studies on the haemoglobins of Australian lampreys. Australian Journal of Experimental Biology and Medical Science 46:639–641.

Potter, I. C., and T. S. Osborne. 1975. Systematics of British larval lampreys. Journal of Zoology 176:311–329.

Potter, I. C., E. S. Robinson, and S. N. Walton. 1968. The mitotic chromosomes of *Mordacia mordax* (Agnatha: Petromyzonidae). Experientia 24:966–967.

Purvis, H. A. 1970. Growth, age at metamorphosis, and sex ratio of northern brook lamprey in a tributary of southern Lake Superior. Copeia 1970:326–332.

Purvis, H. A. 1980. Effects of temperature on metamorphosis and the age and length at metamorphosis in sea lamprey (*Petromyzon marinus*) in the Great Lakes. Canadian Journal of Fisheries and Aquatic Sciences 37:1827–1834.

Raney, E. C. 1952. A new lamprey, *Ichthyomyzon hubbsi*, from the upper Tennessee River system. Copeia 1952:93–99.

Regan, C. T. 1911. A synopsis of the marsipobranchs of the order Hyperoartii. Annals and Magazine of Natural History 7:193–204.

Reighard, J., and H. Cummins. 1916. Description of a new species of lamprey of the genus *Ichthyomyzon*. Occasional Papers of the Museum of Zoology University of Michigan 31:1–12.

Reitzel, A. M., C. M. Miles, A. Heyland, J. D. Cowart, and L. R. McEdward. 2005. The contribution of the facultative feeding period to echinoid larval development and size at metamorphosis: a comparative approach. Journal of Experimental Marine Biology and Ecology 317:189–201.

Renaud, C. B. 1982. Revision of the lamprey genus *Eudontomyzon* Regan, 1911. Master's thesis. University of Ottawa, Ottawa.

Renaud, C. B. 1997. Conservation status of Northern Hemisphere lampreys (Petromyzontidae). Journal of Applied Ichthyology 13:143–148.

Renaud, C. B., M. F. Docker, and N. E. Mandrak. 2009a. Taxonomy, distribution, and conservation of lampreys in Canada. Pages 293–309 *in* L. R. Brown, S. D. Chase, M. G. Mesa, R. J. Beamish, and P. B. Moyle, editors. Biology, management, and conservation of lampreys in North America. American Fisheries Society, Symposium 72, Bethesda, Maryland.

Renaud, C. B., H. S. Gill, and I. C. Potter. 2009b. Relationships between the diets and characteristics of the dentition, buccal glands and velar tentacles of the adults of the parasitic species of lamprey. Journal of Zoology 278:231–242.

Renaud, C. B., and J. Holčík. 1986. *Eudontomyzon danfordi* Regan, 1911. Pages 146–164 *in* J. Holčík, editor. The freshwater fishes of Europe, volume 1, part 1: Petromyzontiformes. AULA-Verlag, Wiesbaden, Germany.

Renaud, C. B., and J. Holčík. 1988. *Lampetra* (*Eudontomyzon*) *gracilis*, a synonym of *Eudontomyzon danfordi*. Environmental Biology of Fishes 23:127–130.

Richards, J. E., R. J. Beamish, and F. W. H. Beamish. 1982. Descriptions and keys for ammocoetes of lampreys from British Columbia, Canada. Canadian Journal of Fisheries and Aquatic Sciences 39:1484–1495.

Richardson, J. 1836. Fauna Boreali-Americana; or the zoology of the northern parts of British America: containing descriptions of the objects of natural history collected on the late northern land expeditions under command of Captain Sir John Franklin, R.N, part 3: the fish. J. Murray, London.

Richardson, J. 1846. The zoology of the voyage *H.M.S.* Erebus and Terror under the command of Captain Sir James Clark Ross, R.N., F.R.S., during the years 1839–43, volume 2: reptiles, fishes, crustacea, insects, mollusca. E.W. Janson, London.

Riva-Rossi, C., M. A. Pascual, J. A. Babaluk, M. Garcia-Asorey, and N. M. Haldenk. 2007. Intrapopulation variation in anadromy and reproductive life span in rainbow trout introduced in the Santa Cruz River, Argentina. Journal of Fish Biology 70:1780–1797.

Robins, C. R., R. M. Bailey, C. E. Bond, J. R. Brooker, E. A. Lachner, R. N. Lea, and W. B. Scott. 1980. A list of common and scientific names of fishes from the United States and Canada, 4th edition. American Fisheries Society, Special Publication 12, Bethesda, Maryland.

Robins, C. R., R. M. Bailey, C. E. Bond, J. R. Brooker, E. A. Lachner, R. N. Lea, and W. B. Scott. 1991. A list of common and scientific names of fishes from the United States and Canada, 5th edition. American Fisheries Society, Special Publication 20, Bethesda, Maryland.

Robinson, E. S., and I. C. Potter. 1969. Meiotic chromosomes of *Mordacia praecox* and a discussion of chromosome numbers in lampreys. Copeia 1969:824–828.

Robinson, E. S., I. C. Potter, and C. J. Webb. 1974. Homogeneity of holarctic lamprey karyotypes. Caryologia 27:443–454.

Rohde, F. C., and R. G. Arndt. 1975. Records of the freshwater lampreys, *Lampetra lamottenii* and *Okkelbergia aepyptera*, from the Delmarva Peninsula (East Coast, United States). Chesapeake Science 16:70–72.

Salewski, V. 2003. Satellite species in lampreys: a worldwide trend for ecological speciation in sympatry? Journal of Fish Biology 63:267–279.

Savvaitova, K. A., and V. A. Maksimov. 1979. The spawning of the Arctic lamprey of the genus *Lampetra* in connection with the problem of the taxonomic status of small forms. Journal of Ichthyology 18:555–560.

Schreiber, A., and R. Engelhorn. 1998. Population genetics of a cyclostome species pair, river lamprey (*Lampetra fluviatilis* L.) and brook lamprey (*Lampetra planeri* Bloch). Journal of Zoological Systematics and Evolutionary Research 36:85–99.

Schuldt, R. J., and R. Goold. 1980. Changes in the distribution of native lampreys in Lake Superior tributaries in response to sea lamprey (*Petromyzon marinus*) control, 1953–1977. Canadian Journal of Fisheries and Aquatic Sciences 37:1872–1885.

Scott, W. B., and E. J. Crossman. 1973. Freshwater Fishes of Canada. Fisheries Research Board of Canada Bulletin 184.

Seagle, H. H., and J. W. Nagel. 1982. Life cycle and fecundity of the American brook lamprey, *Lampetra appendix*, in Tennessee. Copeia 1982:362–366.

Selkoe, K. A., and R. J. Toonen. 2006. Microsatellites for ecologists: a practical guide to using and evaluating microsatellite markers. Ecology Letters 9:615–629.

Shaffer, H. B., and S. R. Voss. 1996. Phylogenetic and mechanistic analysis of a developmentally integrated character complex: alternate life history modes in ambystomatid salamanders. American Zoologist 36:24–35.

Silver, M. R., H. Kawauchi, M. Nozaki, and S. A. Sower. 2004. Cloning and analysis of the lamprey GnRH-III cDNA from eight species of lamprey representing the three families of Petromyzoniformes. General and Comparative Endocrinology 139:85–94.

Sower, S. A. 2003. The endocrinology of reproduction in lampreys and applications for male lamprey sterilization. Journal of Great Lakes Research 29 (Supplement 1):50–65.

Sower, S. A., and H. Kawauchi. 2001. Update: brain and pituitary hormones of lampreys. Comparative Biochemistry and Physiology 129B:291–302.

Sower, S. A., S. Moriyama, M. Kasahara, A. Takahashi, M. Nozaki, K. Uchida, J. M. Dahstrom, and H. Kawauchi. 2006. Identification of sea lamprey GTH beta-like cDNA and its evolutionary implications. General and Comparative Endocrinology 148:22–32.

Strathmann, R. R., L. Fenaux, and M. F. Strathmann. 1992. Heterochronic developmental plasticity in larval sea urchins and its implications for evolution of nonfeeding larvae. Evolution 46:972–986.

Taborsky, M. 1998. Sperm competition in fish: 'bourgeois' males and parasitic spawning. Trends in Ecology and Evolution 13:222–227.

Tagliavini, J., R. Tizzi, F. Conterio, P. Mariottini, and G. Gandolfi. 1994. Mitochondrial DNA sequences in three genera of Italian lampreys. Bolletino Zoologica 61:331–333.

Takahashi, M. K., and M. J. Parris. 2008. Life cycle polyphenism as a factor affecting ecological divergence within *Notophthalmus viridescens*. Oecologia 158:23–34.

Taylor, E. B. 1999. Species pairs of north temperate freshwater fishes: evolution, taxonomy, and conservation. Reviews in Fish Biology and Fisheries 9:299–324.

Taylor, E. B., J. W. Boughman, M. Groenenboom, M. Sniatynski, D. Schluter, and J. L. Gow. 2006. Speciation in reverse: morphological and genetic evidence of the collapse of a three-spined stickleback (*Gasterosteus aculeatus*) species pair. Molecular Ecology 15:343–355.

Taylor, E. B., C. J. Foote, and C. C. Wood. 1996. Molecular genetic evidence for parallel life-history evolution within a Pacific salmon (sockeye salmon and kokanee, *Oncorhynchus nerka*). Evolution 50:401–416.

Thrower, F., C. Guthrie, J. Nielsen, and J. Joyce. 2004. A comparison of genetic variation between an anadromous steelhead, *Oncorhynchus mykiss*, population and seven derived populations sequestered in freshwater for 70 years. Environmental Biology of Fishes 69:111–125.

Tilesius von Tilenau, W.G. 1811. Piscium camtschaticorum descriptions et icones. Mémoires de l'Académie Impériale des Sciences de St. Pétersbourg 3:225–285.

Vladykov, V. D. 1951. Fecundity of Quebec lampreys. Canadian Fish Culturist 10:1–14.

Vladykov, V. D. 1955. *Lampetra zanandreai*, a new species of lamprey from northern Italy. Copeia 1955:215–223.

Vladykov, V. D. 1973. *Lampetra pacifica*, a new nonparasitic species of lamprey (Petromyzonidae) from Oregon and California. Journal of the Fisheries Research Board of Canada 30:205–213.

Vladykov, V. D., and W. I. Follett. 1958. Redescription of *Lampetra ayresii* (Günther) of western North America, a species of lamprey (Petromyzontidae) distinct from *Lampetra fluviatilis* (Linnaeus) of Europe. Journal of the Fisheries Research Board of Canada 15:47–77.

Vladykov, V. D., and W. I. Follett. 1965. *Lampetra richardsoni*, a new nonparasitic species of lamprey (Petromyzonidae) from western North America. Journal of the Fisheries Research Board of Canada 22:139–158.

Vladykov, V. D., and E. Kott. 1976a. A new nonparasitic species of lamprey of the genus *Entosphenus* Gill, 1862 (Petromyzonidae) from south central California. Bulletin of the Southern California Academy of Science 75:60–67.

Vladykov, V. D., and E. Kott. 1976b. A second nonparasitic species of *Entosphenus* Gill, 1862 (Petromyzonidae) from Klamath River system, California. Canadian Journal of Zoology 54:974–989.

Vladykov, V. D., and E. Kott. 1976c. The taxonomic significance of velar tentacles in Holarctic lampreys (Petromyzonidae). Revue des Travaux de L'Institut des Peches Maritimes 40:787–789.

Vladykov, V. D., and E. Kott. 1978a. A new nonparasitic species of the holarctic lamprey genus *Lethenteron* Creaser and Hubbs, 1922, (Petromyzontidae) from Hokkaido, Japan. Canadian Journal of Zoology 56:1792–1800.

Vladykov, V. D., and E. Kott. 1978b. A new nonparasitic species of the holarctic lamprey genus *Lethenteron* Creaser and Hubbs, 1922, (Petromyzontidae) from northwestern North America with notes on other species of the same genus. Biological Papers of the University of Alaska 19:1–74.

Vladykov, V. D., and E. Kott. 1979. Satellite species among the holarctic lampreys (Petromyzonidae). Canadian Journal of Zoology 57:860–867.

Vladykov, V. D., and E. Kott. 1980. Description and key to metamorphosed specimens and ammocoetes of Petromyzonidae found in the Great Lakes region. Canadian Journal of Fisheries and Aquatic Sciences 37:1616–1625.

Vladykov, V. D., C. B. Renaud, E. Kott, and P. S. Economidis. 1982. A new nonparasitic species of Holarctic lamprey, genus *Eudontomyzon* Regan 1911 (Petromyzontidae), from Greece. Canadian Journal of Zoology 60:2897–2915.

Voss, S. R. 1995. Genetic basis of paedomorphosis in the axolotl, *Ambystoma mexicanum*: a test of the single-gene hypothesis. Journal of Heredity 86:441–447.

Voss, S. R., and J. J. Smith. 2005. Evolution of salamander life cycles: a major-effect quantitative trait locus contributes to discrete and continuous variation for metamorphic timing. Genetics 170:275–281.

Waldman, J. R., C. Grunwald, N. K. Roy, and I. I. Wirgin. 2004. Mitochondrial DNA analysis indicates sea lampreys are indigenous to Lake Ontario. Transactions of the American Fisheries Society 133:950–960.

Ward, R. D. 2009. DNA barcode divergence among species and genera of birds and fishes. Molecular Ecology Resources 9:1077–1085.

Ward, R. D., R. Hanner, and P. D. N. Hebert. 2009. The campaign to DNA barcode all fishes, FISH-BOL. Journal of Fish Biology 74:329–356.

Yamazaki, Y., and A. Goto. 1997. Morphometric and meristic characteristics of two groups of *Lethenteron reissneri*. Ichthyological Research 44:15–25.

Yamazaki, Y., and A. Goto. 1998. Genetic structure and differentiation of four *Lethenteron* taxa from the Far East, deduced from allozyme analysis. Environmental Biology of Fishes 52:149–161.

Yamazaki, Y., and A. Goto. 2000. Breeding season and nesting assemblages in two forms of *Lethenteron reissneri*, with reference to reproductive isolating mechanisms. Ichthyological Research 47:271–276.

Yamazaki, Y., H. Sugiyama, and A. Goto. 1998. Mature dwarf males and females of the arctic lamprey, *Lethenteron japonicum*. Ichthyological Research 45:404–408.

Yamazaki, Y., R. Yokoyama, M. Nishida, and A. Goto. 2006. Taxonomy and molecular phylogeny of *Lethenteron* lampreys in eastern Eurasia. Journal of Fish Biology 68 (Supplement B):251–269.

Youson, J. H. 1980. Morphology and physiology of lamprey metamorphosis. Canadian Journal of Fisheries and Aquatic Sciences 37:1687–1710.

Youson, J. H. 2003. The biology of metamorphosis in sea lampreys: endocrine, environmental, and physiological cues and events, and their potential application to lamprey control. Journal of Great Lakes Research 29 (Supplement 1):26–49.

Youson, J. H., and R. J. Beamish. 1991. Comparison of the internal morphology of adults of a population of lampreys that contains a nonparasitic life-history type, *Lampetra richardsoni*, and a potentially parasitic form, *L. richardsoni* var. *marifuga*. Canadian Journal of Zoology 69:628–637.

Youson, J. H., M. Docker, and S. A. Sower. 1995. Concentration of gonadotropin-releasing hormones in brain of larval and metamorphosing lampreys of two species with different adult life histories. Page 83 *in* F. W. Goetz and P. Thomas, editors. Proceedings of the 5th International Symposium on the Reproductive Physiology of Fish. University of Texas Press, Austin.

Youson, J. H., J. A. Heinig, S. F. Khanam, S. A. Sower, H. Kawauchi, and F. W. Keeley. 2006. Patterns of proopiomelanotropin and prooplocortin gene expression and of immunohistochemistry for gonadotropin-releasing hormones (1GnRH-I and III) during the life cycle of a nonparasitic lamprey: relationship to this adult life history type. General and Comparative Endocrinology 148:54–71.

Zanandrea, G. 1959. Speciation among lampreys. Nature (London) 184:360.

Zimmerman, C. E., and G. H. Reeves. 2002. Identification of steelhead and resident rainbow trout progeny in the Deschutes River, Oregon, revealed with otolith microchemistry. Transactions of the American Fisheries Society 131:986–993.

American Fisheries Society Symposium 72:115–124, 2009

# Passage Considerations for Anadromous Lampreys

Mary L. Moser*
*NOAA Fisheries, Northwest Fisheries Science Center*
*2725 Montlake Boulevard East, Seattle, Washington 98112, USA*

Matthew G. Mesa
*U.S. Geological Survey, Columbia River Research Laboratory*
*5501 Cook-Underwood Road, Cook, Washington 98605, USA*

*Abstract.*—Both juvenile and adult anadromous lampreys encounter a variety of obstacles to passage during migration. Hydropower dams can delay or obstruct adults, and turbine entrainment or screen impingement can injure or kill juveniles. Lampreys also face less dramatic obstacles such as culverts, irrigation diversion dams and screens, weirs, and other low-elevation structures. The extent to which most structures affect juvenile and adult movements is not known. However, recent research on lamprey swimming performance and migration behavior has provided new insight into the physical and performance factors that may limit lamprey movements. We drew from both field and laboratory studies to review lamprey swimming performance and migration behavior and the effects of water velocity, attachment sites, light, and temperature on lamprey movements. Based on this review, we identified the following as primary research needs for all lamprey life stages: determination of cues lampreys use to orient during migration, documentation of lamprey endurance limits, and evaluation of energy expenditure during migration and its effects on lamprey fitness.

## Introduction

With declines in anadromous lamprey populations throughout their native range in coastal North America, conservation of these species has become of increasing importance to fisheries managers. This is particularly true on the west coast of the United States where anadromous lampreys are of cultural importance to Native Americans and still support tribal fisheries (Close et al. 2002). Concerns for conservation of lampreys in the Pacific Northwest resulted in a petition to list the anadromous Pacific lamprey *Entosphenus tridentatus* (formerly *Lampetra tridentata*) and river lamprey *L. ayresii* under the Endangered Species Act. Now, with no recovery in sight, calls for action to reverse downward trends in lamprey abundance have become increasingly urgent.

As is the case for other anadromous species of concern, eliminating obstacles to passage for both adult and juvenile life stages of lamprey is a formidable recovery goal. This is particularly true in the Columbia and Snake River basins where lampreys must pass up to nine main-stem hydropower dams and many additional low-elevation structures in order to reach spawning areas. Removing impediments to anadromous lamprey passage was therefore identified as the highest priority recovery objective for lampreys in the Columbia River basin (www.fws.gov/columbiariver/lampreywg/docs/CritUncertFinal.pdf). However, research on lamprey migration performance and behavior has lagged behind

* Corresponding author: mary.moser@noaa.gov

the need to establish passage criteria for all life stages of these fish.

Lampreys are anguilliform swimmers and therefore differ from most teleosts in both swimming capability and behavior. Lampreys even show differences in swimming performance when compared to catadromous anguillid eels because adult lampreys are able to use their oral disk to attach to surfaces and do not have pectoral fins. Because of these differences, it is important to consider the abilities and limitations of lampreys when designing fishways, weirs, dams, and culverts that might be used by anadromous or smaller-bodied adfluvial forms. We reviewed the available literature and our own recent work on swimming and migration behavior of anadromous lampreys to describe some of the problems these fish face during passage, potential solutions, and needs for future research.

In general, anadromous lampreys spawn and rear in freshwater streams. Larval lampreys (ammocoetes) burrow into stream sediments soon after hatching and are largely sedentary for the next 4–6 years (Hardisty and Potter 1971a; Beamish and Levings 1991). Although ammocoetes can make short upstream excursions, they move downstream more often and over greater distances (Quintella et al. 2005). After this period of freshwater residence, juvenile lampreys metamorphose (becoming macrophthalmia) and migrate to the sea. The parasitic lampreys remain in the marine environment for 1–3 years and then embark on a free-swimming migration to freshwater spawning habitats (Hardisty and Potter 1971b; Beamish 1980). In the Columbia River drainage, adult Pacific lampreys may migrate over 700 km to reach tributary streams in the upper Columbia and Snake rivers.

During juvenile and adult freshwater migrations, lampreys may encounter a variety of obstacles to passage, such as large hydropower dams. In the Columbia River drainage, adult Pacific lamprey passage success through fishways at main-stem dams is poor relative to salmonids (Moser et al. 2002a). Moreover, while adult Pacific lampreys normally are capable of moving at rates of 11–20 km/d (Robinson and Bayer 2005), they are significantly delayed during passage at dams (Moser et al. 2005). This may reduce lamprey fitness by causing excessive use of energetic reserves, exposure to suboptimal temperature, and/or direct mortality by predators at the base of dams. Large hydropower dams also can injure or kill migrating juvenile lampreys as a result of turbine entrainment or screen impingement (Moursund et al. 2003; Dauble et al. 2006).

Less dramatic obstacles to lamprey passage may occur along their migration routes. These include, but are not limited to, culverts, irrigation diversion dams, screens, weirs, and other low-head structures (Kostow 2002). For example, sea lampreys Petromyzon marinus in the River Modego, Portugal, were delayed by block weirs of less than 1 m in height and these fish expended a great deal of energy surmounting these obstacles (Almeida et al. 2002, 2005). Similarly, adult Pacific lampreys exhibit poor passage success at low-elevation irrigation diversion dams in the Umatilla River, Oregon (A. Jackson, Confederated Tribes of the Umatilla Indian Reservation, unpublished data).

Research on lamprey swimming performance and migration behavior at barriers has provided insight into physical factors that can limit lamprey movements. In this paper, we drew from field and laboratory studies to identify factors that should be considered when designing or modifying structures to facilitate the passage of anadromous lampreys in their native range. This information is provided for the freshwater phase of both lamprey adults and juveniles (including larvae and macrophthalmia), as most barriers to native lamprey movements are in freshwater streams and rivers. We caution that the passage considerations presented here should not be viewed as design criteria. To develop such criteria, adequate testing must be done using the lamprey population and life stage that the structure is designed to accommodate.

## Swimming Performance

### Adults

Information on the swimming performance of adult lampreys is necessary for development of structures that allow upstream passage of prespawning fish. Several studies have examined the swimming performance of sea lampreys (Beamish 1974; Hanson 1980; Bergstedt et al. 1981; McCauley 1996) and Pacific lampreys (Close et al. 2003; Mesa et al. 2003). In general, this research has shown that anguilliform swimming is probably less efficient and powerful than other swimming modes. However, many anguilliform swimmers undertake long migrations without feeding (e.g., eels or lampreys; van Ginneken and van den Thillart 2000) or swim constantly (e.g., some sharks; Donley and Shadwick 2003), which is inconsistent with a mode of swimming that is inefficient. Indeed, recent research indicates that anguilliform swimming may be more efficient and powerful than previously thought (Tytell and Lauder 2004) and that the physiological cost of migration for eels was relatively low (van Ginneken and van den Thillart 2003). Whether the same is true for migrating lampreys is unknown.

Beamish (1974) showed that the distance (m) sea lampreys could swim declined with an increase in swimming speed between 20 and 60 cm/s. Swimming speed was positively related to temperature (range 5–15°C). For example, the maximum speed that adult sea lampreys could sustain for 1 min was about 35 cm/s at 15°C, but was only 23 cm/s at 2°C. The total distances traveled were 21.3 and 13.8 m. In contrast, adult Chinook salmon *Oncorhynchus tshawytscha*, sockeye salmon *O. nerka*, and steelhead *O. mykiss* at similar temperatures and velocities were able to ascend an experimental endless fishway for distances exceeding 300 m without fatiguing (Collins et al. 1962). Bergstedt et al. (1981) tested swimming of adult land-locked sea lampreys in a tunnel respirometer at 6°C and 10°C and reported that these fish could not swim for more than 1 min at about 1.4 m/s. Hanson (1980) tested large numbers of these fish in a flume at different temperatures and a velocity range of 1.5–4 m/s. At temperatures below 15°C, few fish attempted to swim upstream in the 3-m-long flume, regardless of velocity. At 16–24°C, however, 17% (98 of 590) of the sea lampreys attempted to ascend the flume. He noted that the lampreys were able to endure velocities of about 4 m/s for an average of 1.3 s. McCauley (1996) allowed adult land-locked sea lampreys to volitionally swim up a 30-m-long flume at a variety of water velocities and temperatures. At velocities of about 1.5 m/s, fish were able to swim for up to 50 s, but at 3 m/s they could swim for only 2–3 s. In comparison, Slatick (1971) found that from 78% to 100% of adult Chinook salmon, sockeye salmon, coho salmon *O. kisutch*, and steelhead passed through a 31.4-m-long inclined pipe at velocities of 1.22 m/s within 3–13 min.

Mesa et al. (2003) reported that the mean (±SD) critical swimming speed (as defined by Beamish 1978) of adult Pacific lampreys was 86.2 ± 7.5 cm/s at 15°C. In these experiments, the fish swam to exhaustion and exhibited significant physiological changes that lasted for 1–4 h (e.g., increased blood lactate levels, decreases in blood pH). Critical swimming speeds of lampreys are thus much lower than those of adult Pacific salmon that commonly use fishways, such as Chinook salmon (range = 155–165 cm/s; Geist et al. 2003), pink salmon *O. gorbuscha* (range = 101–118 cm/s; Williams and Brett 1987), and sockeye salmon (mean = 143 cm/s; Brett 1965). On the other hand, the critical swimming speeds of lampreys were similar to other fish in the Pacific Northwest, including bull trout *Salvelinus confluentus* (range = 48–74 cm/s; Mesa et al. 2004) and northern pikeminnow *Ptychocheilus oregonensis* (mean = 72 cm/s; Kolok and Farrell 1994). Close et al. (2003) examined swimming of radio-tagged and untagged adult Pacific lampreys in a 4-m-long flume for up to 1 h at 40 cm/s. The times that both groups of fish were able to swim at this speed ranged from about 100 to 3,600 s. In general, conducting tests in small swim chambers with free-swimming fish is difficult, and therefore these estimates of lamprey swimming performance are probably conservative (Peake 2004). Also, comparisons of swimming speeds between species is complicated by a variety of factors, including study protocol, experimental conditions (e.g., temperature and dissolved oxygen), sex, stage of maturity, level of training, rearing conditions, and physiological status (Geist et al. 2003).

### *Juveniles*

Information on the swimming performance of juvenile lampreys is sparse. For Pacific lampreys, the only information available is from research conducted to develop bypass screens at dams that allow safe lamprey passage (Moursund et al. 2003). In these studies, maximum burst speed of Pacific lamprey macrophthalmia (mean length 136 mm) ranged from 55 to 94 cm/s with a mean of 71 cm/s (Dauble et al. 2006). Metamorphosed sea lampreys of similar length (130 mm) were able to sustain swimming speeds of 60 cm/s for periods of 10–21 s, depending on temperature (Beamish 1974). Maximum sustained swimming speed for metamorphosed Pacific lamprey was 36 cm/s (Dauble et al. 2006) or 2.4 body lengths/s at 12°C. In comparison, the same measure for adult Pacific lamprey was 1.4 body lengths/s, even though these experiments were conducted at higher water temperature (15°C; Mesa et al. 2003).

Dauble et al. (2006) noted that swimming endurance of Pacific lamprey macrophthalmia decreased rapidly at water velocities greater than 46 cm/s. In addition, endurance of ammocoetes is probably even less, due to their greater dependence on anaerobic metabolism (Paton et al. 2001). Thus, the ability of juvenile lampreys to avoid diversion or barrier screens will be greatly reduced when average perpendicular velocity at the face of these structures exceeds 0.4 m/s (Dauble et al. 2006).

## Attachment and Rheotaxis

### *Adults*

When confronted with rapid current velocities, adult Pacific lampreys orient into the current and use their oral disk to attach to the substrate, presumably resting between bouts of burst swimming. This saltatory mode of movement is most pronounced in current velocities greater than 60 cm/s (Daigle et al. 2005). Consequently, it is critical that adult lampreys are provided with adequate attachment surfaces in fishways or culverts where they might encounter high current velocity. Recent laboratory research using sea lampreys has shown that these fish must expend more energy to remain attached on surfaces where they cannot achieve a good seal (U. Reinhardt, Eastern Michigan University, personal communication). Consequently, the best surfaces for lamprey attachment are probably smooth and nonporous.

Grating, wire mesh, and perforated plate are all materials that prevent lamprey attachment. For example, in a flume that generated water velocities greater than 2 m/s, lamprey passage was reduced through areas with grates on the floor (2.5 × 10.0 cm grating of 0.5-cm thick aluminum) relative to passage when the flooring was a continuous, smooth surface (Daigle et al. 2005). In these same experiments, attaching a 30.5-cm wide metal plate over the grate allowed lamprey to attach near an orifice opening with greater than 2.4 m/s velocity and successfully pass through. Field observations of radio-tagged lampreys confirmed that metal grating in the floor of fishways can result in poor lamprey passage efficiency (Ocker et al. 2001; Moser et al. 2003).

While smooth surfaces, such as polished metal and glass, would apparently provide a preferred surface for holding, lamprey are able to achieve suction on roughened, wetted concrete. Experiments on sea lampreys revealed that the flexibility of the oral disk varied among individuals (U. Reinhardt, personal communication). That is, some fish were able to conform their fimbriae to nooks and crannies more effectively than others. In some cases, the inside of tightly corrugated metal pipe could challenge lamprey attachment to culvert walls or floors. Natural substrate in culverts is needed to ensure that adult lamprey have adequate attachment areas during periods of high flows.

Sharp angles prevent lampreys from being able to stay attached as they move around a corner in high velocities. Both laboratory and field experiments at Bonneville Dam, the lowermost dam on the Columbia River, indicate that Pacific lamprey have difficulty negotiating 90° corners in high velocity situations near the entrances of fishways. When bulkheads were rounded, lamprey achieved better entrance success (Moser et al. 2002a). Moreover, laboratory testing indicated that lampreys approaching rounded corners (20.3 cm diameter) had significantly higher passage success than those that encountered square corners (Daigle et al. 2005).

Similarly, vertical surfaces and overhanging ledges at the top of weirs or other structures may prevent lamprey from staying attached as they attempt to pass over. For example, at Willamette Falls on the Willamette River near Portland, Oregon, adult Pacific lampreys attempting to migrate upstream were limited by vertical, wooden flashboards installed around the perimeter of the falls to control upstream water levels (Magie et al. 2007). These boards were vertical to the water surface, did not provide an attachment surface for lamprey at their upper edge, had high current velocities passing over the top, and became covered with algae during the summer and early fall. All of these factors contributed to the boards being a major impediment to upstream passage of lamprey. Sea lamprey managers have employed overhanging ledges or lips to exclude prespawning adult fish from tributary streams of the Great Lakes. The 1977 task force on sea lamprey barriers recommended a 15–30-cm overhanging lip of sheet metal at the top of a barrier to preclude lamprey passage over the structure (Hunn and Youngs 1980). Based on these observations, it is likely that perched culverts are also impediments to migrating adult lamprey.

### *Juveniles*

Like anadromous salmonids, lamprey show broad behavioral changes with metamorphosis and the onset of seaward migration. Ammocoetes rapidly burrow into substrate when exposed, and rates of burrowing are affected by both ammocoete size and substrate coarseness (Quintella et al. 2007). In contrast, macrophthalmia clearly exhibit a preference for horizontal movement and are reluctant to move vertically (M. L. Moser, unpublished data).

With development of the suctorial disk, macrophthalmia also show extended periods of attachment to the substrate (Dauble et al. 2006; Moser, unpublished data). The role of attachment for macrophthalmia is likely as important in their migration as it is for adults. Dauble et al. (2006) found that it was necessary to line a swim chamber with 5-mm square nylon mesh to prevent lampreys from attaching during sustained swimming trials. As is the case for adults, macrophthalmia may require attachment surfaces to rest between bouts of movement.

## Environmental Effects on Behavior

### *Adults*

Adult Pacific and sea lampreys typically show higher levels of activity during night than day (Haro and Kynard 1997; Kelso and Gardner 2000; Moser et al. 2002b; Binder and McDonald 2007). Other lamprey species also show negative phototaxis (Tuunainen et al. 1980; Ullen et al. 1997). Therefore, it seems likely that both artificially high light intensity and abrupt changes in lighting could affect lamprey movements. Experiments in the field and laboratory at Bonneville Dam found no evidence that lighting at fish counting stations (1–3 lux at the channel bottom) was avoided by adult Pacific lampreys (Moser et al. 2002c). However, both blinded and control sea lampreys respond to direct illumination, indicating that light avoidance is mediated by dermal photoreceptors (Binder and McDonald 2007). These results suggest that while adult lampreys will move through a lighted areas, they may be obstructed by very bright or abruptly changing light conditions, for example at culvert entrances or exits.

Water temperature also seems to play an important role in adult lamprey migration and should be considered when evaluating passage at dams or other structures (Binder and McDonald 2008). For example, radio-tagged adult Pacific lampreys in the Columbia River showed less delay and moved faster through the tailrace and fishways at Bonneville and The Dalles Dams as water temperatures increased (Moser et al. 2005). Also, Magie et al. (2007) noted that passage of Pacific lampreys at Willamette Falls dam was strongly correlated with temperature during the summer and declined as temperatures cooled in the early fall. Similarly, lamprey movements in the John Day River, Oregon, ceased in fall when water temperature dropped below 20°C (Robinson and Bayer 2005). In laboratory trials, Binder and McDonald (2008) also found that temperature mediated both diel activity and rheotaxis in adult sea lampreys.

Flow in the tailraces of dams and around the entrances to fishways is another important environmental factor. Recent work at Willamette Falls Dam has shown that the entrance efficiency of lampreys at a fishway near the powerhouse was higher when all turbine units were functioning compared to when they were off (M. G. Mesa, unpublished data). Similarly, adult Pacific lampreys approaching Bonneville Dam were attracted to the powerhouse generating the greatest discharge (Moser et al. 2003). In both cases, discharge from the turbines created an attraction flow that lampreys used to orient. Thus, as for salmonids, lampreys may require a certain amount of flow to efficiently locate and enter fishways or other aids to passage.

### *Juveniles*

Both ammocoetes and macrophthalmia are nocturnal, showing higher levels of activity during night than day (Quintella et al. 2005; Dauble et al. 2006, Moser, unpublished data). Laboratory observations indicated strong negative phototaxis by Pacific lamprey macrophthalmia exposed to abrupt changes in lighting (R. Moursund, Pacific Northwest National Laboratory, personal communication). Consequently, movements of juvenile lampreys could be affected by changes in lighting that are produced at culverts. In addition, their propensity to avoid light might be exploited to guide juvenile lamprey away from areas where they could become impinged (e.g., irrigation diversion screens, fishway bypass screens, etc.).

Juvenile lamprey move downstream primarily at night, but they are also profoundly affected by flow. The abundance of both ammocoetes and macrophthalmia in the water column is highest during freshets when stream or river discharge is greatest (Applegate 1950; R. Tudor, Washington Department of Fish and Wildlife, unpublished data). Surprisingly, Applegate (1950) found little effect of temperature on juvenile sea lamprey movement. Thus, periods of time when juvenile lamprey are most likely to migrate past barriers to downstream areas probably coincide with peak water discharge periods.

## Research Needs

### *Adults*

Impediments to passage are critical limiting factors for native lampreys and research is needed to improve lamprey passage success at dams, culverts, weirs, and other structures. Most fishways that lampreys encounter were designed for more powerful swimmers and test the limits of lamprey performance. More information is needed on the basic biology of lamprey migration to aid in development of fishways that are effective for lampreys. For example, little is known about the migration behavior of lampreys in their native range and the cues that they use to orient and navigate. Extensive research on adult sea lampreys in the Great Lakes indicates that these fish orient to olfactory cues both early in migration (larval pheromones) and on the spawning grounds (male sex pheromones; reviewed in Dittman 2005). However, the relative roles of pheromones, other olfactory cues, current velocity, piloting, or other means of navigating are not known for any native lamprey species.

In addition to determining the cues that lampreys use during migration, basic information on their swimming performance and behavior is still needed. Studies on the volitional swimming performance of adult lampreys in large flumes or test culverts would be helpful in assessing their true performance capabilities. Also, further work on the physiological and bioenergetic effects of swimming at different speeds would help to define limits to lamprey endurance, identify conditions in the wild that are stressful, and determine the effects of energy expenditure on lamprey fitness.

Adult lamprey passage could be improved in traditional fishways by making modifications that are known to be effective (e.g., rounded corners or placing plates over grating). To develop effective lamprey passage at other, relatively unstudied impediments to migration (e.g., culverts and low-elevation dams), specific information about lamprey behavior around these structures is needed. Moreover, the roles of lighting, water temperature, salinity, flow, or other abiotic factors that may influence lamprey passage are largely unknown. For example, the effect of high total dissolved gas levels around dams has never been studied.

### *Juveniles*

While there is very little data available on the freshwater migratory behavior of adult lampreys, the lack of information on juvenile lamprey movements is even more alarming. To what extent do ammocoetes move upstream? Is movement of juveniles largely active or passive? What is the extent of losses at dams and irrigation diversions? How can these losses be mitigated? One of the largest impediments to our understanding of juvenile lamprey behavior is the fact that currently available radio, acoustic, and passive integrated transponder (PIT) tags are too large for use with these small fish. Some recent work has indicated that the use of very small PIT tags holds promise for observation of juvenile lamprey movements (Moursund et al. 2003; Quintella et al. 2005). However, only the largest juveniles can be fitted with these tags and surgical implantation methods are tedious (Mueller et al. 2006). Therefore, development of new and novel methods for assessing juvenile lamprey migration behavior is critically needed.

## Conclusion

As interest in lamprey conservation and biology increases, providing safe and effective passage alternatives for lampreys will become a higher priority. This is essential for maintaining healthy populations of these ancient fish. The development of sound passage alternatives for all lamprey species and life stages will require sophisticated information on cues that guide lamprey movements, the limits of lamprey endurance, and the ultimate effects of energy expenditure during migration on lamprey fitness.

## Acknowledgments

We thank the Columbia River Lamprey Technical Workgroup for their initial interest in this work and their review of an early draft. In particular, Doug Young provided enthusiastic and persistent encouragement throughout this process. This work was partially funded by the U.S. Army Corps of Engineers, Portland and Walla Walla Districts. We also thank two anonymous reviewers for helpful suggestions to improve the organization and content of this review.

## References

Almeida, P. R., B. R. Quintella, and N. M. Dias. 2002. Movement of radio-tagged anadromous sea lamprey during the spawning migration in the River Mondego (Portugal). Hydrobiologia 483:1–8.

Almeida, P. R., B. R. Quintella, A. Koed, and N. Andrade. 2005. Using electromyogram telemetry to study the spawning migration of sea lamprey (*Petromyzon marinus* L). Pages 3–13 *in* M. T. Marmulla and G. Lembo, editors. Aquatic telemetry: advances and applications. Food and Agriculture Organization of the United Nations and COISPA TECONOLOGIA & RICERCA, Rome.

Applegate, V. C. 1950. Natural history of the sea lamprey (*Petromyzon marinus*) in Michigan. U.S. Fish and Wildlife Service Special Scientific Report Fisheries 55.

Beamish, F. W. H. 1974. Swimming performance of adult sea lamprey, *Petromyzon marinus*, in relation to weight and temperature. Transactions of the American Fisheries Society 103:355–358.

Beamish, F. W. H. 1978. Swimming capacity. Pages 101–187 *in* W. S. Hoar and D. J. Randall, editors. Fish physiology, volume VII. Academic Press, New York.

Beamish, R. J. 1980. Adult biology of the river lamprey (*Lampetra ayresi*) and the Pacific lamprey (*Lampetra tridentata*) from the Pacific coast of Canada. Canadian Journal of Fisheries and Aquatic Sciences 37:1906–1923.

Beamish, R. J., and C. C. Levings. 1991. Abundance and freshwater migrations of the anadromous parasitic lamprey, *Lampetra tridentata*, in a tributary of the Fraser River, British Columbia. Canadian Journal of Fisheries and Aquatic Sciences 48:1250–1263.

Bergstedt, R. A., D. V. Rottiers, and N. R. Foster. 1981. Laboratory determination of maximum swimming speed of migrating sea lampreys: a feasibility study. U.S. Fish and Wildlife Service, Great Lakes Fishery Laboratory, Administration Report No. 81-3, Ann Arbor, Michigan.

Binder, T. R., and D. G. McDonald. 2007. Is there a role for vision in the behavior of sea lampreys (*Petromyzon marinus*) during their upstream spawning migration? Canadian Journal of Fisheries and Aquatic Sciences 64:1403–1412.

Binder, T. R., and D. G. McDonald. 2008. The role of temperature in controlling diel activity in upstream migrant sea lampreys (*Petromyzon marinus*). Canadian Journal of Fisheries and Aquatic Sciences 65:1113–1121.

Brett, J. R. 1965. The relation of size to rate of oxygen consumption and sustained swimming speed of sockeye salmon (*Oncorhynchus nerka*). Journal of the Fisheries Research Board of Canada 22:1491–1501.

Close, D. A., M. S. Fitzpatrick, and H. W. Li. 2002. The ecological and cultural importance of a species at risk of extinction, Pacific lamprey. Fisheries 27:19–25.

Close, D. A., M. S. Fitzpatrick, C. M. Lorion, H. W. Li, and C. B. Schreck. 2003. Effects of intraperitoneally implanted radio transmitters on the swimming performance and physiology of Pacific lamprey. North American Journal of Fisheries Management 23:1184–1192.

Collins, G. B., J. R. Gauley, and C. H. Elling. 1962. Ability of salmonids to ascend high fishways. Transactions of the American Fisheries Society 91:1–7.

Daigle, W. R., C. A. Peery, S. R. Lee, and M. L. Moser. 2005. Evaluation of adult Pacific lamprey passage and behavior in an experimental fishway at Bonneville Dam. Report to U.S. Army Corps of Engineers, Portland District, Portland, Oregon.

Dauble, D. D., R. A. Moursund, and M. D. Bleich. 2006. Swimming behavior of juvenile Pacific lamprey, *Lampetra tridentata*. Environmental Biology of Fishes 75:167–171.

Dittman, A. 2005. Chemical cues for sea lamprey migration. Nature Chemical Biology 1:316–317.

Donley, J. M., and R. E. Shadwick. 2003. Steady swimming muscle dynamics in the leopard shark *Triakis semifasciata*. Journal of Experimental Biology 206:1117–1126.

Geist, D. R., R. S. Brown, V. I. Cullinan, M. G. Mesa, S. P. Vanderkooi, and C. A. McKinstry. 2003. Relationships between metabolic rate, muscle electromyograms and swim performance of adult Chinook salmon. Journal of Fish Biology 63:970–989.

Hanson, L. H. 1980. Study to determine the burst swimming speed of spawning-run sea lampreys (*Petromyzon marinus*). U.S. Fish and Wildlife Service Research Completion Report, Millersburg, Michigan.

Hardisty, M. W., and I. C. Potter. 1971a. The behaviour, ecology , and growth of larval lampreys. Pages 85–125 *in* M. W. Hardisty and I. C. Potter, editors. The biology of lampreys, volume 1. Academic Press, London.

Hardisty, M. W., and I. C. Potter. 1971b. The general biology of adult lampreys. Pages 249–277 *in* M. W. Hardisty and I. C. Potter, editors. The biology of lampreys, volume 1. Academic Press, London.

Haro, A., and B. Kynard. 1997. Video evaluation of passage efficiency of American shad and sea lamprey in the modified Ice Harbor fishway. North American Journal of Fisheries Management 17:981–987.

Hunn, J. B., and W. D. Youngs. 1980. Role of physical barriers in the control of sea lamprey (*Petromyzon marinus*). Canadian Journal of Fisheries and Aquatic Sciences 37:2118–2122.

Kelso, J. R. M., and W. M. Gardner. 2000. Emigration, upstream movement, and habitat use by sterile and fertile sea lampreys in three Lake Superior tributaries. North American Journal of Fisheries Management 20:144–153.

Kolok, A. S., and A. P. Farrell. 1994. Individual variation in the swimming performance and cardiac performance of northern squawfish, *Ptychocheilus oregonensis*. Physiological Zoology 67:706–722.

Kostow, K. 2002. Oregon lampreys: natural history, status, and analysis of management issues. Oregon Department of Fish and Wildlife, Portland.

Magie, R. J., E. S. Copeland, and M. G. Mesa. 2007. Passage and behavior of radio-tagged adult Pacific lamprey (*Lampetra tridentata*) at Willamette Falls Dam, Oregon. Report to the Pacific General Electric Company, Portland, Oregon.

McCauley, T. C. 1996. Development of an instream velocity barrier to stop sea lamprey migrations in the Great Lakes. Master's thesis. University of Manitoba, Winnipeg.

Mesa, M. G., J. M. Bayer, and J. G. Seelye. 2003. Swimming performance and physiological responses to exhaustive exercise in radio-tagged and untagged Pacific lampreys. Transactions of the American Fisheries Society 132:483–492.

Mesa, M. G., L. K. Weiland, and G. Barbin-Zydlewski. 2004. Critical swimming speeds of wild bull trout. Northwest Science 78:59–65.

Moser, M. L., A. L. Matter, L. C. Stuehrenberg, and T. C. Bjornn. 2002a. Use of an extensive radio receiver network to document Pacific lamprey (*Lampetra tridentata*) entrance efficiency at fishways on the lower Columbia River, USA. Hydrobiologia 483:45–53.

Moser, M. L., P. A. Ocker, L. C. Stuehrenberg, and T. C. Bjornn. 2002b. Passage efficiency of adult Pacific lampreys at hydropower dams on the lower Columbia River, USA. Transactions of the American Fisheries Society 131:956–965.

Moser, M. L., L. C. Stuehrenberg, W. Cavender, S. G. McCarthy, and T. C. Bjornn. 2002c. Radiotelemetry investigations of adult Pacific lamprey migration behavior: evaluation of modifications to improve passage at Bonneville Dam, 2000. Report to U.S. Army Corps of Engineers, Portland District, Portland, Oregon.

Moser, M. L., D. A. Ogden, S. G. McCarthy, and T. C. Bjornn. 2003. Migration behavior of adult Pacific lamprey in the lower Columbia River and evaluation of Bonneville Dam modifications to improve passage, 2001. Report to U.S. Army Corps of Engineers, Portland District, Portland, Oregon.

Moser, M. L., R. W. Zabel, B. J. Burke, L. C. Stuehrenberg, and T. C. Bjornn. 2005. Factors affecting adult Pacific lamprey passage rates at hydropower dams: using "time to event" analysis of radiotelemetry data. Pages 61–70 *in* M. T. Spedicato, G. Marmulla, and G. Lembo, editors. Aquatic telemetry: advances and applications. Food and Agriculture Organization of the United Nations and COISPA TECONOLOGIA & RICERCA, Rome.

Moursund, R. A., D. D. Dauble, and M. Langeslay. 2003. Turbine intake diversion screens, investigating effects on Pacific lamprey. Hydro Review XXII:40–46.

Mueller, R. P., R. A. Moursund, and M. D. Bleich. 2006. Tagging juvenile Pacific lamprey with passive integrated transponders: methodology, short-term mortality, and influence on swimming performance. North American Journal of Fisheries Management 26:361–366.

Ocker, P. A., L. C. Stuehrenberg, M. L. Moser, A. L. Matter, J. J. Vella, B. P. Sandford, T. C. Bjornn, and K. R. Tolotti. 2001. Monitoring adult Pacific lamprey (*Lampetra tridentata*) migration behavior in the lower Columbia River using radiotelemetry, 1998–99. Report to U.S. Army Corps of Engineers, Portland District, Portland, Oregon.

Paton, K. R., M. H. Cake, and I. C. Potter. 2001. Muscle glycogen, lactate and glycerol-3.-phosphate concentrations of larval and young adult lampreys in response to exercise. Comparative Biochemistry and Physiology B 129:759–766.

Peake, S. 2004. An evaluation of the use of critical swimming speed for determination of culvert water velocity criteria for smallmouth bass. Transactions of the American Fisheries Society 133:1472–1479.

Quintella, B. R., N. O. Andrade, N. M. Dias, and P. R. Almeida. 2007. Laboratory assessment of sea lamprey larvae burrowing performance. Ecology of Freshwater Fish 16:177–182.

Quintella, B. R., N. O. Andrade, R. Espanhol, and P. R. Almeida. 2005. The use of PIT telemetry to study movements of ammocoetes and metamorphosing sea lampreys in river beds. Journal of Fish Biology 66: 97–106.

Robinson, T. C., and J. M. Bayer. 2005. Upstream migration of Pacific lampreys in the John Day River, Oregon: behavior, timing, and habitat use. Northwest Science 79:106–119.

Slatick, E. 1971. Passage of adult salmon and trout through an inclined pipe. Transactions of the American Fisheries Society 100:448–455.

Tytell, E. D., and G. V. Lauder. 2004. The hydrodynamics of eel swimming. I. Wake structure. Journal of Experimental Biology 207:1825–1841.

Tuunainen, P., E. Ikonen, and H. Auvinen. 1980. Lamprey and lamprey fisheries in Finland. Canadian Journal of Fisheries and Aquatic Sciences 37:1953–1959.

Ullen, F., T. G. Deliagina, G. N. Orlovsky, and S. Grillner. 1997. Visual pathways for postural control and negative phototaxis in lamprey. Journal of Neurophysiology 78:960–976.

van Ginneken, V. J. T., and G. E. E. J. M. van den Thillart. 2000. Eel fat stores are enough to reach the Sargasso. Nature (London) 403:156–157.

Williams, I. V., and J. R. Brett. 1987. Critical swimming speed of Fraser and Thompson River pink salmon (*Oncorhynchus gorbuscha*). Canadian Journal of Fisheries and Aquatic Sciences 44:348–356.

American Fisheries Society Symposium 72:125–138, 2009


# Ability of Adult Sea Lamprey to Climb Inclined Surfaces

Ulrich G. Reinhardt*
*Eastern Michigan University, Biology Department*
*316 Mark Jefferson Hall, Ypsilanti, Michigan 48103, USA*

Thomas Binder and D. Gordon McDonald
*Department of Integrative Biology, University of Guelph,* Guelph, Ontario N1G 2W1, Canada

*Abstract.*—Low-head barriers against invasive sea lampreys *Petromyzon marinus* in the Great Lakes are designed to maintain a minimum crest height of 30 cm and have a lip on the crest to prevent them from climbing over the barrier. We tested the ability of migratory-phase sea lampreys to scale inclined ramps with shallow (0.7–5 cm) water depth. We predicted that sea lampreys would jump the barrier and that their ability to attach would increase passage success. A recirculating flume and ramp with a vertical height of 10–30 cm and an inclination between 30° and 60° were used to evaluate lamprey climbing ability. Lampreys trying to scale the ramp were monitored by passive integrated transponder tag readers and low-light video cameras. No lampreys were observed jumping out of the water to scale a barrier. Independent of ramp angle, no fish passed over a 30-cm ramp. Lampreys often attached themselves to the ramp, but without a gain of vertical height between repeated attempts. The success rate at lower ramp heights varied between 0% (15 cm height, 30° angle) and 63% (10 cm height, 60° angle). Only ramps shorter than half the body length of the lampreys could be surmounted. Apparently, the lampreys had to have their dorso-ventral fins fully submerged in the downstream pool to create enough propulsion to scale a ramp in burst-swimming mode. An analysis of 1,300 passage attempts in a field-validation experiment showed a greater apparent motivation to move up a ramp but reconfirmed our laboratory findings on passage technique and maximum performance. We conclude that sea lamprey barrier height could be further reduced and that an overhanging lip is not necessary as sea lampreys neither climb nor jump over barriers. A ramp with a shallow inclination and moderate vertical height and water flow is a new design suggestion for a barrier that blocks sea lampreys and may allow other fish species to pass.

## Introduction

Sea lampreys *Petromyzon marinus* have caused great damage to Great Lakes fishery resources since their accidental introduction to the upper Great Lakes after opening of the Welland canal that bypasses the Niagara Falls (Pearce et al. 1980). The exotic parasite severely harmed native fish species such as lake trout *Salvelinus namaycush* and lake whitefish *Coregonus clupeaformis*. Since the 1950s, sea lamprey numbers have been reduced by a variety of control methods. While injury levels to native fish have been reduced through widespread lampricide treatment of streams and rivers, the Great Lakes Fishery Commission (GLFC) is committed to achieving ecologically and economically sound lamprey management through a more effective use of alternative control technologies (GLFC

* Corresponding author: ureinhard@emich.edu

2001). These include more effective methods of lamprey trapping and improvements in the design and operation of physical barriers.

The GLFC operates 69 constructed barriers to block adult sea lampreys from access to stream habitat for spawning and larval rearing (McLaughlin et al. 2007). Many of these barriers are designed to pass fish species capable of jumping (e.g., trout, salmon). Based on observations by Youngs (1979) on passage probability of sea lampreys over vertical steps, the crest height (vertical distance from tailwater to headwater) of low-head barriers should be a minimum of 30 cm to block all migrating sea lampreys. At present, the maximum crest height does not exceed ~45 cm and a deep pool is typically incorporated on the downstream side of the barrier to provide a "run-up" for jumping species. Although lampreys can jump (e.g., Applegate 1950), their ability to do so is thought to be less than that of other jumping species.

Although low-head barriers are, for the most part, effective, several lamprey barriers have been unsuccessful in blocking lampreys either for a part or for all of the time they have been in operation (Lavis et al. 2003). The exact reasons for the failure of barriers are largely unknown (McLaughlin et al. 2007). The hypothesized reasons for variations in the effectiveness of barriers include temporal variations in water temperature and hydraulic conditions around the barrier (McLaughlin et al. 2007). For example, there may be conditions at barriers (e.g., flooding and loss of crest height) that favor passage by jumping or swimming up the nappe (the sheet of water flowing over the crest) of the ramp. Another possibility is that lampreys scale the barrier or its abutments by exploiting their ability to periodically attach to smooth surfaces using their oral sucker. This attachment is commonly observed in submerged sea lampreys as a means of holding station in current. Furthermore, a cycle of "attach-burst-attach" has been observed in the wild and in laboratory flumes as a means of making horizontal progress against current (e.g., Hanson 1980; Quintella et al. 2004). A much more dramatic example of this cycle has been observed in Pacific lampreys *Lampetra tridentata*, which are able to ascend an inclined wetted aluminum surface over dozens of meters by exploiting an attach-twitch-attach cycle (e.g., Moser et al. 2005; Reinhardt et al. 2008).

In addition to less than 100% effectiveness in blocking lampreys, low-head barriers have some negative impacts on native fish species, with some being absent or reduced in numbers upstream of barrier locations (McLaughlin et al. 2006). This is probably a result of blocked or impeded movement of some species across barriers (Porto et al. 1999) and therefore isolation of their upstream populations and resulting lowered viability. An ideal sea lamprey barrier would successfully block sea lamprey migration while allowing passage of all other native species (McLaughlin et al. 2007).

Apart from limited studies on the ability of sea lampreys to pass over vertical obstacles (Youngs 1979) and some qualitative observations (Applegate 1950; Youngs 1979), there is no published scientific study, as far as we are aware, on the mechanism of sea lamprey passage over vertical barriers. In order to identify reasons for failure of sea lamprey barriers and to develop improved lamprey barrier designs that selectively allow improved passage of native fish species, we studied the mode by which adult sea lamprey attempt to pass over inclined and near-vertical obstacles. We hypothesized that sea lampreys may be able to jump over barriers when crest height is compromised or may use oral sucker attachment in conjunction with muscular effort (twitch, burst, or jump) to pass barriers. We used detailed observation of adult sea lampreys attempting upstream passage in the laboratory and in the field to test our hypotheses.

## Methods

### *Experimental animals*

A total of 270 adult spawning-phase sea lampreys were used in 2005 for laboratory experiments and a further 110 in 2007 for field studies. The sex ratio in 2005 was 1:1 (in 2007, 2:3

male:female), and the mean total body length (BL) was 49.9 ± 0.57 SE cm (50.8 ± 0.47 SE cm in 2007) with less than 0.8 cm difference in male and female mean length. The difference in mean length between years was not significant ($t$-test, $P > 0.05$). The range of BL in both years combined was 33–59.5 cm. Lampreys were obtained from the Toronto Region Conservation Authority, which collects lampreys during the spawning run from traps on Duffins Creek (Ajax, Ontario) and Humber River (Toronto, Ontario). Lampreys were transported to the Hagen Aqualab (University of Guelph), where they were housed in circular tanks (1.85 m diameter, total volume 2.6 $m^3$) at 7°C, under a 12-h photoperiod (0700–1900 hours). Freshwater flowed into the tanks at a rate of one exchange per 24 h. All lampreys were implanted with a passive integrated transponder (PIT) tag (Texas Instruments Radio Frequency Identification tag, 23 mm long, 134.2 kHz) inserted into the peritoneal cavity under general anesthesia. At the end of the experiment, it was observed that in males, the PIT tag had shifted in the body cavity a few centimeters toward the tail, whereas in females, the egg mass prevented the shifting of the tag. This shift might have caused a small bias in the PIT tag reading by the antenna toward recording greater apparent number of attempts by females on the ramp (see Results).

### *Experimental apparatus*

Laboratory studies were conducted in a 6.4-m-long swim flume (total volume = 3,629 L) consisting of an upstream and downstream tank (1,819 and 1,088 L respectively), connected by a 2.5-m-long swim channel with a wire screen at the upstream end (see Figure 1). A 5-hp, 60-Hz centrifugal pump, controlled by a calibrated variable frequency drive (WF2 series drive; TB Wood's, Chambersburg, Pennsylvania), pumped 14.0°C water from a 6,500-L reservoir at three flow rate settings. The downstream tank of the flume was covered in 1-$cm^2$ plastic mesh to prevent the lampreys from attaching. Mesh also prevented the fish from swimming underneath the ramp. A 40-cm-long and 45-cm-wide

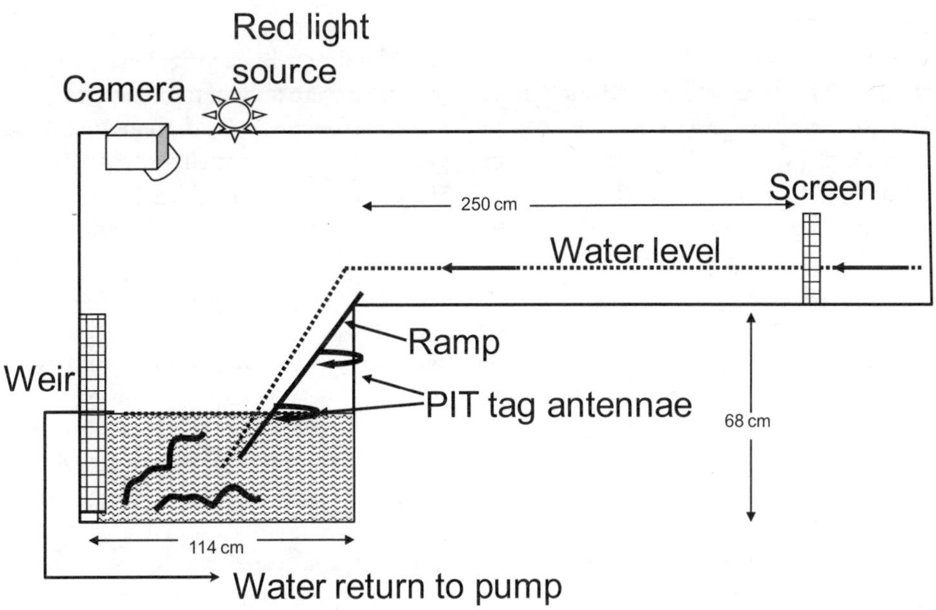


Figure 1. Schematic drawing of the experimental set-up in the laboratory, showing the recirculating flume with two lampreys in the downstream pool, the ramp with passive integrated transponder tag antennae, and the video camera. Dimensions not drawn to scale.

ramp of 0.5-cm-thick white acrylic with painted markings for size reference was positioned at the downstream entrance to the swim channel.

Activity on the ramp was monitored under dim red light to simulate nighttime conditions. An infrared video camera (black and white CCD with infrared emitters, CCTV Wholesalers, model # KG-190) connected to a VCR was mounted above the ramp so that the optical plane was about parallel to the ramp. Two PIT-tag antennae (controlled with a Texas Instruments Series 2000 reader unit; R. Moroz Ltd., Markham, Ontario), attached to the underside of the ramp at the water level and ~15 cm above water level, recorded the identity of each individual as it attempted to traverse the ramp.

## Experimental series

*Experiment lA. Influence of ramp angle on climbing behavior.*—This experiment evaluated the climbing behavior of sea lampreys on a wetted ramp ("low flow," water depth in downstream tank 60 cm) at three different angles of inclination (30°, 45°, 60°). The water depth on the ramp was 0.7 cm in the center. Exposed ramp length was 35–40 cm at all ramp angles. Lampreys ($N$ = 8 per trial) were placed in the downstream tank (1700–1830 hours) and their activity at the ramp was monitored for 4 h. A total of eight replicates were run (two at 30°, four at 45°, and two at 60°). The observations for each angle were pooled for analyses.

*Experiment 1B. Influence of flow on climbing behavior.*—The climbing behavior of lampreys on the 30° inclined ramp, at "low flow" (mean water depth of 0.7 cm, see above) was compared to that at "medium flow" (mean water depth on ramp 3 cm) and "high flow" (mean water depth 5 cm). The exposed ramp length was 35 cm. The nappe was thickest at the top of the ramp where water velocity was slowest and considerably thinner in the higher velocity toward the bottom of the ramp. Water velocity in the center of the ramp ranged between 50 cm/s at low flow and 150 cm/s at high flow. Two additional replicates were run at medium flow with a 45° angle ramp (creating a 110 cm/s velocity in the center of the ramp). Each flow regime was created by varying pump discharge and repeated in two replicates with each replicate using eight naive lampreys. The results from each flow were pooled for analysis.

*Experiment 2. Influence of exposed ramp length on climbing behavior.*—This experiment tested the effect of exposed length of the ramp on the passage success of sea lamprey. Three ramp angles (30°, 45°, and 60°) and two vertical heights (10 and 15 cm from upper to lower water level) were used to produce a total of six different exposed ramp lengths (12.1, 14.9, 16.7, 20.5, 21.0, and 29.0 cm). Water flow was held constant (water depth 3 cm, velocity 130 cm/s at the center of the ramp). Behavior was recorded as in experiment 1. Starting times varied between 0930 and 1930 hours with about equal numbers of replicates starting before noon, in the afternoon (1200–1730), and early evening (1800–1930 hours). Since pilot observations showed that lampreys started to move up the ramp very quickly after lights were dimmed, irrespective of time of day, no attempt was made to adjust the photoperiod of the experimental animals prior to each replicate. Each ramp length was repeated three times, using eight naive lampreys in each replicate, and the three replicates were pooled for analysis.

*Experiment 3. Field validation.*—Field experiments were carried out in May 2007 at Shelter Valley Creek (SVC), a sea lamprey barrier stream near Grafton, Ontario. These trials were designed to create similar experimental conditions to those in the laboratory, but with a shorter holding time in captivity than in the laboratory experiments. The animals used (N = 110) were all caught in traps at the SVC sea lamprey barrier. Two-thirds of those were SVC resident lamprey, which were tagged on site as described above and allowed a minimum of 24 h rest before further use. We assumed that they were fully motivated to migrate as they had entered into the river voluntarily and proceeded to the barrier. The rest were PIT-tagged animals originating from Duffins Creek and Humber River (collected and handled as described above) that were brought from our holding facility and seeded downstream of the barrier into the river to increase the number of moving lamprey available to a different study.

The acrylic-lined ramp was rectangular (190 × 26 cm, length × width) angled at 45° and had 25-cm sidewalls. Water from a 150-L head tank drained at flow rates of 10 and 1.6 L/s via the ramp into a wire mesh cage (60 × 200 × 50 cm) in the stream. At the low-flow rate, the ramp was covered with a thin (1–2 mm) film of water. This was chosen to attract lamprey into a climbing mode. At the 10 L/s flow rate, the water depth on the ramp (1 cm) was comparable to the low-flow treatment in experiment 1 while velocity (160 cm/s in the middle of the ramp) was close to the high-flow treatment in experiment 1. Water velocity and depth on the ramp were adjusted to resemble conditions in experiments 1 and 2 by installing velocity-reducing small weirs and flow-smoothing brushes on the upper portion of the ramp. Water velocity was measured by filming a floating piece of wood on the ramp and measuring its velocity on the recording. The ramp had PIT-tag antennas mounted at about 10 and 25 cm above the water line.

Ten lampreys were held several hours before each trial in a plastic container within the downstream cage. At 2200 hours every night between May 18 and 28, the lampreys were released into the cage and their behavior on the ramp was recorded by PIT-tag readers and a video camera under infrared illumination. Seven high-flow and four low-flow replicates were carried out with 10 fish each at water temperatures between 11.5°C and 17.2°C (measured at 2400 hours).

### Data Analysis

Video analysis was used to assess the ability of lampreys to pass over a barrier by swimming, jumping, or climbing. Criteria for classifying behavioral categories were "passage"—the whole body of the fish has crossed to the upstream side of the ramp; "climbing"—an attempt to pass that included repeated attachment to the ramp by the oral sucker (>3 s attached to the ramp = "attachment"), followed by release and reattachment higher up; "jumping"—a rapid vertical or near-vertical movement out of the water that exposed more than three-quarters of the length of the fish to the air; "swimming"—an attempt to pass upstream that did not include jumping; "attempt to pass"—any movement toward the barrier where the head of the lamprey, as seen on the video recordings, reached 10 cm or more onto the ramp. An attempt ended with the lamprey either passing upstream or resubmerging in the lower pool.

For experiment 1, each attempt at passage was analyzed for successful passage, incidence of jumping, number and duration of attachments, and maximum height (along the ramp plane) reached by the tip of the snout. In the few cases that an animal reached beyond the top of the ramp in experiment 1, the height was scored as the maximum ramp height, which may have biased the calculated mean height to lower values than the actual mean height. The identity of an individual was determined by synchronizing the video with the PIT-tag records. This was sometimes not possible when several animals attempted to climb simultaneously because the PIT-tag antennae are unable to record more than one signal at a time. Attempts without individual identification were excluded from the statistical analysis. Repeated attempts by the same individual were pooled and the quantified observation (e.g., maximum height on ramp) averaged into one observation. This was done instead of taking the highest value for each fish and thereby bias the data by the number of attempts (because selecting maxima from a larger sample size yields greater values). For experiment 2, the only metric drawn from the videos was success or failure of passage attempts by tag-identified individuals. The field experiment (experiment 3) was analyzed in a similar fashion as experiment 1, but only the maximum height of attempts was determined. For the first three replicates of experiment 3, we determined how many attempts were observed that had no accompanying PIT-tag antenna record. This was done to get an estimate of a possible bias between measurements of apparent motivation between the field and the laboratory experiments, caused by possible differences in the detection efficiency of the PIT-tag antennae between the field and the laboratory experiments.

Heights achieved on the ramp were log-transformed after testing for normality and analyzed by either *t*-test or analysis of variance with post hoc comparisons done by Tukey tests. As observations of attachment time (experiment 1) or lag time until first observed attempt (experiment 2) could not be normalized with a simple transformation, a Kruskal–Wallis test was done on ranked observations. A $P_{\alpha}$ value of less than 0.05 was considered statistically significant in all tests. To evaluate the relative importance of ramp angle and ramp length in experiment 2 separately, we assessed the probability of passage in a logistic regression analysis using success and failure of the attempts as input parameters. The ramp length was transformed into proportion of total body length of an observed individual. For example, a 20-cm-long ramp would be scored as 0.5 relative ramp length for a 40-cm lamprey, and 0.66 relative ramp length for a 60-cm individual. The effect of the ramp parameters (relative length, angle) on the passage probability were tested by an effect likelihood ratio test.

## Results

### *Influence of Ramp Angle and Velocity on Climbing Behavior (Experiment 1)*

A total of 684 attempts at passage were made by 70 individuals, but no animal was able to successfully pass over the 35–40-cm ramp. In addition, no lampreys were observed jumping out of the water to pass the barrier. Instead, a passage attempt typically consisted of an individual emerging from the downstream pool and vigorously undulating its lower body in an effort to swim up the ramp and over the barrier. When it could not pass, the individual would either drop back into the pool or attach by its oral sucker to the ramp. After a variable time, it would release suction and either drop back into the pool or try again to reach higher on the ramp.

The number of lampreys attempting passage up the ramp ranged from five to seven (out of eight) fish in each trial, and the number of attempts per individual ranged among trials from a mean of 1.6 ± 0.5 (SD) to 12 ± 9.4 over the 4 h of observation. Because of the variability among replicates, there was no clear trend to suggest that a particular angle or water flow attracted more attempts. Females appeared to attempt passage more often than males (overall mean 6.9 versus 5.1 over 4 h), but this difference was not significant ($t = 1.21$, df = 84, $P_{\alpha} = 0.11$). Thirty-eight percent of all attempts seen on video had no accompanying PIT-tag identification.

The maximum height that a lamprey's head reached up on the ramp (not the vertical height) varied with ramp angle and water flow (Figure 2). Average height achieved decreased as ramp steepness increased. At low flow, mean maximum height decreased from 27.6 ± 1.1 (SE) cm (0.57 ± 0.08 BL) at 30° incline to 21.6 ± 0.7 cm (0.44 ± 0.04 BL) at 60° incline. The difference between angles was significant ($F_{2,47} = 19.97$, $P_{\alpha} < 0.001$), with performance on the steep angle significantly less than that on the other two angles (Tukey test). Water flow also significantly influenced maximum achieved height ($F_{2,30} = 4.48$, $P_{\alpha} < 0.05$), with low flow allowing a greater maximum height than medium flow (Tukey test). Also, there was a significant effect of body length on the maximum height reached on the ramp. A linear regression of body length on maximum height achieved for the pooled data from all treatments resulted in a best-fit maximum height of about 22 cm for a 40-cm-long individual and 27 cm for a 60-cm individual ($F_{1,81} = 16.28$, $P_{\alpha} < 0.001$, $R^2 = 0.17$).

Seventy-seven percent of all animals tested attached at least once to the ramp while attempting to pass. The maximum height reached on the ramp was on average about 2 cm (<10%) greater in attempts with attachments compared to those without attachment.

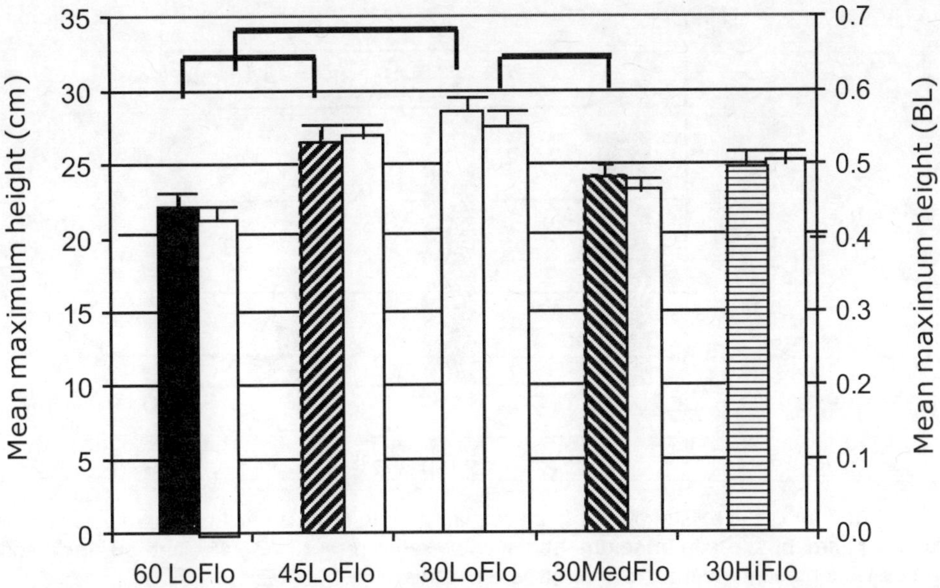


Figure 2. Mean maximum height (+SE) expressed in centimeters (left columns) and body lengths (right axis, right columns) sea lampreys reached with the tip of their head on inclined ramps with varying angles (60°, 45°, and 30°) and three water flows (LoFlo = low, MedFlo = medium, HiFlo = high) in experiment 1. Brackets denote significant differences in Tukey tests.

Using the pooled data from the three angles (at low flow), this difference was statistically significant (paired t-test on mean maximum height an individual achieved in attempts with and without attachment, $t = 2.81$, df = 35, $P_\alpha < 0.05$). However, an analysis of the height gain between subsequent attachments (31 cases where an individual attached more than once during an attempt) showed an average small height loss (mean = −1.3 ± 7.6 cm). This loss was not significantly different from zero ($t = 0.9$, df = 30, $P_\alpha = 0.38$).

The mean time of attachment varied greatly among individuals and within an individual from one attempt to the next. The mean attachment time per attachment event was calculated as the total attachment time divided by the total number of attempts for each individual. The medians for this observation ranged from a low of 146 s for the steep ramp treatment to 60° s for the 45° ramp at medium flow. However, no significant trends among ramp treatments were found in mean attached time per event (for different angles: $H = 2.02$, $n = 37$, $P_\alpha = 0.36$; for different flows: $H = 4.53$, $n = 24$, $P_\alpha = 0.10$).

### *Influence of Exposed Ramp Length on Climbing Behavior (Experiment 2).*

As in experiment 1, no attempts to jump were observed in more than 1,100 recorded attempts at passage. However, fish passage by swimming up the ramp was successful in some of the treatments (Figure 3). No fish passed over the ramps with a 15-cm vertical drop when the ramp angle was 30° or 45°. Lampreys had some success (8.3% of all attempts) in passing a 15-cm drop ramp when the angle was 60°. The highest success rate was 62% at the 60° angle and 10-cm vertical drop. Lampreys that passed the ramp were prevented from proceeding upstream and usually returned to the downstream pool to attempt passage again. There was no apparent effect of start time of the experiment on the time lag until the first observed attempt to pass the barrier (start before

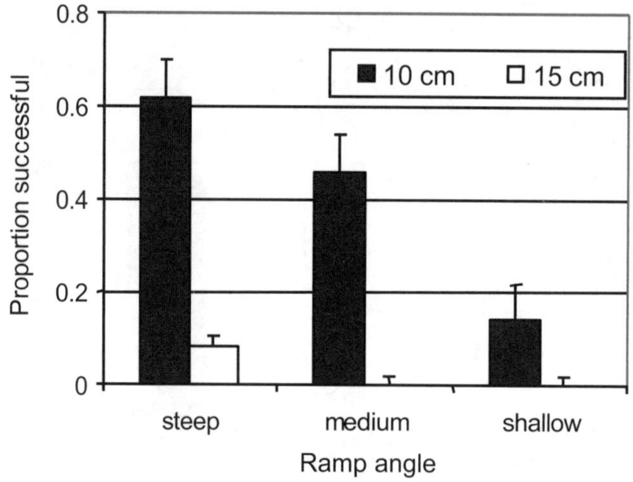


Figure 3. Proportion of successful attempts by sea lamprey to scale an inclined barrier of two vertical heights in experiment 2. Steep, medium, and shallow ramp refer to 60°, 45°, and 30° inclination, respectively. Bars represent means (+SE) of three replicates.

1200 hours, mean lag of 6.5 min; 1200–1730 hours, 3.5 min; after 1730 hours, 5 min; $H$ = 1.95, $n$ = 18, $P_\alpha = 0.38$).

One hundred twenty-nine individuals were included in the logistic regression analysis (Figure 4). The $R^2$ of the model was 0.387 when relative ramp length and angle were included and 0.345 when only relative ramp length was included. However, both parameters were significant (for relative length: $\chi^2 = 48.72$, $P_\alpha < 0.001$; for angle: $\chi^2 = 6.89$, $P_\alpha < 0.01$). According to the regression model, the probability of successful passage was zero when ramp length exceeded half of the lamprey body length. Shallow ramps allowed for slightly easier passage; a 0.5 probability of passage per attempt was reached at ramp lengths of between 37% (shallow angle) and 29% (steep angle) of lamprey body length.

## Field Validation (Experiment 3)

Of 110 animals used in the field validation experiment, 98% tried at least once to scale the ramp during a 4-h observation period. As in the laboratory experiments, often several individuals attempted to pass the ramp simultaneously. In the first three replicates, 50, 29, and 30% of all attempts had no accompanying PIT-tag record, for a mean of 36% (38% in the laboratory). More than 1,300 attempts could be attributed to identified individuals and were used in the analysis. The general behavior observed during attempts to pass was very similar to that observed in laboratory experiments. The mean maximum height obtained on the ramp was 28.9 ± 0.8 cm (SE) in the low-flow treatment and 25.1 ± 0.6 cm in the higher flow treatment (Figure 5). These numbers are quite similar to the mean maximum height achieved at the 45° ramp in experiment 1: 27.1 ± 0.5 cm in the low flow treatment and 24.2 ± 0.8 cm in the medium flow treatment, (termed "high-flow" in Figure 5). Comparing experiments 1 and 3 revealed that the higher flow rate led to significantly lower achieved heights in both, but there was no overall effect of field versus laboratory setting (Table 1). The number of observed attempts per individual, however, was much higher in the field experiment. While the mean number of attempts in the laboratory experiment at the 45° ramp was 4.3 over 4 h, in the field a mean of 12 attempts was recorded (mean 13.5 ± 11.3 for low flow and 11.1 ± 6.7 for high flow). Both experimental location and flow significantly impacted the number of attempts to pass the ramp (Table 1).

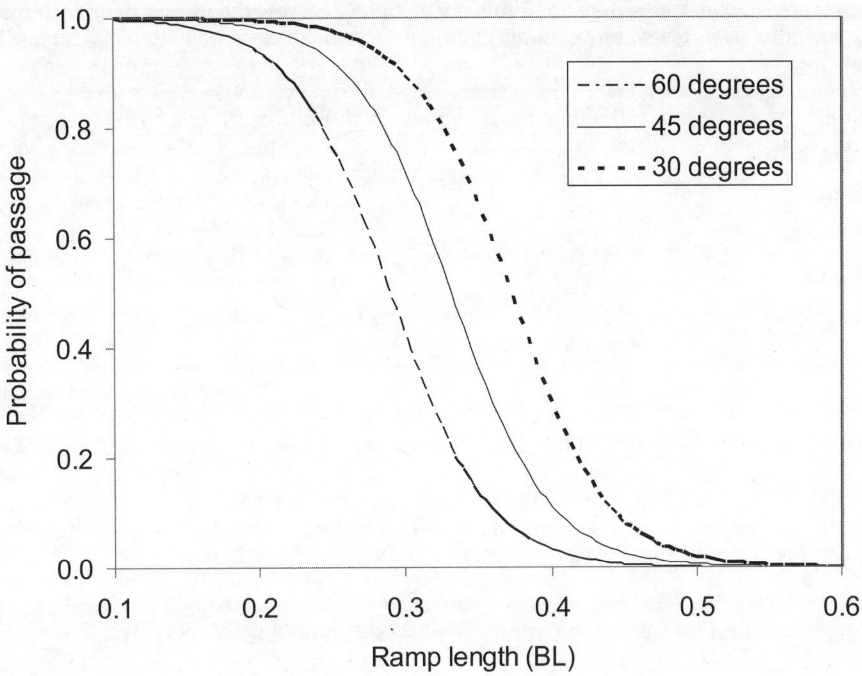


Figure 4. Results of a logistic regression analysis using the ramp passage success rate of 129 pooled individuals in experiment 2. The *x*-axis shows the length of the ramp, expressed as mean body length of lampreys used in the study. Each line represents a different ramp angle. See text for details.

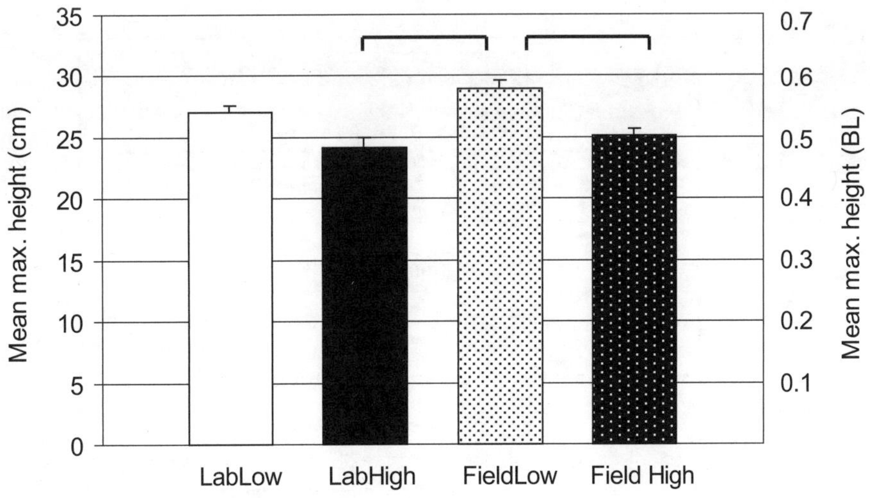


Figure 5. Comparison of mean maximum heights (+SE) in centimeters achieved by sea lampreys attempting to pass a 45° ramp in the laboratory and under field conditions. "Low" and "High" refer to relative flow rate (see text for details). The brackets denote significant differences in posthoc comparisons (Tukey test). The right axis (body lengths) is provided for illustrative purposes, where 25 cm = 0.5 body lengths based on a mean length of 50 cm for lampreys used in the experiments.

Table 1. Analysis of variance comparison of maximum height achieved and number of attempts by sea lampreys attempting to pass 45° ramps under laboratory ($N = 41$) and field ($N = 109$) conditions. NS = not significant.

| Dependent variable | Factor | Means | *F*-value | *P*-value |
|---|---|---|---|---|
| Maximum height (cm) | | | | |
| | Flow rate | Low: 28.2 High: 24.9 | 20.18 | <0.001 |
| | Lab-field | Lab: 26.1 Field: 26.5 | 3.16 | 0.08 NS |
| | Interaction | | 0.29 | 0.58 NS |
| Attempts per 4 h | | | | |
| | Flow rate | Low: 9.4 High: 10.3 | 4.81 | <0.05 |
| | Lab-field | Lab: 4.3 Field: 12.0 | 42.90 | <0.001 |
| | Interaction | | 2.93 | 0.09 NS |

Temperature was another factor in the sea lamprey's ability to swim up the ramp (Figure 6). Most of the replicates were carried out at a temperature similar to the one in the laboratory, plus two replicates below 13°C and one above 15°C. Linear regression analysis of the 69 animals observed in the high flow treatment revealed a statistically significant ($R^2 = 0.224$, $t = 4.36$, $P_\alpha < 0.001$) positive effect of temperature on the mean maximum height achieved on the ramp, with 22.5-cm maximum height calculated from the best-fit regression model at 11.5°C and 29 cm at 17.2°C.

## Discussion

This is the first detailed analysis of the behavioral mechanisms involved in sea lamprey passage over barriers. Our analysis of more than 3,000 passage attempts by more than 300 animals suggests that adult sea lamprey from the Great Lakes region have a limited ability to pass over smooth inclined barriers with restricted water flow.

We found no evidence, in either a laboratory or field setting, that sea lampreys can climb up wetted surfaces using an attach-twitch-attach

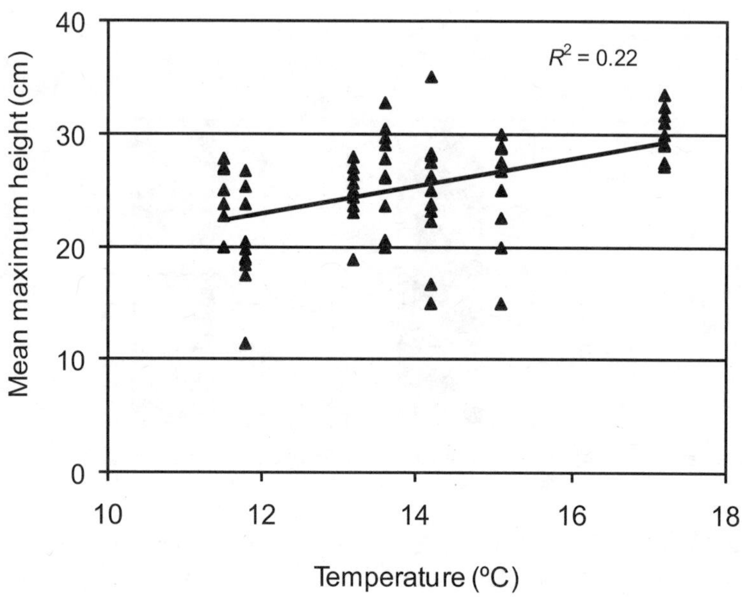


Figure 6. Effect of temperature on mean maximum height in centimeters reached while attempting to pass a ramp in experiment 3 with least-square regression and $R^2$ value.

strategy. Instead, when they detach, they either fall back or struggle to hold position. In the latter case, they gain no height on the ramp, only managing on average to return to their previous height. It thus appears that attachment to a barrier serves sea lampreys solely for the purpose of resting and preventing loss of gained ground in rapid water (Hanson 1980; Quintella et al. 2004) and is not an integral part of the technique by which this species passes over barriers. Youngs (1979), in his study of lamprey passage over low-head barriers, also remarked that sea lamprey did not gain vertical height from attachment to a barrier, but he did not make methodical observations to underpin his conclusion. The attach-twitch-attach cycle described in Pacific lampreys (Moser et al. 2005; Reinhardt et al. 2008) has never been witnessed in sea lampreys by us or by a number of sea lamprey experts we consulted (A. Haro, Conte Anadromous Fish Laboratory; S. Gephard, Connecticut Fisheries Agency; and R. Bergstedt, U.S. Geological Survey, Hammond Bay Laboratory; personal communications) and does not appear to be part of the behavioral repertoire of sea lampreys. Since anecdotal accounts by fishery professionals report that sea lampreys attached high out of the water on inclined dams or waterfalls, it is possible that small surface structure (e.g., ledges or crevices) facilitates vertical movement of lampreys out of the water. Blob et al. (2006) reported that surface roughness facilitates climbing in juveniles of some species of waterfall-climbing Hawaiian gobiids.

Our sea lampreys also did not attempt to jump over the barrier. Accounts from other researchers (e.g., Applegate 1950) and our own occasional observations in the holding tanks show that sea lampreys are capable of jumping clear out of the water. Our impression of startled lamprey jumping from their tanks is that the direction of jumps is rather erratic. Youngs (1979) also reported that sea lampreys were only seen jumping up vertically, and apparently not in an angle or direction that would help them pass over a barrier. Applegate (1950) reported "occasional" instances of sea lamprey jumping at the base of falls but did not mention if this behavior led to successful passage. We conclude that sea lamprey jumping at the base of barriers is not a directed movement and that passing a barrier by jumping from the plunge pool would be a rare event.

The only method of passing over the ramps we observed was burst swimming. Applegate (1950) described this method of swimming at barriers as "short, wriggling thrusts." Youngs (1979) similarly described lamprey passage behavior at low-head barriers as "exerted swimming rather than jumping." The analysis of passage success in experiment 2 suggests that at least half of the lamprey body must be immersed in the downstream pool for the fish to gain enough momentum to pass over the barrier. The posterior half of the lamprey body has the dorsal and caudal fins that are the likely source of most of the lateral leverage against the water, and thus thrust. It appears that when a substantial part of the lamprey is exposed to air, because the fish rears up at the ramp or tries to swim up in a thin nappe, it cannot get enough thrust for successful passage. It is possible that a thicker nappe on the ramps would have made passage easier, despite an accompanying greater water resistance, which we showed reduced achieved height. We observed both in the laboratory and in the field that lamprey occasionally interfered with each other during their attempts to pass the barrier, but since "bunching" at barriers is commonly seen in migrating sea lampreys and is believed to sometimes facilitate passage (Applegate 1950), we considered it a realistic situation and did not try to remove any group effect by observing animals singly.

The maximum distance sea lampreys can scale up on a smooth vertical or angled barrier is not more than half of their body length. Since the maximum length of adult sea lamprey in the Great Lakes region is less than 60 cm (Applegate 1950), we conclude that sea lamprey barriers longer or taller than 30 cm are insurmountable. The effect of body size in our experiment was such that a 10-cm increase in body length yielded a height advantage on the ramp of only 2.5 cm. As in other fish species, swim performance of sea lampreys is positively related to body size (Beamish 1974). Ocean-going sea lampreys are much larger at maturity (e.g., 95 cm mean total BL in one European population, Quintella et al. 2004) and thus most likely possess a greater ability to pass over vertical and inclined barriers than the populations we studied. Using Figure 4 as a guideline, we would expect 50% of lampreys 95 cm in BL to pass a steep (60°) barrier of about 24 cm vertical height (27.5-cm ramp length). However, further work is needed to determine if there are qualitative or quantitative differences in climbing ability between landlocked and anadromous sea lampreys.

Water temperature significantly influenced the ability of lampreys to swim up inclined wetted barriers. At the highest stream temperature (17.2°C), sea lampreys could reach about a third higher than at the lowest observed temperature (11.5°C). While this was not directly tested in our experiments, we expect that up to around 20°C, the rate of successful passage over barriers and other challenging stream obstructions would increase with temperature. This contention is supported by the positive relationship between temperature and swim performance of sea lampreys, described by Beamish (1974) and McAuley (1996). Although Beamish (1974) did not measure burst performance, the mode of swimming used by the lampreys in ascending the ramp, he did show that endurance in these animals nearly doubles between 5°C and 15°C.

Our field validation experiment showed that the motivation to move over barriers was lower under the laboratory conditions. On average, three times more attempts were recorded over the observation period in the field experiment. The overall ratio of successful to missed individual identification of an attempting animal by PIT tag was similar in the laboratory and in the field, indicating that the difference in motivation was real and not a result of differences in the detection efficiency of the equipment. One reason for the lower apparent motivation of fish we used in the laboratory may have been their being held in captivity for a longer time than the ones we used in the field and their sudden transfer from a lower holding temperature that suppressed activity levels (Binder and McDonald 2008); this may have reduced their drive to migrate upstream. Moreover, the laboratory set-up did not allow use of the sea lamprey migratory pheromone, a strong guiding factor in sea lamprey migration (Wagner et al. 2006). We are not certain if migratory pheromone was present in the field experiment. However, we are confident that our laboratory results are representative of the ability of sea lamprey in this size range to scale ramps. This is because we observed the same quality of behavior (i.e., absence of jumps and suction-aided climbing) and quite similar achieved heights under the various conditions. We interpret this to mean that a lower motivation to move is not necessarily tied to a lower maximum performance of the moving fish. Even if one acknowledges the fact that a number of diverging factors (e.g., flow regimes, temperature range, and handling of the animals) makes a direct comparison of field and laboratory results somewhat uncertain, our results show that laboratory experiments are useful and valid in sea lamprey behavioral studies.

The current low-head dam design guidelines prescribe a minimum crest height of 30 cm to assure that Great Lakes sea lamprey will not pass and include a lip to prevent climbing. Our results suggest that the usual lip is an unnecessary design detail because sea lampreys do not climb. Youngs (1979) came to a similar conclusion, as he did not find an effect of an overhanging lip on passage success. Youngs (1979) found that no lamprey passed over vertical barriers of 30 cm height, and less than 1% managed to pass over barriers greater than 15.7 cm. These results are similar to our observation of no passage at 30 cm vertical height and an 8% success rate at 14.5 cm vertical height and a near-vertical ramp angle. T. Binder and D. G. McDonald (unpublished data) found in laboratory experiments that a vertical barrier of 16 cm was sufficient to block all sea lampreys. Taken together, these three studies show that 30 cm is indeed an insurmountable barrier to the Great Lakes sea lamprey. In addition, some further reduction in barrier height seems possible without compromising full blockage of the lampreys, provided that the technical difficulty of maintaining minimum crest height in varying stream flows can be solved.

Reducing the barrier height might allow more nontarget species to pass over lamprey barriers. Furthermore, our results suggest that an inclined barrier with limited water flowing over it is a suitable barrier against sea lamprey passage. Based on our results, the vertical barrier height could be reduced to 15 cm, as long as distance on the ramp exceeds 30 cm. Some recent studies (Voegtle et al. 2002; Yasuda et al. 2004; Baker and Boubée 2006) have evaluated the suitability of ramps as fish passage devices and confirmed their usefulness for some smaller fishes, particularly if the surface is roughened in some way. Further work is necessary to determine if a ramp with limited flow could be a useful addition to a lamprey low-head barrier to allow passage of small or juvenile native fish species.

This study has demonstrated that unlike Pacific lampreys, sea lampreys lack the ability to scale smooth vertical surfaces and, though probably capable of jumping over barriers, they seem to lack the behavioral programming to do so. Our results provide an impetus for rethinking the design of low-head barriers. However, observational studies on "leaky" low-head barriers are needed to find out the reasons for failure of these barriers.

## Acknowledgments

We would like to thank the following individuals for technical support during the project: Jeremy Roos, Stacey Lee-Jenkins, Emily Smith, Marcie Ninness, Sarah Friedl, Scott Hawes, Keith Stamplecoskie, Chunfang Wang, and the technical staff of the Hagen Aqualab at University of Guelph. This research has been made possible by funding from the Great Lakes Fishery Commission. All procedures were approved by the Animal Care committee at the University of Guelph and followed Canadian Council of Animal Care guidelines.

## References

Applegate, V. C. 1950. Natural history of the sea lamprey, *Petromyzon marinus*, in Michigan. U.S. Fish and Wildlife Service, Special Scientific Report: Fisheries 55, Washington, D.C.

Baker, C. F., and J. A. T. Boubée. 2006. Upstream passage of inanga *Galaxias maculatus* and redfin bullies, *Gobiomorphus huttoni* over artificial ramps. Journal of Fish Biology 69:668–681.

Beamish, F. W. H. 1974. Swimming performance of adult sea lamprey, *Petromyzon marinus*, in relation to weight and temperature. Transactions of the American Fisheries Society 103:355–358.

Binder, T. R., and D. G. McDonald. 2008. The role of temperature in controlling diel activity in upstream migrant sea lampreys (*Petromyzon marinus*). Canadian Journal of Fisheries and Aquatic Sciences 65:1113–1121.

Blob, R. W., R. Rai, M. L. Julius, and H. L. Schoenfuss. 2006. Functional diversity in extreme environments: effects of locomotor style and substrate texture on the waterfall-climbing performance of Hawaiian gobiid fishes. Journal of Zoology 268:315–324.

GLFC (Great Lakes Fishery Commission). 2001. Strategic vision of the Great Lakes Fishery Commission for the first decade of the new millennium. Great Lakes Fishery Commission, Ann Arbor, Michigan.

Hanson, L. H. 1980. Study to determine the burst swimming speed of spawning-run sea lampreys (*Petromyzon marinus*). U.S. Fish and Wildlife Service, Research Completion Report, Millersburg, Michigan.

Lavis, D. S., A. Hallett, E. M. Koon, and T. C. McAuley. 2003. History of and advances in barriers as an alternative method to suppress sea lampreys in the Great Lakes. Journal of Great Lakes Research 29 (Supplement 1):362–372.

McAuley, T. C. (1996). Development of an instream velocity barrier to stop sea lamprey (*Petromyzon marinus*) migrations in Great Lake streams. Master's thesis. University of Manitoba, Winnipeg.

McLaughlin, R. L., L. Porto, D. L. G. Noakes, J. R. Baylis, L. M. Carl, H. R. Dodd, D. B. Hayes, J. D. Goldstein, and R. Randall. 2006. Effects of low-head barriers on stream fishes: taxonomic affiliations and morphological correlates of sensitive species. Canadian Journal for Fisheries and Aquatic Sciences 63:766–779.

McLaughlin, R. L., A. Hallett, T. C. Pratt, L. M. O'Connor, and D. G. McDonald, 2007. Research to guide use of barriers, traps and fishways to control sea lamprey. Journal of Great Lakes Research 33 (Special Issue 2): 7–19.

Moser, M. L., D. A. Ogden, B. J. Burke, and C. A. Peery. 2005. Evaluation of lamprey collector in the Bradford Island makeup water channel, Bonneville Dam, 2003. Report to U.S. Army Corps of Engineers, Portland District, Contract E96950021, Portland, Oregon.

Pearce, W. A., R. A. Braem, S. M. Dustin, and J. J. Tibbles. 1980. Sea lamprey (*Petromyzon marinus*) in the lower Great Lakes. Canadian Journal of Fisheries and Aquatic Sciences 37:1802–1810.

Porto L. M., R. L., Mclaughlin, and D. L. G. Noakes. 1999. Low-head barrier dams restrict the movements of fishes in two Lake Ontario streams. North American Journal of Fisheries Management 19:1028–1036.

Quintella, B. R., N. O. Andrade, A. Koed, and P. R. Almeida. 2004. Behavioural patterns of sea lampreys' spawning migration through difficult passage areas, studied by electromyogram telemetry. Journal of Fish Biology 65:961–972.

Reinhardt, U. G., L. Eidietis, S. E. Friedl, and M. L. Moser. 2008. Pacific lamprey climbing behavior. Canadian Journal of Zoology 86:1264–1272.

Voegtle, B, M. Larinier, and P. Bosc. 2002. Experimental study of the climbing capabilities of the goby *Sicyopterus Lagocephalus* (Pallas, 1770) for the design of upstream facilities at the Salazie diversion water intakes (Reunion Island). Bulletin Français de la Peche et de la Pisciculture 364:109–120.

Wagner, C. M., M. L. Jones, M. B. Twohey, and P. W. Sorensen. 2006. A field test verifies that pheromones can be useful for sea lamprey (*Petromyzon marinus*) control in the Great Lakes. Canadian Journal of Fisheries and Aquatic Sciences 63:475–479.

Yasuda, Y., I. Ohtsu, and M. Takahashi. 2004. New portable fishway design for existing trapezoidal weirs. Journal Environmental Engineering Sciences 3:391–401.

Youngs, W. D. 1979. Evaluation of barrier dams to adult sea lamprey migration. Great Lakes Fishery Commission, Project Completion, Report, Ann Arbor, Michigan.

American Fisheries Society Symposium 72:139–151, 2009


# Predation on Lampreys

Philip A. Cochran*
*Biology Department, Saint Mary's University of Minnesota*
*700 Terrace Heights, Winona, Minnesota 55987, USA*

*Abstract.*—Reported predators of lamprey include a variety of fishes, amphibians, reptiles, birds, and mammals, and predation on lamprey is known in both marine and freshwater habitats. Although lampreys are not typically prominent in the reported diets of predators, it does not follow that predation is not an important source of lamprey mortality. Concentrations of migrating and spawning lampreys may be especially vulnerable. Assemblages of predators on lampreys have changed through human activities such as stocking and harvest of fishes. In southeastern Minnesota, for example, most of the 1,145 km in 139 streams that currently are managed for trout now support brown trout *Salmo trutta*, an exotic species that has been reported to prey on several species of lamprey. Prior to its establishment, relatively few fish in these streams would have been capable of feeding on large ammocoetes or adult lampreys.

## Introduction

Interest in the ecology of lampreys, including their predators, has stemmed in large part from two perspectives. While the adverse effects of the sea lamprey *Petromyzon marinus* in the Great Lakes have fostered a negative attitude toward this species, with an emphasis on reducing its numbers, "native" lampreys have been viewed in a more positive light in many cultures, especially where they have been used as food. In particular, concern about threatened or endangered lampreys (e.g., Close et al. 2002) has stimulated interest in ways to enhance their populations.

In what he claimed was the first published report on the natural enemies of lampreys, Surface (1899, cited by Wigley 1959) suggested that lampreys were eaten by many predators, including raccoons, muskrats, rats, minks, weasels, foxes, hawks, and owls. However, his evidence in most cases was circumstantial (e.g., tracks along the shoreline or interpretations of wounds on dead lampreys). Information on the predators of most species of lampreys that is based on direct evidence (e.g., direct observation of predation events or analysis of stomach contents) remains scanty and scattered. The purpose of this paper is to review predation on lampreys and to examine trends in predation with respect to predator taxa, habitat, and lamprey life history stage.

## Methods

I have tried to compile information on lamprey predators in a form that will be easy to amend in the future (Appendix A). Parasites of lampreys were not included because they have been reviewed previously (Appy and Anderson 1981). Whenever possible, I have cited original rather than secondary reports, although I have relied on some of the latter (Scott and Crossman 1973; Holčík 1986a). A number of accounts of predation on lampreys have appeared in works of fiction or speculation. These have been reviewed separately (Cochran, in press).

When it was possible to interpret lamprey life history stage from published accounts,

* Corresponding author: pcochran@smumn.edu

I included that information in Appendix A. Lamprey life histories include parasitic and nonparasitic trajectories. All species spawn in freshwater, typically in shallow, gravel-bottomed stream habitat, and the adults die after spawning. Eggs hatch into burrowing, filter-feeding larvae (ammocoetes) that inhabit areas with finer sediments and quieter water than those used for spawning. The larval stage may last for 4 or 5 years or longer. The adults of nonparasitic species engage in spawning shortly after completing metamorphosis from the ammocoetes stage. In parasitic species, metamorphosis is followed by a period of parasitic feeding on fish or cetaceans. Depending on the lamprey species, parasitic-stage individuals may move downstream from the larval habitat to seek hosts in large rivers, lakes, or marine habitat. This necessitates an upstream migration prior to spawning. For the purposes of this review, all postmetamorphic lampreys were referred to as "adults."

## Results and Discussion

### *New data*

I have recorded some instances of predation on lampreys during studies of other aspects of their biology (e.g., Cochran et al. 1992). In addition, because of my interest in lampreys, colleagues and students have provided me with observations of predation on several lamprey species.

*Hornyhead chub.*—I have often observed cyprinids in the vicinity of spawning lampreys, presumably feeding on eggs. Close observation on May 26, 2006 confirmed that juvenile hornyhead chubs *Nocomis biguttatus* were consuming the eggs of the southern brook lamprey *Ichthyomyzon gagei* in the Yellow River, Washburn County, Wisconsin.

*Brown trout.*—Predation by brown trout *Salmo trutta* on northern brook lampreys *I. fossor* occurred in Rocky Run Creek, Columbia County, Wisconsin (J. Lyons, Wisconsin Department of Natural Resources, personal communication). Stomach contents of six trout (total length [TL]: 23–28 cm) and three yellow perch *Perca flavescens* (TL: 13–18 cm) were collected on June 6, 1995 over 500 m of the stream. A 23-cm trout collected at the lower end of the reach at 1000 hours contained a well-digested lamprey 100 mm long. A 25-cm trout collected at the upper end of the reach at 1300 hours contained a well-digested 100-mm lamprey and a smaller, mature male northern brook lamprey that was still alive. The other fish had eaten invertebrates but contained no lampreys. A brown trout (TL: 41 cm) caught at midday on August 5, 1997 in Wedde Creek, Marquette County, Wisconsin (A. Leisten, Saint Norbert College, personal communication) contained fragments of two lampreys, which, based on the locality, were probably northern brook lampreys (Lyons et al. 1997). At this time of year, it would be expected that only ammocoetes would be present, and my examination revealed in the less-digested specimen a well-developed digestive tract that contained fine-gained sediment. Cochran et al. (1992) also reported a case of predation by a brown trout on a northern brook lamprey.

One of 14 brown trout (TL: 33 cm) collected on May 8, 2003 in Rollingstone Creek, Winona County, Minnesota, contained an adult 16-cm American brook lamprey (A. K. Schumacher, Saint Mary's University, personal communication). I have found no previous accounts of predation by fish on this species. However, anglers may be less likely to report predation on American brook lampreys because they spawn earlier than other lamprey species, often before the start of the fishing season (trout season in Minnesota starts on April 15, whereas in Wisconsin it starts on May 6).

*Walleye.*—Several walleye *Sander vitreus* containing silver lampreys *I. unicuspis* have been collected in the St. Clair River in Michigan, with as many as three to four small lampreys in the same fish (M. V. Thomas, Michigan Department of Natural Resources, personal communication). In 2002, a walleye caught in the Mississippi River in Wabasha County, Minnesota, contained a silver lamprey approximately 26 cm long.

*Spotted sandpiper.*—Spotted sandpipers *Actitus macularis* have been observed digging ammocoetes *Lampetra* sp. out of mud banks in the Pacific Northwest (A. F. Brumo, Confederated Tribes of Warm Springs, personal communication).

## Previous reports

*Predator taxa and habitats.*—A wide variety of predators from all major taxa of vertebrates have been reported to feed on lampreys (Appendix A). Although predation has been reported in both marine and freshwater habitats, accounts for marine habitats are more sporadic, perhaps reflecting the fact that lampreys themselves in marine habitats are more scattered. A notable exception is the consumption of large numbers of pouched lampreys *Geotria australis* by albatrosses *Thalassarche* spp., which Potter et al. (1979) interpreted to reflect their concentration near the surface.

Even in freshwater, most reports of lamprey as prey are anecdotal in nature and involve small numbers of lampreys, but some trends are apparent. Among fishes, brown trout, northern pike *Esox lucius*, burbot *Lota lota*, and walleye have been reported as predators of multiple species of lampreys (Appendix A). These species share with lampreys a preference for cool or cold water, they tend to occur in habitats where lampreys are concentrated during upstream migration or spawning, and, in the case of burbot and walleye, they are adapted to feed at low light levels, when lampreys tend to be most active (Cochran 1986). Among birds, gulls *Larus* spp. and mergansers *Mergus* spp. are among the most widely reported predators (Appendix A) and tend to take advantage primarily of concentrations of upstream migrant, spawning, or spent lampreys. Mergansers in particular provide one of the few examples in which lampreys form a significant fraction of a predator's diet (Sjöberg 1985, 1988, 1989) to the extent that the timing of their seasonal and diel activity patterns is related to the activity of the lampreys. The ability of the mergansers to exploit nocturnal lampreys effectively is related to the favorable light regimes that occur seasonally at high latitudes (Sjöberg 1985), a factor that may benefit other lamprey predators in similar regions (Hardisty 2006). Among mammals, pinnipeds are the most notable predators of lampreys, usually in estuaries or the lower reaches of large rivers (Roffe and Mate 1984).

I have not included in Appendix A accounts of humans as direct consumers of lampreys. Neither have I included accounts of lampreys used as bait. However, their historical or contemporary importance as bait for a variety of fishes in Eurasia (e.g., the species accounts in Holčík 1986a) and North America (e.g., Vladykov 1949; Scott and Crossman 1973) is indicative of the potential importance of lampreys as prey. Most often, ammocoetes are used as bait because they can be harvested in greater numbers than adults, but upstream migrant or spawning adults have been harvested for bait, and small adult silver lampreys have been used in Wisconsin (Becker 1983). Artificial lamprey imitations have also been produced for use in fly fishing (e.g., Whitlock 1994). In addition to their use in recreational and commercial fisheries, lampreys have been used as bait by researchers, notably for white sturgeon *Acipenser transmontanus* (e.g., Elliott and Beamesderfer 1990).

*Lamprey life history stage.*—It has often been suggested that cyprinids, cottids, or other small fishes prey upon eggs or young ammocoetes, although fewer studies have actually reported observations of eggs being eaten or recovered among gut contents (Appendix A), and Brumo's (2006) observations suggested the possibility that salmonids fed on insects dislodged by spawning lampreys. Brumo (2006) found speckled dace *Rhinichthys osculus* feeding in Pacific lamprey *Lampetra tridentata* nests at densities as high as approximately 100 individuals/m$^2$, but only at temperatures greater than 14°C. Egg predation did not have a significant effect on relative survival of ammocoetes cohorts. Although Brumo (2006) acknowledged that some eggs may not have been safely buried in the substrate even if they had not been consumed by speckled dace, he did observe speckled dace actively burrowing into gravel within nests in search of eggs.

The ammocoetes phase is the longest stage in the lamprey life cycle, ammocoetes often occur at high density relative to other stages (with the exception of spawning adults), and ammocoetes are relatively sedentary and predictable in their locations within stream systems. For these reasons, the ammocoetes stage was targeted as most vulnerable by biologists attempting to control sea lampreys in the Great Lakes. It is surprising, therefore, that relatively few accounts have recorded observations of predation on ammocoetes (Appendix A). Even so, representatives of at least four vertebrate classes have been reported as predators of ammocoetes. That sturgeons have been reported as predators suggests the possibility that other benthic fishes may also feed on ammocoetes. For example, common carp *Cyprinus carpio* and many catostomids are known to remove invertebrates from bottom sediments and could probably feed just as easily on ammocoetes, although their pharyngeal teeth might cause larval lampreys to be mechanically processed beyond easy recognition.

Adult lampreys are most concentrated during upstream migrations and spawning, and parasitic-phase adults are most dispersed during the time that they are actively feeding. Thus, accounts of predation on parasitic-phase lampreys within their feeding ranges tend to be anecdotal, whereas some of the most consistently reported lamprey predators are those that target upstream migrants in estuaries or rivers. These include mergansers, seals, and sea lions (Appendix A). In addition, observations of predation on nonparasitic lampreys tend to occur during their spawning periods, when nesting by groups of adults occurs during daylight periods as well as at night.

*Why is predation on lampreys not observed more frequently?*—Lampreys are a potential source of food that comes in a range of sizes. They have no obvious physical defenses and they are conveniently shaped for predators that swallow their food whole (but see Wolf and Jones 1989). They may also be relatively rich in energy relative to other prey types; reported energy concentrations (kJ/g wet mass) range from 2.35 to 6.01, depending on life history stage, size, and time since cessation of feeding (Beamish and Legrow 1983; Whyte et al. 1993; Cochran et al. 2003). Why have lampreys not been reported as prey more frequently?

Some lamprey predation may go undetected. Many diet surveys occur at times of the year when parasitic lampreys are widely scattered. Even though an individual predator might readily consume any lamprey that it encountered during these times, its chance of being included in a sample for diet analysis would be low. Moreover, because lampreys lack bone or other hard structures resistant to digestion, they are less likely to be detected among gut contents. This is especially true for eggs and early ammocoetes. Roffe and Mate (1984) noted that because lampreys are relatively small, pinnipeds are more likely to consume them underwater rather than return with them to the surface. Thus, studies of pinniped diets based on direct observations of surface-feeding animals are likely to underestimate the importance of lampreys as prey.

Lampreys may not be completely defenseless against predators. Some experiments have indicated that salmonines reject Pacific lampreys and western brook lampreys *L. richardsoni* because of distasteful secretions released from the skin (Pfeiffer and Pletcher 1964; Scott and Crossman 1973). Moreover, flexible anguilliform bodies and smooth, slippery texture, which allow lampreys to burrow as ammocoetes and exploit interstices in the substrate as adults, may render them difficult for some predators to grasp or manipulate. I have observed small lampreys in laboratory tanks that were ingested by salmonids only to emerge from beneath their opercular flaps.

Most importantly, many aspects of lamprey life history minimize their exposure to predators. Spawning in freshwater may have evolved at a time when relatively few predators occurred there relative to marine habitats (Wallace 2007). Ammocoetes spend much of their time buried in areas within stream systems that are relatively inaccessible to the types of large predators that might be capable of sifting or excavating them from the sediments. At least some species of parasitic lampreys tend to remain under cover when they are not attached to hosts (Cochran and Gripentrog 1992). Many activities engaged in by lampreys occur primarily at night (Cochran 1986) when they are less vulnerable to visually hunting predators. An exception is spawning, but this activity tends to occur relatively early in the year when relatively low water temperatures limit the metabolic and food consumption rates of some ectothermic predators. Several species of lampreys at least occasionally spawn in cavities beneath boulders or other cover objects where both the adults and eggs may be less vulnerable to at least some types of predators, but the northern brook lamprey, which spawns relatively late in the spring or early summer, is especially prone to do so (Cochran and Gripentrog 1992).

Luiselli et al. (2007) examined the diet of the aquatic snake *Natrix tessellata* in central Italy and compared observed prey species with potential prey species that were not consumed (including ammocoetes of the European brook lamprey *L. planeri*. The snakes were significantly less likely to eat prey that used mud substrate, were diurnal, or were not active foragers.

*The importance of predation on lampreys to humans and vice versa.*—It is conceivable that humans benefit when lampreys are preyed upon. It has been suggested that lampreys in tributaries to the Pacific coast of North America buffer more highly valued salmonids against predators (Landeen and Pinkham 1999; Close et al. 2002). In the Great Lakes, it might be concluded that any predation on the sea lamprey is a benefit, but no one has suggested that predation can regulate sea lamprey populations there.

It is more likely that humans have affected lampreys by changing assemblages of predators via stocking and harvest. Some of the cases of predation on lampreys listed in Appendix A involve nonnative predators, including the Sacramento pikeminnow *Ptychocheilus grandis*, brown trout, and rainbow trout *Oncorhynchus mykiss*. In some situations, these new predators might negatively affect lamprey populations. In southeastern Minnesota, for example, most of the 1,145 km in 139 streams that currently are managed for trout now support brown trout (Thorn et al. 1997). Prior to its establishment, relatively few fishes in these streams would have been capable of feeding on large ammocoetes or adult lampreys, especially at temperatures too warm for native brook trout *Salvelinus fontinalis*. It is unknown whether the somewhat spotty distribution of American brook lampreys and northern brook lampreys in this region reflects at least in part the past impact of this predation, although other human activities in addition to stocking must have played a role (Mundahl et al. 2006). Elsewhere, Kappus et al. (1995) expressed concern about the stocking of catchable brown and rainbow trout in areas inhabited by the endangered European brook lamprey after observing these species aggregating near spawning areas and feeding on adult lampreys.

Because few predators have been reported to feed extensively on lampreys, it might be tempting to dismiss concerns about predation on lampreys of threatened or endangered status. However, although lampreys are not typically prominent in the reported diets of predators, it does not follow that predation is not an important source of lamprey mortality. Concentrations of migrating and spawning lampreys may be especially vulnerable.

## Acknowledgments

I thank all biologists indicated in the text who shared with me their observations of predation on lampreys. Cindy Paszkowski and William Poly supplied me with copies of obscure articles from the literature. The reviewers provided many useful comments on a previous draft of this manuscript.

## References

Allin, A. E. 1951. Records of the sea lamprey and the silver lamprey from the Canadian waters of the western end of Lake Superior. Canadian Field-Naturalist 65:184–185.

Applegate, V. C. 1950. Natural history of the sea lamprey, *Petromyzon marinus*, in Michigan. U.S. Fish and Wildlife Service, Special Scientific Report: Fisheries No. 55, Washington, D.C.

Appy, R. G., and R. C. Anderson. 1981. The parasites of lampreys. Pages 1–42 *in* M. W. Hardisty and I. C. Potter, editors. The biology of lampreys, volume 3. Academic Press, London.

Beamish, F. W. H. 1980a. Biology of the North American anadromous sea lamprey, *Petromyzon marinus*. Canadian Journal of Fisheries and Aquatic Sciences 37:1924–1943.

Beamish, F. W. H., and M. Legrow. 1983. Bioenergetics of the southern brook lamprey, *Ichthyomyzon gagei*. Journal of Animal Ecology 52:575–590.

Beamish, R. J. 1980b. Adult biology of the river lamprey (*Lampetra ayresi*) and the Pacific lamprey (*Lampetra tridentata*) from the Pacific coast of Canada. Canadian Journal of Fisheries and Aquatic Sciences 37:1906–1923.

Becker, G. C. 1983. Fishes of Wisconsin. University of Wisconsin Press, Madison.

Bonnot, P. 1951. The sea lions, seals and sea otter of the California coast. California Fish and Game 37:371–389.

Braswell, A. L. 2005. *Necturus lewisi* Brimley, 1924. Pages 868–870 *in* M. Lanoo, editor. Amphibian declines: the conservation status of United State species. University of California Press, Berkeley.

Brown, E. E. 1958. Feeding habits of the northern water snake, *Natrix sipedon sipedon* Linnaeus. Zoologica 43:55–71.

Brown, L. R., and P. B. Moyle. 1997. Invading species in the Eel River, California: successes, failures, and relationships with resident species. Environmental Biology of Fishes 49:271–291.

Browne, P., J. L. Laake, and R. L. deLong. 2002. Improving pinniped diet analysis through identification of multiple skeletal structures in fecal samples. Fishery Bulletin 100:423–433.

Brumo, A. F. 2006. Spawning, larval recruitment, and early life survival of Pacific lamprey in the South Fork Coquille River, Oregon. Master's thesis. Oregon State University, Corvallis.

Buchwald, D. G. 1968. The Arctic lamprey of Great Slave Lake, N.W.T. Master's thesis. University of Alberta, Edmonton, Alberta, Canada.

Bury, R. B. 1986. Feeding ecology of the turtle, *Clemmys marmorata*. Journal of Herpetology 20:515–521.

Chan, M. D., and G. R. Parsons. 2000. Aspects of brown madtom, *Noturus phaeus*, life history in northern Mississippi. Copeia 2000:757–762.

Churchill, W. S. 1945. The brook lamprey in the Brule River. Transactions of the Wisconsin Academy of Sciences, Arts and Letters 37:337–346.

Close, D. A., M. S. Fitzpatrick, and H. W. Li. 2002. The ecological and cultural importance of a species at risk of extinction, Pacific lamprey. Fisheries 27:19–25.

Close, D. A., M. S. Fitzpatrick, H. W. Li, B. Parker, D. Hatch, and G. James. 1995. Status report of the Pacific lamprey (*Lampetra trzdentata* [sic]) in the Columbia River basin. U.S. Department of Energy, Bonneville Power Administration, Division of Fish and Wildlife, Project Number 94–026, Washington, D.C.

Cochran, P. A. 1986. The daily timing of lamprey attacks. Environmental Biology of Fishes 16:325–329.

Cochran, P. A. In press. Fictional accounts of predation on lampreys. American Currents.

Cochran, P. A., and A. P. Gripentrog. 1992. Aggregation and spawning by lampreys (genus *Ichthyomyzon*) beneath cover. Environmental Biology of Fishes 33:381–387.

Cochran, P. A., J. Y. Hodgson, and A. P. Kinziger. 2003. Change in energy density of the sea lamprey (*Petromyzon marinus*) during its parasitic phase: implications for modeling food consumption and growth. Journal of Great Lakes Research 29 (Supplement 1):297–306.

Cochran, P. A., A. A. Leisten, and M. E. Sneen. 1992. Cases of predation and parasitism on lampreys in Wisconsin. Journal of Freshwater Ecology 7:435–436.

Collis, K., D. Ruby, D. P. Craig, S. Adamany, J. Y. Adkins, and D. E. Lyons. 2002. Colony size and diet composition of piscivorous waterbirds on the lower Columbia River: implications for losses of juvenile salmonids to avian predation. Transactions of the American Fisheries Society 131:537–550.

Couch, S. L. L., and M. M. Lance. 2004. Diet composition and breeding success of Brandt's cormorants in the Columbia River estuary. Northwestern Naturalist 85:62–66.

Davis, R. M. 1967. Parasitism by newly transformed anadromous sea lampreys on landlocked salmon and other fishes in a coastal Maine lake. Transactions of the American Fisheries Society 96:11–16.

DeLong, J. 2000. Walking in 100-year-old footsteps in southern Oregon. American Currents 26:1–10.

Denoncourt, C. E., and J. R. Stauffer, Jr. 1993. Feeding selectivity of the American eel *Anguilla rostrata* (LeSueur) in the Upper Delaware River. American Midland Naturalist 129:301–308.

Ebert, D. A. 1986. Biological aspects of the sixgill shark, *Hexanchus griseus*. Copeia 1986:131–135.

Elliott, J. C., and R. C. Beamesderfer. 1990. Comparison of efficiency and selectivity of three gears used to sample white sturgeon in a Columbia River reservoir. California Fish and Game 76:174–180.

Feltham, M. J. 1990. The diet of red-breasted mergansers (*Mergus serrator*) during the smolt run in N.E. Scotland: the importance of salmon (*Salmo salar*) smolts and parr. Journal of Zoology, London 222:285–292.

Gregory, R. S., and C. D. Levings. 1998. Turbidity reduces predation on migrating juvenile Pacific salmon. Transactions of the American Fisheries Society 127:275–285.

Griffiths, D. E. 1980. The Slave River pelicans. Alberta Naturalist 10:138–143.

Hagelin, L. 1959. Further aquarium observations on the spawning habits of the river lamprey (*Petromyzon fluviatilis*). Oikos 10:50–64.

Hardisty, M. W. 1986. *Lampetra planeri* (Bloch, 1784). Pages 279–304 *in* J. Holčík, editor. The freshwater fishes of Europe, volume 1, part I. Petromyzontiformes. AULA-Verlag GmbH, Wiesbaden, Germany.

Hardisty, M. W. 2006. Lampreys: life without jaws. Forrest Text, Cardigan, UK.

Heard, W. R. 1966. Observations on lampreys in the Naknek River system of southwest Alaska. Copeia 1966:332–339.

Himes, J. G. 2003. Diet composition of *Nerodia sipedon* (Serpentes: Colubridae) and its dietary overlap with, and chemical recognition of *Agkistrodon piscivorus* (Serpents: Viperidae). Amphibia-Reptilia 24:181–188.

Holčík, J., editor. 1986a. The freshwater fishes of Europe, volume 1, part I. Petromyzontiformes. AULA-Verlag GmbH, Wiesbaden, Germany.

Holčík, J. 1986b. *Lethenteron japonicum* (Martens, 1968). Pages 198–219 *in* J. Holčík, editor. The freshwater fishes of Europe, volume 1, part I. Petromyzontiformes. AULA-Verlag GmbH, Wiesbaden, Germany.

Holčík, J. 1986c. *Lethenteron kessleri* (Anakin, 1905). Pages 220–236 *in* J. Holčík, editor. The freshwater fishes of Europe, volume 1, part I. Petromyzontiformes. AULA-Verlag GmbH, Wiesbaden, Germany.

Holčík, J., and C. B. Renaud. 1986. *Eudontomyzon mariae* (Berg, 1931). Pages 165–185 *in* J. Holčík, editor. The freshwater fishes of Europe, volume 1, part I. Petromyzontiformes. AULA-Verlag GmbH, Wiesbaden, Germany.

Hubbs, C. L. 1967. Occurrence of the Pacific lamprey, *Entosphenus tridentatus*, off Baja California and in streams of southern California; with remarks on its nomenclature. Transactions of the San Diego Society of Natural History 14:301–312.

Hubbs, C. L. 1971. *Lampetra* (*Entosphenus*) *lethophaga*, new species, the nonparasitic derivative of the Pacific lamprey. Transactions of the San Diego Society of Natural History 16:125–164.

Hubbs, C. L., and M. B. Trautman. 1937. A revision of the lamprey genus *Ichthyomyzon*. University of Michigan Museum of Zoology Miscellaneous Publication No. 35.

Jameson, R. J., and K. W. Kenyon. 1977. Prey of sea lions in the Rogue River, Oregon. Journal of Mammalogy 58:672.

Jang, M.-H., and M. C. Lucas. 2005. Reproductive ecology of the river lamprey. Journal of Fish Biology 66:499–512.

Jordan, D. S., and C. H. Gilbert. 1899. The fishes of the Bering Sea. Pages 433–492 *in* D. S. Jordan, editor. The fur-seals and fur seal islands of the North Pacific Ocean. Reports of the fur-seal investigations 1896–1897, Washington (part 3). Government Printing Office, Washington, D.C.

Kappus, B., W. Jansen, P. Fok, and H. Rahmann. 1995. Threatened lamprey populations of the Danube basin within Baden-Württemberg, Germany. Miscellanea Zoologica Hungarica 10:85–98.

Laake, J. L., P. Browne, R. L. DeJong, and H. R. Huber. 2002. Pinniped diet composition: a comparison of estimation models. Fishery Bulletin 100:434–447.

Lagler, K. F., and J. C. Salyer, II. 1947. Food and habits of the common watersnake, *Natrix s. sipedon*, in Michigan. Papers of the Michigan Academy of Sciences, Arts and Letters 31:169–180.

Landeen, D., and A. Pinkham. 1999. Salmon and his people: fish & fishing in Nez Perce culture. Confluence Press, Lewiston, Idaho.

Leonard, J. W., and D. S. Shetter. 1936. Studies on merganser depredations in Michigan trout waters. Transactions of the American Fisheries Society 66:335–337.

Little, A. S., W. M. Tonn, R. S. Tallman, and J. D. Reist. 1998. Seasonal variation in diet and trophic relationships within the fish communities of the lower Slave River, Northwest Territories, Canada. Environmental Biology of Fishes 53:429–445.

Ludwig, J. P., C. N. Hull, M. E. Ludwig, and H. J. Auman. 1989. Food habits and feeding ecology of nesting double-crested cormorants in the upper Great Lakes, 1986–1989. The Jack-Pine Warbler 67:117–129.

Luiselli, L., D. Capizzi, E. Filippi, C. Anibaldi, L. Rugiero, and M. Capula. 2007. Comparative diets of three populations of an aquatic snake (*Natrix tessellata*, Colubridae) from Mediterranean streams with different hydric regimes Copeia 2007:426–435.

Lutterbie, G. W. 1976. The darters (Pisces: Percidae: Etheostomatinae) of Wisconsin. Master's thesis. University of Wisconsin-Stevens Point, Stevens Point.

Lyons, J., P. A. Cochran, and M. E. Sneen. 1997. Taxonomic status and distribution of the lamprey *Ichthyomyzon cf. gagei*. American Midland Naturalist 138:69–76.

Matson, T. O. 2005. *Necturus maculosus* (Rafinesque, 1818). Pages 870–871 *in* M. Lanoo, editor. Amphibian declines: the conservation status of United State species. University of California Press, Berkeley.

McCafferty, D. J. 2005. The dietary response of otters (*Lutra lutra*) to introduced ruffe (*Gymnocephalus cernuus*) in Loch Lomond. Journal of Zoology, London 266:255–260.

McPhail, J. D., and C. C. Lindsey. 1970. Freshwater fishes of northwestern Canada and Alaska. Fisheries Research Board of Canada Bulletin 173.

Migdalski, E. C. 1962. Fresh water sport fishes of North America. Ronald Press Company, New York.

Moore, G. A., and J. M. Paden. 1950. The fishes of the Illinois River in Oklahoma and Arkansas. American Midland Naturalist 44:76–95.

Mundahl, N. D., G. Sayeed, S. Taubel, C. Erickson, A. Zaletel, and J. Cousins. 2006. Densities and habitat of American brook lamprey (*Lampetra appendix*) larvae in Minnesota. American Midland Naturalist 156:11–22.

Nickerson, M. A., and R. E. Ashton, Jr. 1983. Lampreys in the diet of hellbender *Cryptobranchus alleganiensis* (Daudin), and the Neuse River waterdog *Necturus lewisi* (Brimley). Herpetological Review 14:10 (note erratum concerning authorship 14:48).

Orr, A. J., A. S. Banks, S. Mellman, H. R. Huber, R. L. DeLong, and R. F. Brown. 2004. Examination of the foraging habits of Pacific harbor seal (*Phoca vitulina richardsi*) to describe their use of the Umpqua River, Oregon, and their predation on salmonids. Fishery Bulletin 102:108–117.

Park, E. 1971. The world of the otter. J.B. Lippincott Company, Philadelphia.

Peterson, C. L., J. W. Reed, and R. F. Wilkinson. 1989. Seasonal food habits of *Cryptobranchus alleganiensis* (Caudata: Cryptobranchidae). Southwestern Naturalist 34:438–441.

Pfeiffer, W., and T. F. Pletcher. 1964. Club cells and granular cells in the skin of lamprey. Journal of the Fisheries Research Board of Canada 21:1083–1088.

Pike, G. C. 1950. Stomach contents of whales caught off the coast of British Columbia. Fisheries Research Board of Canada, Progress Reports of the Pacific Biological Stations 83:27–28.

Poe, T. P., H. C. Hansel, S. Vigg, D. E. Palmer, and L. A. Prendergast. 1991. Feeding of predaceous fishes on out-migrating juvenile salmonids in John Day Reservoir, Columbia River. Transactions of the American Fisheries Society 120:405–420.

Potter, I. C., P. A. Prince, and J. P. Croxall. 1979. Data on the adult marine and migratory phases in the life cycle of the southern hemisphere lamprey, *Geotria australis* Gray. Environmental Biology of Fishes 4:65–69.

Raney, E. C. 1939. The breeding habits of *Ichthyomyzon greeleyi* Hubbs and Trautman. Copeia 1939:111–112.

Roffe, T., and B. Mate. 1984. Abundances and feeding habits of pinnipeds in the Rogue River, Oregon. Journal of Wildlife Management 48:1262–1274.

Rohde, F. C., and R. G. Arndt. 1981. Life history of a Coastal Plain population of the mottled sculpin, *Cottus bairdi* (Osteichthyes: Cottidae), in Delaware. Brimleyana 7:69-64.

Salyer, J. C., II, and K. F. Lagler. 1940. The food and habits of the American merganser during winter in Michigan, considered in relation to fish management. Journal of Wildlife Management 4:186–219.

Scott, W. B., and E. J. Crossman. 1973. Freshwater fishes of Canada. Fisheries Research Board of Canada Bulletin 184.

Scott, W. B., and S. N. Tibbo. 1968. Food and feeding habits of swordfish, *Xiphias gladius*, in the western North Atlantic. Journal of the Fisheries Research Board of Canada 25:903–919.

Sjöberg, K. 1980. Ecology of the European river lamprey (*Lampetra fluviatilis*) in northern Sweden. Canadian Journal of Fisheries and Aquatic Sciences 37:1974–1980.

Sjöberg, K. 1985. Foraging activity patterns in the goosander (*Mergus merganser*) and the red-breasted merganser (*M. serrator*) in relation to patterns of activity in their major prey species. Oecologia 67:35–39.

Sjöberg, K. 1988. Food selection, food-seeking patterns and hunting success of captive goosanders *Mergus merganser* and red-breasted mergansers *M. serrator* in relation to the behavior of their prey. Ibis 130:79–93.

Sjöberg, K. 1989. Time-related predator/prey interactions between birds and fish in a northern Swedish river. Oecologia 80:1–10.

Surface, H. A. 1899. Removal of lampreys from the interior waters of New York. Pages 191–245 *in* Fourth Annual Report, Commission of Fisheries, Game and Forests of the State of New York. Wynkoop Hallenbeck Crawford Co., Albany, New York.

Tabor, R., J. Chan, and S. Hager. 1998. Predation of sockeye salmon fry by cottids and other predatory fishes in the Cedar River and southern Lake Washington, 1997. U.S. Fish and Wildlife Service, Western Washington Office, Aquatic Resources Division, Lacey, Washington.

Tabor, R., S. Hager, A. Hird, and R. Piaskowski. 2001. Predation of juvenile salmonids by predatory fishes in the lower Cedar River, 1999. U.S. Fish and Wildlife Service, Western Washington Office, Fisheries and Watershed Assessment Division, Lacey, Washington.

Tabor, R. A., E. J. Warner, K. L. Fresh, B. A. Footen, and J. R. Chan. 2007. Ontogenetic diet shifts of prickly sculpin in the Lake Washington basin. Transactions of the American Fisheries Society 136:1801–1813.

Tate, W. H. 1949. Growth and food habit studies of smallmouth black bass in some Iowa streams. Iowa State College Journal of Science 23:343–354.

Thomas, J. D. 1962. The food and growth of brown trout (*Salmo trutta* L.) and its feeding relationships with the salmon parr (*Salmo salar* L.) and the eel (*Anguilla anguilla* (L.)) in the River Teify, West Wales. Journal of Animal Ecology 31:175–205.

Thorn, W. C., C. S. Anderson, W. E. Lorenzen, D. L. Hendrickson, and J. W. Wagner. 1997. A review of trout management in southeast Minnesota streams. North American Journal of Fisheries Management 17:860–872.

Tuunainen, P., E. Ikonen, and H. Auvinen. 1980. Lampreys and lamprey fisheries in Finland. Canadian Journal of Fisheries and Aquatic Sciences 37:1953–1959.

Vladykov, V. D. 1949. Quebec lampreys (Petromyzonidae). I. List of species and their economical importance. Department of Fisheries, Province of Quebec, Contribution No. 26.

Vladykov, V. D., and W. I. Follett. 1958. Redescription of *Lampetra ayresii* (Günther) of western North America, a species of lamprey (Petromyzontidae) distinct from *Lampetra fluviatilis* (Linnaeus) of Europe. Journal of the Fisheries Research Board of Canada 15:47–77.

Wahl, D. H., and L. A. Nielsen. 1985. Feeding ecology of the sauger (*Stizostedion canadense*) in a large river. Canadian Journal of Fisheries and Aquatic Sciences 42:120–128.

Wallace, D. R. 2007. Neptune's ark: from ichthyosaurs to orcas. University of California Press, Berkeley.

Whitlock, D. 1994. A guide to aquatic trout food. Swan Hill Press, Shrewsbury, UK.

Whyte, J. N. C., R. J. Beamish, N. J. Ginther, and C. E. Neville. 1993. Nutritional condition of the Pacific lamprey (*Lampetra tridentata*) deprived of food for periods of up to two years. Canadian Journal of Fisheries and Aquatic Sciences 50:591–599.

Wigley, R. L. 1959. Life history of the sea lamprey of Cayuga Lake, New York. U.S. Fish and Wildlife Service Fishery Bulletin 59:561–617.

Wolf, B. O., and S. L. Jones. 1989. Great blue heron deaths caused by predation on Pacific lamprey. The Condor 91:482–484.

# Appendix A.

Reported predators of lampreys are listed below by taxonomic groups. When known, the species of lamprey and life history stage are indicated.

**Invertebrates**

Insect larvae – lampreys (Tuunainen et al. 1980: ammocoetes)

**Vertebrates**

Cephalaspidomorphi

Petromyzontidae: sea lamprey *Petromyzon marinus* – sea lamprey (Davis 1967: adults)

Chondrichthyes

Hexanchidae: bluntnose sixgill shark *Hexanchus griseus* – Pacific lamprey *Lampetra tridentata* (Migdalski 1962, Ebert 1986)

Actinopterygii

Acipenseridae: Siberian sturgeon *Acipenser baeri* – Siberian brook lamprey *Lethenteron kessleri* (Holčík 1986c: ammocoetes and adults)
Acipenseridae: white sturgeon *A. transmontanus* – lampreys (Landeen and Pinkham 1999); Pacific lamprey (Close et al. 1995: all stages)
Anguillidae: European eel *Anguilla anguilla* – European brook lamprey *Lampetra planeri* (Hardisty 1986); sea lamprey (Thomas 1962: ammocoetes)
Anguillidae: American eel *A. rostrata* – sea lamprey (Denoncourt and Stauffer 1993: ammocoetes)
Anguillidae: eels – (Close et al. 1995: eggs and ammocoetes)
Balitoridae: stone loach *Barbatula barbatula* (stone loach) – European river lamprey *L. fluviatilis* (Jang and Lucas 2005: eggs); European brook lamprey (Hardisty 1986)
Cyprinidae: central stoneroller *Campostoma anomalum* – mountain brook lamprey *Ichthyomyzon greeleyi* (Raney 1939: eggs)
Cyprinidae: minnows – Pacific lamprey (Close et al. 1995: eggs and ammocoetes)
Cyprinidae: hornyhead chub *Nocomis biguttatus* – southern brook lamprey *I. gagei* (this study: eggs)
Cyprinidae: *Notropis* sp. (shiners) – sea lamprey (Surface 1899 [cited by Wigley 1959]: eggs)
Cyprinidae: Eurasian minnow *Phoxinus phoxinus* – European river lamprey (Jang and Lucas 2005: eggs); European brook lamprey (Hardisty 1986)
Cyprinidae: Sacramento pikeminnow *Ptychocheilus grandis* – *Lampetra* sp. (Brown and Moyle 1997: ammocoetes)
Cyprinidae: Northern pikeminnow *P. oregonensis* – Pacific lamprey (Close et al 1995: adults; Landeen and Pinkham 1999: adults); lamprey – (Poe et al. 1991)
Cyprinidae: *Rhinichthys* sp. (dace) – sea lamprey (Surface 1899 [cited by Wigley 1959]: eggs)
Cyprinidae: blacknose dace *R. atratulus* – sea lamprey (Wigley 1959: eggs)
Cyprinidae: speckled dace *R. osculus* – Pacific lamprey (Brumo 2006: eggs)
Ictaluridae: channel catfish *Ictalurus punctatus* – Pacific lamprey (Close et al. 1995: adults); lamprey (Poe et al. 1991)
Ictaluridae: brown madtom *Noturus phaeus* – lamprey (Chan and Parsons 2000: ammocoetes)
Siluridae: wels *Silurus glanis* – Ukrainian brook lamprey *Eudontomyzon mariae* (Holčík and Renaud 1986: adults)
Esocidae: northern pike *Esox lucius* – Arctic lamprey *L. camtschatica* (Buchwald 1968: ammocoetes and adults; McPhail and Lindsey 1970: ammocoetes; Little et al. 1998); Siberian brook lamprey (Holčík 1986c: ammocoetes and adults); sea lamprey: (Applegate 1950: adult)
Salmonidae: salmonids – western brook lamprey *L. richardsoni* (Scott and Crossman 1973: eggs); Pacific lamprey (Brumo 2006: eggs)
Salmonidae: taimen *Hucho hucho taimen* – Siberian brook lamprey (Holčík 1986c: ammocoetes and adults)

Salmonidae: rainbow trout *Oncorhynchus mykiss* – northern brook lamprey *Ichthyomyzon fossor* (Churchill 1945: adults?); European brook lamprey (Kappus et al. 1995: adults); Pacific lamprey (Close et al. 1995: eggs and ammocoetes)

Salmonidae: brown trout *Salmo trutta* – northern brook lamprey (Cochran et al. 1992: adults; this study: ammocoetes and adults); American brook lamprey *L. appendix* (this study: adult); European brook lamprey (Hardisty 1986: including small ammocoetes; Kappus et al. 1995: adults); sea lamprey: (Applegate 1950: adult; Thomas 1962: ammocoetes)

Salmonidae: inconnu *Stenodos leucichthys* – Arctic lamprey (Buchwald 1968); Siberian brook lamprey (Holčík 1986c: ammocoetes and adults)

Gadidae: Atlantic cod *Gadus morhua* – Arctic lamprey (Holčík 1986b)

Gadidae: burbot *Lota lota* – Arctic lamprey (Buchwald 1968: ammocoetes); European river lamprey (Sjöberg 1980; Tuunainen et al. 1980: ammocoetes); Siberian brook lamprey (Holčík 1986c: ammetes and adults)

Gasterosteidae: threespine stickleback *Gasterosteus aculeatus* – European brook lamprey (Hardisty 1986: small ammocoetes)

Moronidae: striped bass *Morone saxatilis* – sea lamprey (Beamish 1980a)

Centrarchidae: rock bass *Ambloplites rupestris* – northern brook lamprey (Hubbs and Trautman 1937: adult)

Centrarchidae: smallmouth bass *Micropterus dolomieu* – lamprey (Tate 1949)

Percidae: rainbow darter *Etheostoma caeruleum* – American brook lamprey (Lutterbie 1976: eggs)

Percidae: Eurasian perch *Perca fluviatilis* – European river lamprey (Tuunainen et al. 1980: ammocoetes); Siberian brook lamprey (Holčík 1986c: ammocoetes and adults)

Percidae: logperch *Percina caprodes* – Pacific lamprey (Close et al. 1995: eggs and ammocoetes)

Percidae: sauger *Sander canadensis* – least brook lamprey *L. aepyptera* (Wahl and Nielsen 1985)

Percidae: zander *S. lucioperca* – Ukrainian brook lamprey (Holčík and Renaud 1986: adults)

Percidae: walleye *S. vitreus* – silver lamprey *Ichthyomyzon unicuspis* (this study: adults); Arctic lamprey (Buchwald 1968: ammocoetes; Little et al. 1998); sea lamprey (Applegate 1950: adult)

Hexagrammidae: lingcod *Ophiodon elongatus* – river lamprey *L. ayresii* (Vladykov and Follett 1958: adult; Scott and Crossman 1973: adult)

Cottidae: *Cottus* sp. (sculpins) – Arctic lamprey (Heard 1966: possibly on eggs); Pacific lamprey (Close et al. 1995: eggs and ammocoetes)

Cottidae: coastrange sculpin *C. aleuticus* – lamprey (Tabor et al. 1998)

Cottidae: prickly sculpin *C. asper* – river lamprey (Tabor et al. 2007: ammocoetes and adults); western brook lamprey (Tabor et al. 2007: ammocoetes and adults); Pacific lamprey (Gregory and Levings 1998: ammocoetes); lamprey (Tabor et al. 1998)

Cottidae: mottled sculpin *C. bairdi* – least brook lamprey (Rohde and Arndt 1981: egg)

Cottidae: European bullhead *C. gobio* – European brook lamprey (Hardisty 1986)

Cottidae: riffle sculpin *C. gulosus* – lamprey (Tabor et al. 2001)

Cottidae: torrent sculpin *C. rhotheus* – western brook lamprey (Scott and Crossman 1973: eggs); lamprey (Tabor et al. 1998)

Xiphiidae: swordfish *Xiphias gladius* – sea lamprey (Scott and Tibbo 1968: adults)

## Amphibia

Cryptobranchidae: hellbender *Cryptobranchus alleganiensis* – lampreys (Peterson et al. 1989: ammocoetes); northern brook lamprey (Nickerson and Ashton 1983: adults)

Proteidae: Neuse River waterdog *Necturus lewisi* (Neuse River waterdog) – a report that this species will eat lampreys (Braswell 2005) is actually based on laboratory trials with adult least brook lamprey (Nickerson and Ashton 1983)

Proteidae: mudpuppy *Necturus maculosus* – lampreys (Matson 2005: ammocoetes disabled by lampricide treatments); silver lamprey (Allin 1951: adult)

## Reptilia

Emydidae: western pond turtle *Actinemys* (= *Clemmys*) *marmorata* – Pacific lamprey (Bury 1986: adult – scavenged?)

Colubridae: *Nerodia* (= *Natrix*) sp. (water snake) – chestnut lamlprey *Ichthyomyzon castaneus* (Moore and Paden 1950: adult?)

Colubridae: midland water snake *Nerodia* (= *Natrix*) *sipedon pleuralis* – lampreys (Himes 2003: ammocoetes)
Colubridae: northern water snake *Nerodia* (= *Natrix*) *sipedon sipedon* – lampreys (Lagler and Salyer 1947: ammocoetes?); chestnut lalmprey (Lagler and Salyer 1947: adult); American brook lamprey (Lagler and Salyer 1947); sea lamprey (Surface 1899 [cited by Wigley 1959]: adults; Brown 1958: adults; Wigley 1959: adult)

Aves

Anatidae: common merganser or goosander *Mergus merganser* – lamprey (Leonard and Shetter 1936: ammocoetes; Salyer and Lagler 1940); European river lamprey (Sjöberg 1980, 1985, 1989: adults)
Anatidae: red-breasted merganser *M. serrator* – European river lamprey (Sjöberg 1980, 1985, 1989: adults); European brook lamprey (Feltham 1990)
Ardeidae: heron *Ardea cinerea* – Euopean brook lamprey (Kappus et al. 1995)
Ardeidae: great blue heron *A. herodias* – Pacific lamprey (Wolf and Jones 1989: adults; Close et al. 1995: adults); sea lamprey (Applegate 1950:adult)
Ardeidae: green heron *Butorides virescens* – American brook lamprey (Surface 1899 [cited by Wigley 1959])
Corvidae: common raven *Corvus corax* – western brook lamprey (Scott and Crossman 1973: adults)
Diomedeidae: grey-headed albatross *Thalassarche chrysostoma* – pouched lamprey *Geotria australis* (Potter et al. 1979: adults)
Diomedeidae: black-browed albatross *T. melanophrys* – pouched lamprey (Potter et al. 1979: adults)
Laridae: gulls *Larus* spp. – silver lamprey (Cochran et al. 1992: adult); Arctic lamprey (Buchwald 1968); European river lamprey (Hagelin 1959: adults); Pacific lamprey (Landeen and Pinkham 1999: adults); sea lamprey (Applegate 1950: adults)
Laridae: herring gull *L. argentatus* – European river lamprey (Sjöberg 1980, 1989: adults)
Laridae: California gull *L. californicus* – Pacific lamprey (Close et al. 1995: ammocoetes)
Laridae: common gull *L. canus* – European river lamprey (Sjöberg 1980, 1989: adults); Siberian brook lamprey (Holčík 1986c: ammocoetes and adults)
Laridae: ringbill gull *L. delawarensis* – Pacific lamprey (Close et al. 1995: ammocoetes)
Laridae: lesser black-backed gull *L. fuscus* – European river lamprey (Sjöberg 1980, 1989: adults)
Laridae: westerm gull *L. occidentalis* – Pacific lamprey (Close et al. 1995: ammocoetes)
Laridae: terns *Sterna* spp. – lampreys (Tuunainen et al. 1980: adults); Pacific lamprey (Landeen and Pinkham 1999: adults)
Laridae: Caspian tern *S. caspia* – *Lampetra* sp. (Collis et al. 2002)
Laridae: Forster's tern *S. forsteri* – Pacific lamprey (Close et al. 1995: ammocoetes)
Pelecanidae: American white pelican *Pelecanus erythrorhynchos* – *Lampetra* sp. (Griffiths 1980); Pit-Klamath brook lamprey *L. lethophaga* – (Hubbs 1971: adults)
Phalacrocoracidae: cormorant *Phalacrocorax* sp. – Pacific lamprey (Landeen and Pinkham 1999: adults)
Phalacrocoracidae: double-crested cormorant *P. auritus* – sea lamprey (Ludwig et al. 1989: adults)
Phalacrocoracidae: Brandt's cormorant *P. penicillatus* – river lamprey (Couch and Lance 2004: adults)
Podicipedidae: western grebe *Aechmophorus occidentalis* – *Lampetra* sp. (DeLong 2000)
Scolopacidae: spotted sandpiper *Actitus macularius* – *Lampetra* sp. (this study: ammocoetes)

Mammalia

Canidae: domestic dog *Canis familiaris* – sea lamprey (Applegate 1950: adults)
Mustelidae: river otter *Lutra canadensis* – sea lamprey (Park 1971)
Mustelidae: Eurasian otter *L. lutra* – *Lampetra* spp. (McCafferty 2005: adults)
Mustelidae: American mink *Mustela vison* – Pacific lamprey (Beamish 1980b); sea lamprey (Applegate 1950: possibly on an adult)
Procyonidae: Raccoon *Procyon lotor* – sea lamprey (Applegate 1950: adults)
Otariidae: northern fur seal *Callorhinus ursinus* – Arctic lamprey (Jordan and Gilbert 1899 [cited by Heard 1966]: adult); Pacific lamprey (Hubbs 1967: adults)
Otariidae: Steller sea lion *Eumetopias jubatus* – Pacific lamprey (Bonnot 1951: adults; Jameson and Kenyon 1977: adults; Roffe and Mate 1984: adults; Close et al. 1995: adults)
Otariidae: California sea lion *Zalophus californianus* – Pacific lamprey (Jameson and Kenyon 1977: adults; Roffe and Mate 1984: adults; Close et al. 1995: adults)

Phocidae: harbour seal *Phoca vitulina* – *Lampetra* sp. (Laake et al. 2002: adults); river lampry (Browne et al. 2002: adults; Orr et al. 2004: adults); Pacific lamprey (Hubbs 1967: adult; Roffe and Mate 1984: adults; Close et al. 1995: adults; Browne et al. 2002: adults; Orr et al. 2004: adults)
Physeteridae: sperm whale *Physeter catodon* – Pacific lamprey (Pike 1950: adults; Hubbs 1967: adults; Close et al. 1995: adults)

American Fisheries Society Symposium 72:153–164, 2009


# Diel Feeding by Larval Northern Brook Lampreys in Two Northern Michigan Streams

Trent M. Sutton*,[1] and Stephen H. Bowen[2]
*Michigan Technological University, Department of Biological Sciences Houghton, Michigan 49931, USA*

*Abstract.*—Larval lampreys are burrowing suspension feeders in streams that efficiently use organic detritus as their primary nutrient source. Although larval lampreys often exhibit daily patterns in activity (being most active at night), diel variations in food habits and feeding behavior have not been investigated. The objective of this study was to determine whether larval northern brook lampreys *Ichthyomyzon fossor* exhibit diel periodicity in diet composition, assimilation efficiency, or gut fullness. Larvae were collected at 4-h intervals over a 24-h period from the Pike and Pilgrim rivers, Michigan. Organic detritus composed most of the diet ash-free-dry-mass (AFDM) for each sampling period, and averaged 94.7% and 97.4% for the Pike and Pilgrim rivers, respectively. Assimilation efficiency of AFDM averaged 61.3% (range, 55.7–68.9%) for the Pike River and 87.5% (range, 80.8–94.3%) for the Pilgrim River. Gut fullness in the Pike and Pilgrim rivers was low, averaging 0.23 (range, 0.18–0.30) and 0.20 (0.14–0.25) mg diet AFDM per grams larvae, respectively. Although gut fullness did vary slightly on a diel basis, larval northern brook lampreys feed continuously but slowly in order to achieve a high rate of assimilation from a low-quality food resource.

## Introduction

The northern brook lamprey *Ichthyomyzon fossor*, a nonparasitic species, is endemic to tributaries of Hudson Bay, the Mississippi River, and the Laurentian Great Lakes (Scott and Crossman 1973; Becker 1983; Lanteigne 1992). In recent years, there has been growing concern regarding the population status of northern brook lampreys, particularly in the Great Lakes basin. For example, this species is currently listed as state threatened in Pennsylvania and state endangered in Indiana, Illinois, and Vermont (Nettles et al. 2001; B. Fisher, Indiana Department of Natural Resources, personal communication). While a host of anthropogenic factors negatively impacts stream ecosystem structure and function (Hynes 1970; Waters 1995), the inadvertent killing of larval northern brook lampreys with lampricides in streams containing larval sea lampreys *Petromyzon marinus* appears to be the largest current impediment to the long-term sustainability of this species (Purvis 1970; Vladykov 1973; Johnson 1987). As a consequence, there is a need to better understand the life history of this species, particularly larval life stages that are most vulnerable to sea lamprey control efforts.

During the larval (ammocoete) phase, which lasts from 3 to 13 years, lampreys live in burrows in the fine sediments of streams (Hardisty and Potter 1971; Randall 1972; Moore and Beamish 1973; Scott and Crossman

* Corresponding author: tmsutton@alaska.edu
[1] Current address: University of Alaska Fairbanks, School of Fisheries and Ocean Sciences, 245 O'Neill Building, Fairbanks, Alaska 99775, USA
[2] Current address: Emory University, Oxford College, 100 Hamil Street, Oxford, Georgia 30054, USA

1973; Potter 1980; Beamish and Lowartz 1996). Within the burrow, larvae feed on suspended particles of biofilm by drawing a respiratory current from the overlying water through a small connecting tube (Yap and Bowen 2003). Although these biofilm particles include algae and bacteria as well as organic detritus, it is the detrital fraction that serves as the primary food source (Moore and Beamish 1973; Moore and Potter 1976a, 1976b; Sutton and Bowen 1994; Mundahl et al. 2005).

Larval lampreys have the capacity to meet their nutritional needs by feeding on organic detritus, even though it is a low-quality food resource (Bowen 1980; Mundahl and Wissing 1988; Sutton and Bowen 1994; Mundahl et al. 2005). Specific physiological attributes, such as efficient digestion and assimilation and a low metabolic rate, have been identified as adaptations that allow larval lamprey to effectively use this food source (Moore and Beamish 1973; Moore and Mallatt 1980; Sutton and Bowen 1994; Yap and Bowen 2003; Mundahl et al. 2005). Seasonal- and habitat-related variations in the consumption and utilization of organic detritus by larval lampreys have been documented within the context of life history adaptations to detritivory (Sutton and Bowen 1994; Yap and Bowen 2003; Mundahl et al. 2005). However, no studies to date have examined whether larval lampreys exhibit diel (i.e., 24 h) periodicity in food and feeding habits.

Several studies have demonstrated that some fish species exhibit diel patterns in movements, habitat use, and feeding behaviors and success (Hobson 1979; Cerri 1983; Beers and Culp 1990; McMahon and Holanov 1995; Brodeur et al. 2000; Yu and Peters 2002; Schabetsberger et al. 2003; Daugherty and Sutton 2005). As a consequence, there are often differences in the catchability of those fishes over diel periods (Mendelson 1975; Craig 1977; Kelso 1978; Helfman 1981; Adams et al. 1988). Therefore, knowing the optimum time to collect fish is not only essential for analyses of food habits and feeding intensity, but the results of these studies will also increase our understanding of fish life history and ecology. However, few researchers have included these kinds of comparisons in their studies despite the importance of diel changes in behavior on fish distribution (Harris et al. 1992; Shuler et al. 1994; Jakober et al. 2000; Yu and Peters 2002; Schabetsberger et al. 2003; Daugherty and Sutton 2005).

The objective of this study was to determine whether northern brook lamprey ammocoetes exhibited diel periodicity in diet, assimilation efficiency, or feeding rate. We hypothesized that larval northern brook lampreys exhibit diel periodicity in (1) the percentage of algae, bacteria, and organic detritus comprising the diet, (2) the assimilation efficiency of consumed food material, and (3) the fullness of the digestive tract. Based on these study results, we expected to gain a better understanding of how larval lampreys meet their nutritional needs by feeding on organic detritus, a low-quality food resource apparently not of value to most fishes.

## Methods

### *Larval lamprey collections*

Northern brook lamprey ammocoetes were collected using backpack electrofishing (model ABP-2, Instrumentation Systems Center, University of Wisconsin, Madison) at 4-h intervals over a 24-h period from two Lake Superior tributaries in the Upper Peninsula of Michigan. The Pike River (47°52'N 88°50'E) was sampled from June 3 to June 4, 1992, while the Pilgrim River (47°02'N 88°50'E) was sampled from August 2 to August 3, 1992. Both rivers are second-order and ranged from 4 to 9 m in width at the collection sites, with a riparian canopy that ranged from closed (Pike River) to partly closed (Pilgrim River; Yap and Bowen 2003). For each sampling event, larval lampreys were randomly collected from pool habitats until the desired sample size was reached ($N$ = 10 individuals) for that collection period.

During each sampling period, 10 freshly caught ammocoetes from each river were held on wet ice until quiescent and euthanatized with tricaine methanesulfonate (MS-222; Sigma-Aldrich Corporation, St. Louis, Missouri). This procedure was followed because it prevented ingested food material from being regurgitated (Doxtater 1963), which was not observed for larval lampreys (T. M. Sutton, personal observation). Following euthanization, all ammocoetes were measured for total length (to the nearest 1 mm) and wet weight (to the nearest 0.01 g). Larvae were then fixed in a 10% buffered formalin solution for 24 h and stored in Carosafe postfixation preservative (Carolina Biological Supply Company, Burlington, North Carolina) pending subsequent laboratory analyses.

## Laboratory analyses

The ammocoete digestive tract is a straight, undifferentiated tube measuring 45% of total body length (Sutton and Bowen 1994). Because there is no mixing or differential movement of ingested diet contents in the digestive tract of lamprey ammocoetes, it is assumed that consumed food material moves uniformly through the intestine as plug flow. As a result, diet contents from the anterior end of the digestive tract are representative of recently consumed food items (Sutton 1993).

Gut contents were removed from the anterior one-tenth of the digestive tract for each larval northern brook lamprey collected during each sampling period for both rivers. To determine diet composition, two 50-μL subsamples from a 10-mL solution were examined for each ammocoete by quantitative microscopy following procedures described by Ahlgren (1990a). One subsample was used to estimate algal biovolume, which was then converted to ash-free-dry-mass (AFDM) using an empirically determined weight/volume conversion factor (0.3633 mg AFDM/mm$^3$) similar to ratios reported by Nalewajko (1966). The other subsample was to determine the contribution of bacteria to total diet AFDM by staining the subsample with acridine orange (0.1 mg/mL in 0.1 M potassium phosphate buffer, pH 7.5), as described by Hobbie et al. (1977). Bacterial biomass per sample (mg AFDM/mL) was calculated as (cells/mL) × (mean biovolume [1 μm$^3$/cell; vanDuyl and Kop 1990]) × (2.2 × 10$^{-10}$ mg C/μm$^{-3}$; carbon conversion factor of Bratback and Dundas 1984) × (0.5 AFDM/mg C). The remainder of the sample was dried to a constant weight at 110°C and total sample AFDM was calculated as the weight loss after combustion at 550°C for 2 h. The AFDM of organic detritus in the diet was calculated as the total AFDM (as determined from combustion of organic material) minus microorganism (algae and bacteria) AFDM (Bowen 1979a).

Assimilation efficiency (%AE), the net amount of food assimilated through the gut wall expressed as a percentage of the amount ingested, was estimated for total organic matter in the gut contents of each larval northern brook lamprey, as described by Conover (1966). For each ammocoete, AFDM in the gut contents taken separately from the first and last tenths of the digestive tract were compared, using diet ash as an unassimilated reference material. The assumption that ash is not assimilated is reasonable because most of the ash appeared to be small sand grains and diatom frustules (Sutton, personal observation).

Gut fullness of the anterior one-tenth of the digestive tract was measured for each larval northern brook lamprey. Although gut fullness is a static measure of the mass of food consumed by a larval lamprey immediately prior to sampling, it also serves as an index of feeding rate. Estimates of gut fullness were measured as the mg diet AFDM per gram wet mass of lamprey ammocoete.

## Data analyses

To test for differences in larval northern brook lamprey diet composition, assimilation efficiency, and gut fullness among 4-h sampling periods in each river, data were analyzed using one-way analysis of variance. Our null hypothesis was that there were no significant differences ($\alpha$ = 0.05) in diet composition (percent algae, bacteria, and detritus), assimilation efficiency, or gut fullness of ammocoetes among sampling periods in each river. When significant differences among sampling periods were detected, a

Student-Neuman-Keuls multiple comparisons test was used to separate the means. Percentage data were arcsine transformed prior to statistical analysis. All statistical analyses were conducted using SigmaStat 3.5 software (Systat Software, Inc., San Jose, California) following procedures as described in Zar (1990).

## Results

Seventy larval northern brook lampreys (10 larvae for each of the seven sampling periods) were collected from each of the Pike and Pilgrim rivers. Larvae captured from the Pike River ranged from 68 to 115 mm in total length (mean = 91.74 mm; SE = 2.18) and 0.19–2.84 g in wet body mass (mean = 1.53 g; SE = 0.12). Similarly, larvae captured from the Pilgrim River ranged from 76 to 120 mm in total length (mean = 92.43 mm; SE = 1.62) and 0.64–3.12 g (mean = 1.57 g; SE = 0.09) in wet body mass. There were no significant differences in the mean total length or wet body mass of larval lampreys among sampling periods for the Pike River (length: $F = 1.16$, $P = 0.35$; weight: $F = 1.72$, $P = 0.15$) or Pilgrim River (length: $F = 1.71$, $P = 0.16$; weight: $F = 1.75$, $P = 0.19$).

### *Diet composition*

Algal and bacterial abundance in the diet of larval northern brook lampreys followed no discernable diel pattern in the Pike River. The percentage of diet AFDM composed of algae was low (3.45–4.77%), and did not differ significantly among sampling periods ($F = 0.70$, $P = 0.63$; Figure 1A). Diet AFDM composed of bacteria was also low, ranging from 1.06% to 1.46% (Figure 1B). Although there were significant differences in the percentage of bacteria in the diet AFDM among sampling periods ($F = 4.49$, $P < 0.01$), there was no clear diel trend in this dietary component.

Similarly, the algal and bacterial component of the diet AFDM for larval northern brook lampreys also did not follow a clear diel trend in the Pilgrim River. The abundance of algae in the diet was low (1.26–2.95%). There were significant differences in the percentage of the diet AFDM composed of algae among sampling periods ($F = 5.31$, $P < 0.001$; Figure 2A), with a steady decline from 1600 to 0800 hours. The percentage of the diet AFDM that was composed of bacteria was also low (0.10–0.75%). Although there were significant differences among sampling periods ($F = 3.66$, $P < 0.01$; Figure 2B), there was no clear diel trend in the percentage of bacteria in the diet AFDM.

The utilizable portion of the diet for northern brook lamprey larvae consisted of mostly organic detritus in both the Pike and Pilgrim rivers, regardless of sampling period. Organic detritus ranged from 94.02% to 95.32% of the total diet AFDM in the Pike River, with no significant differences in this dietary component among sampling periods ($F = 1.93$, $P = 0.11$; Table 1). Similarly, the percentage of the diet AFDM composed of organic detritus in the Pilgrim River ranged from 96.39% to 98.54% of the diet AFDM and did not differ significantly among sampling periods ($F = 0.93$, $P = 0.49$; Table 1). For both rivers, there was no discernable diel trend in the consumption of organic detritus by larval northern brook lampreys (Table 1).

### *Assimilation efficiency and gut fullness*

There were no significant diel trends in the assimilation efficiency of the organic fraction of the diet and gut fullness by larval northern brook lampreys in the Pike River. Assimilation efficiency ranged from 55.77% to 68.88%, with no significant differences in %AE among sampling periods ($F = 1.01$, $P = 0.44$; Figure 3A). Gut fullness ranged from 0.18 to 0.30 mg diet AFDM per gram larvae (Figure 3B). Although there were no significant differences in gut fullness among sampling periods ($F = 0.68$, $P = 0.67$), values were, on average, slightly higher during early morning (0700 hours; mean = 0.23 mg diet AFDM per gram larvae) than during the day (1500 and 1900 hours; mean = 0.20 mg diet AFDM per gram larvae)

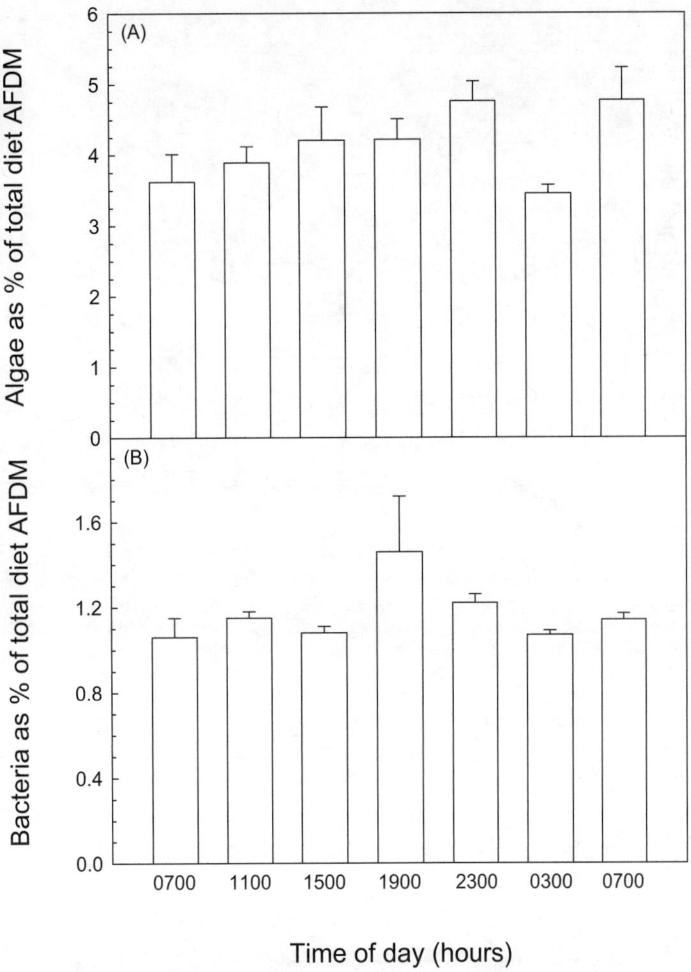


Figure 1. Daily variation in the contribution of (A) algae and (B) bacteria ingested in northern brook lamprey ammocoete diets as a percentage of total ash-free-dry-mass (AFDM) for the Pike River. Values for each 4-h sampling period are means (SE) of five individuals.

Although we observed no trends among sampling periods in the assimilation efficiency of diet AFDM for larval northern brook lampreys in the Pilgrim River, there was a significant diel trend in gut fullness. Assimilation efficiency ranged from 80.83% to 94.29%, with no significant differences in %AE among sampling periods ($F = 2.30$, $P = 0.44$; Figure 4A). Gut fullness of lamprey ammocoetes ranged from 0.14 to 0.25 mg diet AFDM per gram larvae (Figure 4B). There was a significant difference in gut fullness among sampling periods ($F = 6.52$, $P < 0.001$), with higher values during the evening (2000–0400 hours; mean = 0.26 mg diet AFDM per gram larvae) than during the day (1200 hours; mean = 0.19 mg diet AFDM per gram larvae).

## Discussion

There was no diel periodicity in the diet composition of northern brook lamprey ammocoetes among sampling periods in the Pike and Pilgrim rivers. Diets of larval lampreys were dominated by detritus, which ranged from 94% to 99% of the organic fraction of the gut

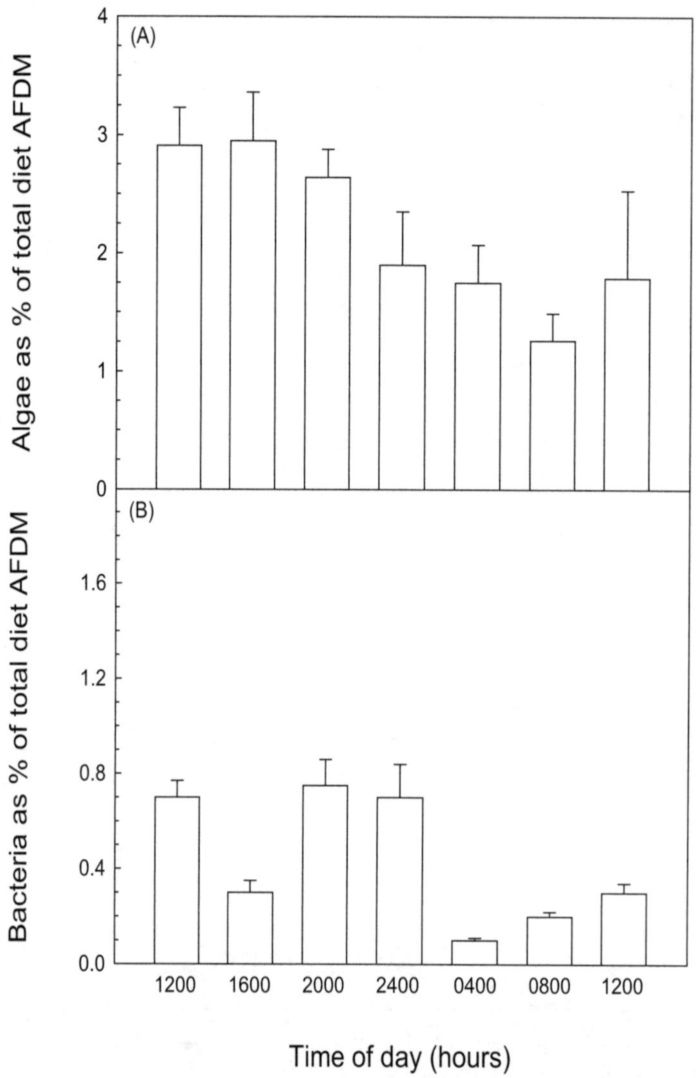


Figure 2. Daily variation in the contribution of (A) algae and (B) bacteria ingested in northern brook lamprey ammocoete diets as a percentage of total ash-free-dry-mass (AFDM) for the Pilgrim River. Values for each 4-h sampling period are means (SE) of five individuals.

contents. The remainder of the contents in the digestive tract was composed of algae and bacteria, which in sum contributed less than 6% of the total diet AFDM. Similar observations of organic detritus-dominated diets have been reported for larval northern brook lampreys, sea lampreys, and American brook lampreys *Lampetra appendix* during summer months (Sutton and Bowen 1994; Mundahl et al. 2005). Our study results suggest that the organic biofilm fragments consumed by larval lampreys do not vary with any diel periodicity. Other studies have shown that the quality and quantity of stream biofilm that is available for ammocoete consumption varies on a predictable annual basis and is regulated by a suite of abiotic factors that include stream discharge, riparian shading, hydrological characteristics of specific habitats, primary and microbial productivity, biofilm particle age, and other

Table 1. Means (SE) for percent ash-free-dry-mass (AFDM) in the diet for northern brook lamprey ammocoetes collected from the Pike River (June 3–4, 1992) and Pilgrim River (August 2–3, 1992) during each 4-h sampling period.

| Pike River | | Pilgrim River | |
|---|---|---|---|
| Time of day (hours) | Percent detritus of total diet AFDM | Time of day (hours) | Percent detritus of total diet AFDM |
| 0700 | 95.32 (0.40) | 1200 | 96.39 (1.01) |
| 1100 | 94.96 (0.25) | 1600 | 96.75 (0.58) |
| 1500 | 94.71 (1.48) | 2000 | 96.61 (0.57) |
| 1900 | 94.32 (0.48) | 2400 | 97.40 (0.51) |
| 2300 | 94.02 (0.88) | 0400 | 98.15 (0.30) |
| 0300 | 95.48 (0.13) | 0800 | 98.54 (0.33) |
| 0700 | 94.09 (0.59) | 1200 | 97.91 (0.91) |

anthropogenic influences (Decho 1990; Sutton and Bowen 1994; Ledger and Hildrew 1998; Yap and Bowen 2003; Mundahl et al. 2005).

Assimilation efficiencies of larval northern brook lampreys also did not vary on a diel basis. Values for assimilation efficiency ranged from 56% to 69% (Pike River) and 81–94% (Pilgrim River), which were similar to %AE values reported for northern brook and sea lamprey ammocoetes in June (48–68%) and August (63–94%), respectively, in Great Lakes tributaries (Sutton and Bowen 1994). Similar values for %AE have also been reported for American brook lampreys in Minnesota streams (Mundahl et al. 2005). These results suggest that variations in assimilation efficiency are regulated by water temperature and the digestibility of detrital aggregate (Bowen 1979b; Decho 1990; Sutton and Bowen 1994). For example, the higher %AE values observed for ammocoetes in the Pilgrim River relative to the Pike River in our study were likely due to warmer stream water temperatures and older, more digestible detritus particles. Because both factors change seasonally with ambient air temperature, riparian shading, and the age and quality of the detrital aggregate, it is unlikely that assimilation efficiency of larval lampreys would vary on a diel basis.

Gut fullness of larval northern brook lampreys was low but followed a distinct seasonal cycle. However, gut fullness also appeared to vary to some extent on a diel basis. According to Malmqvist and Brönmark (1981) and Sutton and Bowen (1994), gut fullness and feeding rate of larval lampreys are most closely associated with water temperature. Our results for the Pike River in June and the Pilgrim River in August were consistent with the pattern and range in gut fullness values that have been reported for larval lampreys during the same time of year (Sutton and Bowen 1994; Mundahl et al. 2005). Values for gut fullness were slightly higher during evening (Pilgrim River) and early morning hours (Pike River) than at any other time of day. Although larval lampreys are suspension feeders that spend most of their time burrowed in the substrate of stream bottoms (Moore and Beamish 1973), this behavior may serve as a predator-avoidance strategy. However, larval lampreys are also negatively phototactic, and respond to light by retracting deeper into their burrows or by vigorously burrowing into the substrate if exposed at the

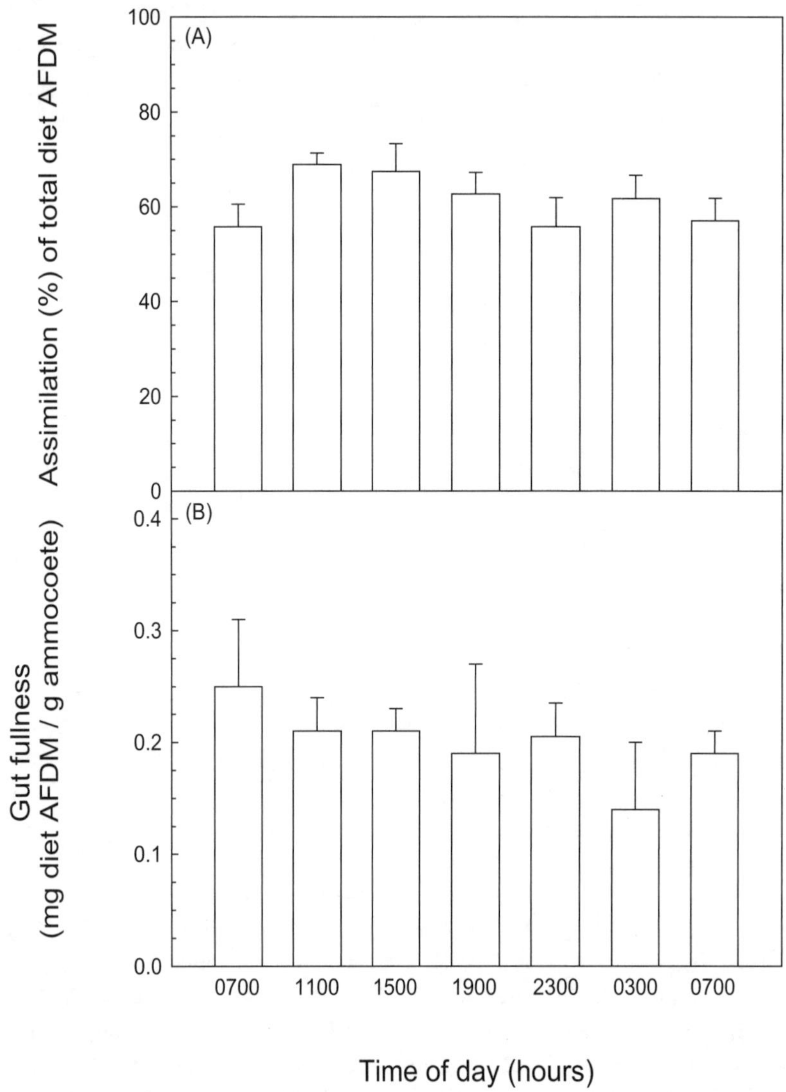


Figure 3. Daily variation in the (A) assimilation and (B) gut fullness of organic matter ingested by northern brook lamprey ammocoetes for the Pike River. Estimates for each 4-h sampling period consist of means (SE) for five pooled ammocoetes for assimilation efficiency and means (SE) of five individuals for gut fullness.

surface (Ullén et al. 1995; Sutton, personal observation). As a result, the higher gut fullness values and subsequently higher feeding rates observed during night or crepuscular periods may simply reflect their aversion to light. Additional evaluations are needed to determine whether ammocoetes exhibit diel periodicity in feeding rate.

Although most detritivorous fishes possess unique morphological or physiological adaptations to effectively digest and assimilate organic detritus, larval lampreys appear to have no specialized adaptations for consuming this low-quality food resource (Hardisty and Potter 1971; Bowen 1980; Mundahl and Wissing 1988; Ahlgren 1990b). However, ammocoetes can achieve high assimilation through a long period of digestion that occurs via a slow rate

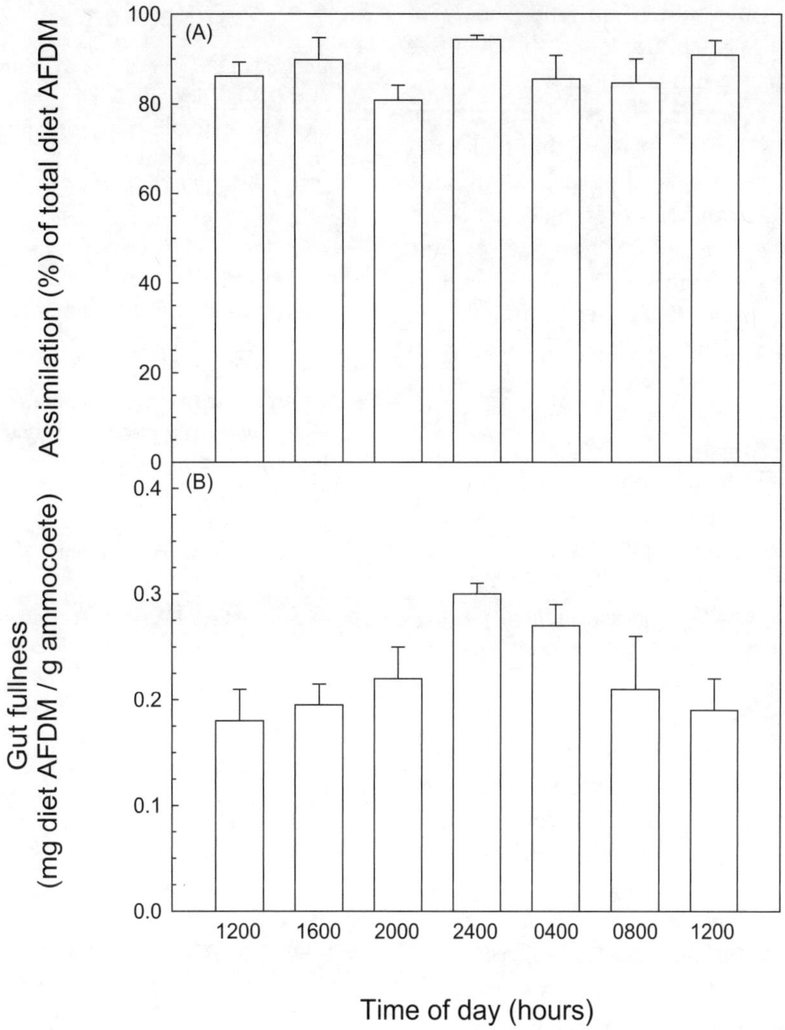


Figure 4. Daily variation in the (A) assimilation and (B) gut fullness of organic matter ingested by northern brook lamprey ammocoetes for the Pilgrim River. Estimates for each 4-h sampling period consist of means (SE) for five pooled ammocoetes for assimilation efficiency and means (SE) of five individuals for gut fullness.

of food passage through the digestive tract (Sutton and Bowen 1994). Yap and Bowen (2003) demonstrated that larval northern brook lampreys selectively ingested detrital particles that were higher in AFDM and amino acids, which further facilitates high digestibility. Because digestibility is strongly linked to primary and microbial productivity and biofilm particle age, both of which vary seasonally rather than on a diel basis in streams (Decho 1990; Ledger and Hildrew 1998; Yap and Bowen 2003), it is unlikely that the quality and availability of organic detritus in streams fluctuates consistently or predictably over a 24-h period. Further, larval lampreys have a lower metabolic rate than other fishes, which is linked to their slow growth rate and the protracted duration of this life stage (Hill and Potter 1970; Mallatt 1982). This suggests that it would be adaptive for larval lampreys to feed consistently throughout the day and night to maintain a constant supply of organic detritus passing through the digestive tract to enhance feeding efficiency.

Our study results suggest that northern brook lamprey larvae do not exhibit diel periodicity in food and feeding habits. This life history attribute can be interpreted as an adaptation to a low-quality food resource (organic detritus) that varies on a seasonal rather than a diel basis (Decho 1990; Sutton and Bowen 1994; Ledger and Hildrew 1998; Yap and Bowen 2003). Because natural selection has favored a unique suite of ecological adaptations for the efficient use of organic detritus, larval lampreys are tightly coupled to biofilm production and processing in aquatic ecosystems (Hynes 1970; Cummins 1974). Although ammocoetes have low metabolic requirements and exhibit slow growth, they must feed continuously on detrital aggregate to meet their nutritional demands. Consequently, studies on food and feeding habits of larval lampreys should not be influenced by the time of day of sampling. By gaining a better understanding of the relationship between organic detritus as a food source and larval lamprey life history and production, strategies might also be identified that lead to more biologically effective means of control for sea lamprey populations in the Great Lakes basin.

## Acknowledgments

We are thankful to J. Weisser, D. Lavis, and S. Morkert (U.S. Fish and Wildlife Service Sea Lamprey Control); W. Swink (U.S. Geological Survey Hammond Bay Biological Station); and J. Weise (Canadian Department of Fisheries and Oceans Sea Lamprey Control Centre) for their expertise on larval lamprey collections and ecology. Thanks to B. Young, M. Yap, and M. Lemke for their assistance during field collections and processing of laboratory samples. Constructive comments on earlier drafts by D. Daugherty, H. Patrick, and W. Swink improved this manuscript. This research was funded by the Great Lakes Fishery Commission, Michigan Technological University, and the Michigan Chapter of Trout Unlimited.

## References

Adams, N. J., D. R. Barton, R. A. Cunjak, G. Power, and S. C. Riley. 1988. Diel patterns of activity and substrate preference in young Arctic char from the Koroc River, northern Quebec. Canadian Journal of Zoology 66:2500–2502.

Ahlgren, M. O. 1990a. Diet selection and the contribution of detritus to the diet of the juvenile white sucker (*Catostomus commersoni*). Canadian Journal of Fisheries and Aquatic Sciences 47:41–48.

Ahlgren, M. O. 1990b. Nutritional significance of facultative detritivory to the juvenile white sucker (*Catostomus commersoni*). Canadian Journal of Fisheries and Aquatic Sciences 47:49–54.

Beamish, F. W. H., and S. Lowartz. 1996. Larval habitat of American brook lamprey. Canadian Journal of Fisheries and Aquatic Sciences 53:693–700.

Becker, G. C. 1983. Fishes of Wisconsin. The University of Wisconsin Press, Madison.

Beers, C. E., and J. M. Culp. 1990. Plasticity in foraging behaviour of a lotic minnow (*Rhinichthys cataractae*) in response to different light intensities. Canadian Journal of Zoology 68:101–105.

Bowen, S. H. 1979a. A nutritional constraint in detritivory by fishes: the stunted population of *Sarotherodon mossambicus* in Lake Sibya, South Africa. Ecological Monographs 49:17–31.

Bowen, S. H. 1979b. Determinants of the chemical composition of periphytic detrital aggregate in a tropical lake (Lake Valencia, Venezuala). Archivos de Hydrobiologia 87:166–177.

Bowen, S. H. 1980. Detrital nonprotein amino acids are the key to rapid growth of tilapia in Lake Valencia, Venezuela. Science 207:1216–1218.

Bratback, G., and I. Dundas. 1984. Bacterial dry matter content and biomass estimations. Applied Environmental Microbiology 48:755–757.

Brodeur, R. D., M. T. Wilson, and L. Ciannelli. 2000.: Spatial and temporal variability in feeding and condition of age-0 walleye pollock (*Theragra chalcogramma*) in frontal regions of the Bering Sea. ICES Journal of Marine Science 57:256–264.

Cerri, R. D. 1983. The effect of light intensity on feeding of predator and prey behaviour in cyprinid fish: factors that influence prey risk. Animal Behaviour 31:736–742.

Conover, R. J. 1966. Assimilation of organic matter by zooplankton. Limnology and Oceanography 11:338–345.

Craig, J. F. 1977. Seasonal changes in the day and night activity of adult perch, *Perca fluviatilis* L. Journal of Fish Biology 11:161–166.

Cummins, K. W. 1974. Structure and function of stream ecosystems. BioScience 24:631–641.

Daugherty, D. J., and T. M. Sutton. 2005. Diel movement patterns and habitat use of flathead catfish in the lower St. Joseph River, Michigan. Journal of Freshwater Ecology 20:1–8.

Decho, A. W. 1990. Microbial exopolymer secretions in ocean environments: their role. (s) in food webs and marine processes. Oceanography and Marine Biology Annual Reviews 28:73–153.

Doxtater, G. D. 1963. Use of ice water to prevent regurgitation of stomach contents of fish. Transactions of the American Fisheries Society 92:68.

Hardisty, M. W., and I. C. Potter. 1971. The behaviour, ecology, and growth of larval lampreys. Pages 85–125 *in* M. W. Hardisty and I. C. Potter, editors. The biology of lampreys, volume 1. Academic Press, London.

Harris, D. D., W. A. Hubert, and T. A. Wesche. 1992. Habitat use by young-of-year brown trout and effects on weighted usable area. Rivers 3:99–105.

Helfman, G. S. 1981. Twilight activities and temporal structure in a freshwater fish community. Canadian Journal of Fisheries and Aquatic Sciences 38:1405–1420.

Hill, B. J., and I. C. Potter. 1970. Oxygen consumption in ammocoetes of the lamprey *Ichthyomyzon hubbsi* Raney. Journal of Experimental Biology 53:47–57.

Hobbie, J. E., R. J. Daley, and S. Jasper. 1977. Use of nucleopore filters for counting bacteria by fluorescence microscopy. Applied Environmental Microbiology 33:1225–1228.

Hobson, E. S. 1979. Interactions between piscivorous fishes and their prey. Pages 231–242 *in* R. H. Stroud and H. Clepper, editors. Predator-prey systems in fisheries management. Sport Fishing Institute, Washington, D. C.

Hynes, H. B. N. 1970. The ecology of running water. University of Toronto Press, Toronto, Ontario.

Jakober, M. J., T. E. McMahon, and R. F. Thurow. 2000. Diel habitat partitioning by bull charr and cutthroat trout during fall and winter in Rocky Mountain streams. Environmental Biology of Fishes 59:79–89.

Johnson, J. E. 1987. Protected fishes of the United State and Canada. American Fisheries Society, Bethesda, Maryland.

Kelso, J. R. M. 1978. Diel rhythm in activity of walleye, Stizostedion vitreum. Journal of Fish Biology 12:593–599.

Lanteigne, J. 1992. Status of the northern brook lamprey, *Ichthyomyzon fossor*, in Canada. Canadian Field-Naturalist 106:7–13.

Ledger, M. E., and A. H. Hildrew. 1998. Temporal and spatial variation in the epilithic biofilm of an acid stream. Freshwater Biology 40:655–670.

Mallatt, J. 1982. Pumping rates and particle retention efficiencies of the larval lamprey, an unusual suspension feeder. Biological Bulletin 163:197–210.

Malmqvist, B., and C. Brönmark. 1981. Filter feeding in larval Lampetra planeri: effects of size, temperature and particle concentration. Oikos 38:40–46.

McMahon, T. E., and S. H. Holanov. 1995. Foraging success of largemouth bass at different light intensities: implications for time and depth of feeding. Journal of Fish Biology 46:759–767.

Mendelson, J. 1975. Feeding relationships among species of *Notropis* (Pisces: Cyprinidae) in a Wisconsin stream. Ecological Monographs 45:199–230.

Moore, J. W., and F. W. H. Beamish. 1973. Food of larval sea lamprey (*Petromyzon marinus*) and American brook lamprey (*Lampetra lamottei*). Journal of the Fisheries Research Board of Canada 30:7–15.

Moore, J. W., and J. M. Mallatt. 1980. Feeding of larval lamprey. Canadian Journal of Fisheries and Aquatic Sciences 37:1658–1664.

Moore, J. W., and I. C. Potter. 1976a. A laboratory study on the feeding of the brook lamprey *Lampetra planeri* (Bloch). Journal of Animal Ecology 45:81–90.

Moore, J. W., and I. C. Potter. 1976b. Aspects of feeding and lipid deposition and utilization in the lampreys, *Lampetra fluviatilis* (L.) and *Lampetra planeri* (Bloch). Journal of Animal Ecology 45:699–712.

Mundahl, N. D., C. Erickson, M. R. Johnston, G. A. Sayeed, and S. Taubel. 2005. Diet, feeding rate, and assimilation efficiency of American brook lamprey larvae. Environmental Biology of Fishes 72:67–72.

Mundahl, N. D., and T. E. Wissing. 1988. Selection and digestive efficiencies of gizzard shad feeding on natural detritus and two laboratory diets. Transactions of the American Fisheries Society 117:480–487.

Nalewajko, C. 1966. Dry weight, ash, and volume data for some freshwater planktonic algae. Journal of the Fisheries Research Board of Canada 23:1285–1288.

Nettles, D. C., N. R. Staats, B. D. Chipman, and L. J. Nashett. 2001. A long-term program, of sea lamprey control in Lake Champlain. U.S. Fish and Wildlife Service, Final supplemental environmental impact statement, Burlington, Vermont.

Potter, I. C. 1980. Ecology of larval and metamorphosing lampreys. Canadian Journal of Fisheries and Aquatic Sciences 37:1641–1657.

Purvis, H. A. 1970. Growth, age at metamorphosis, and sex ratio of northern brook lamprey in a tributary of southern Lake Superior. Copeia 1970:326–332.

Randall, D. J. 1972. Respiration. Pages 287–306 *in* M. W. Hardisty and I. C Potter, editors. The biology of lampreys, volume 2, Academic Press, London.

Schabetsberger, R., M. Sztatecsny, G. Drozdowski, R. D. Brodeur, G. L. Swartzman, M. T. Wilson, A. G. Winter, and J. M. Napp. 2003. Size-dependent, spatial, and temporal variability of juvenile walleye pollack (*Theregra chalcogramma*) feeding at a structural front in the southeast Bering Sea. Marine Ecology 24:141–164.

Scott, W. B., and E. J. Crossman. 1973. Freshwater fishes of Canada. Fisheries Research Board of Canada, Ottawa., Ontario.

Shuler, S. W., R. B. Nehring, and K. D. Fausch. 1994. Diel habitat selection by brown trout in the Rio Grande River, Colorado, after placement of boulder structures. North American Journal of Fisheries Management 14:99–111.

Sutton, T. M. 1993. The importance of organic detritus in the diets of larval sea lamprey (*Petromyzon marinus*) and northern brook lamprey (*Ichthyomyzon fossor*) in the Great Lakes basin. Master's thesis. Michigan Technological University, Houghton.

Sutton, T. M., and S. H. Bowen. 1994. Significance of organic detritus in the diet of larval lampreys in the Great Lakes basin. Canadian Journal of Fisheries and Aquatic Sciences 51:2380–2387.

Ullén, F., T. G. Deliagina, G. N. Orlovsky, and S. Grillner. 1995. Spatial orientation in the lamprey. II. Visual influence on orientation during locomotion and in the attached state. The Journal of Experimental Biology 198:675–681.

vanDuyl, F. C., and A. J. Kop. 1990. Seasonal patterns of bacterial production and biomass in intertidal sediments of the western Dutch Wadden Sea. Marine Ecology Progress Series 59:249–261.

Vladykov, V. D. 1973. North American nonparasitic lampreys of the family Petromyzontidae must be protected. Canadian Field-Naturalist 87:235–239.

Waters, T. F. 1995. Sediment in streams: sources, biological effects and control. American Fisheries Society Monograph 7, Bethesda, Maryland.

Yap, M. R., and S. H. Bowen. 2003. Feeding by northern brook lamprey (*Ichthyomyzon fossor*) on sestonic biofilm fragments: habitat selection results in ingestion of a higher quality diet. Journal of Great Lakes Research 29 (Supplement 1):15–25.

Yu, S.-L., and E. J. Peters. 2002. Diel and seasonal habitat use by red shiner (*Cyprinella lutrensis*). Zoological Studies 41:229–235.

Zar, J. H. 1990. Biostatistical analysis, 4th edition. Prentice Hall, Upper Saddle River, New Jersey.

American Fisheries Society Symposium 72:165–172, 2009


# A Comparison of Native and Exotic Hosts for the Silver Lamprey

Philip A. Cochran*
*Saint Mary's University of Minnesota, Biology Department*
*700 Terrace Heights, Winona, Minnesota 55987, USA*

*Abstract.*—Silver lamprey *Ichthyomyzon unicuspis* has been most strongly associated with native host species that are relatively large and have naked skin or relatively small scales, including paddlefish *Polyodon spathula,* lake sturgeon *Acipenser fulvescens,* lake trout *Salvelinus namaycush,* ictalurid catfishes, and esocids. In many regions, however, and especially in the southern part of the silver lamprey's geographic range, these host species have declined in abundance through a combination of human impacts, and the most abundant large fish species is now the common carp *Cyprinus carpio,* an exotic species with large scales. Silver lampreys removed from paddlefish in the Wisconsin River and allowed to feed on carp in the laboratory displayed very little growth relative to lampreys in the river. Although paddlefish lack scales, skin densities of carp and paddlefish, measured as dry mass per unit area, were similar. However, dried samples of paddlefish skin displayed a substantial lipid residue, and skin density of paddlefish expressed as ash weight was less than that of carp. It is likely that increased handling time contributes to reduced growth by silver lampreys on carp. The importance of paddlefish skin lipids to silver lamprey feeding ecology should be investigated.

## Introduction

The biology of the silver lamprey *Ichthyomyzon unicuspis* has been reviewed by Cochran and Marks (1995) and Cochran and Lyons (2004), with additional information provided by Cochran (2004). Silver lampreys have been most strongly associated with native host species that are relatively large and have naked skin or relatively small scales (Cochran and Lyons 2004). These include paddlefish *Polyodon spathula,* lake sturgeon *Acipenser fulvescens,* lake trout *Salvelinus namaycush,* ictalurid catfishes, and esocids (e.g., Wagner 1908; Becker 1983; Vladykov 1985; Lyons 1993, Runstrom et al. 2001; Renaud 2002; Cochran et al. 2003). Cochran (1994) provided a theoretical basis for these trends. In many regions, however, and especially in the southern part of the silver lamprey's geographic range, these host species have declined in abundance through a combination of anthropogenic effects (Becker 1983), and the most abundant large fish species is now the common carp *Cyprinus carpio,* an exotic species with large scales with which the silver lamprey shares no coevolutionary history (Cochran and Lyons 2004). Silver lampreys are known to attack carp (e.g., Becker 1983; Lubinski et al. 1986). Cochran (2004) reported that a commercial fisher on the Mississippi River in Wisconsin associated parasitic lampreys (primarily silver lampreys, but possibly including chestnut lampreys *I. castaneus*) most strongly with carp and paddlefish but saw more lampreys on carp simply because his operation caught carp more frequently and in much greater numbers. However, Cochran (1991) questioned the ability of silver lampreys

* Corresponding author: pcochran@smumn.edu

to form puncture wounds through the carp's heavy scale layer.

The data reported herein were collected in association with a study of silver lamprey and paddlefish interactions below the Prairie du Sac dam, the first dam on the Wisconsin River upstream from its confluence with the Mississippi River (Lyons 1993). Paddlefish are aggregated at this site throughout the year and display high rates of parasitism by silver lampreys (Lyons 1993; Runstrom et al. 2001; Cochran et al. 2003), and very few fishes other than paddlefish have been observed with lampreys attached. I provide in this account details of an experiment in which silver lampreys from Prairie du Sac readily attached to carp in the laboratory but displayed little growth relative to those on paddlefish in the field. One hypothesis for this difference in growth is that the skin and scale layer of carp is more substantial than that of paddlefish and takes longer for a lamprey to penetrate. I report here preliminary data on the skin density of the two species.

## Methods

Silver lampreys attached to paddlefish were captured by electrofishing in the tailwaters of the Prairie du Sac dam. Lampreys used in laboratory experiments were transported in insulated coolers. They were maintained as described by Cochran (1989) in Living Streams (Frigid Units, Toledo, Ohio) filled to approximately 520 L with filtered, recirculating water under a light cycle of 12 h:12 h light:dark. Water temperature was maintained at 20°C (±1°C). Common carp (total length: 31–53 cm) from the Fox River in Brown County, Wisconsin were used as hosts.

A group of 11 lampreys collected on June 15 (water temperature: 18°C) was held without feeding until they were measured on June 24 and supplied with carp. They were remeasured on July 26 and compared to a group of 10 lampreys removed from paddlefish on July 18 (water temperature: 24°C) and held without feeding until measurement on July 20. Although this was an unplanned experiment, the "captive" lampreys feeding on carp and "wild" lampreys with access to paddlefish experienced similar periods with access to food (32 versus 33 d, respectively) and similar if not identical temperature regimes. It is assumed that both groups of lampreys were of similar mean size on June 15.

Lengths referred to herein, unless otherwise specified, were total lengths, and "body mass" refers to wet live mass. Measurements of living lampreys were obtained after anesthesia with tricaine methanesulfonate. Relationships between body mass and length were assessed by calculating simple linear least-squares regressions of the natural logarithm of body mass on the natural logarithm of length (hereafter referred to as "mass-length regressions"), and they were compared by analysis of covariance.

Samples used for analysis of skin density were obtained from three paddlefish (eye-to-fork lengths: 88, 99, and 104 cm) collected for other studies in the tailwaters of the Prairie du Sac dam on April 4 and May 10, 1990 and a specimen from the Mississippi River at Wabasha, Wabasha County, Minnesota, collected in August, 2005 (eye-to-fork length: 88 cm). Carp (46–71 cm) used for analysis of skin density included two collected from the Fox River on July 3, 1990 and two collected on September 27, 2005 in Upper Lake Winona, Winona County, Minnesota. Samples of skin of known area were excised with a cork borer or scalpel from ventral sites posterior to the pectoral fins, dorsal sites along the dorsal midline anterior to the dorsal fin, and lateral sites approximately halfway in between. After the skin layer was separated from the underlying musculature, they were dried to a constant weight at 60°C. Samples from the Minnesota fish were further processed in a muffle furnace at 500°C to obtain ash weight.

## Results

Growth rates of silver lampreys confined with carp were negligible. The 11 lampreys collected on June 15 were measured on June 24 (mean length: 144.2 mm, SE = 7.5 mm; mean mass: 5.25

g, SE = 0.89 g) and supplied with carp. At least two attachments occurred by the following day, a minimum of 29 attachments occurred during the period June 24–July 21, and, during the same period, an average of 5.9 (SE = 0.3, range = 2–8, $n$ = 21) lampreys were attached when observations were made (no more than one observation on any given date). One lamprey (129 mm, 4.57 g) died on July 13. The others, when remeasured on July 26 (mean length: 146.7 mm, SE = 7.9 mm; mean mass: 5.36 g, SE = 0.91 g), were not significantly different in mean length ($T$ = 0.232, df = 19, $P$ = 0.819) or wet mass ($T$ = 0.086, df = 19, $P$ = 0.933) from when they were originally measured on June 24 (comparisons of group means are presented here because it was not possible to measure the growth of individual lampreys). The captive lampreys, however, were significantly shorter ($T$ = 2.224, df = 18, $P$ = 0.039) and lighter ($T$ = 2.892, df = 18, $P$ = 0.010) than the 10 lampreys removed from paddlefish on July 18 and held without feeding until measurement on July 20 (mean length: 169.8 mm, SE = 6.8 mm; mean mass: 9.89 g, SE = 1.28 g). Furthermore, comparisons of the intercepts of mass–length regressions failed to reveal a significant difference between the captive lampreys measured on June 24 and remeasured on July 26 (Figure 1, $F_{1,18}$ = 0.497, $P$ = 0.490), whereas the wild lampreys measured on July 20 were significantly heavier at a given length than the captives (Figure 2, $F_{1,17}$ = 8.649, $P$ = 0.009).

Although it was expected that skin density of carp in terms of dry mass would be greater than that of paddlefish, values for the two species were similar in magnitude (Figure 3). However, while the dried skin density of carp was greater on the dorsal and lateral surfaces than ventrally, samples from ventral locations on paddlefish were greater than those from dorsal or lateral sites. Paddlefish skin samples, especially those from ventral sites, retained a substantial oily residue when dried, but this was largely combusted in the muffle furnace, and skin density in terms of ash weight was lower for paddlefish than for carp (Figure 4).

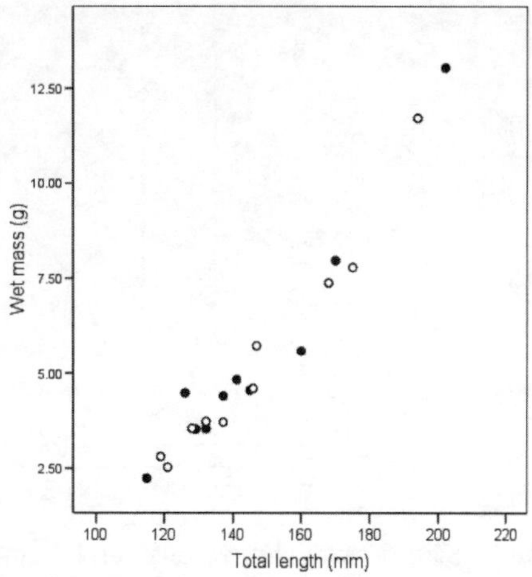


Figure 1. Wet mass (g) versus total length (mm) of silver lampreys before (closed circles) and after (open circles) a period of captive feeding on common carp.

## Discussion

This study compares silver lamprey growth rates in the field on one host, the native paddlefish, with growth rates in the laboratory on another host, the introduced common carp. Thus, two factors, host species and setting, vary

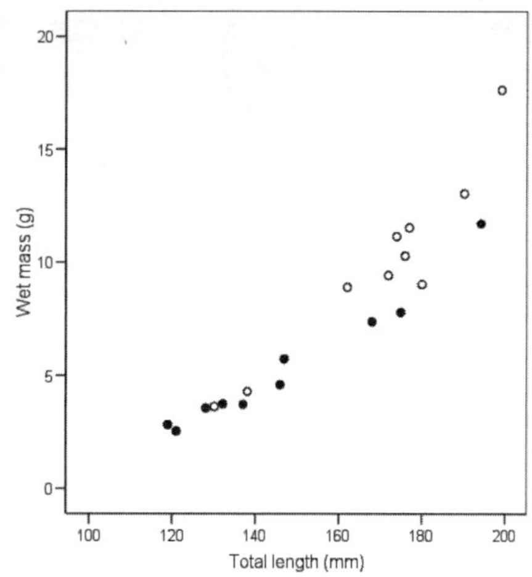


Figure 2. Wet mass (g) versus total length (mm) for a group of silver lampreys after a period of captive feeding on common carp (closed circles) and another group of silver lampreys collected on paddlefish (open circles).

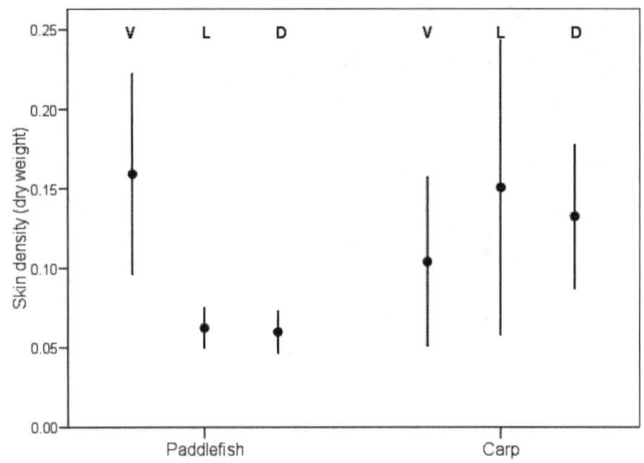


Figure 3. Skin density of paddlefish and common carp in terms of dry mass (g/cm$^2$) for samples taken from ventral (V), lateral (L), and dorsal (D) locations. Means and 95% confidence limits are indicated.

in the experimental design, but only two of four possible treatment combinations are included. A fully balanced design would be very challenging to accomplish given the apparent scarcity of lampreys feeding on carp below the Prairie du Sac dam and the logistical difficulty of acquiring and maintaining large paddlefish in the laboratory. Differences in growth can be attributed to the effect of host species only if it can be assumed that handling and maintenance of captive lampreys did not significantly detract from their growth. However, Roy (1973) reported rapid growth by captive silver lampreys that had attached primarily to the longnose sucker *Catostomus catostomus* and the white sucker *C. commersonii*, and Cochran (1989) successfully maintained other species of parasitic lampreys using the same protocol as in the present study.

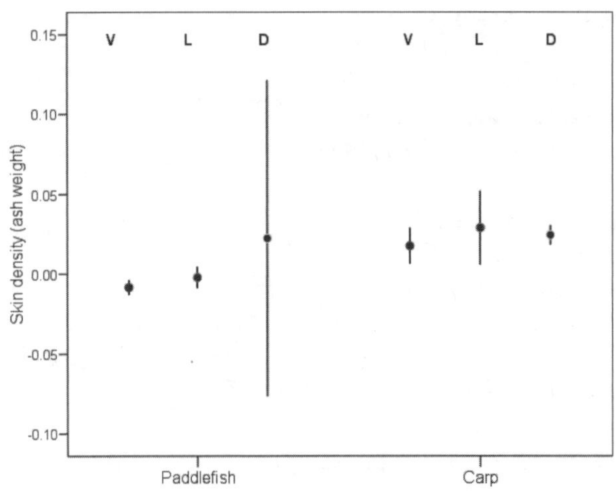


Figure 4. Skin density of paddlefish and carp in terms of ash weight (g/cm$^2$) for samples taken from ventral (V), lateral (L), and dorsal (D) locations. Means and 95% confidence limits are indicated. The confidence limits for the paddlefish dorsal mean reflect the inclusion of an outlier. Slight negative values for paddlefish ventral and lateral sites reflect ash weights of essentially zero (less than the measurement error of the balance); statistical conclusions would be unchanged if negative values were replaced by zeroes.

Although captive silver lampreys attached to carp readily during this study, they did not grow well on this host species. This is in contrast to Roy's (1973) results with *Catostomus*. As noted by Cochran (1991) and Cochran and Lyons (2004), captive silver lampreys seemed unable to form well-defined puncture wounds through the formidable scales of carp. Silver lampreys might grow more quickly on *Catostomus* than on carp because the former have much smaller scales, especially anteriorly. It is of interest that Cochran (2004) found silver lampreys in the field attached to hosts with formidable scale layers, including carp and longnose gar *Lepisosteus osseus* and that in some cases the lampreys had created wounds that appeared to penetrate between adjacent scales; it seems likely that they can more readily do so as they get larger. Moreover, parasitic lampreys can conceivably obtain at least some blood from the living tissue that overlies the scales.

The silver lamprey has been described on the basis of morphology as the most primitive of the extant lampreys (Hubbs and Potter 1971; Potter and Hilliard 1987), although in a more recent study its basal position was not as clear (Gill et al. 2003). It has also been characterized as a blood feeder that forms a puncture wound through the outer surface of its host rather than a flesh feeder that feeds primarily on muscle and other solid tissue (Potter and Hilliard 1987). However, Blatchley (1938) raised the additional possibility that silver lampreys feed primarily on surface secretions: "It is probable that their natural food is the slime or mucous exuded in abundance from the pores of their host, but they frequently retain their hold upon a spot until they have eaten through to the flesh, and deep ulcerous cavities occasionally result from the sore." A lamprey that fed on mucus would be an intermediate step consistent with Mallatt's (1984) scenario for the evolution of ectoparasites from epilithic algal feeders. What seems clear is that the silver lamprey has shared a long coevolutionary history with the paddlefish in the Mississippi River system (Hubbs and Potter 1971). Although it definitely feeds on blood to a large extent, it may not have faced strong selection pressure to refine its ability to penetrate thick scale layers, unlike lamprey species in other geographic areas.

I anticipated that the substantial scales of carp would result in a greater dried skin density than the essentially naked skin of paddlefish. A complete understanding of skin density would require measurements over a broad range of body sizes (e.g., Webb and Skadsen 1979), but the data presented here are representative of carp and paddlefish of sizes parasitized by silver lampreys in the field. In contrast to my expectation, dried skin densities of paddlefish and carp were similar (Figure 3), and both were greater than the values reported by Webb and Skadsen (1979) for rainbow trout *Oncorhynchus mykiss* 35 cm long (0.0252–0.0299 g/cm$^2$). Moreover, silver lampreys tend to attack paddlefish ventrally on paddlefish in the field and dorsally to carp in the laboratory (Wagner 1908; Cochran and Lyons 2004), in both cases where dried skin density is greatest. Cochran (1986) suggested that dorsal attachments to benthic hosts such as carp may minimize the chance of injury or detachment through abrasion against the substrate and also found that dorsal attachments were more prevalent in small laboratory tanks. Paddlefish, however, tend to occur well off the bottom in deep water (e.g., Lyons 1993), and it has been hypothesized that attachments to pelagic hosts tend to occur ventrally because the skin layer is less formidable there (Cochran 1986).

The substantial lipid content of paddlefish skin evident when skin samples were dried, especially noticeable in ventral samples, helps to explain discrepancies between observed and expected patterns in dried skin density. The oil was burned during combustion in the muffle furnace so that paddlefish skin density in terms of ash weight per unit area was significantly less than that of carp. The possibility that the oil itself constitutes a significant source of nutrition for silver lampreys that feed on paddlefish may explain a tendency for them to attach ventrally where dried skin density is greatest.

Cochran (1994) used a model of feeding by parasitic lampreys to provide insight into factors that might affect host species selection under the assumption that natural selection should favor behavior that maximizes the rate of net energy intake over a feeding bout. If net energy intake by a lamprey from a host can be plotted against feeding time as a curve with an upward trajectory that is initially steep but that levels off or begins to decline (Cochran and Kitchell 1986, 1989), then four general quantities influence the overall rate of net energy intake in positive or negative ways: (1) the initial slope of the energy intake curve (+), (2) the maximum or peak value attained by the energy intake curve (+), (3) the time elapsed between feedings (–), and (4) the energy expended during the time between feedings (–). In a comparative assessment of the suitability of paddlefish and carp as hosts for silver lampreys, the latter two factors seem most relevant. The initial slope of the energy intake curve is likely to be maximized for lampreys feeding on relatively large hosts of either species, although it might be affected by differences in blood chemistry. The maximum value attained by the energy intake curve is theoretically greater for paddlefish because they reach larger sizes, but field evidence suggests that silver lampreys do not remain attached to paddlefish long enough to make a difference. Previous observations below the Prairie du Sac dam have indicated that the number of lampreys attached per paddlefish is consistently far exceeded by the number of wounds (e.g., Cochran et al. 2003), and at least some lamprey attachments are terminated through the breaching behavior of the paddlefish (Wagner 1908; P. Cochran, personal observations). The relative time it takes for a silver lamprey to find a paddlefish or carp will depend on the geographic location, but paddlefish in the tailwaters of the Prairie du Sac dam are aggregated to the extent that a silver lamprey knocked off one individual is probably capable of finding another fairly quickly. Moreover, it seems clear that the time and energy expended in forming a puncture wound (i.e., the handling time) is greater for a silver lamprey on a carp. Indeed, the results of the present study suggest the possibility that a silver lamprey acquires significant energy via skin lipids while forming a puncture wound on a paddlefish, and it may not be appropriate to consider this time as part of the interval between feedings.

It would be desirable for future investigations to clarify the role of host skin lipids in the feeding biology of the silver lamprey. Further comparisons involving host species that vary in scale thickness would also be useful, and preparations should be made to investigate how silver and chestnut lampreys interact with the additional species of Asian carps that are expanding their ranges up the Mississippi River valley. After this manuscript was first submitted for review, Ehl (2008) reported chestnut lampreys attached to bighead carp *Hypophthalmichthys nobilis* in the Chariton River in Iowa.

While this study has revealed that silver lampreys attacking paddlefish encounter a high concentration of lipid, Wilkie et al. (2004) discussed the necessity for sea lampreys *Petromyzon marinus* attacking basking sharks *Cetorhinus maximus* to deal with the sharks' dermal denticles and body fluids high in urea. These studies illustrate a need for additional research to account not only for diversity among parasitic lampreys with respect to their feeding adaptations (Potter and Hilliard 1987), but also for diversity among host species with respect to their physical and biochemical features and how they affect the lamprey feeding process.

## Acknowledgments

I thank John Lyons of the Wisconsin Department of Natural Resources for arranging surveys of silver lampreys and paddlefish at the Prairie du Sac dam. Doug Dieterman of the Minnesota Department of Natural Resources supplied an additional paddlefish. Paddlefish were handled under a salvage permit from the Wisconsin Department of Natural Resources.

## References


Becker, G. C. 1983. Fishes of Wisconsin. University of Wisconsin Press, Madison.

Blatchley, W. S. 1938. The fishes of Indiana with descriptions, notes on habits and distribution in the state. The Nature Publishing Company, Indianapolis, Indiana.

Cochran, P. A. 1986. Attachment sites of parasitic lampreys: comparisons among species. Environmental Biology of Fishes 17:71–79.

Cochran, P. A. 1989. Maintaining parasitic lampreys in closed laboratory systems. American Biology Teacher 51:115–119.

Cochran, P. A. 1991. Letter to the editor. American Biology Teacher 53:69. (An addendum to Cochran 1989).

Cochran, P. A. 1994. Why lampreys and humans "compete" (and when they don't): toward a theory of host species selection by parasitic lampreys. Pages 329–345 *in* D. J. Stouder, K. L. Fresh, and R. J. Feller, editors. Theory and application in fish feeding ecology. Belle W. Baruch Library in Marine Science, No. 18, University of South Carolina Press, Columbia, South Carolina.

Cochran, P. A. 2004. Historical notes on lampreys in Wisconsin. American Currents 30(4):4–8.

Cochran, P. A., and J. F. Kitchell. 1986. Use of modeling to investigate potential feeding strategies of parasitic lampreys. Environmental Biology of Fishes 16:219–223.

Cochran, P. A., and J. F. Kitchell. 1989. A model of feeding by parasitic lampreys. Canadian Journal of Fisheries and Aquatic Sciences 46:1845–1852.

Cochran, P. A., and J. Lyons. 2004. Field and laboratory observations on the ecology and behavior of the silver lamprey (*Ichthyomyzon unicuspis*) in Wisconsin. Journal of Freshwater Ecology 19:245–253.

Cochran, P. A., J. Lyons, and M. R. Gehl. 2003. Parasitic attachments by overwintering silver lampreys, *Ichthyomyzon unicuspis*, and chestnut lampreys, *Ichthyomyzon castaneus*. Environmental Biology of Fishes 68:65–71.

Cochran, P. A., and J. E. Marks. 1995. Biology of the silver lamprey, *Ichthyomyzon unicuspis*, in Green Bay and the lower Fox River, with a comparison to the sea lamprey, *Petromyzon marinus*. Copeia 1995:409–421.

Ehl, D. 2008. Living fossil found at dam. Daily Iowegian (April 4).

Gill, H. S., C. B. Renaud, F. Chapleau, R. L. Mayden, and I. C. Potter. 2003. Phylogeny of living parasitic lampreys (Petromyzontiformes) based on morphological data. Copeia 2003:687–703.

Hubbs, C. L., and I. C. Potter. 1971. Distribution, phylogeny and taxonomy. Pages 1–65 *in* M. W. Hardisty and I. C. Potter, editors. The biology of lampreys, volume 1. Academic Press, London.

Lyons, J. 1993. Status and biology of the paddlefish (*Polyodon spathula*) in the lower Wisconsin River. Transactions of the Wisconsin Academy of Sciences, Arts and Letters 81:123–135.

Lubinski, K. S., A. Van Vooren, G. Farabee, J. Janacek, and S. D. Jackson. 1986. Common carp in the upper Mississippi River. Hydrobiologia 136:141–154.

Mallatt, J. 1984. Feeding ecology of the earliest vertebrates. Zoological Journal of the Linnean Society 82:261–272.

Potter, I. C., and R. W. Hilliard. 1987. A proposal for the functional and phylogenetic significance of differences in the dentition of lampreys (Agnatha: Petromyzontiformes). Journal of Zoology (London) 212:713–737.

Renaud, C. B. 2002. The muskellunge, *Esox masquinongy*, as a host for the silver lamprey, *Ichthyomyzon unicuspis*, in the Ottawa River, Ontario/Québec. Canadian Field-Naturalist 116:433–440.

Roy, J.-M. 1973. Croissance, comportement et alimentation de la lamproie du nord (*Ichthyomyzon unicuspis* Hubbs & Trautman) en captivité. Travaux sur les Pêcheries du Québec No. 41. Ministère de l'Industrie et du Commerce, Québec, Canada.

Runstrom, A. L., B. Vondracek, and C. A. Jennings. 2001. Population statistics for paddlefish in the Wisconsin River. Transactions of the American Fisheries Society 130:546–556.

Vladykov, V. D. 1985. Record of 61 parasitic lampreys (*Ichthyomyzon unicuspis*) on a single sturgeon (*Acipenser fulvescens*) netted in the St. Lawrence River (Québec). Naturaliste Canadien 112:435–436.

Wagner, G. 1908. Notes on the fish fauna of Lake Pepin. Transactions of the Wisconsin Academy of Sciences, Arts and Letters 16:23–37.

Webb, P. W., and J. M. Skadsen. 1979. Reduced skin mass: an adaptation for acceleration in some teleost fishes. Canadian Journal of Zoology 57:1570–1575.

Wilkie, M. P., S. Turnbull, J. Bird, Y. S. Wang, J. F. Claude, and J. H. Youson. 2004. Lamprey parasitism of sharks and teleosts: high capacity urea excretion in an extant vertebrate relic. Comparative Biochemistry and Physiology, Part A 138:485–492.

American Fisheries Society Symposium 72:173–189, 2009


# Spawning Distribution and Habitat Use of Adult Pacific and Western Brook Lampreys in Smith River, Oregon

Stephanie L. Gunckel*, Kim K. Jones, and Steven E. Jacobs
*Oregon Department of Fish and Wildlife, 28655 Highway 34, Corvallis, Oregon 97333, USA*

*Abstract.*—Pacific lampreys *Entosphenus tridentatus* (formerly *Lampetra tridentata*) and western brook lampreys *L. richardsoni* are observed in many coastal Oregon basins; however, few data are available to adequately describe their distribution and spawning habitat associations. To document landscape-scale distribution and habitat use of spawning adult lampreys, we conducted biweekly redd surveys at 62 randomly selected sites in Smith River (Umpqua basin, Oregon). Characteristics of lamprey habitat were measured at the redd, habitat unit, and reach scales to quantify available and occupied habitat. Pacific lampreys were found primarily in wider, low-elevation streams, whereas western brook lampreys were more widespread but concentrated in headwater and low-order streams. At the unit scale, unit type and dominant substrate were weakly correlated to presence of spawning lampreys. Both species spawned in gravel rich habitats, predominantly pool tail-outs and low gradient riffles. This study infers habitat associations and broad-scale distribution within a coastal basin to assist in the design of monitoring strategies and population assessments at a regional scale.

## Introduction

Coastal Oregon populations of Pacific lamprey *Entosphenus tridentatus* (formerly *Lampetra tridentata*) and western brook lamprey *L. richardsoni* are considered depressed due to habitat loss and passage problems (Close et al. 2002; Nawa 2003; ODFW 2006). The Pacific lamprey was listed as an Oregon State sensitive species in 1993 and in 1996 was protected through restriction of harvest (ODFW 2006). The western brook lamprey is not protected and has no special state status. Abundance of Pacific lampreys throughout the coast and Columbia River has declined dramatically since the1960s. Dam counts at Winchester (Umpqua River), Bonneville (Columbia River), and Leaburg (McKenzie River) dams show a dramatic decrease from historical levels (Kostow 2002; Nawa 2003; ODFW 2006). In 2003, 11 environmental groups petitioned the U.S. Fish and Wildlife Service to list Pacific, western brook, and two other lamprey species as endangered in the Pacific Northwest and California (Nawa 2003). Even though the petition cited habitat losses due to reduced instream flows, water diversions, dredging, scour and channnelization issues, pollution, and degradation of riparian communities, the U.S. Fish and Wildlife Service determined that the petition did not contain adequate information to warrant a listing (USFWS 2004). The Oregon Department of Fish and Wildlife recently reviewed the status of western brook and Pacific lampreys and found populations to be "at risk" of extinction (ODFW 2006) due to habitat loss, passage barriers and pollution. However, data necessary to conduct a more thorough and detailed assessment are lacking.

Many of the lacking data are critical to the effective management and conservation of Oregon's coastal lamprey species. The Columbia River Basin Lamprey Technical Workgroup

* Corresponding author: stephanie.gunckel@oregonstate.edu

(CRBLTW 2005) and members of Columbia River Inter-Tribal Fish Commission (CRITFC 2004) have identified and prioritized critical data gaps for Pacific lamprey, many of which also apply to western brook lamprey. Among these are (1) methods to assess distribution and abundance of all life stages and appropriate techniques for monitoring population status; (2) population structure and delineation; (3) population dynamics; (4) basic biology, including interspecific and community level relationships; (5) limiting factors and threats, including passage issues, and (6) habitat needs and requirements.

In coastal basins, data describing distribution and abundance of lampreys are rare despite extensive and rigorous sampling since 1998 under the Oregon Plan for Salmon and Watersheds. Survey methods employed to determine status and trends in salmonid populations have not proved applicable for lampreys. Surveyors collecting information on adult steelhead *Oncorhynchus mykiss* during February through May observe adult Pacific and western brook lamprey individuals and redds, but the observations tend to be incidental rather than quantitative. Juvenile western brook and Pacific lampreys are not observed by snorkelers or collected by traditional electrofishing technology used for salmonids. Although out-migrating juvenile lampreys are collected in salmonid migrant traps operated by the Oregon Department of Fish and Wildlife, trap efficiency for lampreys is unknown (D. Jepsen, Oregon Department of Fish and Wildlife, personal communication). Furthermore, these traps are not placed in randomly selected watersheds such that findings at trap sites cannot be extrapolated to other drainages. Habitat surveys conducted under the Oregon plan may be applicable, but habitat specific models are not available. Finally, with the exception of Winchester Dam on the North Umpqua River, counts of adult lampreys that pass dams are largely unavailable because dams in Oregon coastal basins are positioned in the upper portions of drainages. Existing sampling indicates that lampreys are present in coastal basins, but monitoring strategies need to be modified in order to determine an appropriate sample frame and collect data in a quantitative fashion.

Our goal was to identify habitat variables associated with spawning Pacific and western brook lampreys in order to infer distribution throughout coastal Oregon. The objectives of this study were to (1) determine distribution of spawning Pacific and western brook lampreys in the Smith River basin, (2) describe redds of both species, and (3) describe associations of spawning adults in relation to habitat unit- and reach-scale habitat characteristics.

## Methods

### *Study area*

This study was conducted in the Smith River, a tributary of the Umpqua River (Figure 1). The study area consists of 463 km of main-stem and tributary streams upstream of Smith River Falls (river kilometer 48). The basin area above Smith River Falls is approximately 525 km$^2$. The riparian community is dominated by red alder *Alnus rubra* with an understory of sword fern *Polystichum munitum*, salal *Gaultheria shallon*, and vine maple *Acer circinatum*. Smith River above the falls is inhabited by Pacific and western brook lampreys, coho salmon *O. kisutch*, Chinook salmon *O. tshawytscha*, steelhead, coastal cutthroat trout *O. clarkii*, large-scale sucker *Catostomus macrocheilus*, redside shiner *Richardsonius balteatus*, longnose dace *Rhinichthys cataractae*, and Umpqua pikeminnow *Ptychocheilus umpquae*. Land is owned predominately by the U.S. Bureau of Land Management and private forest companies and managed for timber harvest.

### *Sample site selection*

In each of 2004 and 2005, 50 sample sites of 1,000 m were selected from a frame representing the rearing distribution of coho salmon and steelhead above Smith River Falls. Given the lack of available information regarding Pacific and western brook lamprey distributions, the

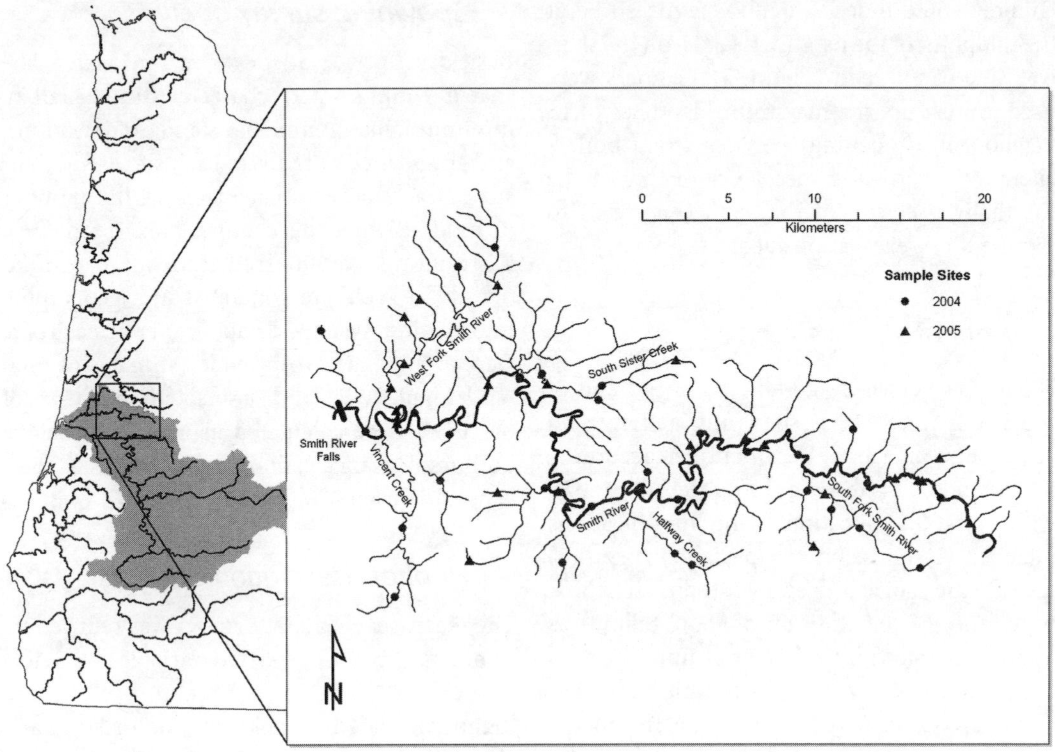


Figure 1. Smith River study area in Oregon and the Umpqua River basin.

salmon and steelhead rearing sample frame offered the best starting point to describe potential lamprey distribution in the basin. Sites were chosen in priority order using the variable probability. Stevens and Olsen (2004) generalized random tessellation stratified algorithm and were randomly selected and spatially balanced ensuring an unbiased and representative sample. Site locations were unique each year.

Geographic coordinates of each site were plotted on U.S. Geological Survey 1:24K topographic maps and uploaded into a handheld Global Positioning System (GPS) unit. Field crews located the precise position of each site, using a map, compass, and the GPS unit. The coordinates represented the downstream end of the sample site, which extended upstream a channel distance of 1,000 m.

We surveyed 31 sites in 2004 and in 2005. Of the original sample draw of 50 sites in 2004, 2 were dry, 2 were not surveyable, and 3 were surveyed by crews targeting steelhead trout where lamprey data were not consistently collected; 1 site was not surveyable in 2005. Field crews were unable to incorporate the remaining sites (12 in 2004 and 18 in 2005) into the biweekly survey rotation. These sites were low priority and eliminating them did not disrupt the spatial balance of the remaining 31 sample sites.

### *Spawning surveys*

All sites were surveyed biweekly from between March 29 and June 24, 2004, and March 28 and June 30, 2005. Newly observed lamprey redds were measured, recorded, and flagged. For each redd, crews noted species; measured length, width, depth of the redd, depth of the water, water velocity, and distance to the nearest cover; and conducted a pebble count. Cover was defined as any structure under which an adult lamprey could hide. Velocity was measured at the midpoint of the lengthwise axis, on the thalweg side of the redd (Figure 2). Calipers were used to measure the maximum diameter of 15 neighboring (touching) pebbles along both a lateral and vertical transect originating at the top-thalwag corner of the redd (Figure 2) to represent pre-excavation substrate.

## Habitat surveys

Habitat surveys were conducted at the end of the season after most spawning activity had ceased. Stream habitat was quantified according to methods described by Moore et al. (2007), with some modifications. The modifications included (1) omission of riparian surveys, (2) measurement of length and width for all habitat units, and (3) the addition of "pool tailout" as a unit separate from other pool units. A total of 54 reach-scale and 38 habitat-unit-scale variables were summarized for each site. In general, descriptions were recorded for channel dimensions, streambed composition, geomorphic reach characteristics, riparian cover, and shade. In order to associate each redd with habitat and reach-scale data, redds were noted within the habitat units in which they occurred.

## Spawning survey analysis

Basic descriptive statistics were used to describe spawn timing, redd characteristics (median, minimum, maximum, and standard deviation), and abundance of redds at each site. Redd density at each site was calculated as the number of redds divided by channel area (redds/m$^2$). Distribution was inferred from maps of redd density at each site generated in a geographic information system. Redd counts at each site were extrapolated to provide estimates of total redd abundance, and associated measures of precision were calculated using a local-neighborhood variance estimator (Stevens 2002; Stevens and Olsen 2003).

## Habitat association analysis

The reach-scale habitat data set was reduced by removing highly correlated variables as determined by correlation analysis. Of two or more highly correlated variables, we retained the variable with the more accurate or precise measurement. Appendix A.1 lists all variables included in the analysis at each scale. We evaluated similarity of reach-scale habitat characteristics between all sites. Similarity was calculated using a relative Sorenson's distance measure, and non-

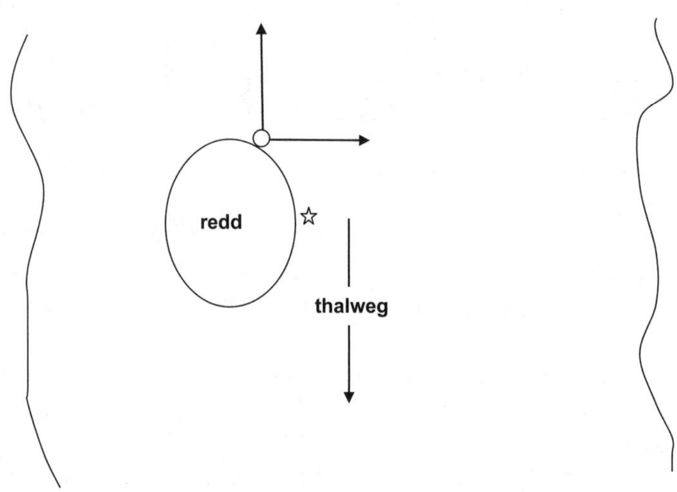


Figure 2. Location of water velocity measurement (☆) and origin of pebble counts (O) in relation to the lamprey redd and thalweg.

metric multidimensional scaling (NMS) ordinations were generated in PC-ORD software (Kruskal 1964; McCune and Mefford 1999). Three hundred iterations were used for each run to evaluate stability, and each analysis was conducted 10 times to ensure the lowest possible stress. The response variables, presence of lamprey redds and redd density, were overlaid on the ordination, and the ordinations were orthogonally rotated to load the maximum variation in lamprey presence on the first axis. Correlation coefficients of habitat variables to axis 1 were used to identify factors related to the presence and density of lamprey redds. Analyses of associations at the habitat unit scale were conducted in the same manner, except only those habitat units consisting of sites where lampreys were detected were included in the analysis.

## Results

### Abundance

In 2004, field crews counted a total of 313 western brook lamprey redds between April 6 and June 7, with the maximum observed within a 2-week period occurring during the May 19 and June 1 survey cycle (Figure 3). In 2005, western brook lamprey spawning activity commenced 1 week later than in 2004. A total of 249 western brook lamprey redds were detected between April 25 and June 16. The maximum number of redds was counted during the May 24–June 6 survey cycle.

A total of 60 Pacific lamprey redds were counted between April 5 and May 5, 2004, with the peak count occurring during the April 7 and April 20 survey cycle (Figure 3). In 2005, Pacific lamprey redds were first detected 1 week later than in 2004; 69 Pacific lamprey redds were counted between April 14 and June 6. Two peak counts were noted in 2005. The first peak occurred between April 26 and May 9 and the second between May 24 and June 6. The later onset of Pacific lamprey spawning activity in 2005 was similar to that observed on the South Fork Coquille River (Brumo and Markle 2006) and may be related to higher stream flows and cooler water temperatures that spring.

We estimated a total of 4,692 (±76%, 95% confidence interval [CI]) western brook lamprey redds (11.25 redds/km) in 2004 and 4,265 (±91%, 95% CI) redds (10.22 redds/km) in 2005. Pacific lamprey redds were estimated to be 504 (±67%, 95% CI) (1.21 redds/km) in 2004 and 459 (±110%, 95% CI) redds (1.10 redds/km) in 2005.

### Redd characteristics

Generally, lamprey redds were identified as round depressions in gravel or cobble substrates. Rocks moved by lampreys to dig the depression typically were piled on all sides of the depression rim and were lighter colored due to exposure of nonalgal-covered surfaces. A plume of finer substrate carried by the current was downstream of the depression as a result of spawning and egg burial; however, in places with low velocities, these finer substrates were not apparent. In many instances, few large rocks were left in the bottom of the depression (Figure 4). Pacific and western brook lamprey redds differed in size and substrate size. Redd characteristics for each species are summarized in Table 1.

Western brook lamprey redds were found in gravel rich and relatively shallow habitats, predominantly pool tailouts (46.3%), low-gradient (slope < 2%) riffles (22.2%), steps over cobble (16.1%), and lateral scour pools (13.1%). Redds were associated with a variety of cover types, most commonly rocky substrates (58.1%), wood (11.9%), and vegetation (8.9%). Because western brook lamprey redds are small, large cobble substrates serve as adequate cover under which spawning adults can hide. Fourteen percent of redds were not associated with any cover.

Pacific lamprey redds were most frequently observed in low-gradient riffles (50.4%), pool tailouts (36.8%), and lateral scour pools (11.2%). Only 43.2% of Pacific lamprey redds were associated with cover, predominately large substrates (26.4%), vegetation (8.0%) and wood (6.5%). The remaining 56.8% of the 125 redds

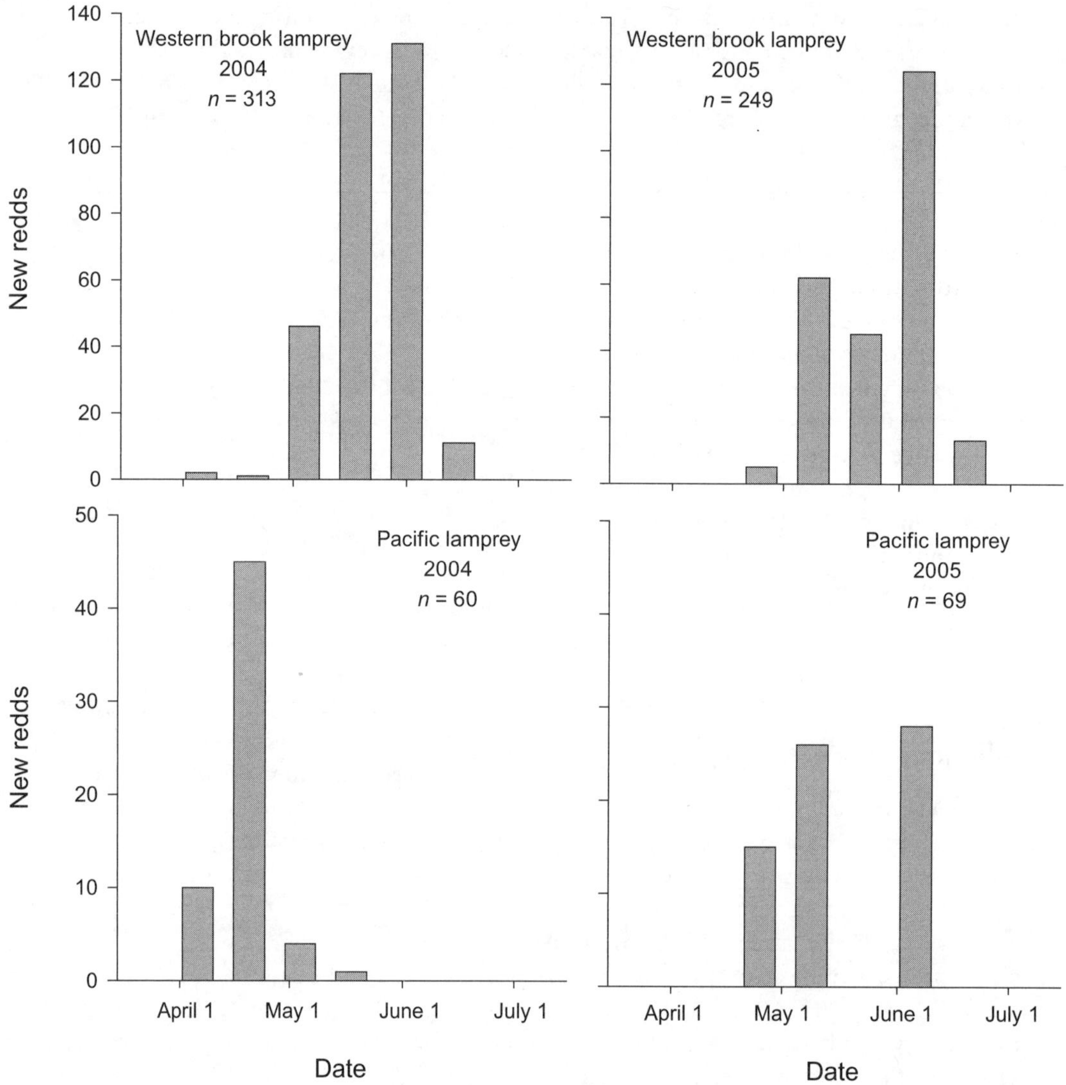


Figure 3. Spawn timing of western brook and Pacific lampreys in 2004 and 2005. Each bar represents a 2-week survey cycle. Surveys were conducted from March 29 through June 24 in 2004 and March 28 through June 30 in 2005.

were observed away from channel features that could serve as cover.

## *Site occupancy and distribution*

Field crews detected western brook lamprey redds in 20 of 31 sites in 2004 and 16 of 31 sites in 2005 (Table 2). Pacific lamprey redds were identified in fewer sites, 8 in 2004 and 9 in 2005.

Generally, western brook lampreys were widespread in the smaller-order tributary streams (Figure 5), whereas Pacific lampreys only occupied large main-stem reaches of Smith River and West Fork Smith River (Figure 6).

## *Habitat associations*

Habitat surveys were completed at 56 of the 62 sites where redd surveys were conducted. Physi-

Figure 4. Pacific lamprey redd (Photo by G. Susac, Oregon Department of Fish and Wildlife).

cal habitat variables were quantified at all sites where lampreys were detected; however, in 2005, habitat surveys at six sites where lampreys were absent were omitted due to time constraints. The data set was too small to describe associations between habitat characteristics and redd density at each site and habitat unit; therefore, our response variable was limited to lamprey presence.

Table 1. Summary of physical characteristics of western brook and Pacific lamprey redds. Redd depth is the distance from the bottom of the depression to the surface of the substrate; water depth is the distance from the substrate to the water surface; substrate is the average maximum diameter of 30 pebbles.

| Variable | Median | Minimum | Maximum | SD |
|---|---|---|---|---|
| | | Western brook lamprey ($n = 539$) | | |
| Length (cm) | 12.0 | 4.0 | 115.8 | 7.8 |
| Width (cm) | 11.0 | 4.5 | 73.2 | 7.8 |
| Redd Depth (cm) | 3.0 | −3.5 | 12.2 | 2.1 |
| Water Depth (cm) | 13.0 | 0.3 | 51.0 | 7.6 |
| Velocity (m/s) | 0.2 | 0.0 | 0.7 | 0.1 |
| Substrate (mm) | 24.0 | 7.8 | 65.2 | 9.0 |
| | | Pacific lamprey ($n = 125$) | | |
| Length (cm) | 39.0 | 21.3 | 210.0 | 18.3 |
| Width (cm) | 36.6 | 20.0 | 270.0 | 23.9 |
| Redd Depth (cm) | 7.6 | −12.0 | 18.3 | 3.8 |
| Water Depth (cm) | 44.0 | 16.0 | 105.0 | 23.2 |
| Velocity (m/s) | 0.6 | 0.2 | 1.0 | 0.2 |
| Substrate (mm) | 48.1 | 27.0 | 88.8 | 12.9 |

Table 2. Occupancy statistics for sites surveyed for lamprey redds in the Smith River basin, 2004 and 2005.

| Year | Sites surveyed | Sites occupied | Minimum number at occupied site | Maximum number at occupied site |
|---|---|---|---|---|
| | | Western brook lamprey redds | | |
| 2004 | 31 | 20 | 1 | 136 |
| 2005 | 31 | 16 | 1 | 114 |
| | | Pacific lamprey redds | | |
| 2004 | 31 | 8 | 1 | 15 |
| 2005 | 31 | 9 | 1 | 48 |

### *Reach-scale analysis*

The reach-scale habitat data set was reduced from 54 to 19 variables (Appendix A.1) and was best described in a two-dimensional ordination that explained 93% of the total variation within the data set. When rotated to load variation associated with presence of western brook lamprey redds on the first axis, the ordination shows that western brook lamprey redds were predominantly detected at higher elevation sites with smaller active channel widths. Axis 1 explains 88% of the variation and is positively correlated with elevation ($r^2$ = 0.849) and negatively correlated with active channel width ($r^2$= 0.772) (Figure 7). However, this relationship is relatively weak given the detection of western brook lamprey redds at some low elevation sites. Figure 8 quantifies the median and range of elevation and active channel width for all sites and those at which western brook lamprey redds occur.

When orthogonally rotated to load variation associated with presence of Pacific lamprey redds on the first axis, the ordination of reach-scale variables shows that Pacific lamprey redds were present in sites at lower elevations and with wider active channels (Figure 9). Axis 1 explained 85% of the variation and was nega-

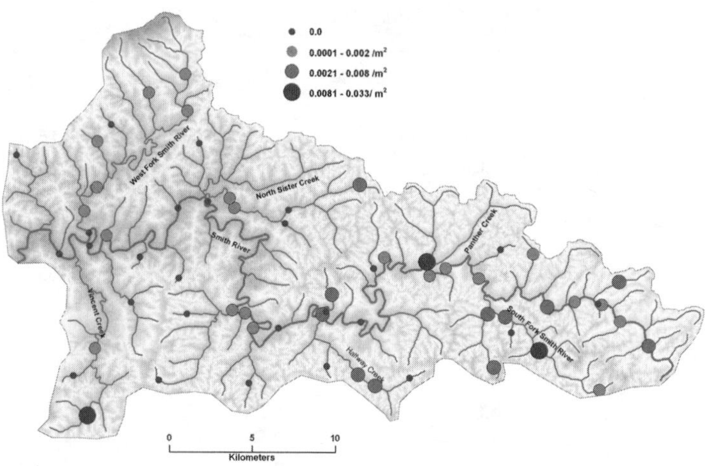


Figure 5. Site occupancy and redd density for western brook lampreys in the Smith River basin, 2004 and 2005.

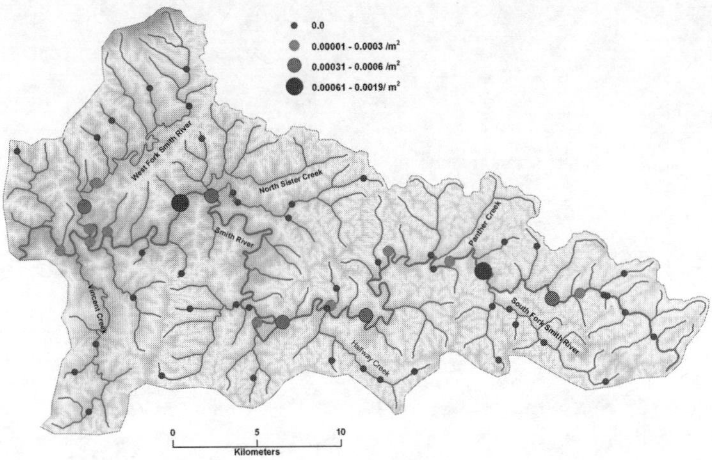


Figure 6. Site occupancy and redd density for Pacific lampreys in the Smith River basin, 2004 and 2005. Figure 6. Site occupancy and redd density for Pacific lampreys in the Smith River basin, 2004 and 2005.

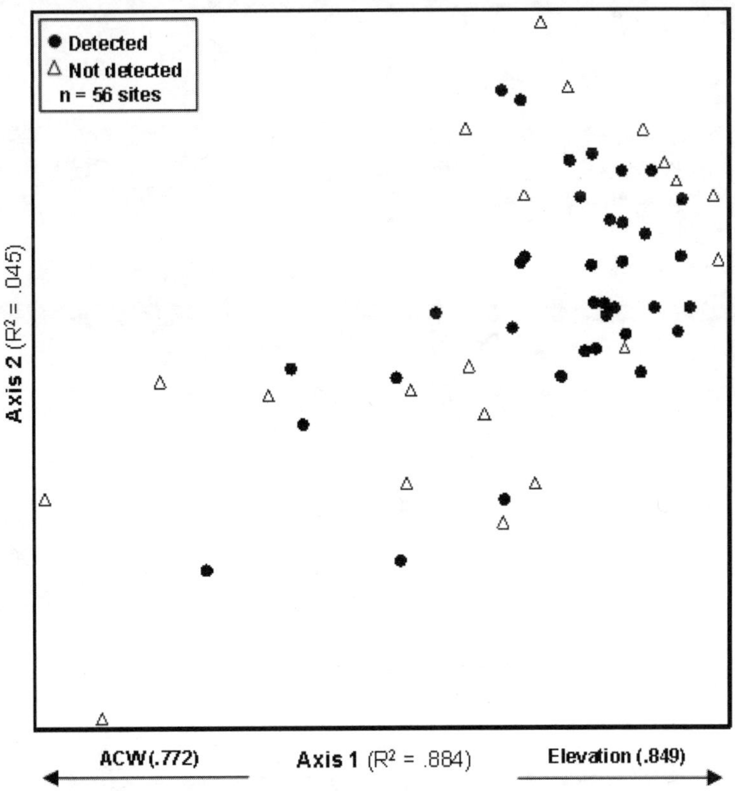


Figure 7. Nonmetric multidimensional scaling ordination of sites as described using reach sale habitat variables. Sites characterized by presence or absence of western brook lamprey redds. Arrows represent direction of correlation. Values in parentheses are correlation coefficients of habitat variables to axis 1. Only variables highly correlated with axis 1 are shown.

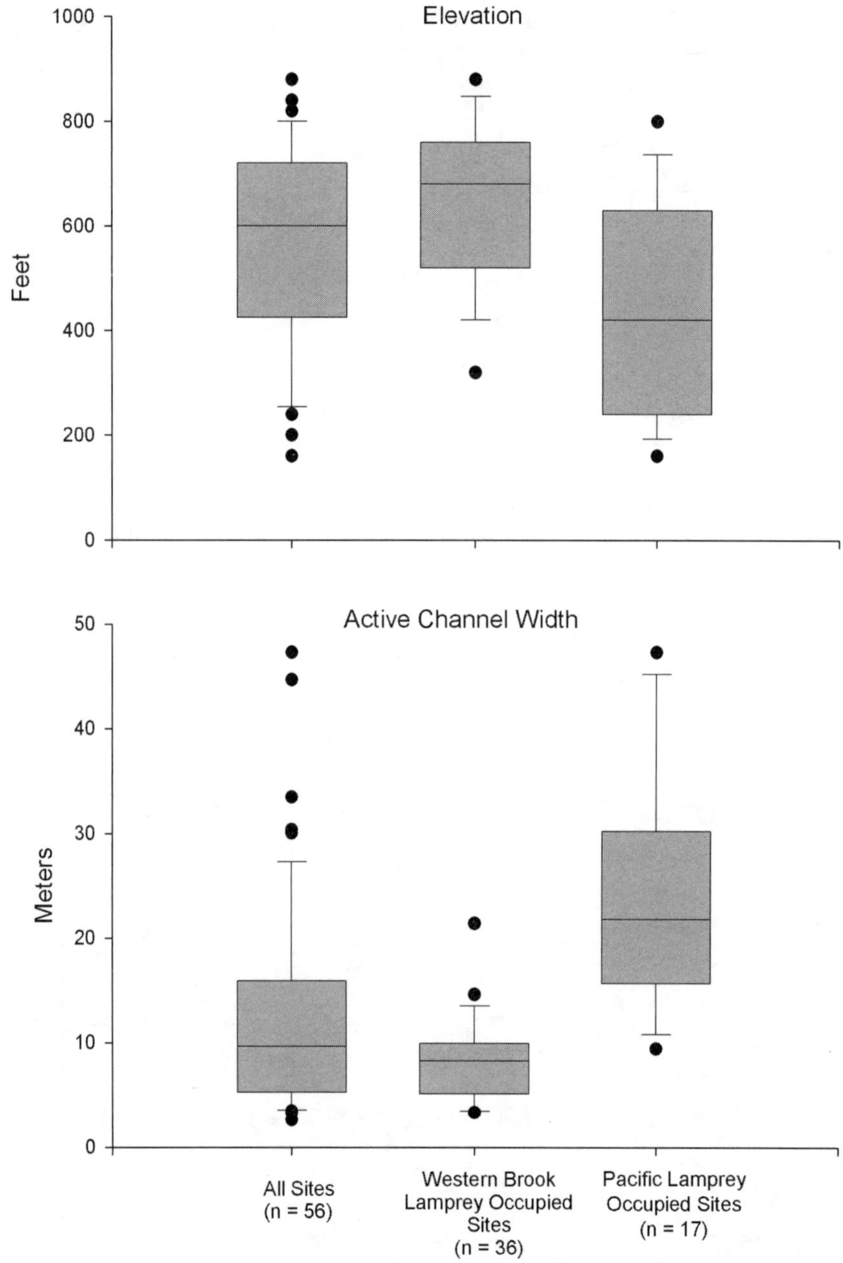


Figure 8. Box and whisker plots of reach variables associated with sites where the presence of western brook and Pacific lamprey redds were detected.

tively associated with elevation ($r^2 = 0.656$) and positively associated with active channel width ($r^2 = 0.851$). The remaining reach-scale variables were weakly correlated with either axis. Figure 8 describes the median and range of these explanatory variables for all sites and those at which Pacific lamprey redds occur.

Results of the multivariate analysis of reach-scale habitat data corresponds well to overall distribution (Figures 5 and 6). Pacific

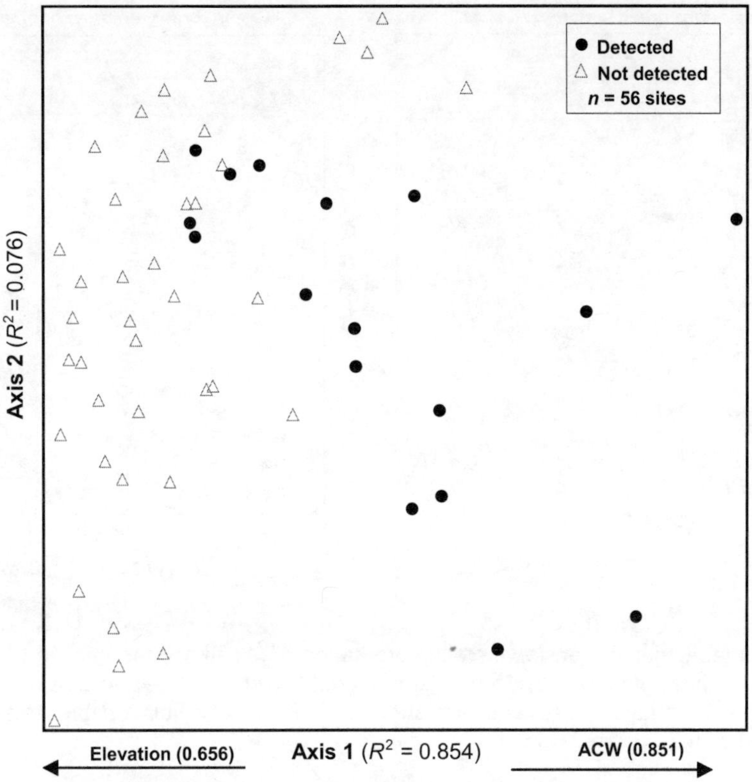


Figure 9. Nonmetric multidimensional scaling ordination of sites as described using reach-scale habitat variables. Sites characterized by presence or absence of Pacific lamprey redds. Arrows represent direction of correlation. Values in parentheses are correlation coefficients of habitat variables to axis 1. Only variables highly correlated with axis 1 are shown.

lampreys are found primarily in wider, high-order streams, whereas western brook lampreys are more widespread but concentrated in headwater and low order streams.

## Habitat unit-scale analysis

The habitat unit-scale data set was reduced from 38 variables to 15 variables (Appendix A.1). A three-dimensional NMS ordination of all habitat units consisting of the 36 sites where western brook lamprey redds were detected explained 75.5% of the variation among units. When rotated to load variation associated with presence of western brook lamprey redds on the first axis, the ordination shows that the presence of redds was weakly correlated with a high percentage of gravel and low percentage of bedrock (Figure 10). Axis 1 explains 61% of the variation associated with all habitat units and was negatively associated with unit area ($r^2$ = 0.427), depth ($r^2$= 0.336), and percent bedrock ($r^2$ = 0.328) and positively associated with percent gravel ($r^2$ = 0.278), though associations were relatively weak. Table 3 describes significant values for these variables for sites with and without western brook lampreys.

To ensure that analysis of all habitat unit types combined did not mask trends within habitat types, we examined each habitat unit type separately and found similar results. For example, a NMS ordination of habitat unit-scale variables of only pool tailouts also suggests that western brook lamprey redds are weakly correlated with substrate. The presence of redds in pool tailouts is positively correlated with

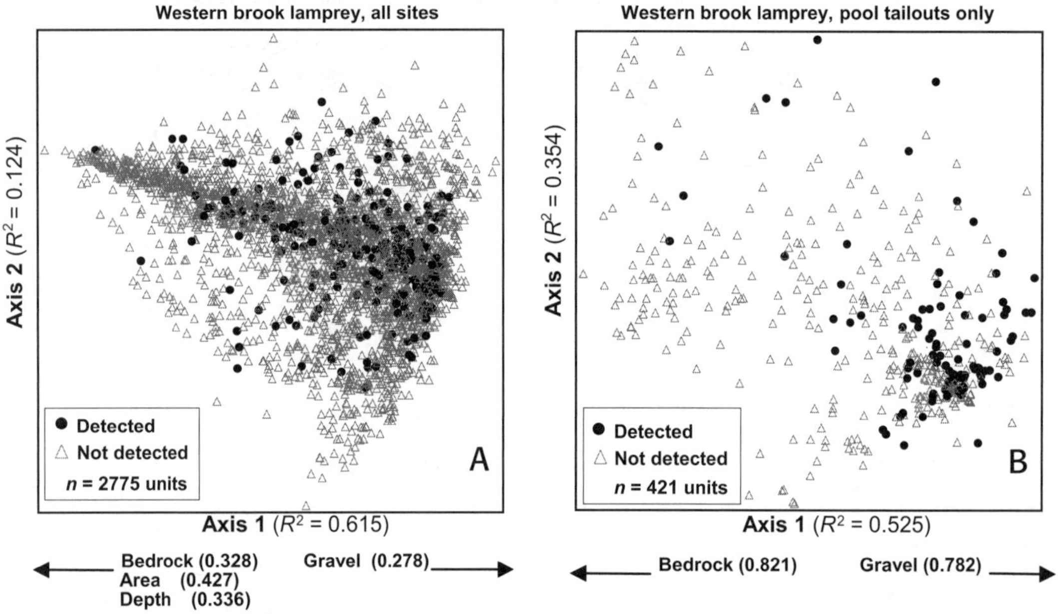


Figure 10. Nonmetric multidimensional scaling ordination of (A) all habitat units and (B) pool tailouts only consisting of sites where western brook lamprey redds were detected. Arrows represent direction of correlation. Values in parentheses are correlation coefficients of habitat variables to axis 1. Only variables correlated with axis 1 are shown.

Table 3. Summary of habitat unit scale variables correlated with sites where presence of western brook lamprey redds was detected.

| | Occupied units | | | Total units | | |
|---|---|---|---|---|---|---|
| | Median | Min | Max | Median | Min | Max |
| | | | All unit types | | | |
| | | $n = 251$ | | | $n = 2775$ | |
| % gravel | 65 | 5 | 100 | 40 | 0 | 100 |
| % bedrock | 0 | 0 | 100 | 0 | 0 | 100 |
| Unit area (m²) | 25.2 | 0.8 | 1,168.0 | 60.6 | 1.5 | 1,284.8 |
| Unit depth (m) | 0.15 | 0.03 | 1.4 | 0.15 | 0.01 | 0.60 |
| | | | Pool tailouts only | | | |
| | | $n = 94$ | | | $n = 421$ | |
| % gravel | 80 | 5 | 100 | 50 | 0 | 100 |
| % bedrock | 0 | 0 | 90 | 0 | 0 | 100 |
| Unit area (m²) | 9.6 | 0.8 | 325.6 | 7.6 | 0.2 | 325.6 |
| Unit depth (m) | 0.15 | 0.05 | 0.60 | 0.15 | 0.03 | 0.60 |

percent gravel ($r^2$ = 0.782) and negatively correlated with percent bedrock ($r^2$ = 0.821) (Figure 10). Axis 1 explains 53% of variation within variables.

Analysis of habitat unit characteristics consisting of the seven sites where Pacific lamprey redds were detected reveals results similar to those for western brook lampreys. The presence of Pacific lamprey redds was weakly correlated with high percentages of gravel and cobble and a low percentage of bedrock. A two-dimensional ordination of all units explained 75.5% of variation among units (Figure 11). When rotated to load the variation associated with presence of Pacific lamprey redds on the first axis, axis 1 explains 63% of the variation. Axis 1 was negatively correlated with percent bedrock ($r^2$ = 0.938) and positively correlated with percent gravel (r2 = 0.518) and cobble ($r^2$ = 0.446). The remaining habitat unit-scale variables were poorly correlated with either axis. Table 4 describes significant values for these variables at sites with and without Pacific lamprey redds.

Examination of unit-scale variables of pool tailouts only show similar results as those for analyses of all unit types combined and separately. A two-dimensional ordination of unit-scale variables describing all pool tailouts, orthogonally rotated to maximize variation associated with the presence of Pacific lamprey redds, shows that axis 1 is loosely correlated with percent gravel ($r^2$ = 0.332) and negatively correlated with percent bedrock ($r^2$ = 0.283) and shade ($r^2$ = 0.314); however, axis 1 explains just 25% of variation within the ordination (Figure 11).

## Discussion

This study described habitat associations, distribution, and redds of Pacific and western brook lampreys while highlighting uncertainties inherent in estimating abundance of lampreys at a landscape scale. In general, we found that Pacific lampreys spawned in lower-elevation sites in the main-stem sections of Smith River, whereas western brook lampreys were more widespread

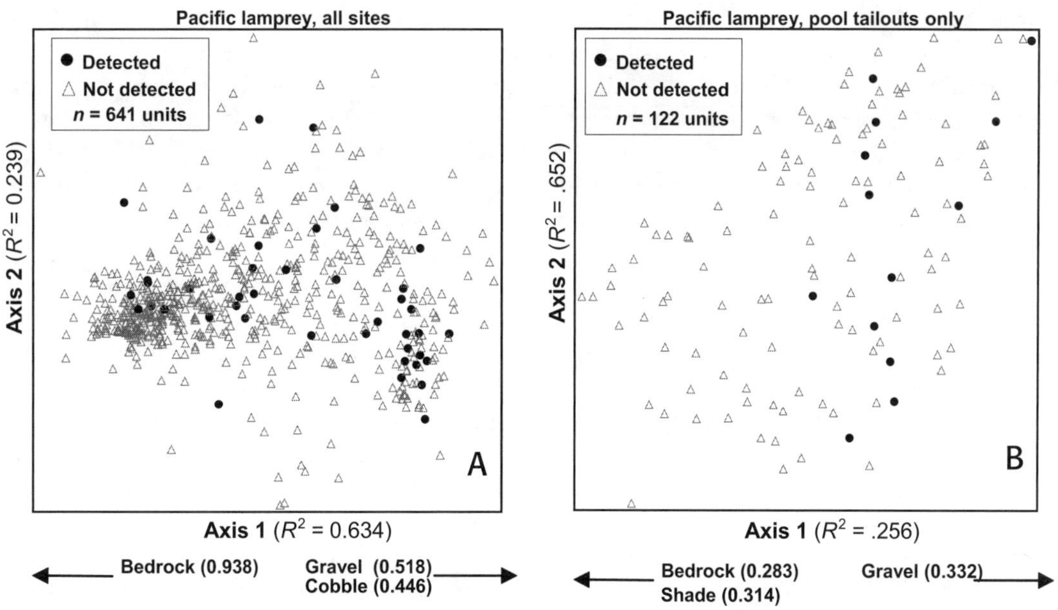


Figure 11. Nonmetric multidimensional scaling ordination of (A) all habitat units and (B) pool tailouts only consisting of sites where Pacific lamprey redds were detected. Arrows represent direction of correlation. Values in parentheses are correlation coefficients of habitat variables to axis 1. Only variables correlated with axis 1 are shown.

Table 4. Summary of habitat unit scale variables correlated with sites where presence of Pacific lamprey redds was detected.

| | Occupied units | | | Total units | | |
|---|---|---|---|---|---|---|
| | Median | Min | Max | Median | Min | Max |
| | | | All unit types | | | |
| | | $n = 41$ | | | $n = 641$ | |
| % gravel | 25 | 4.8 | 90 | 9.9 | 0 | 95 |
| % bedrock | 19.8 | 0 | 90 | 50 | 0 | 100 |
| | | | Pool tailouts only | | | |
| | | $n = 13$ | | | $n = 122$ | |
| % gravel | 60 | 5 | 80 | 20 | 0 | 85 |
| % bedrock | 10 | 0 | 90 | 30 | 0 | 100 |

and distributed throughout the upper mainstem and tributary streams. This pattern of distribution is similar to that observed on Cedar Creek in the Lewis River system where Pacific lamprey spawning locations were in the lower sections of Cedar Creek and western brook lampreys were found in the upper tributaries (Pirtle et al. 2003).

Our a priori estimates of potential distribution (sample frame) and the identification of redds by species increased the amount of uncertainly in the redd abundance estimate. In addition, western brook lampreys may have extended upstream of our sample frame, which was based on the upper extent of steelhead. The precision of redd abundance estimates is less than desired. Our goal was to have a 95% confidence interval of less than 45%, though we preferred one of less than 35%. The large confidence intervals (67–110%) were likely due to the small sample size and high variance between sites. Even though the confidence intervals were wide, the actual estimates for each species and site occupancy remained virtually unchanged from 2004 to 2005. Tailoring the sample frame to the distribution of each species, increasing sample size, and refining the survey protocol may result in more precise estimates and should be considered if redd surveys using similar protocol are to be implemented as a long-term monitoring strategy.

Four factors complicated identification of redds. First, in 27 instances (23 western brook lampreys and 4 Pacific lampreys), redds were identified in large areas of cleaned and disturbed gravel, usually containing two to five depressions. These areas were two to three times the size of a typical redd. It was evident that substrate had been cleared due to lamprey spawning activity, but it was not obvious how many redds these areas represented. Each identifiable depression was counted as an individual red, but measurements were made with considerable uncertainty. These redds were not included in the summaries of physical redd characteristics described in Table 1.

Second, in a few instances field crews observed both western brook and Pacific lampreys filling in newly created redds. Lampreys moved small gravel and cobble from upstream into the depression until it was filled and flush with the substrate surface. In some cases, the depression was filled where it mounded above the surface of the substrate (depicted as negative values in minimum depth measurements in Table 1). This behavior, where common, may make the detection of redds difficult.

Third, in a few instances field crews observed western brook lampreys spawning in redds of Pacific lampreys. Field staff were diligent about revisiting old redds of both species to capture any superimposition by western brook lampreys. Studies in the South Fork Coquille River and Lewis River in Washington have also observed this behavior (Pirtle et al. 2003; Brumo and Markle 2006).

Last, crews were careful to properly identify western brook lamprey redds. In places where redds could be easily confused with large elk tracks, presence of a lamprey positively identified the redd. If a lamprey was not present, then the redd was not counted or measured. Similarly, Pacific lamprey redds could be confused with steelhead redds. Again, field crews made conservative redd calls and only identified redds certain to be created by lampreys.

These complications are likely sources of error for quantitative redd counts and should be taken into consideration if using redd surveys for population monitoring purposes. Similar to redd surveys for bull trout *Salvelinus confluentus* (Dunham et al. 2001) and other salmonids, crew experience is one method to minimize error associated with misidentification, superimposition, and lamprey spawning behavior. In addition, frequent surveys (no less than biweekly surveys) can ensure positive identification of redds and minimize error. In many instances, crews observed redds on one survey that were filled in, superimposed, or otherwise unidentifiable on the subsequent visit.

The multivariate analyses show many unoccupied habitat units that have very similar physical characteristics as those occupied by lampreys, even within the same reach. Thus, at fine scales, other ecological factors may have a stronger influence on spawning distribution than habitat structure. Presence of lampreys at a particular site may influence the behavior of other individuals. It is theorized that Pacific and western brook lampreys are attracted to streams and spawning sites by pheromones released by ammocoetes and mature adults as are sea lampreys *Petromyzon marinus* (Bjerselius et al. 2000; Vrieze and Sorensen 2001; Li et al. 2002). This phenomena may be evident at one particular site on Smith River where 48 Pacific lamprey redds were detected, a value much greater than the mean (Table 2). Habitat characteristics at this site were not significantly different from those of other sites and could not explain the unusually high density of redds. Lamprey pheromones may have a greater influence on distribution than habitat structure. In addition, even though this study quantified a myriad of habitat characteristics, it may not have measured the correct habitat unit-scale variables to fully explain habitat unit selection of spawning lampreys.

Connectivity and distribution of habitats may also have a significant influence on spawning site selection. For example, sites near high-quality rearing habitat for western brook lampreys or Pacific lamprey overwintering habitat may be more likely occupied than sites of similar character, but not near optimal habitats for other life stages. This study did not account for stream network distribution of other habitats important to western brook and Pacific lampreys.

This study found that reach-scale habitat variables correlated with stream size were the best descriptors of potential distribution of spawning lampreys of both species. Pacific lampreys spawned at lower elevation sites in larger stream sizes. Western brook lampreys were associated with higher elevation sites and smaller streams but were relatively widespread. At the habitat unit scale, unit type and dominant substrate were only weakly correlated with presence of spawning lampreys. Both species spawn predominantly in pool tailouts and low-gradient riffles, both gravel-rich habitat types. The association of reach- and unit-scale variables with distribution of adult lampreys helps refine the sample frame for monitoring adult populations and provides context for designing a sample strategy for juvenile lampreys.

## Acknowledgments

This study would not have been possible without the efforts of many dedicated people. We would like to recognize the field crews who collected the data: Michael Scheu, Jeremiah Bernier, Brian Bangs, and Peter Baki. Staci Stein and Brian Bangs completed the data entry and Charlie Stein assisted with the analysis of the habitat data.

## References

Bjerselius, R., W. Li, J. H. Teeter, J. G. Seelye, P. B. Johnsen, P. J. Maniak, G. C. Grant, C. N. Polkinghorne, and P. W. Sorensen. 2000. Direct behavioral evidence that unique bile acids released by larval sea lamprey (*Petromyzon marinus*) function as a migratory pheromone. Canadian Journal of Fisheries and Aquatic Sciences 57:557–569.

Brumo, A.F. and D. F. Markle. 2006. Spawning, larval drift, early mortality, and ecological role of Pacific lampreys in the South Fork Coquille River. Report of Oregon State University Department of Fisheries and Wildlife to Oregon Department of Fish and Wildlife, Corvallis.

CRBLTW (Columbia River Basin Lamprey Technical Workgroup). 2005. Critical uncertainties for lamprey in the Columbia basin: results from a strategic planning retreat of the Columbia River Technical Workgroup, Columbia Basin Fish and Wildlife Authority, Portland OR. Available: www.fws.gov/columbiariver/lampreywg/docs/CritUncertFinal.pdf (June 2006).

CRITFC (Columbia River Inter-Tribal Fish Commission). 2004. Columbia River Basin Pacific Lamprey Summit. Columbia River Inter-Tribal Fish Commission, Portland, OR. Available: www.critfc.org/lamprey_files/lamp_summ.pdf (June 2006).

Close, D. A., M. S. Fitzpatrick, H. W. Li. 2002. The ecological and cultural importance of a species at risk of extinction, Pacific lamprey. Fisheries 27(7)19–25.

Dunham, J., B. Rieman, and K. Davis. 2001. Sources and magnitude of sampling error in redd counts for bull trout. North American Journal of Fisheries Management 21:343–352.

Kostow, K. 2002. Oregon lampreys: natural history, status, and analysis of management issues. Oregon Department of Fish and Wildlife, Portland.

Kruskal, J. B. 1964. Nonmetric multidimensional scaling: a numerical method. Psychometrika 29:115–129.

Li, W., A. P. Scott, M. J. Siefkes, H. Yan, Q. Liu, S. Yun, and D. A. Gage. 2002. Bile acid secreted by male sea lamprey that acts as a sex pheromone. Science 296(5565):138–141.

McCune, B., and M. J. Mefford. 1999. PC_ORD, multivariate analysis of ecological data. MjM Software Design, Gleneden Beach, Oregon.

Moore, K. M. S, K. K. Jones, and J. M. Dambacher. 2007. Methods for stream habitat surveys: aquatic inventories project. Oregon Department of Fish & Wildlife, Information Report 2007–01, Corvallis. Available: http://oregonstate.edu/dept/ODFW/freshwater/inventory/methods_reports.htm. (August 2007).

Nawa, R. 2003. A petition for rules to list Pacific lamprey (*Lampetra tridentata*); river lamprey (*Lampetra ayresi*); western brook lamprey (*Lampetra richardsoni*); and Kern Brook lamprey (*Lampetra hubbsi*) as threatened or endangered under the Endangered Species Act. Letter to the U.S. Fish and Wildlife Service, Washington, D.C.

ODFW (Oregon Department of Fish and Wildlife). 2006. Oregon native fish status report. Oregon Department of Fish and Wildlife, Salem. Available: www.dfw.state.or.us/fish/ONFSR/report.asp (August 2007).

Pirtle, J., J. Stone, and S. Barndt. 2003. Evaluate habitat use and population dynamics of lampreys in Cedar Creek, 2002 Annual Report, Project No. 2000–01400. Bonneville Power Administration, Portland, Oregon.

Stevens, D. L., Jr. 2002. Sampling design and statistical analysis methods for the integrated biological and physical monitoring of Oregon streams. Oregon Department of Fish and Wildlife, Monitoring Program Report Number OPSW-ODFW-2002–7, Portland.

Stevens, D. L., Jr., and A. R. Olsen. 2003. Variance estimation for spatially balanced samples of environmental resources. Environmetrics 14:593–610.

Stevens, D. L., Jr., and A. R. Olsen. 2004. Spatially balanced sampling of natural resources. Journal of the American Statistical Association 99:262–278.

Vrieze, L. A., and P. W. Sorensen. 2001. Laboratory assessment of the role of a larval pheromone and natural stream odor in spawning stream localization by migratory sea lamprey (*Petromyzon marinus*). Canadian Journal of Fisheries and Aquatic Sciences 58:2374–2385.

USFWS (U.S. Fish and Wildlife Service). 2004. Endangered and threatened wildlife and plants: 90-day finding on a petition to list three species of lampreys as threatened or endangered. Federal Register 69/247(27 December 2004):77158–77167.

Table A.1. Redd and habitat variables included in the analyses. Habitat unit and reach variables were measured according to Moore et al. (2007).

| Scale | Variable | Definition |
|---|---|---|
| Redd | Length | Length of redd (cm) |
| | Width | Width of redd (cm) |
| | Depth | Distance from the bottom of the depression to the surface of the substrate (cm) |
| | Water depth | Distance from the substrate to the water surface (cm) |
| | Velocity | Water velocity at the redd (m/s) |
| | Substrate | Average maximum diameter of 30 pebbles (mm) |
| | Cover | Cover type |
| Habitat unit | Unit type | Unit type designation |
| | Unit area | Length times width ($m^2$) |
| | Slope | Unit slope, measured with clinometer (%) |
| | Percent shade | Percent of 180° that topography or vegetation occludes the sky, measured with clinometer |
| | Depth | Maximum depth of pools, modal depth of fast water units (m) |
| | Percent silt/organic | Percent of wetted substrate surface area |
| | Percent sand | Percent of wetted substrate surface area |
| | Percent gravel | Percent of wetted substrate surface area |
| | Percent cobble | Percent of wetted substrate surface area |
| | Percent bedrock | Percent of wetted substrate surface area |
| | Large boulders | Number of boulders > 0.5 m above water surface, roughness index |
| | Percent undercut | Percent of right and left bank |
| | Large wood | Number of large wood pieces |
| | Large wood volume | Large wood volume ($m^3$) |
| | No. of key wood pieces | Wood $\geq 12 \times 0.6$ m |
| Reach | Elevation | Feet above sea level at downstream end of site |
| | Gradient | Gradient measured with clinometer |
| | Channel form | Channel morphology and constraint |
| | Land use | Dominant land use |
| | VWI | Valley width divided by active channel width |
| | Active channel width | Width of active channel (m) |
| | Percent silt | Percent of wetted substrate surface area |
| | Percent sand | Percent of wetted substrate surface area |
| | Percent gravel | Percent of wetted substrate surface area |
| | Percent cobble | Percent of wetted substrate surface area |
| | Percent bedrock | Percent of wetted substrate surface area |
| | Large boulders | Number of boulders > 0. 5 m above water surface, roughness index |
| | Percent undercut | Percent distance of left and right bank |
| | Large wood | Volume of large wood |
| | Width:depth | Mean width divided by mean depth |
| | Percent pool area | Percent of wetted area |
| | No. pools/100 m | Number of pools per 100 m |
| | Mean pool depth | Average pool depth |
| | No. of complex pools | Number of pools associated with wood |

American Fisheries Society Symposium 72:191–201, 2009


# A Field Study to Investigate Repeat Homing in Pacific Lampreys

Douglas R. Hatch* and John M. Whiteaker
*Columbia River Inter-Tribal Fish Commission*
*729 NE Oregon Street, Suite 200, Portland, Oregon 97232, USA*

*Abstract.*—In-season homing of Pacific lampreys *Lampetra tridentata* was investigated using radiotelemetry in the lower Columbia River from 1998 through 1999. A total of 50 Pacific lampreys were captured: 25 at Willamette Falls (river kilometer [rkm] 45 on the Willamette River, a tributary to the Columbia River, with its confluence at rkm 163) and 25 at Bonneville Dam (rkm 238 on the Columbia River). Each fish was fitted with a radio transmitter, transported, and released in the Columbia River approximately 26 km downstream from the confluence of the Willamette River. Movement of the radio-tagged Pacific lampreys was monitored for several months using mobile and fixed receiver stations to observe rates of homing towards the site of original capture. Results indicated that the lampreys exhibited nonsignificant in-season homing fidelity ($p = 0.622$) based on the null expectation that one-half of the total recoveries would home and the other half would stray. Final location classifications were 17 homed, 20 strayed, and 13 undetermined. The undetermined classification included individuals that were not detected upstream of the confluence of the Willamette River or in other Columbia River tributaries. Final location classifications were not influenced by fish length ($p = 0.594$). Although considered weak swimmers, Pacific lampreys were capable of traveling at velocities near 2.5 km/h and sustaining that activity for at least 24 h.

## Introduction

Pacific lamprey *Lampetra tridentata* is endemic to the Columbia River basin and plays an important role in the region's river and marine ecosystems, along with having cultural, medicinal, and religious significance to Native American Indians of the Pacific Northwest. Counts of Pacific lampreys at fishway counting stations in the Columbia River basin are currently a fraction of the numbers that were reported 40–50 years ago, indicating a dramatic decline in lamprey abundance (Close et al. 2002).

Pacific lampreys have a unique anadromous life history. Mating individuals build nests and spawn in headwater streams; larvae live as filter feeders in freshwater for several years prior to their metamorphosis and then migrate to the ocean. Lampreys spend several years feeding in the ocean prior to returning to freshwater to spawn (Beamish 1980). Besides perpetuating the species, these spawning migrations provide a vehicle for deposition of marine nutrients to the streams, as well as a buffer for avian (Merrell 1959), marine mammal (Jameson and Kenyon 1977; Roffe and Mate 1984), and piscivorous (Beamish 1980; Poe et al. 1991) predation on Endangered Species Act-listed Pacific salmon *Oncorhynchus* spp. Since natal homing is common in other anadromous species, it is often assumed that Pacific lampreys also exhibit this life history characteristic.

Natal homing has been described for a

* Corresponding author: hatd@critfc.org

broad range of taxa, including the highly studied salmonids. Natal homing was demonstrated in Gulf of Mexico sturgeon *Acipenser oxrhynchus desotoi* (Stabile et al. 1996), American shad *Alosa sapidissima* (Carscadden and Leggett 1975), and rainbow trout *O. mykiss* (Lindsey et al. 1959), and repeat homing occurs in American shad (Melvin et al. 1986), Windermere char *Salvelinus willughbii* Günther (Frost 1963), brook trout *S. fontinalis* (O'Connor and Power 1973), and brown trout *Salmo trutta* (Tilzey 1977). In-season homing was demonstrated in Colorado pikeminnow *Ptychocheilus lucius* (Tyus 1985), lake trout *Salvelinus namaycush* (Martin 1960), flathead catfish *Pylodictis olivaris* (Hart 1971; Hart and Summerfelt 1974), muskellunge *Esox masquinongy* (Margenau 1994), black-and-yellow rockfish *Sebastes chrysomelas* (Hallacher 1984), mosshead sculpin *Clinocottus globiceps* (Green 1973), Windermere char (Frost 1963), Florida largemouth bass *Micropterus salmoides floridanus* (Mesing and Wicker 1986), cutthroat trout *O. clarkii* (Platts 1959; McCleave 1967; McCleave and Horrall 1970), rainbow trout (Lindsey et al. 1959), brook trout (O'Connor and Power 1973), brown trout (Harcup et al. 1984; Halvorsen and Stabell 1990), largemouth bass *Micropterus salmoides* (Quinn et al. 1978), and American shad (Dodson and Leggett 1973). Prior to this work, homing has not been studied in Pacific lampreys; however, Beamish (1980) speculated that at least some Pacific lampreys homed as evidenced by differences in adult length compositions among rivers that he investigated.

Bergstedt and Seelye (1995) used coded wire tag and recovery methods to reject the hypothesis that Lake Huron sea lampreys *Petromyzon marinus* home to natal streams. However, Wright et al. (1985) used allozyme data to identify statistically significant differences among five sample sites within Lake Huron, which they interpreted to indicate stock structure. For stock structure to exist, Lake Huron sea lampreys must exhibit natal homing.

Gerking (1959) used the term "homing" in a general sense to mean the return of fish after migratory, accidental, or experimental displacement "to a place formerly occupied instead of going to other equally probable places." McCleave (1967) described equally probable places as areas "occupied by other individuals of the same species." In spawning migrations of fishes, three types of homing are recognized (McCleave 1967): (1) natal homing: the return of adults to spawn in the same location in which they were hatched, termed "reproductive, parent stream, or natal homing" by Lindsey et al. (1959); (2) repeat homing: the return of adults to spawn in subsequent breeding seasons at the location of initial spawning; and (3) in-season homing: the return of adults within the same breeding season to the location of initial choice after displacement. This study uses the third definition of homing.

Pacific lamprey homing fidelity has many implications for restoration and management of the species. If lampreys demonstrate high homing fidelity, then substantial population substructuring can be inferred and restoration efforts will need to be focused on multiple distinct populations, as is the case for Pacific salmon. On the other hand, if homing fidelity is low, then a large panmictic metapopulation can be inferred and recovery efforts can be focused at a basin or multibasin scale. The objective of this study was to investigate in-season homing of Pacific lampreys using radiotelemetry.

## Study Area

The study area includes the Columbia River between Columbia City, Oregon (river kilometer [rkm] 137) and Bonneville Dam (rkm 238) and the Willamette River, a tributary of the Columbia from the confluence at rkm 163 to Willamette Falls (rkm 45).

Five fixed receiver sites were located at various points between the upstream locations of capture and the downstream release site (Figure 1). Specific sites were chosen based on the following criteria: usefulness to the study, security from vandals, ambient noise levels,

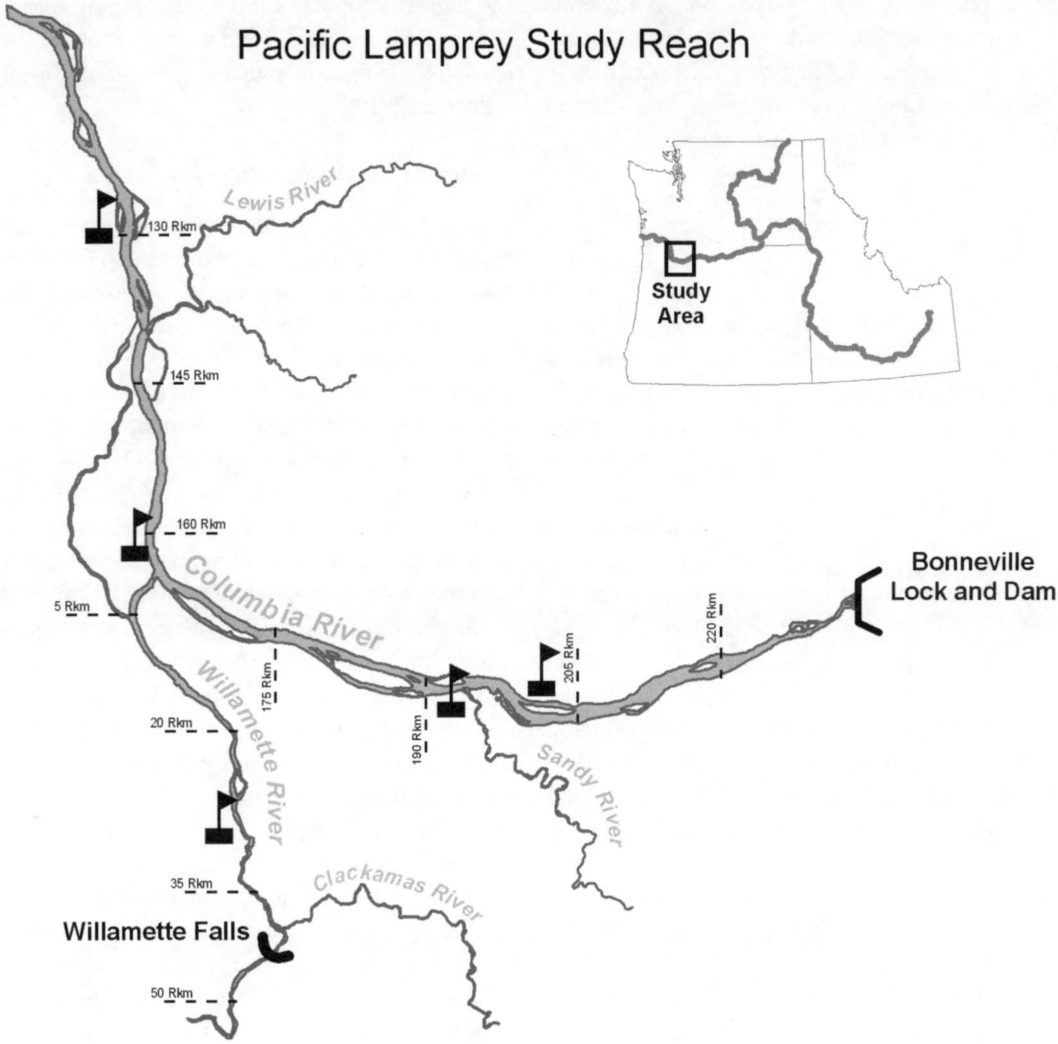


Figure 1. Study area, with locations of the five fixed receiver sites denoted by tower icons. One fixed site was located at river kilometer (rkm) 127 below the release site, one at the confluence of the Willamette and Columbia rivers, one on the Willamette at rkm 27, and two on opposing sides of the Columbia River at rkm 193 on the north shore and at rkm 190 on the south shore.

and elevation. Each fixed receiver site consisted of a Lotek SRX-400 radiotelemetry receiver, a nine-element Yagi antenna, and a deep cycle battery. The SRX-400 receivers stored data to permanent memory (Lotek Engineering 1994) and were set to continuously scan only channel 16 (149.960 MHz). We set the gain levels at each site as high as possible while trying to minimize the number of false detections received.

### *Capture sites*

Adult Pacific lampreys were collected at Bonneville Dam and Willamette Falls at approximately 10-d intervals from June 11 through September 24, 1998. Fish at Bonneville Dam were collected with an adult lamprey trap at the Washington Shore Adult Fish Facility. We collected Willamette River fish by dip net at the Willamette Falls fish ladder while the ladder was in operation and by hand from Willamette Falls after the ladder was dewatered. We immediately transported captured fish to the Abernathy Technology Center near Longview, Washington and held them for 1–4 d until tagging.

## Methods

### Tagging procedure

We surgically implanted uniquely-coded Lotek MCFT-3BM radio tags with 3-s pulse rates and a guaranteed battery life of 163 d in 50 Pacific lampreys (25 from each collection site). The tags had an air weight of 7.7 g and a water weight of 3.7 g. To ensure that the water weight of the tag was less than 1% of the weight of the fish as recommended by Winter et al. (1978), a minimum lamprey length of 620 mm was used.

An effort was made to maintain clean but not necessarily aseptic conditions. All utensils and transmitters were washed in a betadine solution and surgical gloves were worn during surgeries. To ensure the quality of sutures and speed of tagging, all surgeries were conducted by a licensed veterinarian and assisted by a fisheries technician. The adult lampreys were anesthetized in a 150-mg/L solution of MS-222 (Methane-tricaine sulfonate). Throughout the tagging process, the gills were irrigated with a continuous flow of 75-mg/L solution of MS-222; however, a 150-mg/L solution and freshwater were used to maintain a suitable level of anesthesia as needed. A 3–4-cm ventral incision was made just anterior to the second dorsal fin and slightly lateral. Approximately 3 cm posterior to the incision a 16-gauge hypodermic needle was inserted and fed through to the incised area. The whip antenna of the transmitter was inserted through the needle and the needle was removed. The transmitter was then placed into the coelom through the incision, and the incision was sutured with 3/0 nonabsorbable nylon suture. No topical antiseptic was applied because of risk of damage to epithelial cells (Klontz and Smith 1968; Herwig 1979). The Pacific lampreys were placed in freshwater for recovery immediately after suturing. We held the tagged lampreys for 3–5 d prior to release to ensure incisions healed and to allow stress levels of the fish to return to pretagging levels (Close et al. 1996).

### Release and relocation

All radio-tagged Pacific lampreys from the study were released near rkm 137. Relocation was accomplished using a combination of aerial and boat tracking along with the fixed receiver sites. All methods used Lotek SRX-400 receivers. Boat tracking was accomplished in an 18' jet sled fitted with two five-element yagi antennas mounted on 10-ft masts. The antennas were pointed forward and at a 45° angle to the side on both sides of the boat. Aerial tracking was done in a Cessna aircraft with a two-element yagi mounted on each wing pointing down and to the side of the aircraft. Boat tracking was primarily limited to the study area; aerial tracking was conducted within the study area, upstream and downstream of the study area, and in tributaries flowing into the Columbia or Willamette rivers within the study area.

When tracking by boat, we made two to four passes depending on width of the river and the number of islands and side channels present. Both antennas were active when cruising to maximize coverage. When a fish was detected, we disconnected one antenna to determine the direction of the signal and moved closer until we could record the transmitter code and determine the fish's location. When aerial-tracking on the Willamette and Columbia rivers, upstream and downstream passes were made, and on smaller tributaries, a single pass was sufficient to cover the entire river. The transmitter code, date, latitude, and longitude were recorded along with the nearest landmark or channel marker. Fixed receiver stations were downloaded multiple times each week during the lamprey run and every 2–3 weeks in the later stages of the study.

Pacific lampreys were classified as homed, strayed, and undetermined. The last known detection had to be in the river of capture, upstream of the confluence of the Willamette River for fish to be classified as "homed." A "strayed" classification was assigned to an individual that did not return to its river of capture, and the final detection was not in the Columbia River between the confluence of the Willamette River and the release point (rkm 137). If an individual was last detected in the Columbia River between the confluence of the Willamette River and the release point, or in the lower 2 km of the Multnomah Channel, then it was classified as "undetermined."

## Results

Pacific lampreys that were displaced demonstrated low fidelity to capture sites. Of the 50 tagged individuals, 34% ($n$ = 17) homed, 40% ($n$ = 20) strayed, and 26% ($n$ = 13) were undetermined. Pacific lampreys captured in the Columbia had a higher recovery rate and homing rate (20 and 14, respectively) than the Willamette sample (17 and 3, respectively); however, both groups exhibited nonsignificant homing ($p$ = 0.622) based on the null expectation that one-half of the total recoveries would home and the other half would stray. Had all of the individuals that were undetermined (five Columbia and eight Willamette origin) strayed, then the tagged Pacific lampreys would have shown a significant ($p$ = 0.024) tendency to stray. On the other hand, if all of the individuals that were undetermined were found to ultimately home, then the tagged Pacific lampreys would have still exhibited nonsignificant homing ($p$ = 0.157). There does appear to be some temporal variation in the degree of homing. Fewer Pacific lampreys were classified as undetermined from early releases than from later releases (Figure 2).

The duration of time that we were able to track radio tagged individuals averaged 108 d (standard error = 13.0, range 0–254 d, median 143 d). There was no significant difference ($p$ = 0.620) in tracking duration between collections made in the Columbia River and those from the Willamette River. Fish length was not well correlated with tracking duration ($r$ = 0.191), suggesting that the tags did not affect a fish's final classification location.

Using analysis of variance (ANOVA), we found that the final location determination was significantly ($p < 0.001$) affected by the duration of radio tag detections. This effect was entirely due to the undetermined classification having significantly ($p < 0.001$) lower duration than the fish that strayed or homed. In general, most fish that were undetermined were released without subsequent detection. It should be noted though that some of the tagged fish that were classified as undetermined were detected for more than 130 d, but the mean is still significantly lower than the homed or strayed (Figure 3). In addition, a few tagged fish that homed and strayed did so in a very short (2–5 d) period of time.

Pacific lampreys collected at Bonneville Dam (mean 686 mm) were significantly ($p <$ 0.001 $t$ = 4.428, df = 48) longer than those collected from Willamette Falls (total length mean 647 mm). However, we compared the lengths of fish that chose the Willamette River with those that chose the Columbia River following displacement and discovered no significant ($p$ = 0.701) difference in mean total length (672 and 689 mm, respectively). This test should be viewed with some caution as a result of small sample sizes ($n$ = 22 and $n$ = 6, for the Columbia and Willamette rivers). Also, ANOVA results showed that final classifications (Figure 4) were not influenced by length ($p$ = 0.594).

Several interesting migratory behaviors appeared while radio-tracking during this experiment. All of the Pacific lampreys that we continuously tracked migrated upstream approximately 100–150 m from the bank and generally were in water less than 10 m deep. A few fish would cross from one bank to the other, and due to weak or lost signals, we assume that they are in close orientation to the bottom. There was no preference for specific shorelines or any specific circadian pattern to the migrations. We observed that some radio-tagged Pacific lampreys progressed directly to an upstream location while others would move upstream and then backtrack. Some fish that backtrack would

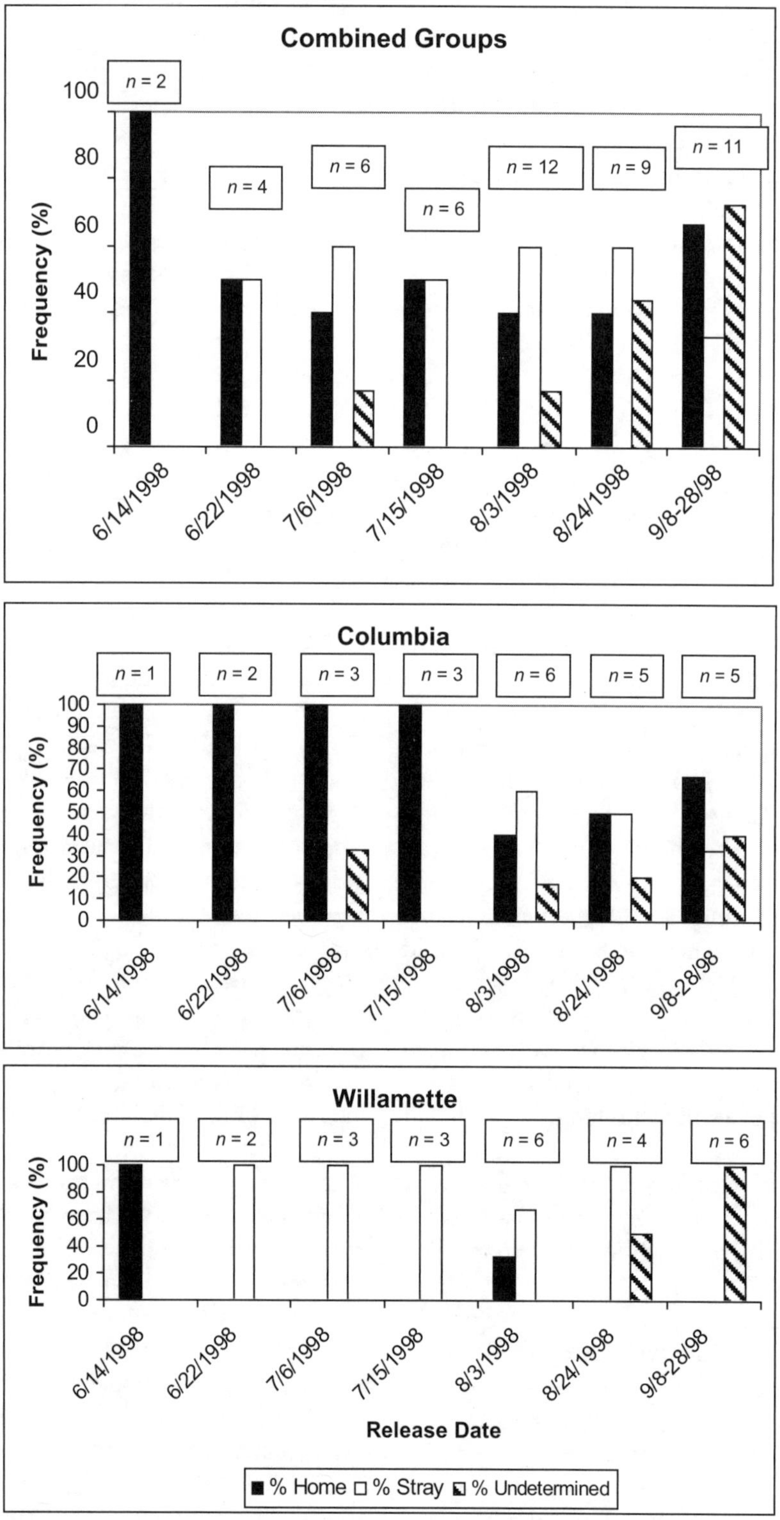


Figure 2. Frequency of final lamprey classifications by release date.

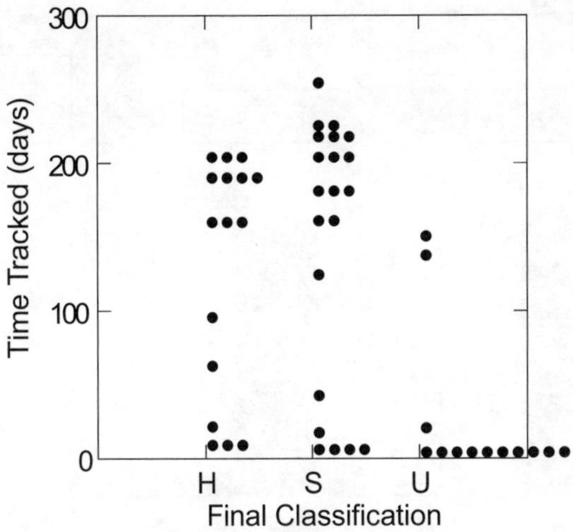


Figure 3. The duration (days) that radio-tagged Pacific lampreys were detected as a function of final location. H = Homed, S = Strayed, and U = Undetermined.

once again continue upstream and others held their position or continued downstream. None of the individuals migrated below the release location. Directionality of movements was not associated with capture location or final classification (Figure 5). The fastest migration speed for Columbia River fish was 2.4 km/h, and it was 2.5 km/h for Willamette River fish. Both groups could maintain these speeds for many hours or days.

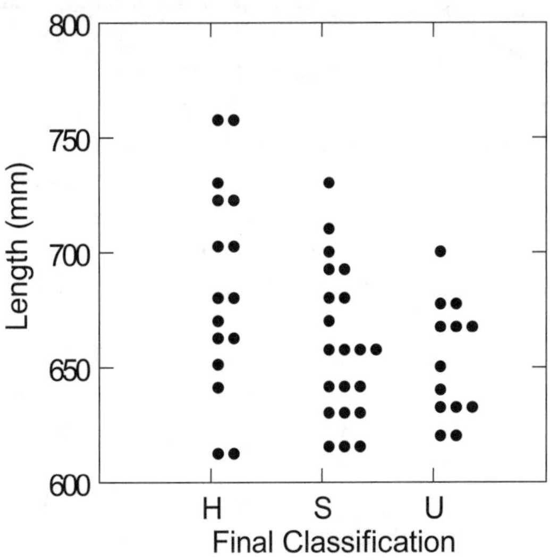


Figure 4. Total length of radio-tagged Pacific lampreys as a function of final classification. H = Homed, S = Strayed, and U = Undetermined.

## Discussion

Pacific lampreys did not exhibit in-season homing. Many factors may have contributed to these findings: trapping and handling effects, in-river environmental factors, or some unknown physiological factor. Nonetheless, the complete lack of homing that Pacific lampreys exhibited has many management implications.

It is possible that the handling stress, the post-surgery holding time, or radio-tag size affected these results. At the time of this study, Pacific lampreys had never been radio-tagged before. Extra care was taken during the surgeries and recovery to ensure that incisions healed and that individuals recovered fully prior to release. Compared to modern radio tags, the tags used in this study were large. This was necessary to achieve a significant battery life. It's possible that the size of the tag impeded swimming performance and ultimately homing, but our data does not support this idea because larger lampreys did not home at a greater rate than smaller lampreys.

We suspected that environmental factors such as river discharge or water temperatures may be important factors influencing whether an individual ultimately homed or strayed. The Columbia River has much higher discharge than the Willamette River, and a higher frequency of Columbia-origin Pacific lampreys did return to the Columbia River than those captured in the Willamette River. However, neither population exhibited a significant homing effect. We did see an increase in the number of individuals that homed earlier in the year (Figure 2), which could be the result of increasing water temperatures (Keefer et al., in press), but we were unable to discern if temperature was the causal effect.

A photophobic response has been observed in upstream migrants of several lamprey species (Hardisty and Potter 1971). Kan (1975) specu-

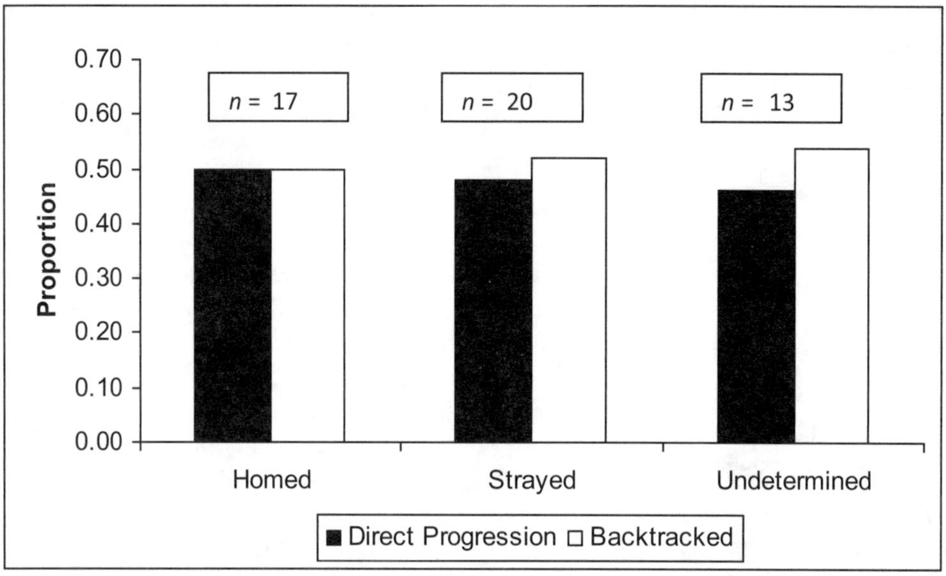


Figure 5. The proportion of radio-tagged Pacific lampreys that were observed to progress directly upstream and those that backtracked by final classification. Samples sizes were 17, 20, and 13 for homed, strayed, and undetermined classifications, respectively.

lated that this response might be true for Pacific lampreys based on observations in fishways where prespawning fish were seen attaching to the shaded side of fishway walls. Based on our telemetry observations, we would conclude that during upstream migrations in the lower Columbia River, adults readily moved during daytime and nighttime periods and did not seem to exhibit any photophobic response. However, this may be more related to water clarity than to phototaxis. Kan (1975) estimated migration time as 4.5 km/d for Pacific lampreys in Clear Creek, a tributary to the John Day River. Our telemetry observations illustrate that although considered weak swimmers (Beamish 1974), Pacific lampreys were capable of traveling at velocities near 2.5 km/h and sustaining that activity for at least 24 h.

Experiments conducted by Bjerselius et al. (2000) showed that adult sea lampreys use larval bile acids as a migratory pheromone to help locate spawning areas during their spawning migrations, and the effect of the pheromone was dependant on river flow, adult sexual maturity, and time of day. In a similar experiment conducted on Pacific lampreys, only one of six individuals exhibited similar responses to juvenile bile acids (Sorensen and Close 1998). Lack of homing in our experiment could be an indication that Pacific lampreys do not follow stream specific migratory pheromones as do sea lampreys. Unlike sea lampreys, Pacific lampreys typically spend at least a year in freshwater prior to spawning (Beamish 1980). It has been shown that olfactory sensitivities in sea lampreys diminish shortly after entering spawning streams and continue to lessen as individuals mature (Sorensen et al. 1995). Therefore, migratory pheromones for Pacific lampreys may only operate at the level sufficient to guide individuals into rivers where lampreys have produced juveniles in the past but may not be as important for in-river homing and detection of specific spawning locations.

Length differences at various collection sites have been taken as evidence supporting a homing hypothesis and hence an inferred stock structure (Beamish 1980; Keefer et al., in press). A similar difference in length was found between our two lamprey populations, but in this case it may be the result of sampling bias since the length composition of fish returning after displacement showed no differences between the Columbia and Willamette rivers. The location of the trap at Bonneville Dam required individuals to traverse the lower (one-fourth) of the fish ladder prior to entering the trap. Moser et al. (2002) demonstrated that Pacific lampreys have difficulty negotiating the Bonneville Dam fish ladders. This phenomenon may be related to size, thus skewing the length composition of Pacific lampreys collected in the trap. Alternatively, the Pacific lampreys collected at Willamette Falls may be biased toward smaller fish as a result of collecting fish by hand.

If Pacific lampreys exhibited a high degree of homing fidelity, one would also expect a high degree of genetic stock structure as seen in Lake Huron sea lampreys (Wright et al. 1985). The lack of in-season homing found in this study combined with no statistical support for genetic population structure or subdivision found in Goodman et al. (2006) supports the hypothesis that Pacific lampreys do not home to natal streams. In our review of the literature, the vast majority of fish species exhibit some level of homing fidelity and it is difficult to consider the idea that an anadromous species with such a large range would not exhibit a similar behavior. Pacific lampreys do have a novel life history that is not fully understood and is considerably different than other fishes. It is possible that lampreys do home at some level and do have some form of population structure that are currently undetectable, but conservation and management of the species must consider the potential implications of managing for a single population.

## Acknowledgments

This study was funded by the Bonneville Power Administration. We thank the U.S. Fish and Wildlife staff at the Abernathy Technology Center for providing a location for surgeries and holding implanted fish. We thank the researchers from the University of Idaho, the U.S. Army Corps of Engineers, and the National Marine Fisheries Service for their assistance with the Bonneville Dam trap. Access to private property for our stationary receivers was provided by Ron Schlumpberger, Marion's Carpets, Staff Jennings, Chinook Landing, and the Camas-Washougal Marina. Field assistance was provided by Bobby Begay, Mark Wishney, Aaron Jackson, and Jed Volkman. Ryan Branstetter, Peter Galbreath, David Graves, Saang-Yoon Hyun, Denise Kelsey, Dale McCullough, Mary Moser, Shawn Narum, Blaine Parker, and Phillip Roger reviewed the document and provided useful comments.

## References

Beamish, F. W. H. 1974. Swimming performance of adult sea lamprey, *Petromyzon marinus*, in relation to weight and temperature. Transactions of the American Fisheries Society 103:355–358.

Beamish, R. J. 1980. Adult biology of the river lamprey *Lampetra ayresi* and the Pacific lamprey *Lampetra tridentata* from the Pacific coast of Canada. Canadian Journal of Fisheries and Aquatic Sciences 37:1906–1923.

Bergstedt, R. A., and J. G. Seelye. 1995. Evidence for lack of homing by sea lampreys. Transactions of the American Fisheries Society 124:235–239.

Bjerselius, R., W. Li., J. H. Teeter., J. G. Seelye., P. B. Johnsen, P. J. Maniak, G. C. Grant, C. N. Polkinghorne, and P. W. Sorensen. 2000. Direct behavioral evidence that unique bile acids released by larval sea lamprey (*Petromyson marinus*) function as a migratory pheromone. Canadian Journal of Fisheries and Aquatic Sciences 57:557–569.

Carscadden, J. E., and W. C. Leggett. 1975. Meristic differences in spawning populations of American shad, *Alosa sapidissima*: evidence for homing to tributaries in the St John River, New Brunswick. Journal of the Fisheries Research Board of Canada 32:653–660.

Close, D. A., M. Fitzpatrick, H. Li, B. Parker, D. Hatch, and G. James. 1996. Status report of the Pacific lamprey *Lampetra tridentata* in the Columbia River basin. Bonneville Power Administration, Portland, Oregon.

Close, D. A., M. Fitzpatrick, and H. Li. 2002. The ecological and cultural importance of a species at risk of extinction, Pacific lamprey. Fisheries 27(7):19–25.

Dodson, J. J., and W. C. Leggett. 1973. Behavior of adult American shad *Alosa sapidissima* homing to the Connecticut River from Long Island Sound. Journal of the Fisheries Research Board of Canada 30:1187–1860.

Frost, W. E. 1963. The homing char *Salvelinus willughbii* (Gunther) in Windermere. Animal Behavior 11:74–82.

Gerking, S. D. 1959. The restricted movement of fish populations. Biological Reviews 34:221–242.

Green, J. 1973. Evidence for homing in the Mosshead sculpin *Clinocottus globiceps*. Journal of the Fisheries Research Board of Canada 30:129–130.

Goodman, D., S. Reid, and M. Docker. 2006. A phylogeographic analysis of the Pacific lamprey *Entosphenus tridentatus*. U.S. Fish and Wildlife Service, Revised Final Report, Contract number 10181-4-M465, Washington, D.C.

Hallacher, L. E. 1984. Relocation of original territories by displaced black-and-yellow rockfish, *Sebastes chrysomelas*, from Carmel Bay, California. California Fish and Game 70(3):158–162.

Halvorsen, M., and O. B. Stabell. 1990. Homing behavior of displaced stream-dwelling brown trout. Animal Behavior 39:1089–1097.

Harcup, M. F., R. Williams, and D. M. Ellis. 1984. Movements of brown trout, *Salmo trutta* L., in the River Gwyddon, South Wales. Journal of Fish Biology 24:415–426.

Hart, L. G. 1971. A telemetric study of homing and home range of flathead catfish, *Pylodictis olivaris* (Rafinesque), in an 850 hectare Oklahoma Reservoir. Master's thesis. Oklahoma State University, Stillwater.

Hart, L. G., and R. C. Summerfelt. 1974. Homing behavior of flathead catfish, *Pylodictis olivaris*, tagged with ultrasonic transmitters. Proceedings of the Annual Conference of Southeast Association of Game and Fish Commissions 27:520–531.

Hardisty, M. W., and I. C. Potter. 1971. The general biology of adult lampreys. Pages 127–206 *in* M. W. Hardisty, and I. C. Potter, editors. The biology of lampreys. Academic Press, London.

Herwig, N. 1979. Handbook of drugs and chemicals used in the treatment of fish diseases: a manual of fish pharmacology and materia medica. Thomas, Springfield, Illinois.

Jameson, R. J., and K. W. Kenyon. 1977. Prey of sea lions in the Rogue River, Oregon. Journal of Mammal 58(4):672.

Kan, T. T. 1975. Systematics, variation, distribution, and biology of lampreys of the genus *Lampetra* in Oregon. Doctoral dissertation. Oregon State University, Corvalis.

Klontz, G. W., and L. S. Smith. 1968. Methods of using fish as biological research subjects. Methods of Animal Experimentation 3:383–385.

Lindsey, C. C., T. G. Nortcote, and G. F. Hartman. 1959. Homing of rainbow trout to inlet and outlet spawning streams at Loon Lake, British Columbia. Journal of the Fisheries Research Board of Canada 16:695–719.

Margenau, T. L. 1994. Evidence of homing of a displaced Muskellunge, *Esox masquinongy*. Journal of Freshwater Ecology 9(3):253–256.

Martin, N. V. 1960. Homing behavior in spawning lake trout. The Canadian Fish Culturist 26:3–6.

McCleave, J. D. 1967. Homing and orientation of cutthroat trout (*Salmo clarki*) in Yellowstone Lake, with special reference to olfaction and vision. Journal of the Fisheries Research Board of Canada 24:2011–2044.

McCleave, J. D., and R. M. Horrall. 1970. Ultrasonic tracking of cutthroat trout (*Salmo clarki*) in Yellowstone Lake. Journal of the Fisheries Research Board of Canada 27:715–730.

Melvin, G. D., M. J. Dadswell, and J. D. Martin. 1986. Fidelity of American shad, *Alosa sapidissima* (Clupeidae), to its river of previous spawning. Canadian Journal of Fisheries and Aquatic Sciences 43:640–646.

Merrell, T. R. 1959. Gull food habits on the Columbia River. Fish Commission of Oregon, Research Briefs 7(1), Portland.

Mesing, C. L., and A. M. Wicker. 1986. Home range, spawning migrations, and homing of radio-tagged Florida largemouth bass in two central Florida lakes. Transactions of the American Fisheries Society 115:286–295.

Moser, M. L., P. A Ocker, L. C. Stuehrenberg, and T. C. Bjornn. 2002. Passage efficiency of adult Pacific lampreys at hydropower dams on the lower Columbia River, USA. Transactions of the American Fisheries Society 131:956–965.

O'Connor, J. F., and G. Power. 1973. Homing of brook trout (*Salvelinus fontinalis*) in Motamek Lake, Quebec. Journal of the Fisheries Research Board of Canada 30:1012–1014.

Platts, W. S. 1959. Homing, movements, and mortality of wild cutthroat trout spawned artificially. Progressive Fish-Culturist 21(1):36–38.

Poe, T. P., H. C. Hansel, S. Vigg, D. E. Palmer, and L. A. Prendergast. 1991. Feeding of predaceous on out-migrating juvenile salmonids in John Day Reservoir, Columbia River. Transactions of the American Fisheries Society 120:405–420.

Quinn, T., G. W. Esch, T. C. Hazen, and J. Whitfield Gibbons. 1978. Long range movement and homing by largemouth bass *Micropterus salmoides* in a thermally altered reservoir. Copeia 1978(3):542–545.

Roffe, T. J., and B. R. Mate. 1984. Abundances and feeding habits of pinnipeds in the Rogue River, Oregon. Journal of Wildlife Management 48(4):1262–1274.

Sorensen, P. W., and D. A. Close1998. The olfactory system of migratory adult Pacific lampreys (*Lampetra tridentate*). Pacific Lamprey Research and Restoration Project, Report to Bonneville Power Administration, BPA Report DOE/BP-00000248-1, Portland, Oregon.

Sorensen, P. W., W. Li , R. Bjerselius, B. Zielinski, and L. Bowdin. 1995. Peripheral olfactory sensitivity of sea lamprey is greatest just prior to their spawning migration and then rapidly deteriorates. Chemical Senses 20(6): abstract #281, p. 782.

Stabile, J., J. R. Waldman, F. Parauka, and I. Wirgin. 1996. Stock structure and homing fidelity in Gulf of Mexico sturgeon *Acipenser oxyrinchus desotoi*, based on restriction fragment length polyorpism and sequence analysis of mitochondrial DNA. Genetics 144:767–775.

Tilzey, R. D. J. 1977. Repeat homing of brown trout (*Salmo trutta*) in Lake Eucumbene, New South Wales, Australia. Journal of the Fisheries Research Board of Canada 34:1085–1094.

Tyus, H. M. 1985. Evidence for homing in Colorado squawfish, *Ptychocheilus lucius*. Proceedings of the Desert Fishes Council 13–15:303–310.

Winter, J. D., V. B. Kuechle, D. B. Siniff, and J. R. Tester. 1978. Equipment and methods for radiotracking freshwater fish. University of Minnesota Agricultural Experiment Station, Miscellaneous Report 152, St. Paul (Available from Communication Resources, Coffey Hall, University of Minnesota, St. Paul.)

Wright, J., C. C. Krueger, P. F. Brussard, and M. C. Hall. 1985. Sea lamprey (*Petromyzon marinus*) populations in northeastern North America: genetic differentiation and affinities. Canadian Journal of Fisheries and Aquatic Sciences 42:776–784.

American Fisheries Society Symposium 72:203–222, 2009


# Approaches for Monitoring Pacific Lamprey Spawning Populations in a Coastal Oregon Stream

ABEL F. BRUMO*
*Stillwater Sciences, 850 G Street, Suite K, Arcata, California 95521, USA*
and
*Department of Fisheries and Wildlife, Oregon State University*
*104 Nash Hall, Corvallis, Oregon 97331, USA*

LEO GRANDMONTAGNE
*Wild Fish for Oregon, 46760 Highway 242, Myrtle Point, Oregon 97458, USA*

STEVEN N. NAMITZ
*Siskiyou National Forest, Powers Ranger District*
*42861 Highway 242, Powers, Oregon 97466, USA*

DOUGLAS F. MARKLE
*Department of Fisheries and Wildlife, Oregon State University*
*104 Nash Hall, Corvallis, Oregon 97331, USA*

*Abstract.*—We evaluated two methods for assessing Pacific lamprey *Lampetra tridentata* spawning populations (visual counts of spawning adults and redds) and one method for assessing larval production (emergent ammocoete counts from drift nets) in the South Fork Coquille River, Oregon in 2004 and 2005. All three methods generally provided similar portrayals of timing, duration, and magnitude of spawning, including greater abundance in 2004 and seasonally bimodal spawning in 2005. We found a linear relationship between adult and redd counts but a high redd to adult ratio that varied seasonally in both years. The high redd to adult ratio can be attributed to short residence time in spawning areas and temperature or habitat-dependent differences in detection of adults, both of which can undermine adult count data. Redds had relatively longer persistence and larger numbers compared to adults and therefore may be a more practical survey method, but variable redd shape, size, and age, as well as superimposition, presented significant counting errors. Both adult and redd counts had no clear-cut way to quantify errors. Sampling emergent ammocoetes in the drift allowed detection of low density early and late season spawning and would be the preferred survey method when surveys of spawning adults and redds are impractical due to river size, visibility, or access. Even when spawning surveys are practical, emergent ammocoete counts may be better for detecting and monitoring small populations. Disadvantages of ammocoete sampling include nighttime work hours, extra laboratory time, and difficulties with species identification. The general absence of a stock–recruit relationship in lampreys means adult and redd counts are poor predictors of ammocoete production and emergent ammocoete abundance is a poor predictor of spawning abundance. The relationship breaks down because of variability in early survival, which is best detected using data from both spawning surveys and larval drift samples.

* Corresponding author: abel@stillwatersci.com

## Introduction

Accurately monitoring trends and abundance of fish spawning populations is crucial for understanding species status and population dynamics, as well as instituting recovery plans and charting their success (Al-chokhachy et al. 2005; CRBLTW 2005; Gallagher and Gallagher 2005). Monitoring primarily involves regular collection of long-term data, but also requires evaluation of data accuracy and relevance (Dunham et al. 2001; Gallagher and Gallagher 2005) and understanding of biological processes that affect interpretation (Neilson and Geen 1981; Morbey 2000). For example, both frequency and scale of spawning surveys can influence conclusions about spawning time and abundance (Isaak and Thurow 2006), particularly for patchily distributed populations (Dunham et al. 2001).

Recently, the Columbia River Basin Lamprey Technical Workgroup documented the need for standardized protocols to monitor population trends and abundance and document life history attributes of lampreys (CRBLTW 2005). In many cases—particularly for lampreys—the choice of a long-term monitoring metric has been based on cost and convenience rather than on systematic evaluation of pros and cons of existing methods. With few exceptions (Graham and Brun 2005; Cochnauer et al. 2006; Luzier et al. 2006; Stone 2006; Gunckel et al. 2009, this volume), monitoring Pacific lampreys *Lampetra tridentata* (also referred to as *Entosphenus tridentatus*) in the northwestern United States has been incidental to salmonid monitoring, often resulting in unreliable data (Beamish 1980; Moser and Close 2003). For instance, dam counts of upstream migrants, designed primarily for monitoring salmon and steelhead *Oncorhynchus mykiss*, have been inconsistent and provide little information on when or where passing fish will spawn (Moser et al. 2002; Moser and Close 2003).

Targeted surveys of spawning adults, redds, or newly emerged larvae (also referred to as emergent ammocoetes) provide more realistic indices of spawning success. Once Pacific lampreys begin redd construction, they are not easily startled on spawning grounds—even when an observer is standing directly over them. For this reason, lampreys offer a unique opportunity for direct observation and enumeration. Redd counts, which are often a convenient byproduct of steelhead monitoring, are one promising method for monitoring lamprey spawning populations (Glenney et al. 2004; Gunckel et al. 2009) but do not always encompass the entire spawning season (Kostow 2002; Glenney et al. 2004; Moser et al. 2007). To date, there has been little effort to develop rigorous, lamprey-specific redd count protocols, and evaluation of the relationship between redd counts and adult abundance is needed before redds can be used for population estimates.

Potential weaknesses of using adults and redds to assess lamprey populations include difficulty locating them in streams with small or fragmented spawning populations and no information on egg survival or ammocoete recruitment. Sampling newly emerged ammocoetes is one possible alternative to adult-oriented monitoring. Because of their high densities, wide distribution, and nighttime drifting behavior, emergent ammocoetes are more vulnerable to quantitative sampling gears and may be sampled to effectively monitor spawning activity and estimate ammocoete production. For marine fishes, it is well known that larval monitoring can detect small populations that might be missed by adult-oriented surveys (Houde 1987; Johnston et al. 1995; McDonald 2004), and this may also be true for some lamprey populations. Despite its potential usefulness, data on newly emerged ammocoetes has rarely been collected (Bennett and Ross 1995; Derosier 2001; White and Harvey 2003).

Our goal, therefore, was to assess three potential metrics for monitoring Pacific lamprey spawning populations: two measures of spawning stock size (adult counts and redd counts) and one measure of production (newly emerged ammocoete counts). Where appropriate, results from each approach were evaluated in terms of congruence with other approaches, and patterns of spawning activity were compared over spatial, seasonal, and annual scales. Supplemental information on timing, distribution, size, and sex of spawning adults was also collected with the intent of improving interpretation of monitoring data. Finally, for each approach, we discuss logistical pros and cons and key error sources.

## Methods

### *Sample Sites*

This study took place during 2004 and 2005 spawning seasons on the South Fork Coquille River in southwest Oregon and included sampling at comparatively large and small spatial scales. Large-scale surveys were carried out in a 9.2-km river section between Yellow Creek (42 56' 57" N, 124 05' 58" W) at river kilometer 90.0 and Baker Creek (42 54' 21" N, 124 06' 40" W) at river kilometer 99.2 (Figure 1). This river segment was floatable in a single day and known to include suitable Pacific lamprey spawning and rearing habitats. Small-scale surveys took place at a single spawning ground (30 × 30 m), designated the focal area, which was located at river kilometer 88.6, roughly 1,000 m from the downstream end of the large-scale section (42 57' 22" N, 124 06' 21" W). The focal area was known to be a heavily used spawning area and had safe and easy access for night sampling.

### *Abiotic Variables*

Water temperature was monitored at 30-min intervals from April 27 to July 13, 2004 and from April 4 to August 3, 2005 using an Onset Optic Stowaway temp logger placed in a shaded area at the river bottom directly downstream of the focal area. From April 6 to 27, 2004, water temperature was recorded daily with a bulb thermometer, and daily mean temperatures were estimated based on bulb temperature at time of observation in relation to diel temperature trends in early May 2004. Daily mean water temperature during periods before April 6 in 2004 and April 4 in 2005 were predicted using year-specific multiple linear regression models with maximum daily air temperature at Powers, Oregon and $\log_e$ of discharge as explanatory variables (2004, $R^2 = 0.863$, $P < 0.0001$; 2005, $R^2 = 0.918$, $P < 0.0001$). Discharge data were obtained from a U.S. Geological Survey gauging station located at Pow-

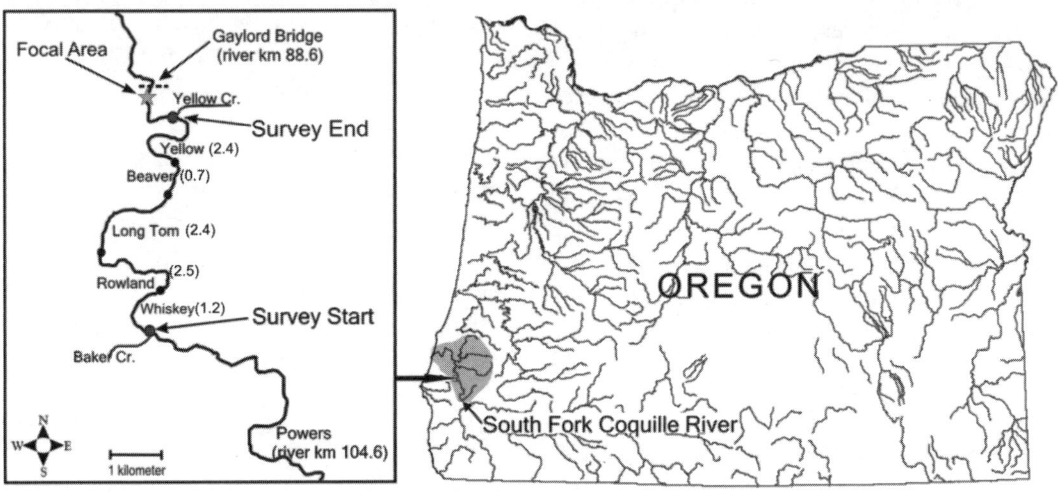


Figure 1. Map showing location of South Fork Coquille River, Oregon. Inset shows location of large-scale survey start and end points, study reaches within the large-scale section, and the focal area. Reach lengths in kilometers are shown in parentheses.

ers, Oregon, 14.5 river kilometers upstream of the focal area (42 53'31" N, 124 4'16" W). No significant tributaries enter the river section between the gauge and the focal area.

## *Large-Scale Surveys*

We used inflatable pontoon boats to conduct adult and redd counts on the large-scale section on a weekly basis from April 29 to June 2, 2004 and from April 21 to June 30, 2005. Surveys normally began around 1000 hours and ended at 1600–1800 hours, though on three occasions in 2004, when adult densities were high, 2 d were required to complete surveys. In 2004, logistical problems dictated large-scale surveys begin roughly 3 weeks after initial observations of spawning at the focal spawning area. In 2005, high-flow events prevented surveys from occurring on a regular 7-d interval.

The large-scale river section was divided into five reaches, as defined by Oregon Department of Fisheries and Wildlife (Figure 1), and all spawning areas in each reach were visually surveyed by floating and wading. All visible, live adult Pacific lampreys were systematically captured by hand with cloth gloves or a modified dip net. Before release, fish were measured and sexed, then tagged with a one-half-inch T-Bar floy tag to avoid recounting in subsequent surveys. Floy tags were colored to denote reach and week of capture to aid in understanding of residence time and movement. All fish captured were mature and in spawning condition. Visible Pacific lamprey carcasses were also recovered, measured, sexed, examined for tags, and cut in half to avoid recounting. Finally, new Pacific lamprey redds, estimated to have been built within the previous week based on redd integrity and color, were counted in each reach.

## *Focal Area Surveys*

We obtained counts of spawning adults at the focal area by systematically wading through the approximately 30 × 30 m spawning ground. Thirty-eight surveys were carried out from March 28 to July 12 in 2004 and 46 from April 5 to July 17 in 2005. Fish were not captured or tagged, in order to avoid disruption of spawning and introduction of error into estimates of ammocoete survival associated with a concurrent study in the area (Brumo 2006). When visibility permitted, counts were made in late afternoon and separated by no more than 4 d. Total redd counts were not made in the focal area because densities were so high in 2004 that individual redds could not be consistently distinguished. Relationships between spawning activity in the large-scale section and the focal area were evaluated for both years using the total number of live adults per weekly large-scale survey, the total number of live adults per reach in each survey, and the mean number of adults per survey in the focal area for each week.

## *Larval Drift Nets*

We used abundance of drifting age-0 ammocoetes in the newly emerged size-class as a measure of larval production and an alternative metric for monitoring spawning. Drift-net samples were taken 27 m downstream of the focal area boundary. The next closest Pacific lamprey spawning area was approximately 300 m upstream of the focal area, immediately below which was a long pool with substantial depositional habitat. We used a 500-μm mesh zooplankton drift net (0.70 × 1.5 m opening; 2.5 m long) that was fastened to rebar posts driven into the substrate. The net was usually set within 2 m of the thalweg (6–12 m from the stream margin), except during periods of high water when it had to be set closer to the stream margin. The net covered most of the water column, which was typically 0.7–0.8 m deep in the thalweg. Loss of drifting ammocoetes due to uneven cobble substrate and flow over the top of the net was assumed to be minimal and consistent between samples.

Drift samples were collected every 3 to 4 d from April 11 to July 14, 2004 and April 25–July 28, 2005. Beginning sample dates were selected to capture initial ammocoete emergence, which was predicted from observed spawning, water temperature, and estimated incubation time. Drift sampling ceased each year after no emergent ammocoetes were caught for 4–5 consecutive samples. Each sample was collected 2–3 h after civil twilight (time at which sun is 6° below horizon; U.S. Naval Observatory Web page: http://aa.usno.navy.mil) in 2004 and 3 h after civil twilight in 2005. This sample time was close to the period of estimated peak daily drift abundance (Brumo 2006). We fished each net for 7–16 min, depending on flow conditions and concentrations of net-clogging matter. A TSK mechanical flowmeter (Tsurumi-Seiki Co. Ltd., Yokohama, Japan) centrally mounted in the net opening was used to estimate water velocity and volume filtered. We assumed that variation in net efficiency across sample dates due to clogging was negligible for age-0 ammocoetes because of their poor swimming ability and short fishing times. High river levels limited access to the focal area on two sample dates in 2004 and five in 2005; consequently, we used an alternate site approximately 200 m upstream. Of these alternate samples, only the last three in 2005 contained age-0 ammocoetes and were included in analyses.

Net samples were rinsed into the cod end using a battery powered wash-down pump. Contents were treated with rose bengal bioactive staining agent to facilitate sorting and preserved in 5% formalin. Age-0 ammocoetes were sorted and counted in the laboratory under magnification and measured to the nearest 0.1 mm total length using a dissecting scope equipped with an ocular micrometer. For samples containing more than 100 age-0 ammocoetes, a random subsample of 100 individuals was measured. Emergent ammocoetes were defined as individuals larger than 7.9 mm and smaller than 9.1 mm (hereafter referred to as 8–9 mm). This size range was based on lengths from early season samples—when all individuals were expected to be emergent—and lengths of larvae hatched in captivity (M. Meeuwig, U.S. Geological Survey, personal communication). A pilot project, in which we reared eggs in situ, also verified emergence at the 8–9 mm size range (Brumo 2006). We assumed that age-0 ammocoetes larger than 9.0 mm did not originate from the focal area, but instead resulted from upstream production. We excluded larvae smaller than 8.0 mm due to the likelihood that they were western brook lampreys *Lampetra richardsoni*. Both mean egg diameter and emergent ammocoete length of western brook lampreys are significantly smaller than those of Pacific lampreys (Pletcher 1963; Meeuwig et al. 2004). A small percentage of larvae included in our emergent 8–9 mm category could have been western brook lamprey, particularly in later season samples, but we believe this influence is inconsequential due to low numbers of spawning western brook lampreys observed and their lower fecundity.

Ammocoete abundance was standardized as the number of individuals drifting past the wetted-width cross section per minute, calculated as individuals/$m^3$ of water filtered by net $\times$ discharge (m3/s) $\times$ 60 s/min. This measure of ammocoete abundance was selected to account for variable stream discharge between samples, though there was a strong correlation between drift rate (individuals/min) and drift density (individuals/$m^3$; $r^2 = 0.963$; $P < 0.0001$). Although we did not quantify actual larval production, we assumed that drift rate reflected relative abundance of emergent ammocoetes over time.

Dates of initial, final, and peak spawning based on emergent larvae in the drift were estimated using an effective degree-days (EDD) temperature–development relationship (Meeuwig et al. 2005). The EDD estimate was based on temperature units required to reach larval stage 18 (Piavis 1961), calculated from a laboratory study of Pacific lamprey egg development at constant water temperatures of 10, 14, 18, and 22°C (Meeuwig et al. 2005). Effective degrees for each day were calculated by subtracting the theoretical temperature for zero development, 4.85°C (Meeuwig et al. 2005), from daily mean water temperature recorded at the focal area. Spawning dates corresponding to each cohort of emergent ammocoetes caught in the drift were estimated by back-calculating from the capture date to the date required to achieve 300.7 EDD.

## Results

### Abiotic Variables

Discharge during the lamprey spawning season in 2004 was marked by a single, mid-April high-flow event, which peaked on April 21 at 137 $m^3/s$ (Figure 2). In 2005, three substantial discharge events (>57 $m^3/s$) occurred early in the spawning season with two smaller events (24 and 29 $m^3/s$) later in the season (Figure 2). During April–July spawning periods, stream discharge was inversely related to water temperature (2004 and 2005 combined; $r^2 = 0.865$; $P < 0.0001$). Average water temperature from April 6–July 13 was warmer in 2004 (16.3°C) than in 2005 (14.0°C) in 2005. In both years, water temperature rose gradually during spring with diel fluctuations of up to 8°C.

### Large-Scale Surveys

We captured, measured, and tagged 446 live adult Pacific lampreys during six weekly large-scale surveys in 2004 (8.1 fish/km) and 140 during nine surveys in 2005 (1.7 fish/km) (Table 1). We saw but missed 54 individuals in 2004 and 31 in 2005, catch rates of 89% and 82%, respectively. Missed individuals were not included in analyses. Peak abundance in 2004 was 169 fish on May 13, and numbers declined dramatically after May 19 (Figure 3). Maximum fish abundance was bimodal in 2005, peaking at 52 on May 5 and 40 on June 1 (Figure 3). Adult density (fish/km) in 2004 was greatest in Whiskey reach, the furthest upstream, and decreased downstream, whereas density was greatest in Long Tom reach in 2005 (Figure 1; Table 1).

We counted 1,759 Pacific lamprey redds over six surveys in 2004 and 1,169 over nine surveys in 2005 (Table 1; Figure 4). New Pacific lamprey redds were detected from April 28 to June 3 in 2004 and from April 21 to June 30 in 2005. In 2004, the peak redd count was 791 on May 5 (Figure 4). In 2005, two peaks occurred: one on May 5 (299) and one on June 1 (384)

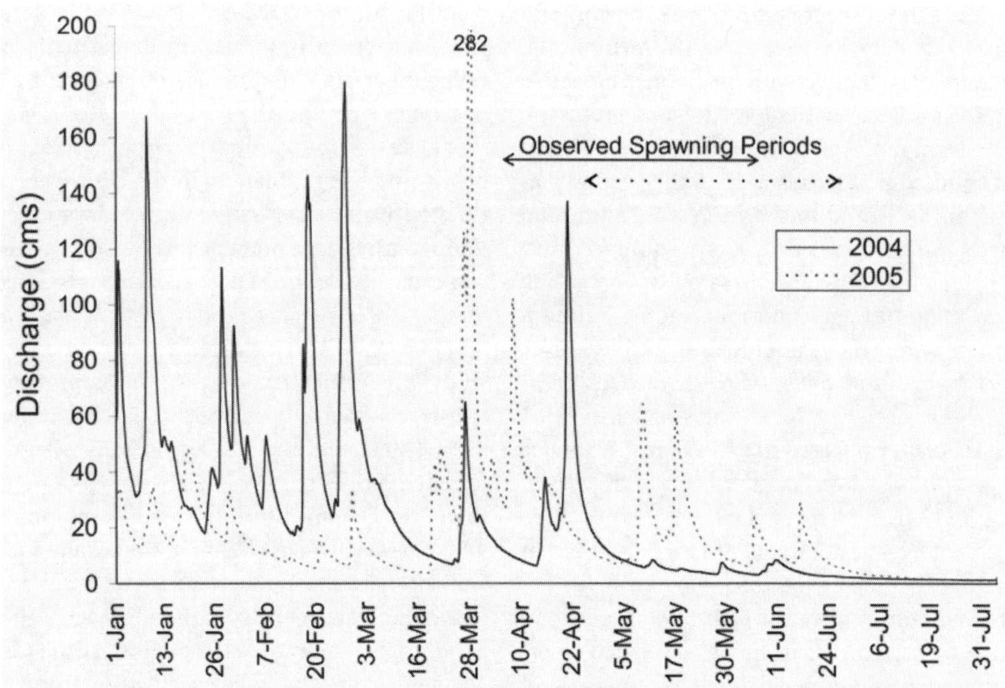


Figure 2. Discharge ($m^3/s$) hydrograph for South Fork Coquille River from January 1 to July 30, 2004 and 2005. Observed spawning periods are based on observations of adults at the focal area.

Table 1. Interannual and spatial patterns in densities of live adult Pacific lampreys and redds for large-scale surveys of the South Fork Coquille River in 2004 and 2005. *N* is total number counted in each year. Reaches are ordered from upstream to downstream.

| | | | Reach | | | | | |
|---|---|---|---|---|---|---|---|---|
| Count | Year | *N* | Whiskey | Rowland | Long Tom | Beaver | Yellow | Total |
| Adults/km | 2004 | 446 | 15.5 | 8.5 | 7.8 | 5.4 | 5.2 | 8.1 |
| Adults/km | 2005 | 140 | 2.3 | 1.5 | 2.6 | 0.8 | 1.3 | 1.7 |
| Redds/km | 2004 | 1,759 | 33.4 | 27.2 | 38.8 | 29.4 | 29.1 | 31.9 |
| Redds/km | 2005 | 1,169 | 16.4 | 13.2 | 18.4 | 11.0 | 10.5 | 14.1 |

(Figure 4). Mean redd density over all surveys was 31.9/km in 2004 and 14.1/km in 2005 (Table 1). Redd density was highest in Long Tom reach in both years (Figure 1; Table 1).

Redd counts and adult counts were highly correlated among surveys in 2004 ($r^2 = 0.867$; $P = 0.0069$) and 2005 ($r^2 = 0.877$; $P = 0.0002$) (Figure 5; Table 1). In 2005, both counts were bimodal. Over all surveys combined, about four redds were counted for every live adult in 2004, whereas more than eight redds were counted per adult in 2005 (Table 1). Excluding an outlier of 73 redds and one adult, the number of redds per adults varied from 2.1 to 5.5 across weekly surveys in 2004, and there was no indication of a temporal pattern (Figure 6). In 2005, the number of redds per adult varied from 5.8 to 24.6 between surveys and appeared to increase seasonally (Figure 6).

Eleven of 446 fish (2.5%) tagged in 2004 were recaptured alive while in 2005, zero of 140 tagged fish were recovered. Ten of 11 live

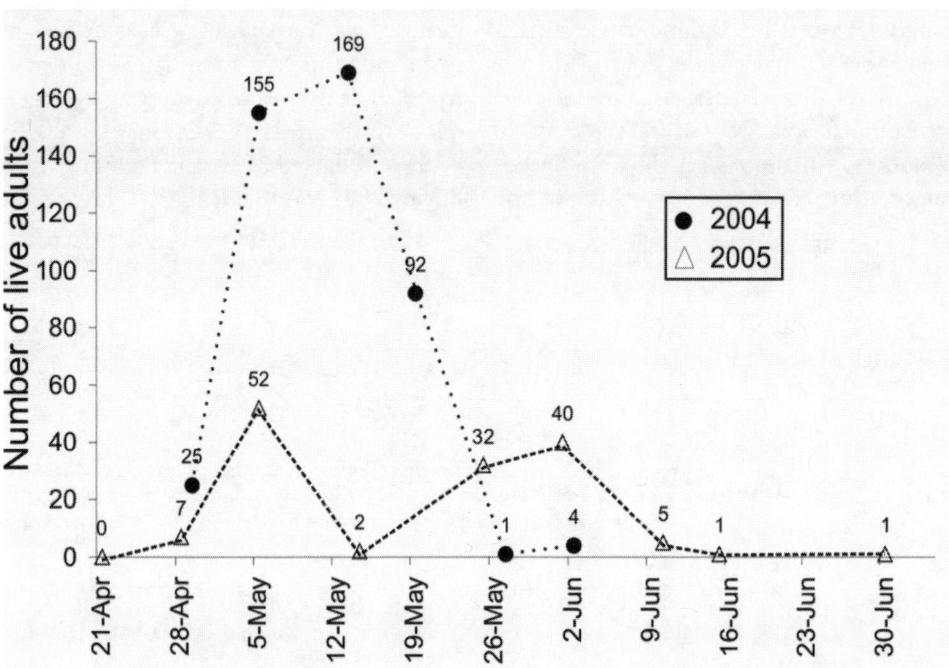


Figure 3. Seasonal patterns in numbers of live adult Pacific lamprey spawners captured in large-scale surveys in 2004 (*N* = 446) and 2005 (*N* =140). Note: figure does not include 54 and 31 individuals seen but not captured in 2004 and 2005, respectively.

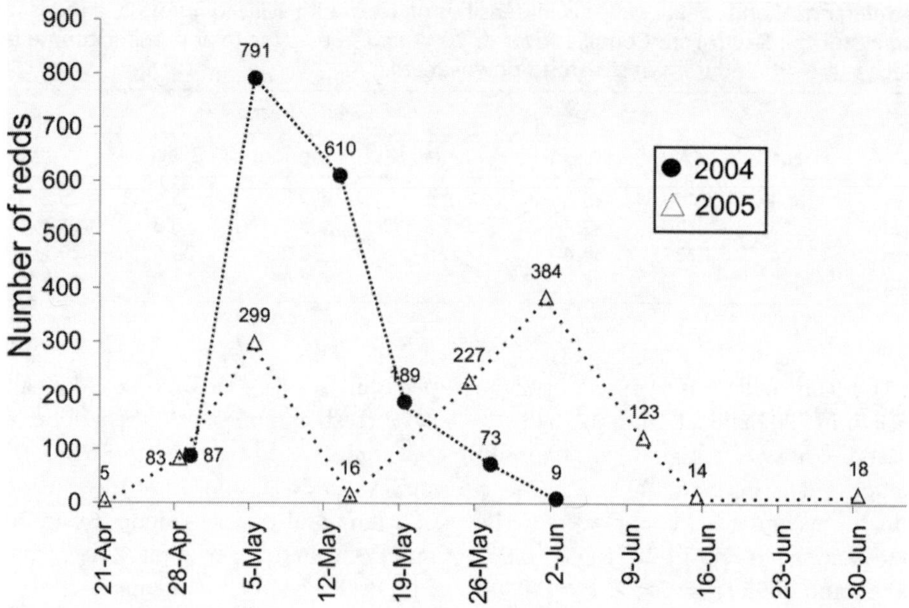


Figure 4. Seasonal patterns in numbers of Pacific lamprey redds counted in large-scale surveys in 2004 ($N = 1{,}759$) and 2005 ($N = 1{,}169$).

recaptures in 2004 were male, all of which were recaptured 1 week after tagging, suggesting a relatively short residence time. Four of the recaptured males moved upstream, two moved downstream, and four were in the same reach. The recaptured female was caught 1 d after being tagged, three reaches downstream of the tagging location. Percent recaptures generally declined as the spawning season progressed, from appoximately 4% in late April/early May to 0% after mid-May, suggesting a longer residence time early in the season.

We collected 373 Pacific lamprey carcasses during 2004 and 17 during 2005 large-scale sur-

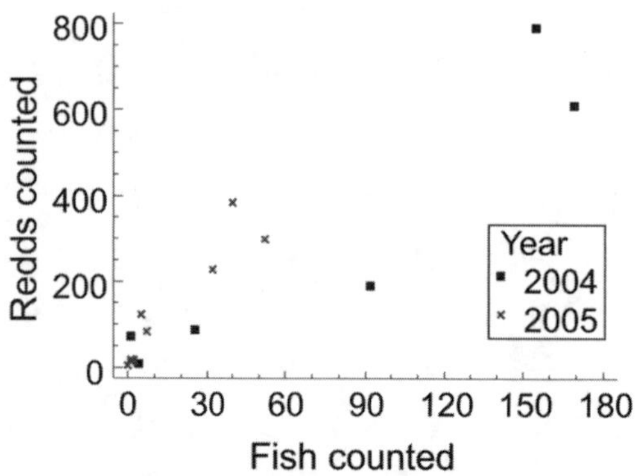


Figure 5. Number of redds counted versus number of live adult Pacific lampreys captured during weekly 9.2 km large-scale surveys in 2004 and 2005.

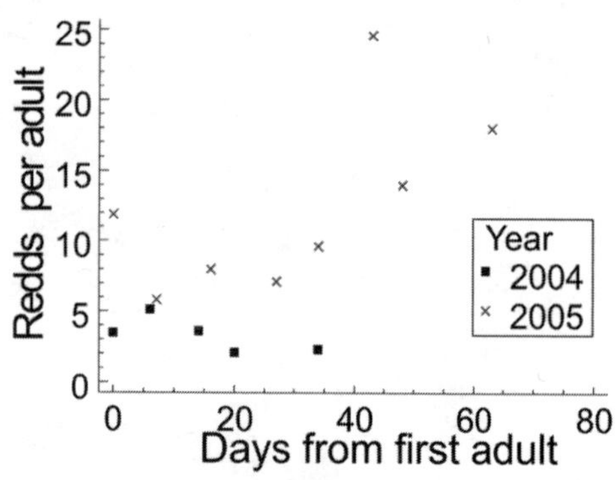


Figure 6. Number of redds counted per adult versus days from first adult observation in large-scale surveys. A 5/27/04 outlier of one fish to 73 redds, 28 d from first adult observation, was omitted.

veys. Few carcasses were found in early surveys in either year, but numbers rose gradually, lagging peak adult counts by about 1–2 weeks in both years. None of the carcasses collected in 2005 contained tags, whereas 35 were recovered with tag intact during 2004. Twenty-seven (77%) of the tagged carcasses were male. Twenty-nine (83%) were found in reaches downstream from their tagging locations. Two, both of which were male, were found upstream while the remaining four were located in the same reach were they were tagged. All tagged carcasses were found within 9 d of their tagging date, again suggesting a short residence time.

In 2004 the sex-ratio of live fish was 0.94:1 (217 male and 229 female), but that of carcasses was 2:1 (233 males and 116 female). Sex could not be determined for 24 of the 373 carcasses. In 2005, the sex ratio of live fish was 1.15:1 (75 male and 65 female) and the sex ratio of dead fish was 1.13:1 (9 male and 8 female). In both years, there was no evidence for a seasonal pattern in sex ratio.

Lengths of mature adults (live and dead) collected from spawning areas in the large-scale section in 2004 ranged from 35.5 to 60.0 cm, and had a mean of 48.0 0 ± 0.147 cm (mean ± SE). Males (50.0 ± 0.173 cm) were significantly longer than females (45.4 ± 0.175 cm) by about 10% in 2004. In 2005 mature adults ranged from 31.0 to 58.0 cm, and had a mean length of 46.9 ± 0.394 cm. Again, in 2005, males (48.3 ± 0.491 cm) were significantly longer than females (45.2 ± 0.574 cm) by approximately 6%. In both years, there was not a significant difference between mean length of live males or females between survey reaches (2004 males: $P = 0.1523$; 2004 females: $P = 0.9312$; 2005 males: $P = 0.9552$; 2005 females $P = 0.1125$) or survey dates (2004 males: $P = 0.6479$; 2004 females: $P = 0.9700$; 2005 males: $P = 0.1395$; 2005 females $P = 0.0562$).

### Focal Area Surveys

We counted 233 adult Pacific lampreys in the focal spawning area in 2004 (8.3/survey), and 85 in 2005 (2.6/survey). Spawning stage adults were seen from April 6–June 3, 2004 (95% by May 17) and from April 25–July 3, 2005 (95% by June 6). Maximum counts were 27, on May 4, 2004 (Figure 7); and 9 on April 26, 2005 (Figure 8). Peak counts in 2005 were bimodal, one in late April and the other from mid-May to early June (Figure 8). High discharge precluded observations from mid to late April in 2004 and five times during 2005 (Figure 2).

In both years, estimates of spawning season and peak spawning period were in general agreement whether based on focal area adult counts, large-scale adult counts, or large-scale redd counts (Table 2). Weekly mean focal area adult counts and weekly large-scale adult counts were significantly correlated over both years combined ($r^2 = 0.690$, $P = 0.0001$) and in 2004 ($r^2 = 0.753$, $P = 0.0250$), but not in 2005 ($r^2 = 0.065$, $P = 0.5069$) when densities were low (Figure 9). Adult counts from each reach in the large-scale section were also significantly correlated with each other, with total large-scale counts, and with weekly mean focal area counts.

### Larval Drift Nets

In 2004, 26 drift samples were taken from April 11–July 14. A total of 6,901 age-0 ammocoetes were captured, of which 55% (3,791) were newly-emerged (8–9 mm), 43% greater than 9 mm, and 2% less than 8 mm. Early season samples in 2004 caught predominately 8–9 mm individuals, and the proportion of older and larger age-0 larvae increased as the season progressed (Figure 10).

Emergent ammocoetes occurred in the drift for 54 d in 2004, from May 6–June 28, 30 d after the first, and 25 d after the last, adult was detected in the focal area (Figures 7 and 10). Emergent ammocoete abundance (fish/min) in 2004 averaged 353.3 ± 141.6 (mean ± SE) and peaked in mid-May at 2,132. The median emergence date was between May 20–24 and 99% were captured by June 21. Abundance of larger (>9 mm) ammocoetes peaked at 610.2 per minute in early June, and remained high throughout the month (Figure 10). A few larvae less than 8 mm, likely western brook lampreys, were caught between

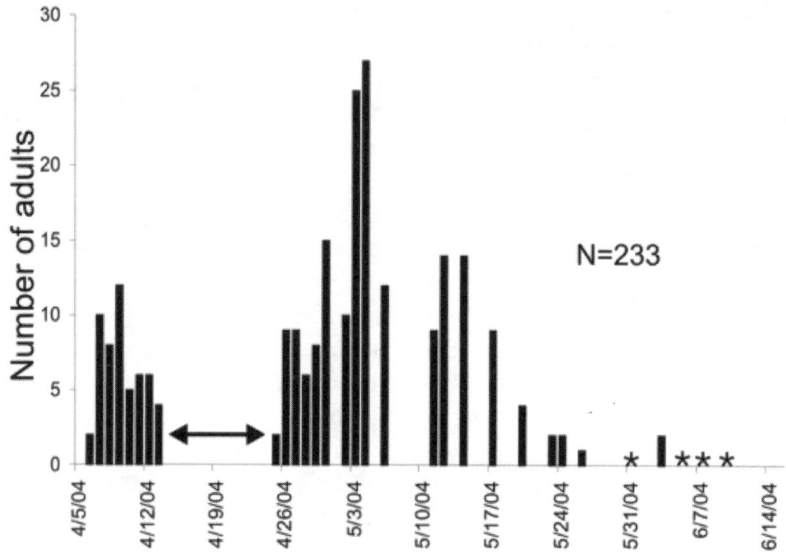


Figure 7. Seasonal patterns of abundance of adult Pacific lamprey spawners in the focal area of the South Fork Coquille River, 2004. Double arrow indicates a period of high discharge when sampling was not possible. Stars represent surveys when no fish were counted.

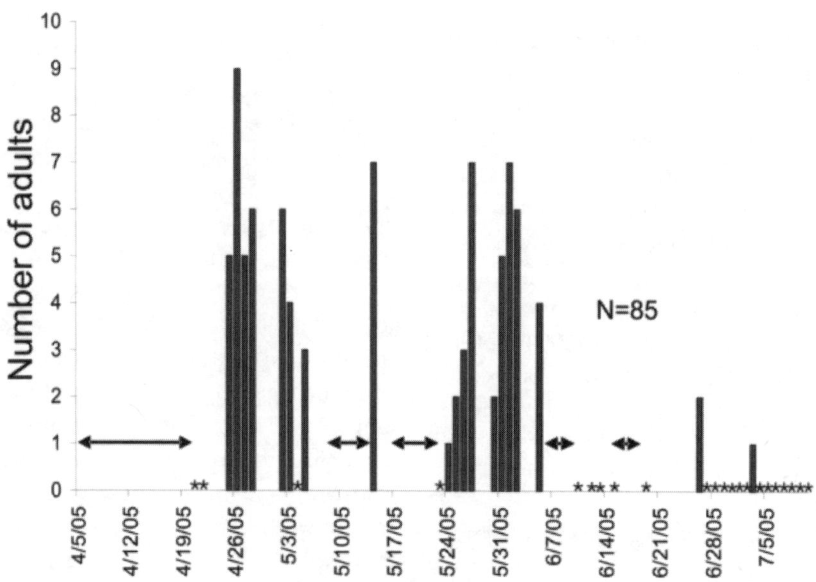


Figure 8. Seasonal patterns of abundance of adult Pacific lamprey spawners in the focal area of the South Fork Coquille River, 2005. Double arrows indicate periods of high discharge when sampling was not possible. Stars represent surveys when no fish were counted.

Table 2. Summary of estimated spawning season start and end dates, peak spawning dates, and annual relative abundance of Pacific lampreys in the South Fork Coquille River based on large-scale adult and redd counts, focal area adult counts, and drift net samples of emergent ammocoetes. Spawning dates from emergent larvae are based on back-calculations as described in text. Asterisks indicate late start of sampling in 2004. The double asterisk indicates an earlier spawning pulse based on presence of older, larger age-0 ammocoetes in drift samples.

| Monitoring approach | Year | Spawning start date | Spawning end date | Peak spawning activity | Relative abundance |
|---|---|---|---|---|---|
| Large-scale weekly adult counts | 2004 | *April 29 | June 3 | May 13 | 8.1 fish/km |
| | 2005 | April 28 | Jun 30 | Bimodal: May 5 (52) and June 1 (40) | 1.7 fish/km |
| Large-scale weekly redd counts | 2004 | *April 29 | June 3 | May 5 | 31.9 redds/km |
| | 2005 | April 21 | June 30 | Bimodal: May 5 (299) and June 1 (384) | 14.1 redds/km |
| Focal area adult counts | 2004 | *April 6 | June 3 | May 4 | 8.3 fish/survey |
| | 2005 | April 25 | July 3 | Bimodal: late April and mid-May–early June | 2.6 fish/survey |
| Drift net samples of emergent ammocoetes | 2004 | March 25 | June 7 | April 11 | 353 emergent ammocoetes/min |
| | 2005 | **March 27 | July 8 | Bimodal: May 13 and June 8 | 112 emergent ammocoetes/min |

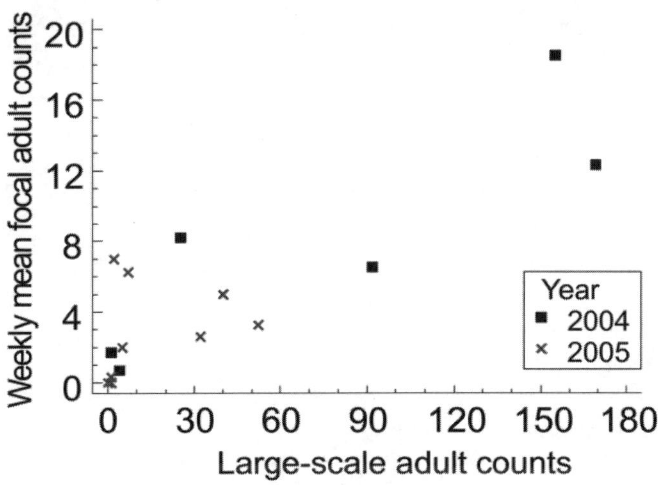


Figure 9. Weekly mean focal area Pacific lamprey adult counts (fish per observation) versus live adults counted during weekly large-scale surveys.

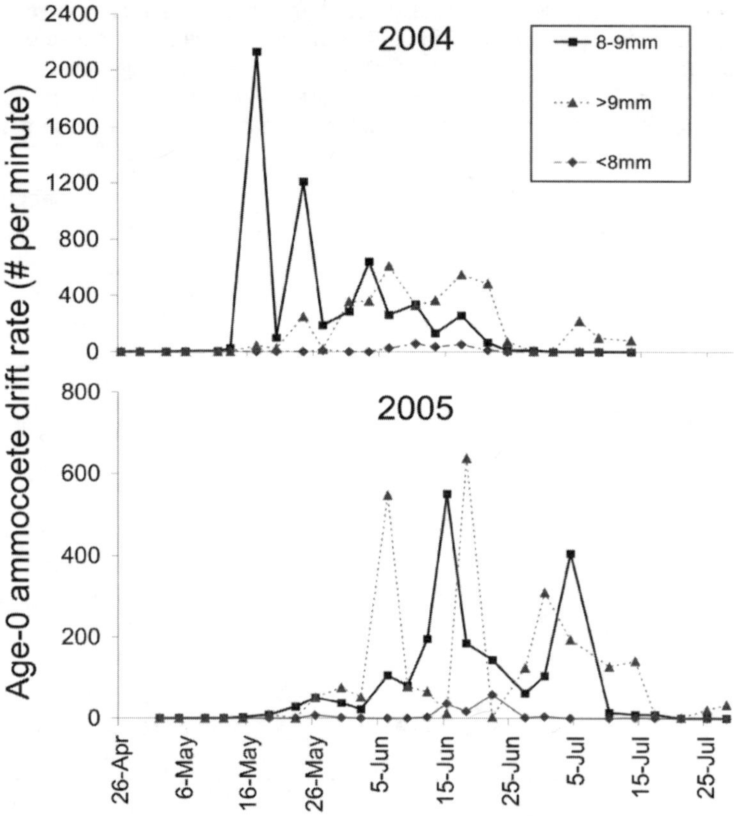


Figure 10. Seasonal patterns of age-0 Pacific lamprey drift rates in the South Fork Coquille River, 2004 and 2005. Note different scales for *y*-axes.

May 27 and June 21, peaking on June 10 at 57.1 fish per minute (Figure 10).

In 2005, 27 samples were taken from April 25 to July 28 (Figure 10). A total of 2,854 age-0 ammocoetes were collected, of which 48% (1,355) were newly emerged, 50% were greater than 9 mm, and 2% were less than 8 mm. Emergent ammocoete abundance during the period of presence averaged 112.1 ± 34.9 individuals/min, about a third of 2004 abundance (Table 2). Duration of ammocoete emergence was much longer in 2005 (71 d) than in 2004 (54 d), likely due to cooler water temperatures and prolonged spawning.

In 2005, emergent ammocoetes were present in the drift from May 15, 20 d after initial adult detection, to July 25, 22 d after final adult detection on the focal area (Figures 8 and 10). As was the case with adult and redd counts, emergent ammocoete abundance was bimodal in 2005, peaking in mid-June (550.9 fish/min) and early July (405.2 fish/min) (Figure 10). The median emergence date occurred between June 12 and June 15. Ammocoetes greater than 9 mm were captured in three main peaks in early, mid-, and late June (Figure 10). In contrast to 2004, a relatively high number of these larger age-0 ammocoetes were present in early season samples—sometimes in numbers greater than the emergent size-class (Figure 10). Again, a few larvae less than 8 mm were caught between May 26 and July 17, and their drift rate peaked at 58.2 fish/min on June 22.

Focal area spawning seasons as estimated from emergent ammocoete presence and the EDD temperature-development relationship were from March 25 to June 7, 2004 and from March 27 to July 8, 2005—about 2–4 weeks earlier than estimates from adult and redd surveys (Table 2). Peak spawning activity based on ammocoete production was estimated to occur around April 11, 2004 and May 13 and June 8, 2005 (Table 2).

## Discussion

### *Annual and Seasonal Trends in Spawning Activity*

All three data types generally gave similar views of annual timing, duration, and magnitude of spawning, including considerably greater spawning activity (multiple of 2.3–4.7) in 2004 and seasonally bimodal activity in 2005. In both years, however, larval drift samples indicated that spawning occurred 2–4 weeks earlier than shown by adult and redd counts, suggesting that low-density, early spawning was missed with adult-oriented surveys. Presence of older age-0 ammocoetes (>9 mm) in early-season drift samples in 2005 also suggested that some upstream spawning occurred earlier than predicted from data on emergent-sized ammocoetes. This early 2005 spawning pulse was likely associated with low flows and unseasonably high temperatures in February and March.

The significant correlations between adult counts in the focal area, the large-scale reaches, and the entire large-scale section suggest that seasonal patterns of spawning were largely independent of spatial scale in our study area—particularly at higher spawner densities as in 2004. Nevertheless, over larger scales and in rivers with greater variability in discharge, temperature, and spawning habitat, such spatial correlations in spawning activity may not persist. For example, studies on salmonids using similar habitats indicate that distribution of spawning is not homogonous but distributed spatially and temporally in relation to habitat availability (Dunham et al. 2001; Isaak and Thurow 2006). Low spawner densities appeared to contribute to the lack of correlation between large-scale and focal area surveys in 2005. Cooler water temperatures in 2005 might have also restricted daily duration of spawning activity more than in 2004. Large-scale surveys typically took place between 1000 and 1500 hours when average water temperatures were below the daily maximum temperature, whereas most focal surveys took place around 1600 hours, a time closer to the daily maximum temperature when fish were more likely to be active.

Estimates of seasonal peaks in spawning activity were largely congruent whether based on large-scale adult and redd counts or focal area adult counts, but estimates of peak activity based on ammocoete production only agreed with adult metrics in 2005. This is not surprising because abundance of spawning lampreys has been shown to be a poor predictor of ammocoete production, likely reflecting seasonal variation in fertilization success or within-redd survival (Jones et al. 2003; Brumo 2006). The disparity between peak spawning as estimated from adult and redd counts and emergent ammocoetes in 2004 could have also resulted from changes in sampling efficiency associated with placement of drift nets further from the thalweg during high flow events or undercounting of adults due to increased turbidity associated with these events. Preliminary studies showed little cross-channel variation in drift rates at the sample site (Brumo 2006), and we expected cross-channel variability in drift abundance to be considerably less than variability between sampling dates. Therefore, it is unlikely that net placement was responsible for the observed pattern. Also, coldwater temperatures and high velocities during high-flow events are expected to deter spawning, suggesting that substantial undercounts during these periods were unlikely. Fertilization success and within-redd survival under varying flow regimes are difficult to evaluate, but our results indicate that these subjects are important for understanding spawning time and larval production.

The late March–July spawning period observed on the South Fork Coquille generally agreed with observations from other coastal systems. Kan (1975) reported coastal Oregon Pacific lampreys spawning from March to May, and on the Smith River in southwest Oregon, Gunckel et al. (2009) detected newly built redds from April to June. In British Columbia, spawning was observed from April to July, but typically peaked in June and July (Beamish 1980; Richards 1980; Farlinger and Beamish 1984). Evidence from the Santa Clara River in Southern California suggests that southern populations spawn as early as January and peak from February to April (Chase 2001), whereas inland and northern populations are thought to initiate spawning considerably later (Kan 1975; Beamish 1980; Richards 1980).

### *Adult versus Redd Counts*

The linear relationship between adult counts and redd counts suggests that both provide similar information about seasonal trends in spawning activity. The expectation of a pair-spawning redd to adult ratio of 1:2, as found by Farlinger and Beamish (1984), or a polygamous ratio (1:>2) was not seen. Instead, redd to adult ratios were much higher in both years (4:1 in 2004; greater than 8:1 in 2005), implying underestimation of the total spawning population with weekly adult counts, or multiple redds per spawning pair. We did not observe multiple redds per spawning pair, but individual spawners were occasionally seen moving rocks in multiple locations within a spawning ground. Other lamprey species have also been observed building redds in several locations (Hardisty and Potter 1971), and some Pacific lamprey redds may be built but never used (Pletcher 1963; Close et al. 2003). Population underestimation using weekly surveys is supported by our tagging results, which indicated short (<1 week) residence times for individuals on spawning grounds, especially females. Farlinger and Beamish (1984) reported that male Pacific lampreys remained on spawning grounds for an average of 6.5 d, versus 4.6 d for females. Beamish (1980) reported that river lamprey *Lampetra ayresii* females died within a few hours after spawning, but males typically survived for about 3 weeks. Such quick turnover of individuals on spawning areas needs to be accounted for in adult count data collected over nondaily scales.

Annual, seasonal, and diel variability in ability to detect adults and redds may also influence observed redd to adult ratios. For example, in 2005, deeper water and lower average visibility made adult detection and capture relatively more difficult than redd observation, which was reflected in an increase in percentage of adults that evaded capture from 2004 to 2005. Cooler water temperatures in 2005 might also have restricted daily duration of spawning activity. Pacific lampreys are thought to require approximately 11°C to initiate spawning (Pletcher 1963; Kan 1975; Close et al. 2003; Brumo 2006), and early in the 2005 spawning season, this temperature was typically not reached until late afternoon. We generally observed greater spawning activity in the late afternoon and at night, especially early in spawning seasons. The influence of such seasonal and diel activity patterns on adult counts requires further investigation, and survey timing should be adjusted accordingly.

The general increase in number of redds counted per adult later in the spawning season, especially in 2005, may be explained by shorter spawner residence time later in the season or counter error due to accumulation of older redds over time. The number of redds constructed per adult and how it varies over time and space need to be more thoroughly understood before redds can be used to accurately assess spawner abundance.

Both adult counts and redd counts present logistical difficulties and have no clear-cut way to quantify errors. These problems include observer variability, movement of lampreys during surveys, greater spawning activity at night, and variable visibility due to rain, wind, turbidity, water depth, sun angle, and discharge. Standardization of both survey types in relation to weather, discharge, and time of day is possible but risks inconsistent data collection over the season and variable sampling frequency, compromising interannual comparability. We expect redd and adult count errors to be more consistent from year to year in shallow river systems with low variation in spring discharge, but such systems are rare over much of the species' range.

Because Pacific lampreys are not easily frightened from spawning areas and can be caught by hand with little trouble, personnel training can be more easily standardized for adult counts, especially in clear, wadeable streams. However, adult counts may only be viable in systems containing relatively high densities of spawners, such as the South Fork Coquille River.

Because of their relative permanence and higher numbers compared to adults, redds may be more useful for detecting spawning activity in streams with low population densities. Yet, variable redd shape, size, and age, as well as superimposition, make consistency in redd counts problematic (Dunham et al. 2001; Al-chokhachy et al. 2005; Gallagher and Gallagher 2005; Stone 2006; Gunckel et al. 2009). Elevated densities of spawners and large areas of disturbed spawning substrate (20 × 5 m) were seen during weekly large-scale floats in May 2004, making individual redds nearly impossible to distinguish. Gunckel et al. (2009) reported similar difficulties. Signs of apparent redd superimposition were also observed in early May of 2004, when more than 25 spawners and particularly high redd concentrations were seen in the focal area. Previous studies have cited anecdotal evidence of superimposed spawning by Pacific lampreys (Pletcher 1963; Kan 1975; Close et al. 2003; Gunckel et al. 2009). Steelhead redd misidentification can also potentially introduce error into redd counts for inexperienced observers (Stone 2006; Gunckel et al. 2009; A. F Brumo, unpublished). In April and May of both years, we observed a small number of spawning steelhead and their redds in spawning areas used by lampreys.

Research on redd count error in bull trout *Salvelinus confluentus* surveys suggests that even with significant training, observer variability is substantial and might be unavoidable (Dunham et al. 2001). We recommend pairing observers unfamiliar with lamprey redds with experienced observers during early surveys to allow corroboration of redd identification and counts. Dunham et al. (2001) also noted that significant counting errors could be attributed to redd age, which may limit the use of redd counts for population monitoring. Marking redds to monitor which are newly built is one solution to this problem. In this case, we did not mark redds, but instead attempted to count all redds perceived to be built within a week of surveys. This judgment was subjective and likely added to survey error. Stone (2006), who marked Pacific lamprey redds, reported that the duration redds remain visible was related to discharge and increased as the spawning season progressed. Until redd longevity is better quantified, marking redds during surveys is recommended.

Adult carcasses may also be a potentially useful monitoring metric, but seasonal and interannual differences in stream turbidity and discharge confound data interpretation. High-discharge events in 2005 likely increased downstream transport and made it more difficult to recover carcasses, as evidenced by the much lower number of carcasses collected per adult counted (17 of 140) compared to 2004 (373 of 446). The low recovery rate of tagged carcasses suggests that most spawners died within a week of tagging and drifted out of the survey reach or into deep pools, or were scavenged by birds or mammals. It is also notable that during large-scale surveys, we commonly observed apparent postspawn or dying Pacific lampreys swimming slowly downstream. Since most (29 of 35) recovered carcasses with tags were found downstream of their tagging, and presumably spawning locations, carcasses likely only provide information on upstream spawning populations.

### *Sex Ratio*

In both years, sex ratio of live adult Pacific lampreys was approximately 1:1. From a sample of 252 fish collected during the spawning season, Farlinger and Beamish (1984) found a sex ratio of around 1.3 males per female. Similarly, Kan (1975) reported a sex ratio of roughly 1.2

males per female for spawning-stage individuals. Hardisty and Potter (1971) reviewed results from studies on various European and North American species and found that there was usually an excess of spawning males in both resident and anadromous species, although sex ratio varied considerably between species and years.

In 2004, the male to female ratio of live fish (1:1) was considerably lower than that of dead fish (2:1). Possible explanations for this discrepancy include differences in postspawning behavior, longevity differences, predator preference, and time of metamorphosis (Kan 1975). The higher proportion of males recaptured (90% live, 77% carcasses) points to a difference in postspawning activity and longer residence time for males. There was not a significant length difference between live and dead fish of either sex, which suggests that they were likely part of the same population (Brumo 2006). Little data on sex ratio of dead Pacific lampreys exists in the literature. Lorion et al. (2000) noted that samples of mature Miller Lake lamprey L. minima usually had a 3:1 male bias, except for a large collection of dead fish where they found a 9:1 female bias.

### *Larval Drift Samples versus Spawning Surveys*

Samples of drifting ammocoetes have advantages and disadvantages relative to adult and redd surveys. When adult surveys are impractical due to river size, visibility, or access, drift nets used in conjunction with water temperature data and known degree-days required for emergence are an effective way to pinpoint spawning times and provide a continuous picture of upstream ammocoete production. Depending on environmental conditions, one important advantage of drift sampling is the approximate 30-d lag between spawning and resultant ammocoete emergence. For example, it is possible to assess presence and timing of spawning during high, turbid flows by capturing drifting ammocoetes 30 d later. In this study, both adult-oriented counts and ammocoete drift samples suffered from data gaps associated with high discharge events, especially during recurrent, mid-season events in 2005. However, with the exception of larger events such as the late April freshet in 2005, when all sampling was impossible, those events were less problematic for drift sampling.

In streams with small or fragmented spawning populations, or in larger water bodies, drift nets could be used to demonstrate upstream spawning presence when visual adult or redd observations might not. For this reason, data on relative ammocoete abundance in the drift can also aid in identification of productive spawning tributaries and assignment of conservation priorities without implementation of traditional surveys. In general, the probability of detection should decrease from age-0 ammocoetes to redds to spawning adults. Assuming minimal error in the temperature-development model used to estimate spawning dates from drift dates, our results also suggest that drift sampling more precisely describes the spawning period than adult-oriented surveys—particularly in early spring when high discharge and low abundance of spawners and redds preclude detection.

Disadvantages of drift samples include nighttime work hours, laboratory time, taxonomic identification problems, lack of information gained on spawning stock or initial egg production, and difficulty quantifying abundance in fluctuating discharge conditions. In this study, drift rates were used primarily to provide estimates of relative ammocoete production from the focal spawning area over time. In order to provide more robust estimates of relative production and to extrapolate drift rates to estimate absolute production from a given area, it would be necessary to better quantify (1) cross-sectional drift patterns and percentage of drifting fish captured (capture efficiency), (2) percentage of daily drift captured during the fishing period and diel periodicity in drift rates and how it varies in relation to environmental factors (e.g., moon phase), (3) day-to-day variation in drift rates related to seasonal variation, and (4) distance drifted by newly emerged ammocoetes before settling into rearing habitats.

For seasonal comparisons of ammocoete production, we assumed that percentage of total drifting individuals caught at the net position was consistent across sampling dates. Exploratory sampling in 2004 using multiple nets set simultaneously across the wetted-width showed no significant differences in drift densities of age-0 ammocoetes between the thalweg and channel margins (Brumo 2006). Site-specific evaluation of cross-channel variation in drift and capture efficiency and how it varies with discharge is necessary for accurate ammocoete production estimates.

Additionally, we assumed our drift samples taken 3 h post-sunset accurately represented relative abundance of the entire sample date. In other words, variation in proportion of total daily drift caught during the period our net was set was considerably lower than variation in total drift abundance among sampling dates. Reichard et al. (2004) found that variation in drift abundance of various cyprinids within a night was typically much less than variation among sampling dates. Replicate drift sampling during each sample period has been recommended by various authors because it provides a more accurate approximation of daily, relative drift abundance (Allan and Russek 1985; Franzin and Harbicht 1992). In this case, we substituted replicate sampling with size by using a drift net (70 × 150 cm opening) approximately eight times larger than a typically sized drift net (30 × 45 cm opening). Other studies on drifting invertebrates (Allan and Russek 1985) and fish (Zitek et al. 2004) have shown that a single sample taken near estimated peak drift time is sufficient to predict total drift from the 24-h period.

We also assumed that observed differences in drift rates between sample dates represented actual seasonality, not short-term fluctuations in emergence timing (variability between sample dates was less than variability between weeks). A test comparing drift-rate variability of nets set on several consecutive nights to variability between weekly samples would help substantiate this assumption.

Whether estimating relative or absolute production, another important consideration when using drift nets is spatial scope of inference. Specifically, what spawning areas did most captured individuals originate from? In this case, some portion of emergent larvae captured likely originated from spawning areas upstream of the focal area. Because seasonal patterns of spawning activity were highly correlated between the focal area and large-scale section upstream, the proportion of emergent ammocoetes originating upstream was assumed to be relatively unchanged over time. For this reason, relative abundance of drifting ammocoetes can be used to draw inferences about spawning time and relative ammocoete production at both areas, assuming early survival was similar between the two.

Quantifying Pacific lamprey emergent ammocoete drift distance over time and describing factors that influence downstream movement would help identify where drifting individuals originate, allowing generation of more spatially explicit conclusions about larval production. Derosier (2001) showed that age-0 sea lampreys can disperse at least 874 m downstream of spawning areas after approximately 3 months from emergence. Manion and McLain (1971) suggested that age-0 sea lampreys remained concentrated near spawning areas during their first year. Proximity of rearing habitats in relation to spawning location is another factor that likely affects drift distance. In the current study, there were several substantial depositional, rearing habitats between the focal spawning area and the nearest upstream spawning area, which may have lessened drift from upstream spawning areas.

## Conclusions

When using emergent ammocoete drift data to monitor spawning, a key issue to understand is the stock–recruit relationship. In lamprey, the relationship is thought to be weak due to high and variable mortality during the incubation period (Jones et al. 2003; Haesker et al. 2003; Brumo 2006). For this reason, spawner stock estimates are poor predictors of ammocoete production and emergent ammocoete abundance a poor predictor of spawner abundance. In spite of this weak relationship, this study demonstrates that both approaches can give similar pictures of timing and duration of spawning, and in some years, similar views of seasonal abundance patterns. The abundance relationship breaks down because of differences in cohort survival, and those differences can only be detected with both spawning stock and ammocoete production data.

Ultimately, the monitoring metric or combination of metrics employed to assess Pacific lamprey spawning populations will depend on availability of personnel and funds, stream size and access, prevailing weather conditions, and information desired. Drawing from our experiences on the South Fork Coquille River, we recommend using a multi-life stage approach to maximize understanding of escapement, ammocoete production, and factors affecting year-class success.

## Acknowledgments

This work was supported by Oregon Department of Fish and Wildlife and U.S. Fish and Wildlife Service. Specimen collections were authorized under a series of Oregon scientific taking permits through OR2004-1365 and OR2005-2310, and an OSU Institutional Animal Care and Use permit LAR-ID 2926. We are grateful for support and assistance from M. Gray, B. Riggers, D. Thompson, B. Alfonse, and M. Jansen. B. Clemens provided critical review of the draft manuscript.

## References

Al-chokhachy, R., P. Budy, and H. Schaller. 2005. Understanding the significance of redd counts: a comparison between two methods for estimation the abundance of and monitoring bull trout populations. North American Journal of Fisheries Management 25:1505–1512.

Allan, J. D., and E. Russek. 1985. The quantification of stream drift. Canadian Journal of Fisheries and Aquatic Sciences 42:210–215.

Beamish, R. J. 1980. Adult biology of the river lamprey (*Lampetra ayresi*) and the Pacific lamprey (*Lampetra tridentata*) from the Pacific coast of Canada. Canadian Journal of Fisheries and Aquatic Sciences 37:1906–1923.

Bennett, R. M., and R. M. Ross. 1995. Stream drift of newly hatched sea lampreys in the upper Delaware River. Journal of the Pennsylvania Academy of Science 69(1):7–9.

Brumo, A. F. 2006. Spawning, larval recruitment, and early life survival of Pacific lampreys in the South Fork Coquille River, Oregon. Master's thesis. Oregon State University, Corvallis.

Chase, S. D. 2001. Contributions to the life history of adult Pacific lamprey (*Lampetra tridentate*) in the Santa Clara river of southern California. Bulletin of the Southern California Academy of Sciences 100:74–85.

Close, D. A., K. Aronsuu, A. Jackson, T. Robinson, J. Bayer, J. Seelye, S. Yun, A. Scott, W. Li, and C. Torgersen. 2003. Pacific Lamprey Research and Restoration Project, Annual Report 2002. Bonneville Power Administration, Project No. 1994-026. Portland, Oregon.

Cochnauer, T., C. Claire., and S. A. Putnam. 2006. Evaluate status of Pacific lamprey in the Clearwater River drainage, Idaho. Annual Report for 2005. Bonneville Power Administration Project Number 200002800. Portland, Oregon.

CRBLTW (Columbia River Basin Lamprey Technical Workgroup). 2005. April 19. Critical uncertainties for lamprey in the Columbia River basin: results from a strategic planning retreat of the Columbia River Lamprey Technical Workgroup. Columbia Basin Fish and Wildlife Authority, Portland, Oregon.

Derosier, A. L. 2001. Early life history of sea lamprey: emergence, movements, and dispersal. Master's thesis. Michigan State University, East Lansing.

Dunham, J., B. Rieman, and K. Davis. 2001. Sources and magnitude of sampling error in redd counts for bull trout. North American Journal of Fisheries Management 21:343–352.

Farlinger, S. P., and R. J. Beamish. 1984. Recent colonization of a major salmon-producing lake in British Columbia by Pacific lamprey. Canadian Journal of Fisheries and Aquatic Sciences 41:278–285.

Franzin, W. G., and S. M. Harbicht. 1992. Tests of drift samplers for estimating abundance of recently hatched walleye larvae in small rivers. North American Journal of Fisheries Management 12:396–405.

Gallagher, S. P., and C. M. Gallagher. 2005. Discrimination of Chinook salmon, coho salmon, and steelhead redds and evaluation of the use of redd data for estimating escapement in several unregulated streams in northern California. North American Journal of Fisheries Management 25:284–300.

Glenney, C., C. Rossell, G. Asbridge, and B. Benjamin. 2004. Fifteenmile Creek basin winter steelhead spawning survey, 2004 Accomplishment Report. Barlow Ranger District, Mt. Hood National Forest, Dufur, Oregon.

Graham, J. C., and C. V. Brun. 2005. Determining lamprey species composition, larval distribution, and adult abundance in the Deschutes River, Oregon, subbasin, 2004–2005 Annual Report. Bonneville Power Administration, Project Number 200201600, Portland, Oregon.

Gunckel, S. L., K. K. Jones, and S. E. Jacobs. 2009. Spawning distribution and habitat use of adult Pacific and western brook lampreys in Smith River, Oregon. Pages 173–189 *in* L. R. Brown, S. D. Chase, M. G. Mesa, R. J. Beamish, and P. B. Moyle, editors. Biology, management, and conservation of lampreys in North America. American Fisheries Society, Symposium 72, Bethesda, Maryland.

Haesker, S. L., M. L. Jones, and J. R. Bence. 2003. Estimating uncertainty in the stock- recruitment relationship for St. Marys River sea lampreys. Journal of Great Lakes Research 29 (Supplement 1):728–741.

Hardisty, M. W., and I. C. Potter. 1971. The general biology of adult lampreys. Pages 127–206 *in* M.W. Hardisty and I. C. Potter, editors. The biology of lampreys. Academic Press, London.

Houde, E. D. 1987. Fish early life dynamics and recruitment variability. Pages 17–29 *in* R. D. Hoyt, editor. 10th Annual Larval Fish Conference. American Fisheries Society, Symposium 2, Bethesda, Maryland.

Isaak, D. J., and R. F. Thurow. 2006. Network-scale spatial and temporal variation in Chinook salmon (*Oncorhyncuhs tshawytscha*) redd distributions: patterns inferred from spatially continuous replicate surveys. Canadian Journal of Fisheries and Aquatic Sciences 63:285–296.

Jones, M. L., R. A. Bergstedt, M. B. Twhoey, M. F. Fodale, D. W. Cuddy, and J. W. Slade. 2003. Compensatory mechanisms in Great Lakes sea lamprey populations: implication for alternative control strategies. Journal of Great Lakes Research 29 (Supplement 1):113–129.

Johnston, T. A., M. N. Gaboury, R. A. Janusz, and L. R. Janusz. 1995. Larval fish drift in the Valley River, Manitoba: influence of abiotic and biotic factors, and relationships with future year-class strengths. Canadian Journal of Fisheries and Aquatic Sciences 52:2423–2431.

Kan, T. T. 1975. Systematics, variation, distribution, and biology of lampreys of the genus *Lampetra* in Oregon. Doctoral dissertation. Oregon State University, Corvallis.

Kostow, K. 2002. Oregon lampreys: natural history, status, and problem analysis. Oregon Department of Fish and Wildlife, Portland.

Lorion, C., D. F. Markle, S. Reid, and M. Docker. 2000. Re-description of the presumed-extinct Miller Lake lamprey, *Lampetra minima*. Copeia 4:1019–1028.

Luzier, C. W., G. Silver, T. A. Whitesel 2006. Evaluate habitat use and population dynamics of lampreys in Cedar Creek, Annual Report 2005. Bonneville Power Administration, Project Number 20000140. Portland, Oregon.

Manion, P. J., and A. L. McLain. 1971. Biology of larval sea lampreys (*Petromyzon marinus*) of the 1960 year class, isolated in the Big Garlic River, Michigan, 1960–65. Great Lakes Fishery Commission Technical Report 16:1–36.

McDonald, L. L. 2004. Sampling rare populations, Pages 11–42 *in* W. L. Thompson, editor. Sampling rare or elusive species: concepts, designs, and techniques for estimating population parameters. Island Press, Washington, D.C.

Meeuwig, M. H., J. M. Bayer, and R. A. Reiche. 2004. Identification of larval Pacific lampreys (*Lampetra tridentata*), river lampreys (*L. ayresi*), and western brook lampreys (*L. richardsoni*), and thermal requirements of early life history stages of lampreys. Amended Final Report for 2002–2004. Bonneville Power Administration, Project Number 20000290 Portland, Oregon.

Meeuwig, M. H., J. M. Bayer, and J. G. Seelye. 2005. Effects of temperature on survival and development of early life stage Pacific and western brook lampreys. Transactions of the American Fisheries Society 134:19–27.

Morbey, Y. 2000. Protandry in Pacific salmon. Canadian Journal of Fisheries and Aquatic Sciences 57:1252–1257.

Moser, M. L., J. M. Butzerin, and D. B. Dey. 2007. Capture and collection of lampreys: the state of the science. Reviews in Fish Biology and Fisheries 17:45–56.

Moser, M. L., and D.A. Close. 2003. Assessing Pacific lamprey status in the Columbia River basin. Northwest Science 77(2):116–125.

Moser, M. L., P. A. Ocker, L. C. Stuehrenberg, and T. C. Bjornn. 2002. Passage efficiency of adult Pacific lampreys at hydropower dams on the lower Columbia River, USA. Transactions of the American Fisheries Society 131:956–965.

Neilson, J. D., and G. H. Geen. 1981. Enumeration of spawning salmon from spawner residence time and aerial counts. Transactions of the American Fisheries Society 110:554–556.

Piavis, G. W. 1961. Embryological stages in the sea lamprey and effects of temperature on development. Fishery Bulletin 61:111–143.

Pletcher, F. T. 1963. The life history and distribution of lampreys in the Salmon and certain other rivers in British Columbia, Canada. Master's thesis. University of British Columbia, Vancouver.

Reichard, M., P. Jurajda, and C. Smith. 2004. Spatial distribution of drifting cyprinid fishes in a shallow lowland river. Archiv für Hydrobiologie 159(3):395–407.

Richards, J. E. 1980. Freshwater biology of the anadromous Pacific lamprey *Lampetra tridentata*. Master's thesis. University of Guelph, Guelph, Ontario.

Stone. J. 2006. Observations of nest characteristics, spawning habitat, and spawning behavior of Pacific and western brook lamprey in a Washington stream. Northwest Science 87:225–232.

White, J. L., and B. C. Harvey. 2003. Basin-scale patterns in drift of embryonic and larval fishes and lamprey ammocetes in two coastal rivers. Environmental Biology of Fishes 67:369–378.

Zitek, A., S. Schmutz, and A. Ploner. 2004. Fish drift in a Danube sidearm-system: II. Seasonal and diurnal patterns. Journal of Fish Biology 65:1339–1357.

American Fisheries Society Symposium 72:223–232, 2009


# Morphological Diagnosis of *Entosphenus* and *Lampetra* Ammocoetes (Petromyzontidae) in Washington, Oregon, and California

Damon H. Goodman*,[1]
*Department of Fisheries Biology, Humboldt State University*
*One Harpst Street, Arcata, California 95521, USA*

Andrew P. Kinziger
*Department of Fisheries Biology, Humboldt State University*
*One Harpst Street, Arcata, California 95521, USA*

Stewart B. Reid
*Department of Fisheries Biology, Humboldt State University*
*One Harpst Street, Arcata, California 95521, USA*
and
*Western Fishes, 2045 East Main Street. Ashland, Oregon, 97520, USA*

Margaret F. Docker
*Department of Biological Sciences, University of Manitoba*
*Winnipeg, Manitoba R3T 2N2, Canada*

*Abstract.*—The objective of this study was to assess the utility of new and previously published morphological characters for identifying lamprey ammocoetes to genus (*Entosphenus* and *Lampetra*) in coastal rivers of California, Oregon, and Washington. We surveyed four morphological characters on 382 ammocoetes (60.0–169.0 mm total length) from 15 localities and compared these against genus assignments using genetic methods. Pigmentation of the caudal fin was dark in *Entosphenus* versus peppered or absent in *Lampetra* and allowed for 95.6% proper genus diagnosis when compared to identification by genetic methods. Pigmentation of the caudal ridge faded from dark to light (anteriorly to posteriorly) in *Entosphenus*, in contrast to a uniformly dark caudal ridge in *Lampetra*, and allowed for 98.7% proper genus diagnosis. Ventral pigmentation and myomere number differed between the two genera but did not allow for diagnosis by themselves. When applied together, the three pigment characters provided 99.7% agreement with genetic identification. We conclude that the morphological characters, alone and in combination, provide valid genus assignments of lamprey ammocoetes within the study area and size range evaluated.

* Corresponding author: damon_goodman@fws.gov
[1] Current address: Fish and Wildlife Service, 1655 Heindon Road, Arcata, California 95521, USA

## Introduction

Little is known about lamprey population status along the Pacific coast of the United States. Data that are available suggest a significant decline in populations (Renaud 1997; Close et al. 2002), which resulted in a petition to list four species of lampreys from this region in 2003 under the U.S. Endangered Species Act (USFWS 2004). Efforts aimed at managing these species, however, are thwarted by improper field identifications, particularly in the ammocoete life history phase. Field data on ammocoete distribution and abundance are often difficult to interpret, due to inclusion of multiple taxa. Thus, the distribution, status, and abundance trends of many Pacific coast lamprey populations remain unclear.

Lamprey identification is complicated by the presence of two distinct life history phases, ammocoete (larva) and adult, each of which requires a different suite of morphological characters for species identification (e.g., Vladykov and Kott 1980; Jenkins and Burkhead 1994). While identification of adult lampreys to species along the Pacific coast is established (Bond 1994; Moyle 2002, Wydoski and Whitney 2003), with a few exceptions (e.g., Klamath basin, Lorion et al. 2000), identification of ammocoetes to species is considerably more difficult (Richards et al. 1982). Since the multi-year ammocoete phase occurs in the freshwater environment and is easily accessible for field collections, it may be a useful life history stage for monitoring distribution as well as trends in abundance and habitat use. Although the links between ammocoete and adult abundance have not been established, monitoring of ammocoetes may still be useful, similar to monitoring of outmigrating juvenile salmonids. Development of methods allowing for identification of ammocoetes is, therefore, critical for effective management of lamprey populations.

The objective of this study was to assess the utility of external morphological characters for identifying lamprey ammocoetes to genus in coastal rivers of Washington, California, and Oregon. Within this region, identification to genus alone is sometimes sufficient for management purposes and can often be used to infer species, due to the established distributions of congeneric taxa. Two lamprey genera occur within the study area, *Entosphenus* and *Lampetra*. In this paper, we follow Gill et al. (2003) and Nelson (2006) in recognizing both as valid genera. *Entosphenus* is represented by four recognized species in Washington, Oregon, and California: Pacific lamprey *E. tridentatus* (formerly *L. tridentata*), Pit-Klamath brook lamprey *E. lethophagus* (formerly *L. lethophaga*), Klamath lamprey *E. similis* (formerly *L. similis*), and Miller Lake lamprey *E. minimus* (formerly *L. minima*), and *Lampetra* is represented by three species: western brook lamprey *L. richardsoni*, river lamprey *L. ayresii*, and Kern brook lamprey *L. hubbsi* (Table 1); Kern brook lamprey was placed within *Entosphenus* (Vladykov and Kott 1976), but mitochondrial DNA sequence suggests that it is more closely related to west coast *Lampetra* species (Docker et al. 1999). This study uses genetic methods to identify ammocoetes to genus, thereby eliminating the need to rear ammocoetes to adulthood to validate identifications.

## Materials and Methods

### Collection of Specimens

Ammocoetes were collected at 15 locations in Washington, Oregon, and California (Figure 1; Table 2). All specimens were preserved in 95% ethanol and have been deposited into the Humboldt State University Fish Collection (HSU). Collections within the known distribution of both genera, *Entosphenus* and *Lampetra*, were preferentially selected for this study since, when comparing allopatric populations, environmentally induced morphological variability can confound interpretations of intergeneric differences. Localities known to support populations of lampreys with limited geographic ranges, specifically Pit-Klamath brook lamprey, Klamath lamprey, and Kern brook lamprey, were targeted in the Pit, Klamath, and San Joaquin drainages.

Table 1. Adult life-history characteristics and geographic distribution of lamprey species occurring in the study area.

| Species | Adult life-history characteristics | Geographic distribution | Reference |
|---|---|---|---|
| Pacific lamprey | Anadromous and predatory | Pacific coast streams from Hokkaido Island, Japan to Baja California, Mexico | Beamish 1980; Rohde 1980 |
| Pit-Klamath brook lamprey | Nontrophic freshwater resident | Pit River, California and upper Klamath River, Oregon | Hubbs 1971 |
| Klamath lamprey | Predatory freshwater resident | Klamath River, California | Vladykov and Kott 1979 |
| Miller Lake lamprey | Predatory freshwater resident | Miller Lake and Klamath River | Bond and Kan 1973; Lorion et al. 2000 |
| Western brook lamprey | Nontrophic freshwater resident | Pacific coast streams from California to British Columbia | Vladykov and Follett 1965 |
| River lamprey | Anadromous and predatory | Larger river systems from Sacramento/San Joaquin to the Fraser | Vladykov and Follett 1958 |
| Kern brook lamprey | Nonparasitic freshwater resident | Sacramento and San Joaquin River | Vladykov and Kott 1976 |

## *Identification Using Genetic Analysis*

Ammocoetes were identified to genus, *Entosphenus* or *Lampetra*, using genetic methods. Docker et al. (1999) found three *Lampetra* species from the west coast of North America (river lamprey, western brook lamprey, and Kern brook lamprey) to differ from four species of the genus *Entosphenus* (Pacific lamprey, Vancouver lamprey, Klamath lamprey, and Pit-Klamath brook lamprey) by more than 8% in their cytochrome b (cyt b) sequences. Out of the 384 base pairs (bp) sequenced at the 5'-end of cyt b, 24 positions were fixed for alternate nucleotides in *Entosphenus* and *Lampetra*. All but two of these diagnostic differences were maintained following inclusion of a fifth *Entosphenus* species, Miller Lake lamprey (Lorion et al. 2000). Two of these 22 diagnostic sites, at positions 96 and 252 of the cyt b molecule, correspond with restriction enzyme recognition sites (*Hae*III and *Bsa*AI) and form the basis of restriction fragment length polymorphism (RFLP) assays to differentiate between the two genera. As explained below, the *Hae*III RFLP assay was used in this study.

DNA was extracted from muscle tissue in Chelex (Part #: Sigma C7901–50G), 0.2 mL distilled water, and 1.2 μL Proteinase K (Part #: Qiagen 19131 [>600 mAU/mL]) incubated at room temperature for 12–24 h and denatured by boiling for 8 min (Miller and Kapuscinski 1993). A 433 bp fragment of the cyt b gene was amplified using the polymerase chain reaction (PCR) with a primer developed for this study 5'-CCA CCM ACC RWC YAT TMT TCG AAA AAC-3' and CB2-H (Palumbi 1996). Amplifications were performed in 25 μL reaction volumes with the following reagent concentrations: 12.5 μL PCR Master Mix (Part #: Promega M7505), 2.0 μL DNA template (<250 ng), 10 μM of each primer (0.8 μM final), and 8.5 μL nuclease-free water. Thermocycling conditions included 35 cycles of 94°C for 1 min, 57°C for 1 min, and 72°C for 1.5 min with a final extension of 72°C for 5 min. The *Hae*III restriction digests were performed in 20 μL reactions with the following reagent concentrations: 4 μL PCR product (0.2–1.5 μg), 0.2 μL BSA (10 μg/μL), 2 μL 10× Buffer, 0.5 μL restriction enzyme *Hae*III (10u/μL), and 13.3 μL nuclease-free water (Part #: Fisher BioReagents *Hae*III BP3368–1). Digests were incubated 12–24 h at 37°C and visualized on 2.0% agarose gels. Based on the known sequence from Docker et al. (1999) and Lorion

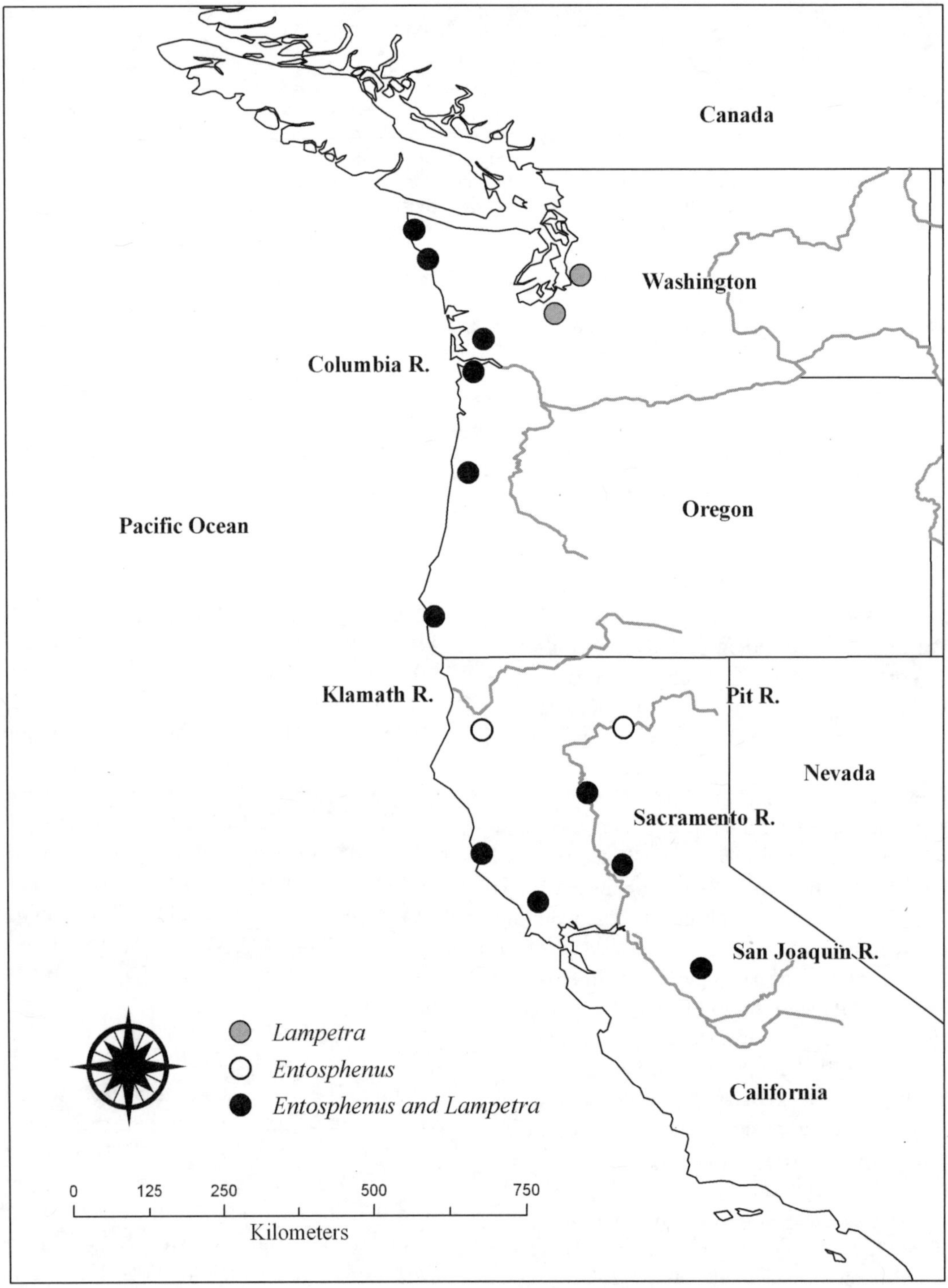


Figure 1. Sampling sites for *Entosphenus* and *Lampetra* ammocoetes in Washington, Oregon, and California. Locality, vouchers, and proportion of *Entosphenus* and *Lampetra* collected at each locality are given in Table 2.

Table 2. Collection localities, Humboldt State University voucher numbers (HSU #), number of individuals (*n*), mean total length (and range), and mean myomere count (and range) by genus for all collections used in the morphological evaluation of lamprey ammocoetes from Washington, Oregon, and California. Latitude and longitude provided in State Plane NAD 1983 coordinate system.

| | | | | | | *Entosphenus* | | | *Lampetra* | | |
|---|---|---|---|---|---|---|---|---|---|---|---|
| Drainage | River | State | Latitude | Longitude | HSU # | *n* | Total length (mm) | Myomere count | *n* | Total length (mm) | Myomere count |
| Lake Ozette | Big River | WA | 48.14 | −124.55 | 3951, 3952, 3953 | 12 | 98 (80–120) | 68 (66–71) | 17 | 110 (88–139) | 60 (58–62) |
| Cedar | Cedar Creek | WA | 47.86 | −124.41 | 3955, 3956 | 11 | 110 (93–129) | 68 (67–71) | 15 | 113 (72–160) | 59 (57–63) |
| Lake Washington | Cedar River | WA | 47.55 | −122.18 | 3927 | 0 | – | – | 27 | 117 (74–169) | 64 (59–68) |
| Nisqually | Nisqually River | WA | 46.95 | −122.49 | 3923, 3924 | 0 | – | – | 28 | 112 (85–143) | 61 (57–64) |
| Willapa | Willapa River | WA | 46.64 | −123.61 | 3963, 3964 | 18 | 108 (90–130) | 68 (66–70) | 10 | 141 (115–163) | 60 (58–62) |
| Columbia | Youngs River | OR | 46.08 | −123.71 | 3858, 3859 | 15 | 88 (69–115) | 67 (65–68) | 10 | 94 (71–122) | 61 (59–63) |
| Yaquina | Yaquina River | OR | 44.72 | −123.85 | 3874, 3875 | 16 | 84 (65–110) | 67 (65–70) | 3 | 83 (78–92) | 62 (62–62) |
| Euchre | Euchre Creek | OR | 42.66 | −124.31 | 3896, 3897, 3898 | 21 | 104 (75–122) | 68 (66–71) | 8 | 107 (93–134) | 59 (57–60) |
| Klamath | Trinity River | CA | 40.92 | −123.64 | 3920 | 20 | 79 (60–96) | 69 (67–71) | 0 | – | – |
| Pit | Hat Creek | CA | 41.02 | −121.51 | 3946 | 30 | 88 (68–133) | 61 (58–64) | 0 | – | – |
| Navarro | Navarro River | CA | 39.14 | −123.64 | 3905, 3906 | 11 | 90 (82–110) | 66 (62–69) | 8 | 97 (77–130) | 61 (59–62) |
| Russian | Mark West Creek | CA | 38.54 | −122.80 | 3976, 3977 | 2 | 86 (84–88) | 68 (66–70) | 27 | 90 (70–108) | 55 (53–57) |
| Sacramento | Mill Creek | CA | 40.11 | −122.25 | 3972 | 20 | 105 (75–126) | 70 (68–72) | 6 | 103 (83–136) | 54 (54–55) |
| | Feather River | CA | 39.03 | −121.65 | 3943, 3974, 3975 | 17 | 108 (92–121) | 69 (66–71) | 11 | 102 (86–120) | 69 (66–71) |
| San Joaquin | Merced River | CA | 37.53 | −120.39 | 3939, 3968, 3969 | 5 | 116 (106–123) | 67 (63–69) | 14 | 110 (86–135) | 54 (50–56) |

et al. (2000) and preliminary assays on adult specimens identified to species (M. F. Docker, University of Manitoba, unpublished data), the restriction endonuclease *Hae*III cleaves the 433 bp cyt b PCR product at one site in *Lampetra* resulting in 167 and 266 bp fragments (but see Results for a second *Lampetra* cut pattern) and cleaves at two sites in *Entosphenus* resulting in fragments 92, 167, and 174 bp in length. The restriction enzyme *Bsa*AI cuts the cyt b PCR product at one site in *Lampetra* resulting in 188 and 245 bp fragments but does not cleave the 433 bp fragment in *Entosphenus*. We used the *Hae*III RFLP assay in this study since anything that would interfere with *Bsa*AI digestion could erroneously lead to diagnosis as *Entosphenus*.

Direct sequencing was performed on the cyt b PCR product from a third previously unknown RFLP type (see Results) using the DTCS Quick Start cycle sequencing kit (Beckman Coulter, Fullerton, California) and CEQ8000 automated sequencer following manufacturer's instructions.

### *Morphological Analysis*

Three pigmentation characters and myomere counts were assessed in each ammocoete. The pigmentation characters examined were pigmentation of the caudal fin, caudal ridge, and ventral surface. Pigmentation of the caudal fin and caudal ridge has been used previously in taxonomic assessments of Pacific coast ammocoetes (Richards et al. 1982), but ventral pigmentation is a novel character. Pigmentation of the caudal fin was scored as (1) dark: uniformly pigmented with melanophores except distal margin which lacks pigmentation; (2) peppered: single melanophores or localized melanophore aggregations; melanophores concentrated at caudal fin base and diffusing towards caudal fin margin becoming completely absent from caudal fin margin; or (3) absent: completely lacking melanophores (Figure 2; Richards et al. 1982). Pigmentation of the caudal ridge, the posterior-most trunk myomeres, was scored as (1) "fade" (i.e., anterior to posterior transition from dark [melanophores present] to light [melanophores absent]); or (2) dark, uniformly pigmented with melanophores (Figure 2; Richards et al. 1982). Pigmentation of the ventral region anterior to the cloaca and posterior to the last branchial pore was scored as (1) absent: completely lacking melanistic pigmentation; or (2) dark: ventral surface uniformly pigmented with melanophores or with localized melanophore aggregations separated by areas lacking melanistic pigmentation (Figure 3). Trunk myomeres were counted according to Hubbs and Trautman (1937) and Vladykov (1955; i.e., myomeres were counted between the last branchial pore and the anterior tip of the cloacal slit). Character states of all characters were developed from specimens preserved in 95% ethanol. To evaluate the ability of each individual character to differentiate *Entosphenus* from *Lampetra*, we generated frequency distributions for each pigmentation character and for myomeres. Two sample *t*-tests were conducted

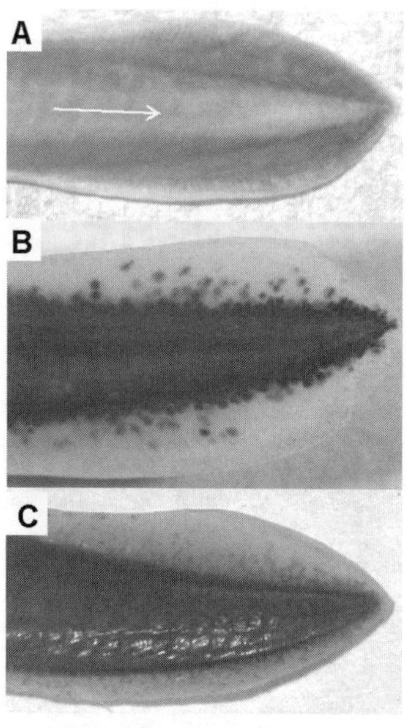


Figure 2. Caudal fin and caudal ridge pigmentation in *Entosphenus* and *Lampetra* ammocoetes: (A) caudal fin dark and caudal ridge fade as in *Entosphenus*; and (B and C) caudal fin peppered and caudal ridge dark as in *Lampetra*. The caudal ridge is identified by the white arrow.

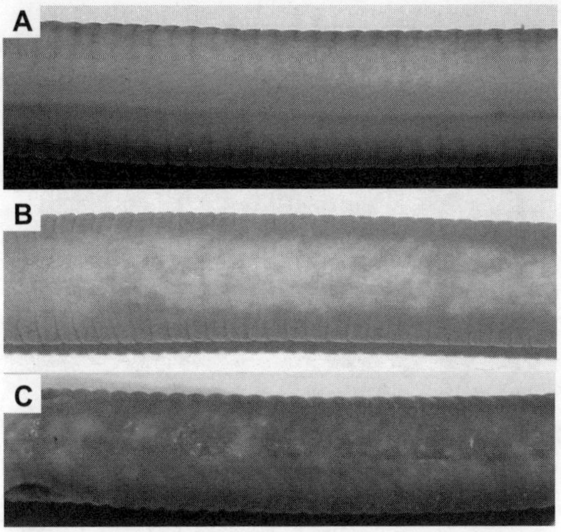


Figure 3. Ventral pigmentation in *Entosphenus* and *Lampetra* ammocoetes: (A) absent as in *Entosphenus*; and (B and C) dark as in *Lampetra*. Ventral surface uniformly pigmented with melanophores (B) or with localized melanophore aggregations separated by areas lacking melanistic pigmentation (C).

to evaluate myomere count differences among genera and populations at $\alpha = 0.05$.

## Results

A total of 382 ammocoetes were examined and the *Hae*III genetic assay identified 198 as *Entosphenus*, 157 as *Lampetra*, and 27 (all from Mark West Creek in California) as a third, previously unknown RFLP type (consisting of two fragments, 92 and 341 bp in length). Direct sequencing showed this third RFLP type to also be *Lampetra* but with substitutions at the *Hae*III recognition sites. *Entosphenus* ammocoetes ranged from 60 to 133 mm total length (TL) with a mean of 96.43 mm, and *Lampetra* ranged from 70 to 169 mm with a mean of 107.62 mm.

Caudal fin and caudal ridge characters provided the highest levels of diagnostic confidence with the identity of 95.6% and 98.7% individuals, corresponding to the genetic identification (Table 3). Ventral pigmentation had a lower level of diagnostic confidence with 77.0% proper identification. When applied together, the three pigment characters provided 99.7% agreement with genetic identification. *Entosphenus* ammocoetes typically had dark caudal fins, with the caudal ridge fading anteriorly to posteriorly and no ventral pigmentation, whereas *Lampetra* ammocoetes typically lacked or had peppered caudal fin pigmentation, a solid dark caudal ridge, and invasion of pigment on the ventral surface (Figures 2 and 3). Misclassifications by each character did not correlate to longitude or other geographic variables. The Navarro River sample was the only individual that was not properly identified when using a combination of the three pigment characters.

Myomoere counts were significantly different between genera ($P < 0.001$). *Entosphenus* myomere counts were typically higher and ranged from 58 to 72 with a mean of 67 ± 3 (±SD); *Lampetra* myomere number ranged from 50 to 71 with mean of 60 ± 4 (Table 2). *Entosphenus* myomere number exhibited a bimodal distribution, whereas *Lampetra* displayed a more complex multimodal distribution (Figure 4). Within *Entosphenus*, the Hat Creek collection myomere counts were significantly lower than other samples ($P = 0.001$) and ranged from 58 to 64, mean 61 ± 2, compared to 62–72, with a mean of 68 ± 2. *Lampetra* collections had higher levels of variability among mean myomere counts and

Table 3. Frequency distribution of caudal fin, caudal ridge and ventral pigmentation in *Entosphenus* and *Lampetra* ammocoetes. See text for character and character state definitions.

| Character | Character state | *Entosphenus* | *Lampetra* |
|---|---|---|---|
| Caudal fin | Dark | 190 | 9 |
| | Peppered | 3 | 159 |
| | Absent | 5 | 16 |
| Caudal ridge | Fade | 198 | 5 |

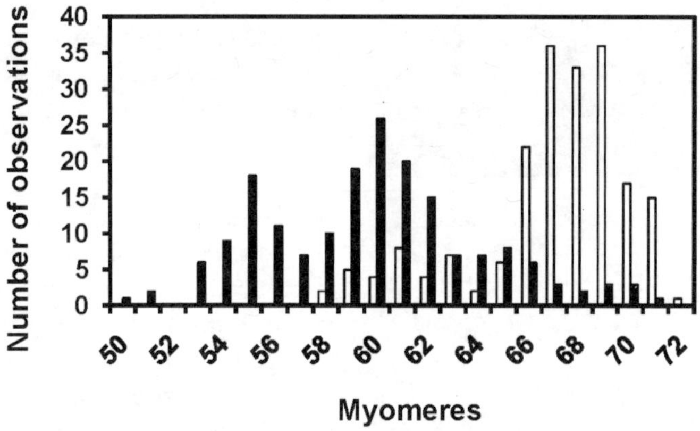


Figure 4. Frequency distribution of myomeres in *Entosphenus* and *Lampetra* ammocoetes. Open bars are *Entosphenus* and filled bars are *Lampetra*.

ranged from 54 to 69. Lower myomere counts occurred in the *Lampetra* individuals in southern collections, including Mark West Creek (55 ± 1), Mill Creek (54 ± 1), and the Merced River (54 ± 2). The highest mean *Lampetra* count was in the Feather River (66–71, 69 ± 2), also one of the southernmost collections. Myomere count ranges overlapped between the genera in two of nine collections that contained both genera: Navarro and Feather rivers. In the Navarro River sample, a single *Lampetra* and a single *Entosphenus* individual shared a myomere count of 62. In the Feather River collection, myomere count ranges were identical for the two genera.

## Discussion

*Entosphenus* and *Lampetra* are sufficiently different in mitochondrial DNA sequences (Docker et al. 1999) that diagnostic restriction fragment length polymorphism (RFLP) assays could be developed to quickly distinguish between them. Interrogating a 433 bp fragment of the cyt b gene with the restriction endonuclease *Hae*III produced a common *Lampetra* pattern (consisting of 167 and 266 bp fragments), a second *Lampetra* pattern (92 and 341 bp) found to date only in ammocoetes from Mark West Creek, California, and a single *Entosphenus* pattern (consisting of 92, 167, and 174 bp fragments).

Using this genetic assay to assign ammocoetes to genus, we found that Entosphenus and Lampetra ammocoetes from 15 localities a sin\ in California, Oregon, and Washington exhibited nearly unambiguous genus assignments using a combination of three pigment characters. Pigmentation of the caudal fin and caudal ridge have been used previously in taxonomic assessments of Pacific coast ammocoetes in southwestern British Columbia, Canada (Richards et al. 1982), and this study confirmed their applicability to a wider geographic extent. Ventral pigment had a lower level of diagnostic confidence but differed among genera. All pigment characters are recognizable in field scenarios within the size range evaluated in this study (60.0–169.0 mm TL), providing the opportunity for valid field identifications within this size range. It is likely that these characters also apply to smaller individuals, although our data are not sufficient to determine this.

Myomere counts were significantly different among the two genera, making it a useful character for genus identification, but count overlap did occur. At collection sites with both genera, myomere ranges did not overlap in 9 of 11 sites, but at one site, myomere counts had the same range. These data indicate that myomere counts are a useful character but do not consistently identify individual ammocoetes to genus. Variability in myomere counts may be related to species level taxonomic differences. The bimodal distribution of *Entosphenus* ammocoetes can be attributed to the low counts in the Hat Creek collection. Pit-Klamath brook lamprey, a species restricted to the upper Klamath and Pit drainages, is the only species of the genus described from Hat Creek. Therefore, the lower myomere number in the Hat Creek collection is presumably related to systematic differences. Removing the Hat Creek sample improves the utility of myomere counts to distinguish between the genera.

Species level diagnosis is possible in some cases when morphological characters are combined with distributional information. For example, Pacific lamprey is the only species of the genus *Entosphenus* that is known to occur outside of Vancouver Island and the Sacramento and Klamath drainages (Beamish 1982; Bond and Kan 1973; Hubbs 1971; Lorion et al. 2000; Vladykov and Kott 1979). Therefore, genus identification presumably provides species diagnosis of Pacific lamprey for a large portion of the study area. Inferring species identification based on geographic distribution may be appropriate in some instances; it is also potentially misleading if range expansions or contractions occur. Although species specific morphological or genetic characters have not yet been identified for some species within these genera (particularly paired species, see Docker 2009, this volume), this study illustrates that the proposed morphological characters allow for accurate assignment of ammocoetes to genus.

## Acknowledgments

For help with genetic data collection we thank Shannon Donohue. Collecting permits were obtained from California Department of Fish and Game, Oregon Department of Fish and Wildlife, and Washington Department of Fish and Wildlife.

## References

Beamish, R. J. 1980. Adult biology of the river lamprey (*Lampetra ayresi*) and the Pacific lamprey (*Lampetra tridentata*) from the Pacific coast of Canada. Canadian Journal of Fisheries and Aquatic Sciences 37:1906–1923.

Beamish, R. J. 1982. *Lampetra macrostoma*, a new species of freshwater parasitic lamprey from the west coast of Canada. Canadian Journal of Fisheries and Aquatic Sciences: 93:736–747.

Bond, C. E. 1994. Keys to Oregon freshwater fishes. Department of Fish and Wildlife, Oregon State University, Corvallis.

Bond, C. E., and T. T. Kan. 1973. *Lampetra* (*Entosphenus*) *minima* n. sp., a dwarfed parasitic lamprey from Oregon. Copeia 3:568–573.

Close, D. A., M. S. Fitzpatrick, and H. W. Li. 2002. The ecological and cultural importance of a species at risk of extinction, Pacific lamprey. Fisheries 27:19–25.

Docker, M. F. 2009. A review of the evolution of nonparasitism in lampreys and an update of the paired species concept. Pages 71–114 *in* L. R. Brown, S. D. Chase, M. G. Mesa, R. J. Beamish, and P. B. Moyle, editors. Biology, management, and conservation of lampreys in North America. American Fisheries Society, Symposium 72, Bethesda, Maryland.

Docker, M. F., J. H. Youson, R. J. Beamish., and R. H. Devlin. 1999. Phylogeny of the lamprey genus *Lampetra* inferred from mitochondrial cytochrome b and ND3 gene sequences. Canadian Journal of Fisheries and Aquatic Sciences 56:2340–2349.

Gill, H. S., C .B. Renaud, F. Chapleau, R. L. Mayden, and I. C. Potter. 2003. Phylogeny of living parasitic lampreys (Petromyzontiformes) based on morphological data. Copeia 4:687–703.

Hubbs, C. L. 1971. *Lampetra* (*Entosphenus*) *lethophaga*, a new species, the nonparasitic derivative of the Pacific lamprey. Transactions of San Diego Society of Natural History 416:125–164.

Hubbs, C. L., and M. B. Trautman. 1937. A revision of the lamprey genus *Ichthyomyzon*. Miscellaneous Publications Museum of Zoology University of Michigan 35.

Jenkins, R. E. and N. M. Burkhead. 1994. Freshwater fishes of Virginia. American Fisheries Society, Bethesda, Maryland.

Lorion, C. M., D. F. Markle, S. B. Reid, and M. F. Docker. 2000. Redescription of the presumed-extinct Miller Lake Lamprey, *Lampetra minima*. Copeia 4:1019–1028.

Miller, L. M., and A. R. Kapuscinski. 1993. Genetic variation in northern pike. Transactions of the American Fisheries Society 125:971–977.

Moyle, P. B. 2002. Inland fishes of California, revised and expanded. University of California Press, Berkley.

Nelson, J. S. 2006. Fishes of the world, 4th edition. Wiley, Hoboken, New Jersey.

Palumbi, S. R. 1996. Nucleic acids II: the polymerase chain reaction. Pages 205–247 *in* D. M. Hillis, C. Moritz, and B. K Mable, editor. Molecular systematics, 2nd edition. Sinauer Associates, Sunderland, Massachusetts.

Renaud, C. B. 1997. Conservation status of Northern Hemisphere lampreys (Petromyzontidae). Journal of Applied Ichthyology 13:143–148.

Richards, J. E., R. J. Beamish, and F. W. H. Beamish. 1982. Descriptions and keys for ammocoetes of lampreys from British Columbia, Canada. Canadian Journal of Fisheries and Aquatic Sciences 39:1484–1495.

Rohde F. C. 1980. *Lampetra tridentata* (Gairdner), Pacific lamprey. Page 34 *in* D. S. Lee, C. R. Gilbert, C. H. Hocutt, R. E. Jenkins, D. E. McAllister, and J. R. Stauffer, Jr., editors. Atlas of North American freshwater fishes. North Carolina State Museum of Natural History, Raleigh.

USFWS (U.S. Fish and Wildlife Service). 2004. Endangered and threatened wildlife and plants; 90-day finding on a petition to list three species of lampreys as threatened or endangered. Federal Register 69:247(27 December 2004):77158–77167.

Vladykov, V. D. 1955. *Lampetra zanandreai*, a new species of lamprey from northern Italy. Copeia 1955:215–223.

Vladykov, V. D., and W. I. Follett. 1958. Redescription of *Lampetra ayresii* (Gunther) of western North America, a species of lamprey (Petromyzonidae) distinct from *Lampetra fluviatilis* (Linnaeus) of Europe. Journal of the Fisheries Research Board of Canada 15:47–77.

Vladykov, V. D. and W. I. Follett. 1965. *Lampetra richardsoni*, a new nonparasitic species of lamprey (Petromyzonidae) from western North America. Journal of the Fisheries Research Board of Canada 22:139–158.

Vladykov, V. D. and E. Kott. 1976. A new nonparasitic species of lamprey of the genus *Entosphenus* Gill, 1862 (Petromyzonidae) from south central California. Bulletin of the Southern California Academy of Sciences 75:60–67.

Vladykov, V. D., and E. Kott. 1979. A new species of the holarctic lamprey genus *Entosphenus* Gill. 1862 (Petromyzonidae) from the Klamath River, in California and Oregon. Canadian Journal of Zoology 57:808–823.

Vladykov, V. D., and E. Kott. 1980. Description and key to metamorphosed specimens and ammocoetes of Petromyzonidae found in the Great Lakes region. Canadian Journal of Fisheries and Aquatic Sciences 37:1616–1625.

Wydoski, R. S. and R. R. Whitney. 2003. Inland Fishes of Washington, 2nd edition. University of Washington Press, Seattle.

American Fisheries Society Symposium 72:233–253, 2009


# Lessons from the Reintroduction of a Noncharismatic, Migratory Fish: Pacific Lamprey in the Upper Umatilla River, Oregon

DAVID A. CLOSE*,[1]
*Tribal Fisheries Program, Department of Natural Resources, Confederated Tribes of the Umatilla Indian Reservation Post Office Box 638, Pendleton, Oregon 97801, USA*

KENNETH P. CURRENS
*Northwest Indian Fisheries Commission 6730 Martin Way East, Olympia, Washington 98516, USA*

AARON JACKSON
*Tribal Fisheries Program, Department of Natural Resources Confederated Tribes of the Umatilla Indian Reservation Post Office Box 638, Pendleton, Oregon 97801, USA*

ANDREW J. WILDBILL
*Tribal Fisheries Program, Department of Natural Resources Confederated Tribes of the Umatilla Indian Reservation Post Office Box 638, Pendleton, Oregon 97801, USA*

JOSH HANSEN
*Oregon Department of Fish and Wildlife, Northeastern Region Pendleton, Oregon 97801, USA*

PRESTON BRONSON
*Tribal Fisheries Program, Department of Natural Resources Confederated Tribes of the Umatilla Indian Reservation Post Office Box 638, Pendleton, Oregon 97801, USA*

KIMMO ARONSUU
*North Ostrobothnia Regional Environment Centre, PL124, 90101, Oulu, Finland*

*Abstract.*—Between 1999 and 2007, more than 2,600 adult Pacific lampreys *Entosphenus tridentatus* (formerly *Lampetra tridentata*) were reintroduced to the Umatilla River, where they had been extirpated by poisoning, from nearby locations in the Columbia River consistent with the International Union for Conservation of Nature and Natural Resources guidelines. Reintroduced adult Pacific lampreys were able

* Corresponding author: d.close@fisheries.ubc.ca
[1] Present address: Fisheries Centre and Department of Zoology, The University of British Columbia, Vancouver, British Columbia, V67 1Z4, Canada

to find suitable spawning habitat, construct nests, and deposit viable eggs (81–93% mean egg viability per nest). Their larvae were able to feed and grow. Median lengths for age 0+, 1+, and 2+ larvae were 19, 63, and 109 mm, respectively. Mean density of larvae in survey plots increased over time from 0.08 to 6.56 larvae/$m^2$. Geographical distribution of larvae in the river increased downstream, but larvae failed to become established in the lower Umatilla River where water flows were regulated for irrigation. Annual abundances of trapped, recently metamorphosed, out-migrating larvae increased during the study from nearly zero to 180,000, but not in all years, which suggests that many might not be surviving migration to the Columbia River, possibly because of irrigation withdrawals. Abundances of trapped, returning adult lamprey also increased from 2003 to 2006, which corresponded with the period when adult lampreys that were the progeny of reintroduced lampreys were expected to return, but long-term monitoring is necessary to confirm that increases were the result of the reintroduction. Our results also demonstrated that even if presumptive causes of extirpation were known and removed before reintroduction, success is not guaranteed. Reintroduction not only assists in redistributing animals to parts of their historical range, but in conjunction with monitoring, it may be essential to identify additional limiting factors that were unknown at reintroduction.

## Introduction

Reintroduction—the attempt to establish a species in a part of its historical range from which it has been extirpated or become extinct (IUCN 1998)—is becoming an increasingly common conservation strategy (Wolf et al. 1996; Fischer and Lindenmayer 2000; Seddon et al. 2007). Reintroductions can increase natural biodiversity, provide keystone components of ecosystems, and create public and political support necessary to restore habitat and implement protection measures (Seddon 1999). Most reintroduction efforts have been focused on mammals, especially charismatic species of *Carnivora* and *Artiodactyla* and birds, especially raptors (Falconiformes) and game birds (Anseriformes, Gruiformes, and Galliformes), although species in these groups are not necessarily more likely to be threatened. In comparison, relatively few reintroductions have occurred for fish (Seddon et al. 2005). Reintroductions of anadromous species that are vulnerable to many different threats during migration and in the ocean or migratory fish species that are not strongly philopatric to localities are especially challenging. In this paper, we report on the initial success and future challenges of reintroducing a noncharismatic anadromous fish with a potentially weak philopatric life history—Pacific lamprey *Entosphenus tridentatus* (formerly *Lampetra tridentata*)—to a tributary of the Columbia River, located in the Pacific Northwest of the United States.

Of the 34 lamprey (Petromyzontidae) species, approximately half are endangered, vulnerable, or extinct at least in parts of their range in the Northern Hemisphere (Renaud 1997). Renaud (1997) and Masters et al. (2006) reported the likely causes were pollution, regulation and diversion of streams, and overharvest. For example, Myllynen et al. (1997) found that high levels of iron and low pH adversely impacted lamprey roe and recently hatched larvae in western Finland. Industrial pollution was identified as a major factor limiting lamprey populations throughout Europe (Kelly and King 2001). Studies in Europe showed that dams constructed for hydroelectricity, flow diversions, and weirs cut off spawning habitat and rearing areas for larvae (Eklund et al. 1983; Kirchhofer 1995; Ojutkangas et al. 1995; Laine et al. 1998; Kelly and King 2001; Almeida and Quintella 2002; Masters et al. 2006; Andrade et al. 2007). In addition, excessive commercial and illegal harvests likely depleted lamprey populations. In some European rivers, river lampreys *Lampetra flu-*

*viatilis* continued to be commercially harvested even where some populations experienced serious declines (Eklund et al. 1983; Ojutkangas et al. 1995; Kelly and King 2001; Masters et al. 2006). In some areas of Portugal, the high value of sea lamprey *Petromyzon marinus* as food (often selling for 45 euros per lamprey) also led to poaching (Andrade et al. 2007).

In the Columbia River basin, Pacific lampreys also experienced sharp declines and in some areas they have been extirpated. In the Willamette River, a tributary to the lower Columbia River, for example, commercial Pacific lamprey harvest peaked in 1946 with approximately 500,000 lampreys (Kostow 2002), a number thought to be approximately 20% of the adult abundance (Mattson 1949). Commercial harvest ended in 2001 when only 12,276 lampreys were harvested and the Oregon Department of Fish and Wildlife became concerned about the status of the fish. At Ice Harbor Dam nearly 720 km upstream on the Snake River, Pacific lamprey abundance decreased exponentially from the 1970s (Close et al. 2002), with a similar trend in lamprey counts at Rocky Reach Dam in the mid-Columbia River. Likewise, in the upper Columbia River basin, surveys of Native Americans, who have extensive traditional ecological knowledge of Pacific lampreys associated with harvest for food, pointed to large declines in lamprey abundance from historical levels (Close et al. 2004).

A growing body of studies suggests that hydroelectric dams, diversions, and low-head irrigations dams may be a principle cause of sharp declines in abundance of lampreys into the interior Columbia and Snake River basins. Large hydroelectric dams were constructed on the main-stem Columbia and Snake rivers from the 1930s through the 1970s. These caused severe declines in anadromous salmonids (Raymond 1979; Raymond 1988) and also likely affected the migration, distribution, and abundance of Pacific lamprey. Recent evidence showed that fewer than 50% of adult Pacific lampreys encountering the lower Columbia River dams were able to pass above the dams using the fishways and they experienced longer delays in migration than those experienced by salmon (Moser et al. 2002a; 2002b).

Low-head irrigation dams may have had similar impacts on migrating adults in the tributaries as the major hydroelectric dams do in the main stem of the Columbia River. Recent radiotelemetry studies showed that adult lamprey passage efficiency was less than 50% at these dams on the Umatilla River (M. Moser, National Marine Fisheries Service, personal communication). In the Columbia River basin, the construction of irrigation diversion dams in the tributary rivers began in the early 1900s and hundreds of these dams now occur throughout the Columbia River basin. Cumulative effects of these may have been significant. Larval recruitment of Pacific lamprey in the tributaries of the Columbia and Snake rivers where these dams occurred, for example, appeared reduced from historical levels (Moser and Close 2003).

Ignorance of the ecological role of lampreys and conflicting social values associated with different species also contributed to the decline. Pacific lampreys are valued by Native American tribes (Close et al. 2002), but until recently, state fish and wildlife agencies treated lampreys as threats to commercial and recreational salmonid fisheries (e.g., Bond and Kan 1973; Farlinger and Beamish 1984). Within the Columbia River basin, for example, the Umatilla River was once a preferred harvest location for Pacific lamprey and salmonids by many Native American tribal members (Close et al. 2002). Lampreys were extirpated from the Umatilla River, however, by poisoning with rotenone treatments in 1967 and 1974 (Close et al. 2004). When restoration of salmon and steelhead populations, which were also declining precipitously, began in the 1980s (Phillips et al. 2000), restoration of lampreys in the Umatilla River and other basins was left to natural recolonization by the rapidly declining numbers of fish that still migrated through the main stem Columbia River. Although occasional pulses of straying lampreys entered the lower Umatilla River in some years, by the 1990s, lampreys still had not recolonized their historical habitat in the basin at abundances that persisted over time.

In the late 1990s, the Confederated Tribes of the Umatilla Indian Reservation initiated research to reintroduce Pacific lampreys to the Umatilla River with the ultimate goal of re-establishing natural production that would provide sustainable and harvestable abundances of adults. The International Union for the Conservation of Nature and Natural Resources (IUCN), whose guidelines we followed, recommends that the causes for population declines must be well understood and removed and that the habitat be restored before species are reintroduced (IUCN 1998). The results we report here, however, suggested that reintroduction and monitoring for species whose life history is not well known may be a necessary intermediate step in identifying key limiting factors to reintroduction that were previously unknown rather than the culminating step. Successful reintroductions also generally occur in three stages: (1) demonstrating that released individuals survive, (2) breeding by the released generation and their offspring, and (3) persistence of the re-established population (Seddon 1999). Judging the ultimate success of the Pacific lamprey reintroductions to re-establish persistent local spawning aggregations will take several decades or more. Assessing the success of a reintroduced cohort takes many years because of the 6–9-year life cycle of the fish. In addition, short-term changes in abundances of adult Pacific lampreys returning to a tributary are potentially misleading because year-to-year ocean survival can be highly variable and that variability may be compounded by potentially weak philopatry (Goodman et al. 2008; Lin et al. 2008a, 2008b). Consequently, here we report results that specifically tested hypotheses for the first two stages of successful reintroduction. These are that reintroduced adult lampreys can (1) build nests in the Umatilla River, (2) produce viable eggs in nests, (3) produce larvae that grow, (4) increase larval abundance, and (5) increase recently metamorphosed lampreys.

# Methods

## *Study area*

The Umatilla River basin is located in northeastern Oregon, in the middle Columbia basin, and drains an area of more than 6,000 km$^2$ (Figure 1). The 185-km river originates in the conifer forests of the Blue Mountains, east of Pendleton, and flows northwest across the semi-arid shrub steppe of the Deschutes-Umatilla Plateau until it enters the Columbia River at approximately river kilometer (rkm) 465. The Umatilla River lies upstream of three of the 19 large Columbia River hydroelectric dams (Bonneville Dam, The Dalles Dam, and John Day Dam). The Umatilla River has nine major tributaries that may have held lampreys: the north and south forks of the Umatilla River, and Meacham, Iskuulpa, Birch, Butter, McKay, Tutuilla and Wildhorse creeks.

Agricultural and rangelands comprise more than 80% of land use in the basin. Forestry (15%) and urban development (3%) comprise most of the remaining land uses. The annual precipitation in the Umatilla basin varies from 22 to 38 cm near the mouth of the Umatilla River and 101–139 cm in the Blue Mountains. Consequently, several irrigation projects (U.S. Department of Interior, Bureau of Reclamation, Washington, D.C.) were completed in the early part of the 1900s to provide water to more arid areas in the west part of Umatilla County. These include two major water-storage reservoirs, McKay Reservoir located on McKay Creek, and Cold Springs Reservoir located about 10 km northeast of Hermiston, Oregon. The projects divert water to the reservoirs from the Umatilla River by the Feed Canal Diversion Dam and Canal. In addition, six large irrigation diversions are located in the lower basin: Three Miles Fall Dam, Boyd Hydropower, Maxwell Irrigation Dam, Dillon Irrigation Diversion Dam, Westland Irrigation Diversion Dam, and Stanfield Irrigation Diversion Dam (Figure 1). Stream flow in the Umatilla River comes mainly from snowmelt in the Blue Mountains, with the largest mean monthly flows in March and April. Lowest flows occur in July, August, and Septem-

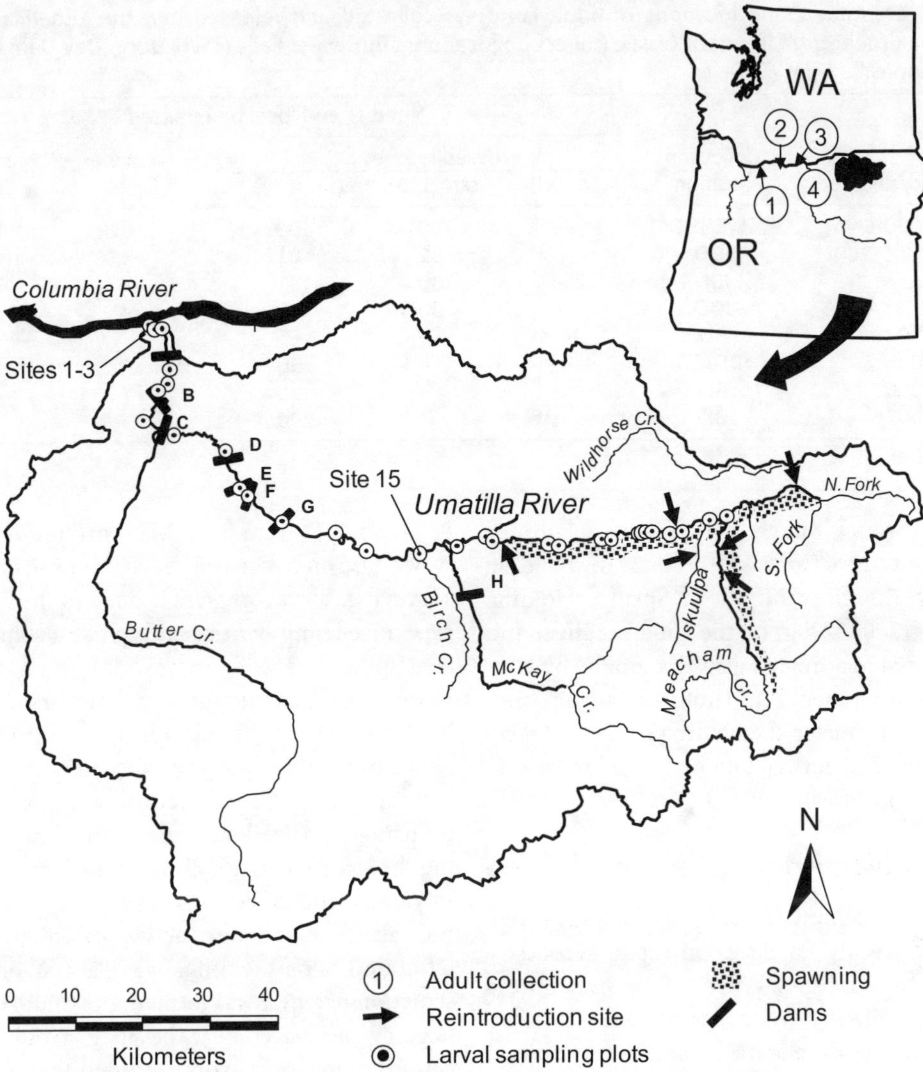


Figure 1. Umatilla River basin with locations where adult migrating lamprey were collected, sites of reintroductions, range of observed spawning, index plots for larval sampling, and locations of dams. Adult collection locations: 1 = Bonneville Dam; 2 = The Dalles Dam; 3 = John Day Dam; 4 = Tumwater Falls, Umatilla River. Dams in the Umatilla River: A = Three Mile Falls; B = Boyd Hydro; C = Maxwell Diversion; D = Dillion; E = Westland Diversion; F = Feed Canal; G = Stanfield Dam; H = McKay; I = Cold Springs Dam.

ber. During summer months, up to 100% of the discharge from the Umatilla River may flow into these diversions.

## *Animal collection and holding*

Pacific lampreys were collected by hand and short-handled nets from Tumwater Falls on the John Day River, the John Day Dam, The Dalles Dam and Bonneville Dam fishways (Figure 1; Table 1). Lampreys were transported to the U.S. Geological Survey Columbia River Research Laboratory (2000 and 2003) or South Fork Walla Walla Acclimation Facility (2002–2006) and maintained in tanks supplied by river water. River water temperatures ranged from 6°C

Table 1. Numbers and locations of adult lampreys collected and released into the Umatilla River, Iskuulpa Creek, and Meacham Creek. Collection locations: Tumwater Falls (TWF), John Day Dam (JDD), and Bonneville Dam (BD).

| Years collected/released | Collection location | Number and location released: Umatilla River (km 98) | Umatilla River (km 118) | Umatilla River (km 140) | Iskuulpa Creek | Meacham Creek |
|---|---|---|---|---|---|---|
| 1999/2000 | TWF, JDD | – | 150 | 300 | – | 150 |
| 2000–2001/2001 | JDD | – | 82 | 81 | – | 81 |
| 2002 | JDD | 150 | 100 | 141 | – | 100 |
| 2002–2003/2003 | JDD | – | 90 | 110 | 54 | 230 |
| 2003/2004 | BD | – | – | 63 | – | 70 |
| 2004/2005 | JDD, TD | – | – | 50 | 15 | 55 |
| 2005/2006 | BD | – | – | 90 | 21 | 87 |
| 2006/2007 | BD | | | 200 | 25 | 169 |

to 8°C through March. Before release, lamprey were moved to Three Mile Falls Dam Holding Facility (2002) or Minthorn Springs Holding Facility (2003–2006) on the Umatilla River and acclimated in a single raceway supplied by Umatilla River water. Temperatures ranged from 4°C to 15°C during the holding period until release. Survival during holding ranged from 96% in 2002 to 100% in 2003–2007.

### *Out-planting*

Beginning in April of each year, lampreys were checked weekly for development of secondary sexual characteristics. When 25% of the lampreys were spermiating or females became distended from the egg development, they were released into the upper Umatilla River at six locations: the main-stem Umatilla River at rkm 98.0, 118.0, and 140.0; Meacham Creek at rkm 10 and 17.5; and Iskuulpa Creek at rkm 5 (Figure 1). The number of adults released each year ranged between 120 and 600 lampreys (Table 1).

### *Nest surveys and egg viability*

Surveys were conducted from 2000 to 2002 to assess the distribution of Pacific lamprey nests in the Umatilla River in the following areas: (1) the main-stem Umatilla River from rkm 90–144, (2) the north fork Umatilla River from rkm 0–4.3, (3) the south fork Umatilla River from rkm 0–4, and (4) Meacham Creek from rkm 0–24. Surveys were from May through August. A two-person crew surveyed 4–5 rkm/d. Prior to surveying, surveyors were trained in the field to recognize lamprey nests and to use a standardized survey method. Surveyors walked downstream along the margins and traversed from bank to bank at the tail out of each pool and above each riffle. The ability to view spawning activity and nest construction was maximized by using polarized sunglasses and only walking the stream when visibility was clear enough to view depths of pools and riffles. Special care was taken not to disturb active spawning. Once a test nest (a nest without eggs), a nest, or construction of a nest was located, pink fluorescent flagging was placed near the site to avoid duplicate counting. Surveyors recorded the number of adults on or near nests, the date, location, and identity of the surveyors. Location was estimated using a hand held Global Positioning System (GPS) unit (Garmin GPS III plus) and mapped with Arcview (GIS version 3.2).

In 2001, 16 nests were sampled to assess egg viability. Approximately 10–20 eggs per nest were collected to determine stage of egg development. Eggs were gently dislodged with a weed picker and were captured in a plankton kick net placed below the nest. The probe sample was used to predict when stage 12 (head formation) was reached (Piavis 1961). Once eggs reached stage 12, a sample of 200 eggs was collected from each nest. Eggs were fixed in

10% formalin and stored in carosafe until the number of viable and unviable eggs could be counted using a dissecting microscope (Nikon). Eggs developing to stages 12–14 were used to calculate percentage of viable eggs. Eggs were classified as unviable if they were covered with fungus or were deformed.

## *Larval abundance and distribution*

From 1998 to 2006, 30 plots from the mouth of the river to the headwaters of the Umatilla River were sampled for larvae by electrofishing to monitor abundance and densities (Figure 2). Years 1998–2000 were prior to any larval production by reintroduced adult lampreys in the Umatilla River. All of the index plots were 7.5 m$^2$ and were located in soft substrate areas of the channel where larvae were typically most abundant. In 2001, one plot within the mainstem Umatilla River was relocated because the river channel shifted. Electrofishing surveys were conducted during August and September, beginning at the index site near the mouth of the river working upstream to site 30. To estimate the larval abundance in each plot, we used the two-pass method (Seber and Le Cren 1967) and the multiple-catch method (Carle and Strub 1978). Using a backpack model Abp-2 electrofishing unit (Engineering Technical Ser-

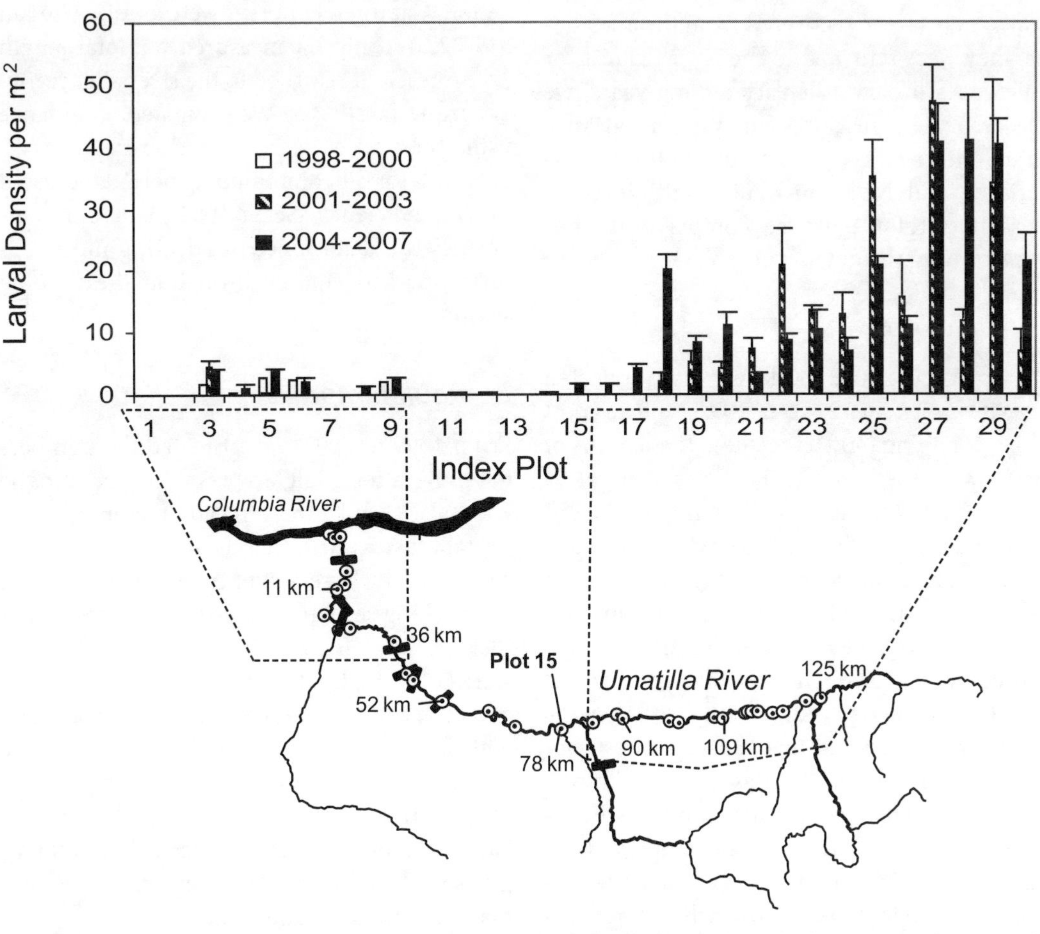


Figure 2. Increases in mean larval density and spatial distribution of Pacific lamprey in the Umatilla River before (1998–2000) and after (2001–2007) reintroduction. Bars show standard error for each period.

vices, University of Wisconsin, Madison), larvae were dislodged from the substrate in two 11.25-min passes. If the catch of the second pass was ≥70% of the first catch, a third pass was conducted. If no larvae were detected in the first pass, only one pass was conducted. The electrofishing unit delivered 3 pulses per second (125 V DC) at 25% duty cycle, with a 3:1 burst train (three pulses on, one pulse off) to remove larvae from the substrate (Weisser and Klar 1990). Once larvae emerged from the substrate, 30 pulses per second were applied to stun and capture larvae. Following collection, larvae were anesthetized in MS-222 solution (50 mg/L tricane methanesulfonate), identified to species by tail pigmentation (Richards et al. 1982), and measured for total length (±1 mm). After larvae recovered from the anesthesia, they were returned to the river. To test for differences in larval density among years, we used repeated measures analysis of variance with Bonferroni's correction for multiple comparisons and Newman-Keuls multiple comparison test for pairwise comparisons after transforming the data by $\ln(1 + x)$.

### Larval growth

During August 2002, 80 larvae were collected from the index sites 22 (rkm 109.1) and 30 (rkm 124.9) for aging (40 larvae/site). The lengths of larvae were measured (±1 mm) and weighed (±0.01 g). The larvae were frozen (–80°C) and stored until statoliths could be extracted. Statoliths, which are analogous to otoliths in teleost fishes, were extracted by removing otic capsules, cleaning them under a dissecting microscope, and cutting them open. Statoliths were transferred randomly to numerically coded wells filled with glycerin so that subsequent estimation of age would not be biased by the knowledge of length of the individual or by an age previously assigned. After 4 weeks, lampreys were aged by counting the number of bands with a dissecting scope (Volk 1986). Each lamprey was aged twice at different times. Larvae with unreadable statoliths (two individuals) were not included in the final data.

### Recently metamorphosed lampreys

We monitored migration of larval and metamorphosed lampreys from October through April using a 1.5-m-diameter rotary-screw trap with a revolving 3.5-mm-mesh cone mounted on aluminum pontoons located 1.9 km from the mouth of the river. Total river width at this location was approximately 75 m, and the river bottom was mainly bedrock with carved channels. A 1.5-m-deep channel on the west bank of the river served as the trap location. The trap was operated 24 h/d, and catch was enumerated twice per day by personnel from Oregon Department of Fish and Wildlife and the Confederated Tribes of the Umatilla Indian Reservation. Captured lampreys were anesthetized in MS-222 (50 mg/L), measured for total length (±1 mm), and returned to the river after recovery from anesthesia. We estimated abundance using the sum of Petersen mark–recapture calculations for different trapping periods adjusted for trap efficiency (Seber 1982), where trap efficiency was determined by releasing and recapturing marked lamprey throughout the trapping season.

### Adult lampreys

From 1999 to 2007, we estimated the numbers of upmigrating adult lampreys entering into the Umatilla River by trapping adult lampreys with portable assessment traps near the mouth of the river. Portable assessment traps are widely used for catching sea lampreys in Great Lakes region (Schuldt and Heinrich 1982). Trap dimensions were 60 cm high, 116 cm long, and 49 cm wide, and each had a 40-cm-deep mouth in both ends with an opening of 12 × 10 cm. Traps were placed just below water surface on both sides of the entrance to the Three Mile Falls Dam fish ladder (Figure 2) and were checked daily. Lampreys captured in the trap were measured and marked for release and recapture.

To examine the relationship among the number of adult lampreys entering the Umatilla River, the number of adults migrating nearby in the Columbia River that could potentially colonize the river, and streamflows in the Umatilla River, which might limit or alternatively attract lampreys, we used daily abundances of Pacific lamprey at the John Day Dam from June 15, 2005–September 15, 2005, where they were counted during two consecutive 8-h shifts from 0500 to 2100 hours (Moser and Close 2003) and daily mean discharge of the Umatilla River. Data were from the Columbia River Fish Passage Center (www.fpc.org).

## Results

### *Nest surveys and egg viability*

Reintroduced adult lampreys were able to locate suitable spawning locations and build nests within the main-stem Umatilla River and Meacham Creek. Spawning activity and nest construction began near the end of May and extended through July. The majority of spawning activity occurred during the first 2 weeks of June in the Umatilla River and Meacham Creek. During the spawning activity in 2001, thermographs recorded mean daily temperatures at 117.3 and 140.0 rkm in the main-stem Umatilla River ranging from 12.7°C to 16.5°C and 9.3°C to 12.9°C, respectively. A thermograph in Meacham Creek at rkm 21 recorded daily mean temperatures that varied between 10.1°C and 13.3°C.

During the 3 years of study, we observed nests from rkm 96.0–144.0 on the Umatilla main stem and rkm 4.3–27.0 in Meacham Creek (Figure 1). During the first year, 81 nests occurred in the Umatilla River and Meacham Creek. In the main-stem Umatilla River, lampreys built 51 nests. In Meacham Creek, 29 nests occurred between rkm 17.0 and 27.0 and one occurred near the mouth of north fork Meacham Creek. Nest distribution was different between the main-stem Umatilla and Meacham Creek, however. Nest distribution was more evenly distributed out along the profile of the main-stem Umatilla than in Meacham Creek, where a majority of the nests occurred less than 1 km from the release site.

Out-planted lampreys were successful in depositing fertilized eggs in the nests and producing viable eggs in the Umatilla River and Meacham Creek. In 2001, 49 viable nests occurred in the main-stem Umatilla River (19 nests) and Meacham Creek (30 nests). In 2002, 67 viable nests were located both in the Umatilla River (21 nests) and Meacham Creek (46). Overall, egg viability ranged from 57.8% to 100.0%, but in 7 of 16 nests, egg viability was greater than 99.0%. In the Umatilla River, mean percent egg viability per nest was 93.4 ± 3.6 ($n$ = 4), and in Meacham Creek, mean percent egg viability was 81.4 ± 5.1 ($n$ = 12). Seventy-five percent of the unviable eggs were covered by fungus, and 25.0% were deformed.

### *Larval abundance and distribution*

Larval abundance and spatial distribution increased dramatically in the Umatilla River following reintroduction (Figure 2). Prior to reintroduction, larvae occurred only rarely and at low densities in the lower 10 km of river where straying migratory adults from the Columbia River may enter and hold in the Umatilla River. Overall the index plots, mean larval density increased 70–80-fold from 0.08 ± 0.05 larvae/m$^2$ before reintroduction to 5.23 ± 1.73 larvae/m$^2$ and 6.56 ± 2.44 larvae/m$^2$ in 2001 and 2002 after reintroduction and remained elevated through 2007 ($F$ = 10.08, df = 9, 29; $P$ < 0.001). Newman-Keuls multiple comparison tests also showed highly significant differences ($P$ < 0.001) in larval densities between pre- and postreintroduction years but no significant differences among postreintroduction densities.

The geographic distribution of increased larval density corresponded to where adults were reintroduced. In 2001–2003, after reintroduction, larvae were found in increasing densities in index plots between rkm 90 and 125, where they had been absent (Figure 2). River kilometer 90 corresponds to the downstream most location of reintroduction and rkm 125 is close to where Meacham Creek, the upstream most location for reintroduction, enters the Umatilla River (Figure 1). Densities were greatest at site 27 (rkm 116), which also corresponded to a reintroduction location (Figures 1 and 2). Between 2004–2007, larval densities increased downstream in index plots 15–18 (rkm 78–90) but no farther (Figure 2). In the lower Umatilla River, lamprey larvae were observed in more locations, but densities did not increase much beyond prereintroduction levels (Figure 2).

### *Larval growth*

Size distributions and statolith analyses of larval lamprey in the upper Umatilla River showed evidence of different year-classes and growth. Larval length-frequency analysis showed a bimodal distribution in larval lengths (Figure 3), with peaks that corresponded with age 1+ and 2+ larvae based on the statolith analysis. In over a year, lampreys grew approximately 40–50 mm. Median length for age 1+ larvae was 63.0 mm (range: 38.0–101.0) and for age 2+ larvae was 109.0 mm (range: 73.0–141.0). Only a single age 0+ larva was aged, and it was 19.0 mm long.

### *Recently metamorphosed lampreys*

The number of migrating metamorphosed lampreys about to enter the Columbia River increased during the 2000–2001 out-migration

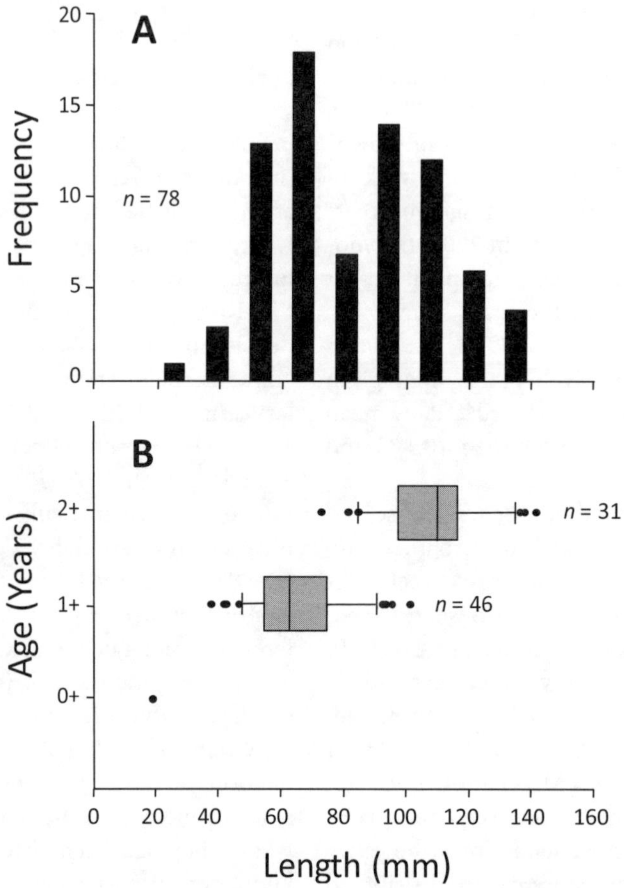


Figure 3. Length-frequency analysis (A) showing bimodal size distribution of larval lamprey and median length at age (B) showing evidence of growth. The box represents 50% of the observations; the vertical line in each box plot represents the median of all the data. Bars outside of the box show the 10th and 90th percentiles of the observations.

(Figure 4), returned to low levels during 2002–2005, and sharply increased again in 2006. Comparison of the lengths of recently metamorphosed lampreys and larvae captured and other age and length data suggested that the rotary screw trap likely selected for larger sizes of lamprey, which makes additional interpretation difficult. Most of the recently metamorphosed lampreys and larvae captured by the trap were of similar size (Figure 5), although larvae might be expected to be smaller (Figure 3). During 2005–2006, the median length of recently metamorphosed lamprey was 143.0 mm (range, 113–180 mm) and the median length of larval lamprey was 145.0 mm (range, 52–182 mm). Regardless, the 2000–2001 peak must have been due to natural production that occurred before the out-planting study began, presumably by strays in the lower river, where we documented the only presence of prereintroduction larvae (Figure 2).

### *Adult lampreys*

Numbers of spawning phase adult lampreys entering the Umatilla River increased from prereintroduction levels in 2003/2004 through the spring of 2007 (Figure 4). All these fish were caught in late winter and early spring, when streamflows in the Umatilla River were highest. Many of the fish showed secondary sexual characteristics, suggesting that they had spent the winter in the main-stem Columbia River. In contrast, we captured no adult lampreys during the peak of the nearby spawning migration at the John Day Dam in July and August, which corresponded to the period when streamflows in the lower Umatilla River were reduced to nearly zero during the summer (Figure 6).

## Discussion

This study showed that reintroduction could be a viable tool for re-establishing locally extirpated aggregations of Pacific lamprey if the life stage specific factors limiting natural production can be reduced or eliminated. Successful reintroductions generally occur in three stages: (1) demonstrating that released individuals survive, (2) breeding by the released generation and their offspring, and (3) persistence of the re-established population (Seddon 1999). To date, we have demonstrated success in the first two stages. In the Umatilla River basin, introduced lampreys were able find suitable spawn-

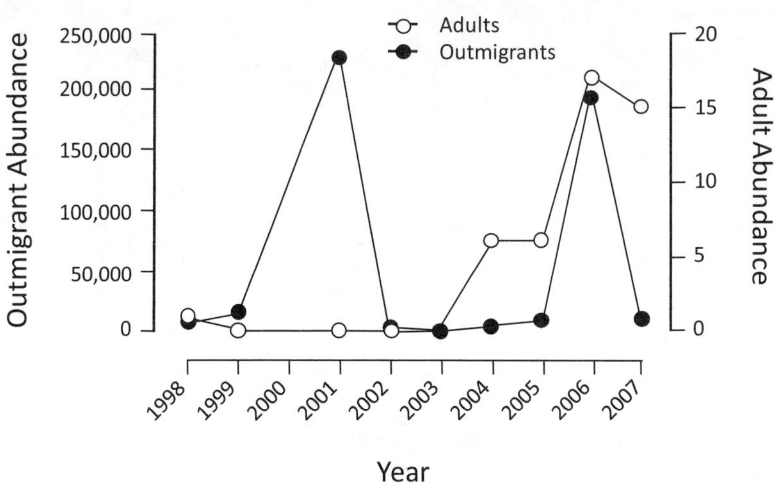


Figure 4. Trends in abundances of out-migrating young lamprey caught by a rotary screw trap near the mouth (1.9 km) of the Umatilla River and the abundances of returning spawning phase adult lamprey entering the Umatilla River during the spring from 1998 to 2007.

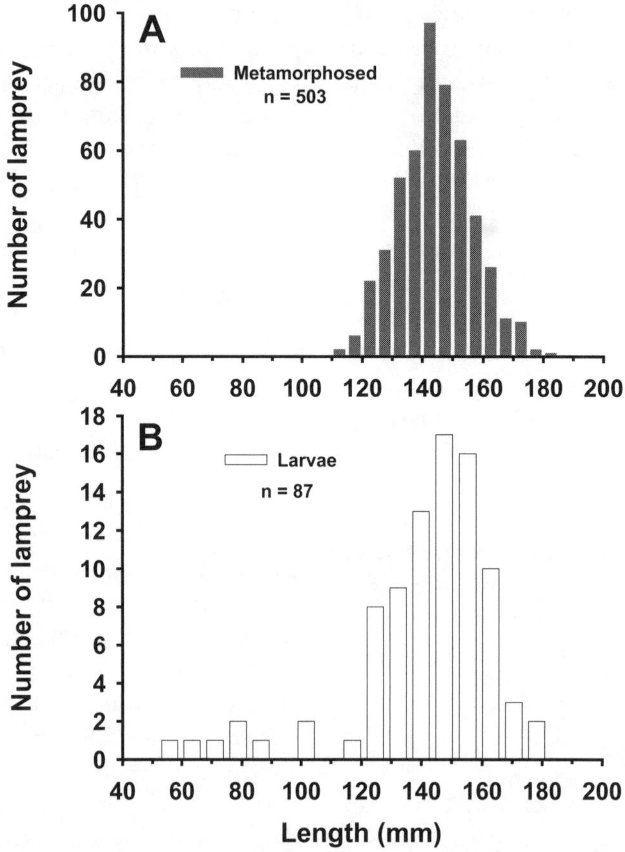


Figure 5. Length–frequency distribution of metamorphosed (A) and larval lamprey (B) captured in a rotary screw trap near the mouth of the Umatilla River in 2005–2006.

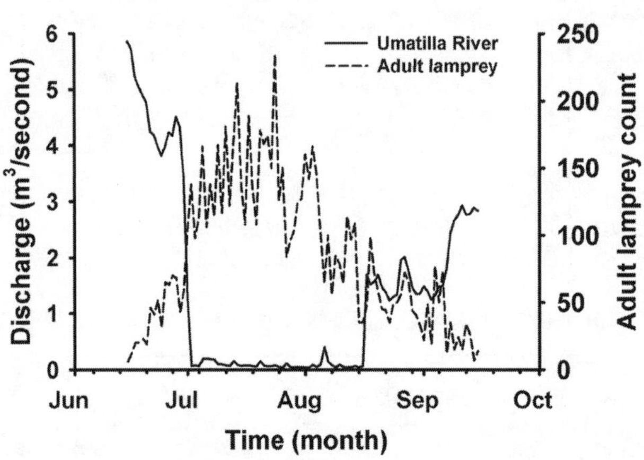


Figure 6. Changes in abundance of lamprey at John Day Dam and the discharge near the mouth of the Umatilla River during the peak of the spawning migration in 2005. The peak migration occurred in July at the John Day Dam, when discharge was lowest in the Umatilla River; nearly all flows are diverted for irrigation.

ing habitat, construct nests, and deposit viable eggs; their larvae were able to feed, grow, and migrate downstream; and the geographical distribution and abundance of larvae expanded in the river. Adults subsequently returned to the river, but it remains unclear if these are the results of the reintroductions.

Transplanted Pacific lampreys were capable of finding suitable habitat for building nests and depositing viable eggs that were favorable for embryological development. We had been concerned that prespawning mortality of released adult lampreys—which we did not observe but which is always a concern with fish potentially stressed because of handling (Wedemeyer et al. 1990)—could contribute to reduced nest building, but this apparently did not occur. The total number of nests was 15.7% (2001 and 2002) of the total number of transplanted lampreys. Life history theory predicts that under the communal behavior and superimposed spawning exhibited by the Pacific lampreys (Pletcher 1963; Kan 1975), nests numbers might be less than that predicted by single pair matings. Our findings were consistent with other studies of natural populations, which suggested that translocation did not significantly affect general spawning behavior. Farlinger and Beamish (1984), for example, found that the total nest abundance for a natural population of Pacific lamprey in Eel Creek, British Columbia was approximately 20% of the total number of spawning adults.

We did observe differences in the distribution of nests relative to release locations between the main-stem Umatilla River and Meacham Creek. One possible explanation may be differences in the distribution of suitable habitat between the two parts of the river. Substrate, current, water depth, water quality, and temperature may affect selection of spawning areas of Pacific lamprey (Pletcher 1963; Kan 1975; Russell et al. 1987). Although we did not examine all the possible parameters of nest selection, we did note correspondence between water temperatures and spawning distribution of Pacific lampreys. Water temperature is an important triggering factor for Pacific lamprey spawning (Pletcher 1963; Kan 1975; Farlinger and Beamish 1984). Given the proximity to the north and south forks of the Umatilla River, for example, we expected introduced lampreys to migrate and use spawning habitat in those forks. However, in the 3 years of surveys, we did not find lamprey nests in either of those streams (Figure 4), which are colder than the main stem. In 2000, the mean daily temperature was 8.3°C during the first 2 weeks in June in the north fork Umatilla River. Pacific lampreys along the coast of Oregon have been observed spawning in May with temperatures between 10°C and 15°C (Kan 1975). Inland populations of Pacific lamprey can spawn through the end of July in the same range of temperatures (Kan 1975; Farlinger and Beamish 1984). In our study, spawning occurred within 2 weeks beginning in June in areas where daily mean temperature was between 9°C and 16°C.

These locations were also favorable for lamprey embryological development in nests and subsequent larval growth. Manion and Hanson (1980) reported average egg viability to be 90% for sea lamprey. In our study, the egg viability levels were 93% and 81% in the Umatilla River and Meacham Creek, respectively, approximately the same as sea lamprey egg viabilities. Likewise, larvae also grew successfully in the upper main-stem Umatilla River (Figure 3). Using length frequency and statolith aging, Morkert et al. (1998) estimated that sea lamprey larvae grew rapidly and metamorphosed at 2+ years in a highly productive river in Michigan. In contrast, based on similar analyses (Figures 3 and 5), we hypothesized that Pacific lampreys do not metamorphose until at least 3–4 years of age in the Umatilla River.

The most striking evidence of success in the first two phases of reintroduction may be the re-establishment of larvae and the increases in larval density over 70 km of stream where they had not occurred for more than 25 years (Figure 2). This confirmed that suitable habitat for lamprey remained in the Umatilla River. Additional research will be needed to determine the full capacity of the stream for larvae, but these results also suggested that adult lamprey that may have theoretically had the opportunity to enter the Umatilla River from the nearby Columbia River have not done so because of ecological or physical barriers to recolonization. Our monitoring showed that although adult Pacific lamprey might otherwise be expected to recolonize the Umatilla River naturally from nearby spawning migrations in the Columbia River, this was unlikely under existing streamflows in the Umatilla River. Adult lamprey were caught entering the Umatilla River when streamflows were high, but no adult lamprey were caught in traps during the peak of the nearby spawning migration at the John Day Dam in July and August, which corresponded to the period when streamflows in the lower Umatilla River were reduced to their lowest levels. It is unclear whether changes in streamflows also prevented adult lampreys entering the Umatilla in early spring from gaining access to the upper watershed.

## Learning from reintroductions

The IUCN recommended that the causes for population declines be well understood and removed and that habitat be restored before the initial releases of the species to allow the species to persist with little subsequent management (IUCN 1998). Identifying and removing the causes of decline is clearly essential for success. Review of reintroduction efforts worldwide showed that none of the programs that acknowledged causes of decline but failed to remove them were successful (Fischer and Lindenmayer 2000). More interesting, however, was that less than one-fourth of the programs that removed the causes of decline were successful. This was not surprising to us. Our results demonstrated that even if the initial causes of the extinctions were well understood and removed (in this case, intentional poisoning), release of the individuals and subsequent monitoring helped reveal watershed specific life-histories of Pacific lamprey in river and potential factors limiting natural production that were unknown at the time of reintroduction.

In contrast to the IUCN, our experience and that of others (Seddon et al. 2007) suggest that species might require postreintroduction management based on learning from reintroductions. This requires a well-designed and implemented monitoring program. This approach may be more realistic and successful than the approach of the IUCN guidelines, especially for reintroduction of species whose life histories and habitat requirements have not been well studied. Where species are rare and poorly understood, opportunities for studying life history requirements are necessarily limited, which can hamper reintroduction efforts if they depended on having comprehensive knowledge of the species. In addition, resources for the necessary research and monitoring necessary for reintroduction may be more available when they are integrated into a reintroduction program.

In our case, monitoring of larval distribution and abundance after reintroduction pointed to factors that might be limiting the survival and distribution of Pacific lamprey in the river. After 7 years of monitoring, the lower distribution in the river did not extend below site 15. Sites 1–15 were all influenced by changes in river flows from irrigation activity. We hypothesize that the failure to re-establish larval distribution in the lower part of the river was associated with highly regulated water flows from the mouth of the river to site 15. McKay Reservoir can release substantial volumes of water from storage that enter the main-stem Umatilla River at the mouth of McKay Creek while the diversion dams below on the main stem remove water for irrigation (Figures 1 and 2). The removal of water occurs during a critical life history stage for larvae. Adult lampreys spawn during the first 2 weeks in June; eggs incubate for approximately 2 weeks before hatching; and after 2 weeks, prolarvae (mean length approximately 20 mm) drift from the gravels downstream to settle in soft substrate to begin their sedentary larval stage. The prolarvae migration downstream corresponds with the periods of the greatest water removal. By July 1, for example, no water remains in the Umatillia River below Westland Diversion Dam (Figure 6).

Larval lampreys have been lost in irrigation canals for more than 100 years in the Umatilla River. In October 1902, the East Oregonian provided a commentary about irrigation and fish, including lamprey, in the Umatilla River:

> As regards the irrigator and his interference with their freedom or their destruction in his ditches, I will venture to say that every one diverting water from this river is as anxious as any lover of good fishing that they be protected and enabled to increase and at the same time is equally as anxious that his lands and crops be free from the presence of all kinds of fish and eel. It may not be generally understood, but fish running down irrigation ditches and dying on the land are extremely obnoxious and positively detrimental to the hay crop because they are frequently raked up and stacked with hay.

Although this commentary occurred before the screening of the canals on the Umatilla River by the U.S. Bureau of Reclamation to prevent salmonid fishes from being diverted into irrigated hay fields, larval lampreys were still being impacted. Larval lampreys were regularly collected in canal traps at Stanfield, Feed, and Westland canals from 1964 to 1974. After rotenone treatments in 1967 and 1974 of the main-stem Umatilla River from the mouth of Meacham Creek downstream, larval lampreys disappeared from the traps completely in 1974 (N. Been, Oregon Department of Fish and Wildlife, retired, personnel communication).

Current screening efforts do not appear to provide protection for lampreys. Since our reintroduction of adult lampreys, lamprey larvae have reappeared and fish salvage operations have retrieved lamprey larvae between the head gates and drum screens at Maxwell and Feed Canals in 2006–2007 (Figure 1; Schwartz 2006; Contor and Wolf 2007). The problem is complex. First, based on electrofishing surveys behind the drum screens located in the Westland Canal, we found that some lamprey larvae (mean length 130 mm; $n = 7$) can pass through the drum screens into the irrigation canals. Those that do not pass may also be lost. Because of the larvae's small size and weak swimming ability, at high flows they can be impinged against screens; at low flows, metamorphosed lamprey may attempt to pass the screens by adhering orally to the screens (a natural behavioral adaptation for climbing waterfalls) where they can be crushed as the screens rotate (Dauble and Moursund 2000; Moursund et al. 2001).

IUCN guidelines for reintroductions (IUCN 1998) include other important tenets: (1) individuals introduced should be taken from genetically similar, wild populations, if possible; (2) habitat and release strategies should be chosen to maximize success and minimize any negative impacts; and (3) evaluation of the reintroduction should include long-term monitoring and modeling of population viability. We chose sources of wild, migrating, adult lampreys that were geographically near the Umatilla River (Table 1), assuming that if adult lampreys return home to areas where they were born, genetic similarity would be greatest between aggregations that were geographically nearby.

Fischer and Lindenmayer (2000) found that the likelihood of success in reintroductions increases when more than 100 animals are released and release strategies and locations target appropriate habitats. We released more than 2,600 lamprey over 7 years (Table 1), recognizing that some lamprey might not breed successfully and that the number and choice of individuals for reintroduction could affect the genetic diversity in the reestablished lamprey (Haig et al. 1990). Recommendations in the conservation biology literature for abundances necessary to maintain genetic diversity ranged from a genetic effective size ($N_e$) of 50 to avoid inbreeding depression (Franklin 1980) to 500 (Lande and Barrowclough 1987; Reed and Bryant 2000) or more (Lynch and Lande 1998), which might be necessary to maintain genetic variability in life history and behavioral traits. No genetic data were available on the ratio of census size to $N_e$ in Pacific lamprey, but assuming overlapping age structure,

3–7-year average life span, and semelparity, we estimated our releases would provide a founding $N_e$ of 250–500 (Waples 1990).

## Declaring success

The final stage of successful reintroduction is persistence of the re-established population (Seddon 1999). Evaluating success of this stage for Pacific lamprey and other species with potentially weak philopatric life history is a unique challenge for two reasons. First, Pacific lampreys are anadromous, which means that the geographical focus for reintroduction is only a small part of their home range. Episodic changes in climate, habitat, water flows, and predators within the Umatilla River, in the main stem of the Columbia River, or in the ocean affect survival from year to year, leading to potentially large variation in abundances. Second, unlike anadromous fishes such as salmonids, Pacific lampreys may not home accurately to their natal streams. If philopatry to natal streams were weak, then reintroduction would be more of a re-expansion of all life history stages into historical habitat than re-establishment of a population and should be evaluated on this basis. The combination of these factors makes evaluating reintroduction efforts more difficult than for a strongly philopatric or nonmigratory species.

Although abundance of out-migrating, metamorphosed lampreys increased following reintroduction of Pacific lampreys in Umatilla River, it is not possible to definitively link this increase to the reintroduction efforts. A longer period of increased out-migrant abundance is needed. A large increase in metamorphosed lampreys occurred during the 2005–2006 (Figure 4), which corresponded with the period when we would have expected increased numbers of out-migrating lamprey, given their life history. Our data collected on the length at age of larvae and the recently metamorphosed lampreys indicated that recently metamorphosed migrating lampreys were 3–4 years old and this peak occurred 4–5 years after the initial reintroductions (Table 1). The largest observed migration of recently metamorphosed lamprey, however, was a result of natural production before our study began (Figure 4). Similarly, we expected increased abundance of out-migrants in 2006–2007 from the continued reintroductions of adults, but abundance did not increase.

Lack of a clear trend of increased abundance of out-migrating, recently metamorphosed lamprey may have at least three possible explanations. First, it is possible that some other factor is affecting survival of lamprey during this portion of their life stage in the Umatilla River. Pacific lampreys metamorphose from July through October (Richards and Beamish 1981). Recently metamorphosed lampreys then begin to move from soft substrate into gravel interstitial spaces (Beamish 1980; Richards and Beamish 1981). This transition may leave them vulnerable to a number of still unknown environmental and ecological challenges, including diversion into the canals during winter months. For example, Feed and Westland canals operate during the winter months in the Umatilla River.

Second, the best methods for getting precise, unbiased estimates of the abundance of out-migrating lampreys remain undeveloped. After metamorphosis, lampreys migrate as streamflows increase during the winter and spring (Beamish and Levings 1991). In the Umatilla River, migration could be patchy and occur in one major flow event, when high flows could limit operating the trap or prevent counting the fish. Although passive traps of different kinds can capture recently metamorphosed lampreys (Moser et al. 2007), we used a rotary screw trap because debris during the high discharges in the Umatilla River precluded using passive traps. Our results suggested, however, that the screw trap was biased for size due to live trap mesh, although this could be confounded by the tendency of larger larvae to drift downstream (Potter 1980; White and Harvey 2003). Improving sampling methods and gear for metamorphosed lamprey will be important for monitoring the success of Pacific lamprey restoration efforts in the Umatilla River and elsewhere.

Finally, detecting clear trends in recently metamorphosed or adult lamprey abundances will require a longer period of monitoring than has occurred so far in the Umatilla River. Annual variation in survival for anadromous fish can be large. Detecting trends in this variation may take many more years of monitoring and an improved understanding the factors affecting this variation. For example, our data were inadequate to show that reintroduction of adult lampreys increased abundance of naturally spawning adult lampreys in the next generation. We did document small increases in the number of adults entering the Umatilla River in the spring 4 or 5 years after the reintroduction began (Figure 4). This may be attributable to our reintroduction, but it is also within the range of natural variation.

Weak philopatry to natal streams would also complicate interpreting evidence for the success of local reintroduction efforts. Evidence from a variety of species indicates that migratory lampreys select rivers due to flow characteristics, migratory pheromones, and possible natal philopatry. For example, increases in river flow induced spawning migration of river lampreys (Asplund and Södergren 1975; Abou-Seedo and Potter 1979; Aronsuu et al. 2002) and rivers with the largest discharges in the Great Lakes attract the largest runs of sea lampreys (Morman et al. 1980). Likewise, evidence also suggests that migratory pheromones released by larval lampreys attract migratory adult lampreys into rivers. Larval sea lampreys produce and release two bile acids; petromyzonol sulfate and allocholic acid, which were detected by the migratory adults (Li et al. 1995). Petromyzonol sulfate has its own specific receptor site in the olfactory epithelium in sea lamprey (Li and Sorensen 1997). Behavioral evidence indicates that these two bile acids function as migratory pheromones in conjunction with other stream odors (Bjerselius et al. 2000; Vrieze and Sorensen 2001). Recently, two new sulfated steroids (petromyzonamine disulfate and petromyzosterol disulfate) were shown to be migratory pheromones in the sea lamprey (Sorensen et al. 2005).

The homing mechanisms for migratory Pacific lampreys have not been as well studied, but they may be similar. Yun et al. (2003) provided evidence that larval Pacific lampreys can produce petromyzonol sulfate and allocholic acid and measured release rates by enzyme linked immunosorbent assay. Fine et al. (2004) provided evidence that petromyzonol sulfate and allocholic acid where detected by an adult Pacific lamprey using electro-olfactogram recordings. One of the authors of this paper (D. A. Close, unpublished data) working with other scientists has recently chemically identified five putative migratory pheromones from larval Pacific lamprey washings using mass spectrometry and they have used electro-olfactogram techniques to develop dose–response curves using the putative pheromones. However, behavioral tests are needed to confirm that these are migratory pheromones. If larval Pacific lampreys are emitting a chemical cue to attract migrating adults, then the lack of larval recruitment in the lower Umatilla River may be a major limiting factor in restoration.

Genetic data also suggested that Pacific lampreys could be weakly philopatric. Goodman et al. (2008) found little evidence of genetic differentiation among Pacific lampreys along the Pacific coast of North America using mitochondrial DNA markers, which could be explained by lack of homing. In contrast, we found that adult Pacific lampreys across their range were genetically differentiated and that that gene flow among aggregations of Pacific lampreys from their natal region to other regions decreased as the distance to those locations increased (Lin et al. 2008a, 2008b). This indicated some greater degree of philopatry. We noted that this level of philopatry could occur through attraction mechanisms such as migratory pheromones or streamflows as long as ocean migration distances were restricted and habitats and aggregations where lampreys historically occurred remained intact (Lin et al. 2008a, 2008b).

## Conclusion

Reintroductions have intuitive appeal for increasing natural biodiversity, providing keystone components of ecosystems and creating public and political support necessary to restore habitat and implement protection measures (Seddon 1999). As used here, reintroduction is also important for also conserving indigenous cultural traditions and traditional ecological knowledge. Local reintroductions of noncharismatic species that are poorly studied, migratory species that cover large home ranges, or species that are weakly philopatric are not easy, however. Successful reintroductions require a multidisciplinary scientific approach (IUCN 1998) involving fisheries and wildlife managers, ecologists, physiologists, and geneticists. In addition, as shown from our example, reintroduction is not only a means of moving species to an area where they historically occurred, but used with well-designed monitoring, it is also a useful tool for identifying previously unknown factors that affect success. Finally, where species may be weakly philopatric, the final stage of success for local reintroductions—persistence—should be evaluated more as a re-expansion and continuing presence of all life history stages into historical habitat rather than just as re-establishment of a unique, closed population. Designing reintroduction projects with these factors in mind would allow managers to collect multifaceted information on both the biological and nonbiological factors affecting success and to develop multiple models for exploring and choosing alternative management scenarios that could lead most rapidly and efficiently to persistence (McLain and Lee 1996).

## Acknowledgments

We thank Brandon Trelor, Jake Settle, and other technicians who assisted in collecting data on this project, the Oregon Department of Fish and Wildlife for assisting in the collection of out-migrant data, and Debbie Docherty for administrative assistance. The Bonneville Power Administration provided funding this project. We are grateful to the tribal elders that inspired this project. *Yox Kalo.*

## References

Abou-Seedo, F. S., and I. C. Potter. 1979. The estuarine phase in the spawning run of the river lamprey (*Lampetra fluviatilis*). Journal of Zoology 188:5–25.

Almeida, P. R., and B. R. Quintella. 2002. Larval habitat of the sea lamprey (*Petromyzon marinus* L.) in the river Mondego (Portugal). Pages 121–130 *in* M. J. Collares-Pereira, M. M. Coelho, and I. G. Cowx, editors. Freshwater fish conservation options for the future. Blackwell Scientific Publications, Oxford, UK.

Andrade, N. O., B. R. Quintella, J. Ferreira, S. Pinela, I. Povoa, S. Pedro, and P. R. Almeida. 2007. Sea lamprey (*Petromyzon marinus* L.) spawning migration in the Vouga River basin (Portugal): poaching impact, preferential resting sites and spawning grounds. Hydrobiologia 582:121–132.

Aronsuu, K. K., E. Ojutkangas and T. J. Marjomäki. 2002. Several environmental factors affecting the timing of river lamprey's (*Lampetra fluviatilis*) spawning migration. Pages 111–113 *in* R. Kamula and A. Laine, editors. Proceedings of the Second Nordic Fish Passage Symposium on Freshwater Fish Migration and Fish Passage Evaluation and Development. Direktoratet for Naturforvaltning, DN-Notat 1-1999, Tronheim, Norway.

Asplund, C., and S. Södergren. 1975. Flodnejonögats (*Lampetra fluviatilis* L.) lekvandring i Rickleån. Zoologisk Revy 36(4):111–119 (In Swedish with English Summary).

Beamish, R. J. 1980. Adult biology of the river lamprey (*Lampetra ayersi*) and the Pacific lamprey (*Lampetra tridentata*) from the Pacific coast of Canada. Canadian Journal of Fisheries and Aquatic Sciences 37:1906–1923.

Beamish, R. J., and C. D. Levings. 1991. Abundance and freshwater migrations of the anadromous parasitic lamprey, *Lampetra tridentata*, in a tributary of the Fraser River, British Columbia. Canadian Journal of Fisheries and Aquatic Sciences 48:1250–1263.

Bjerselius, R., W. Li. J. H. Teeter, J. G. Seelye, P. B. Johnsen, P. J. Maniak, G. C. Grant, C. N. Polkinghorne, and P. W. Sorensen. 2000. Direct behavioral evidence that unique bile acids released by larval sea lamprey (*Petromyzon marinus*) function as a migratory pheromone. Canadian Journal of Fisheries and Aquatic Sciences 57:557–569.

Bond, C. E., and T. T. Kan. 1973. *Lampetra (Entos-*

*phenus) minima* n. sp., a dwarfed parasitic lamprey from Oregon. Copeia 1973:568–574.

Carle, F. L., and M. R. Strub. 1978. A new method for estimating population size from removal data. Biometrics 34:621–630.

Close, D. A., M. S. Fitzpatrick, and H. W. Li. 2002. The ecological and cultural importance of a species at risk of extinction, Pacific lamprey. Fisheries 27(7):19–25.

Close, D. A., A. D. Jackson, B. P. Conner, and H. W. Li. 2004. Traditional ecological knowledge of Pacific lamprey (*Entosphenus tridentatus*) in northeastern Oregon and southeastern Washington from indigenous peoples of the Confederated Tribes of the Umatilla Indian Reservation. Journal of Northwest Anthropology 38:141–161.

Contor, C. R., and J. Wolf. 2007. Umatilla basin project and Umatilla project biological opinion monitoring and evaluation. U.S. Bureau of Reclamation, Portland, Oregon.

Dauble, D. D., and R. A. Moursund. 2000. Effects of extended length bypass screens on juvenile Pacific lamprey. HydroVision 2000 Technical Paper. HCL Publications, Kansas City, Missouri.

Eklund, J., A. Niemi, and E. Ojutkangas. 1983. The river lamprey in two regulated Finnish rivers. Pages 417–426 *in* A. Lillehammer and S. J. Saltveit, editors. Regulated rivers. Universitetsforlaget, Oslo.

Farlinger, S. P., and R. J. Beamish. 1984. Recent colonization of a major salmon-producing lake in British Columbia by Pacific lamprey (*Lampetra tridentata*). Canadian Journal of Fisheries and Aquatic Sciences 41:278–285.

Fine, J. M., L. A. Vrieze, and P. W. Sorensen. 2004. Evidence that petromyzontid lampreys employ a common migratory pheromone that is partially comprised of bile acids. Journal of Chemical Ecology 30:2091–2110.

Fischer, J., and D. B. Lindenmayer. 2000. An assessment of the published results of animal relocations. Biological Conservation 96:1–11.

Franklin, I. R. 1980. Evolutionary changes in small populations. Pages 135–149 *in* M. E. Soule and B. A. Wilcox, editors. Conservation biology: an evolutionary-ecological approach. Sinauer Associates, Sunderland, Massachusetts.

Goodman, D. H., S. B. Reid, M. F. Docker, G. R. Haas, and A. P. Kinziger. 2008. Mitochondrial DNA evidence for high levels of gene flow among populations of a widely distributed anadromous lamprey *Entosphenus tridentatus* (*Petromyzontidae*). Journal of Fish Biology 72:400–417.

Haig, S. M., J. D. Ballou, and S. R. Derrickson. 1990. Management options for preserving genetic diversity: reintroduction of Guam rails to the wild. Conservation Biology 4:290–300.

IUCN (International Union for Conservation of Nature and Natural Resources). 1998. Guidelines for reintroductions. IUCN, IUCN/SSC Re-introduction Specialist Group, Gland, Switzerland and Cambridge, UK.

Kan, T. T. 1975. Systematics, variation, distribution, and biology of lampreys of the genus *Lampetra* in Oregon. Doctoral dissertation. Oregon State University, Corvallis.

Kelly, F. L., and J. J. King. 2001. A review of the ecology and distribution of three lamprey species, *Lampetra fluviatilis* (L.), *Lampetra planeri* (Bloch), and *Petromyzon marinus* (L.): a context for conservation and biodiversity considerations in Ireland. Biology and Environment: Proceedings of the Royal Irish Academy 101B:165–185.

Kirchhofer, A. 1995. Concept of conservation for the European brook lamprey (*Lampetra planeri*) in Switzerland. Fishchökologie 8:93–108.

Kostow, K. 2002. Oregon lampreys: natural history status and analysis of management issues. Oregon Department of Fish and Wildlife, Portland.

Lande, R., and G. F. Barrowclough. 1987. Effective population size, genetic variation, and their use in population management. Pages 87–124 *in* M. E. Soule, editor. Viable populations for conservation. Cambridge University Press, Cambridge, UK.

Laine, A., R. Kamula, and J. Hooli. 1998. Fish and lamprey passage in a combined Denil and vertical slot fishway. Fisheries Management and Ecology 5:31–44.

Li, W., P. W. Sorensen, and D. D. Gallaher. 1995. The olfactory system of migratory adult sea lamprey is acutely and specifically sensitive to unique bile acids released by conspecific larvae. Journal of General Physiology 105:569–587.

Li, W., and P. W. Sorensen. 1997. Highly independent receptor sites for naturally occurring bile acids in the sea lamprey, *Petromyzon marinus*. Journal of Comparative Physiology A 180:429–438.

Lin, B., Z. Zhang, Y. Wang, K. P. Currens, A. Spidle, Y. Yamazaki, and D. A. Close. 2008a. AFLP assessment of genetic diversity of Pacific lamprey. North American Journal of Fisheries Management 28:1182–1193.

Lin, B., Z. Zhang, Y. Wang, K. P. Currens, A. Spidle,

Y. Yamazaki, and D. A. Close. 2008b. Erratum: AFLP assessment of genetic diversity of Pacific lamprey. North American Journal of Fisheries Management 28:1648.

Lynch, M., and R. Lande. 1998. The critical effective size for a genetically secure population. Animal Conservation 1:70–72.

Manion, P. J., and L. H. Hanson. 1980. Spawning behavior and fecundity of lampreys from the three upper Great Lakes. Canadian Journal of Fisheries and Aquatic Sciences 37:1635–1640.

Masters, J. E. G., M-H. Jang, K. Ha, P. D. Bird, P. A. Frear, and M. C. Lucas. 2006. The commercial exploitation of a protected anadromous species, the river lamprey (*Lampetra fluviatilis* (L.)), in the tidal River Ouse, north-east England. Aquatic Conservation: Marine and Freshwater Ecosystems 16:77–92.

Mattson, C. R. 1949. The lamprey fishery at Willamette Falls, Oregon. Oregon Fish Commission Research Briefs 2:23–27.

McLain, R. J., and R. G. Lee. 1996. Adaptive management: promises and pitfalls. Environmental Management 20:437–448.

Morkert, S. B., W. D. Swink, and J. G. Seelye. 1998. Evidence for early metamorphosis of sea lampreys in the Chippewa River, Michigan. North American Journal of Fisheries Management 18:966–971.

Morman, R. H., D. W. Cuddy, and P. C. Rugen. 1980. Factors influencing the distribution of sea lamprey (*Petromyzon marinus*) in the Great lakes. Canadian Journal of Fisheries and Aquatic Sciences 37:1811–1826.

Moser, M. L., P. A. Ocker, L. C. Stuehrenberg, and T. C. Bjornn. 2002a. Passage efficiency of adult Pacific lampreys at hydropower dams on the lower Columbia River, USA. Transactions of the American Fisheries Society 131:956–965.

Moser, M. L., A. L. Matter, L. C. Stuehrenberg, and T. C. Bjornn. 2002b. Use of an extensive radio receiver network to document Pacific lamprey (*Lampetra tridentata*) entrance efficiency at fishways in the lower Columbia River, USA. Hydrobiologia 483:45–53.

Moser, M. L., and D. A. Close. 2003. Assessing Pacific lamprey status in the Columbia River basin. Northwest Science 77:116–125.

Moser, M. L., J. M. Butzerin, and D. B. Dey. 2007. Capture and collection of lampreys: the state of the science. Reviews in Fish Biology and Fisheries 17:45–56.

Moursund, R. A., R. P. Mueller, T. M. Degerman, and D. D. Dauble. 2001. Effects of Dam Passage on Juvenile Pacific Lamprey (*Lampetra tridentata*). U.S. Army Corps of Engineers, Portland, Oregon.

Myllynen, K., E. Ojutkangas, and M. Nikinmaa. 1997. River water with high iron concentration and low pH causes mortality of lamprey roe and newly hatched larvae. Ecotoxicology and Environmental Safety 36:43–48.

Ojutkangas, E., K. Aronen, and E. Laukkanen. 1995. Distribution and abundance of river lamprey (*Lampetra fluviatilis*) ammocoetes in the regulated river Perhonjoki. Regulated Rivers Research and Management 10:239–245.

Phillips, J. L., J. Ory, and A. Talbot. 2000. Anadromous salmonid recovery in the Umatilla River basin, Oregon: a case study Journal of the American Water Resources Association 36:1287–1308.

Piavis, G. W. 1961. Embryological stages in the sea lamprey and effects of temperature on development. Fishery Bulletin 61:111–143.

Potter, I. C. 1980. Ecology of larval and metamorphosing lampreys. Canadian Journal of Fisheries and Aquatic Sciences 37:1641–1657.

Pletcher, T. F. 1963. The life history and distribution of lampreys in the Salmon and certain other rivers in British Columbia, Canada. Master's thesis. University of British Columbia, Vancouver.

Raymond, H. L. 1979. Effects of dams and impoundments on migrations of juvenile Chinook salmon and steelhead from the Snake River, 1966 to 1975. Transactions of the American Fisheries Society 108:505–529.

Raymond, H. L. 1988. Effects of hydroelectric development and fisheries enhancement on spring and summer Chinook salmon and steelhead in the Columbia River basin. North American Journal of Fisheries Management 8:1–24.

Reed, D. H., and E. H. Bryant. 2000. Experimental tests of minimum viable population size. Animal Conservation 3:7–14.

Renaud, C. B. 1997. Conservation status of Northern Hemisphere lampreys (Petromyzontidae). Journal of Applied Ichthyology 13:143–148.

Richards, J. E., and F. W. H. Beamish. 1981. Initiation of feeding and salinity tolerance in the Pacific lamprey *Lampetra tridentata*. Marine Biology 63:73–77.

Richards, J. E., R. J. Beamish, and F. W. H. Beamish. 1982. Descriptions and keys for ammocoetes of lampreys from British Columbia, Canada.

Canadian Journal of Fisheries and Aquatic Sciences 39:1484–1495.

Russell, J. E., F. W. H. Beamish, and R. J. Beamish. 1987. Lentic spawning by the Pacific lamprey, *Lampetra tridentata*. Canadian Journal of Fisheries and Aquatic Sciences 44:476–478.

Schuldt, R. J., and J. W. Heinrich. 1982. Portable trap for collecting adult sea lampreys. Progressive Fish-Culturist 44:220–221.

Schwartz, J. 2006. Umatilla basin project and Umatilla project biological opinion monitoring and evaluation. U.S. Bureau of Reclamation, Portland, Oregon.

Seber, G. A. F. 1982. The estimation of animal abundance and related parameters, 2nd edition. Charles Griffin and Company, London.

Seber, G. A. F., and E. D. Le Cren. 1967. Estimation population parameters from catches large relative to the population. Journal of Animal Ecology 36:631–643.

Seddon, P. J. 1999. Persistence without intervention: assessing success in wildlife re-introductions. Trends in Ecology and Evolution 14:503.

Seddon, P. J., D. P. Armstrong, and R. F. Maloney. 2007. Developing the science of reintroduction biology. Conservation Biology 21:303–312.

Seddon, P. J., P. S. Soorae, and F. Launay. 2005. Taxonomic bias in reintroduction projects. Animal Conservation 8:51–58.

Sorensen, P. W., J. M. Fine, V. Dvornikovs, C. S. Jeffery, F. Shao, J. Wang, L. A. Vrieze, K. R. Anderson, and T. R. Hoye. 2005. Mixture of new sulfated steroids functions as a migratory pheromone in the sea lamprey. Nature Chemical Biology 1:324–328.

Vrieze, L. A., and P. W. Sorensen. 2001. Laboratory assessment of the role of a larval pheromone and natural stream odor in spawning stream localization by migratory sea lamprey (*Petromyzon marinus*). Canadian Journal of Fisheries and Aquatic Sciences 58:2374–2385.

Volk, E. C. 1986. Use of calcareous otic elements (statoliths) to determine age of sea lamprey ammocoetes (*Petromyzon marinus*). Canadian Journal of Fisheries and Aquatic Sciences 43:718–722.

Waples, R. S. 1990. Conservation genetics of Pacific salmon. III. Estimating effective population size. Journal of Heredity 81:277–289.

Wedemeyer, G. A., B. A. Barton, and D. J. McLeay. 1990. Stress and acclimation. Pages 451–489 *in* C. B. Schreck and P. B. Moyle, editors. Methods for fish biology. American Fisheries Society, Bethesda, Maryland.

Weisser, J. W., and G. T. Klar. 1990. Electric fishing for sea lampreys (*Petromyzon marinus*) in the Great Lakes region of North America. Pages 59–64 *in* I. G. Cowx, editor. Developments in electric fishing. Cambridge University Press, Cambridge, U.K.

White, J. L., and B. C. Harvey. 2003. Basin-scale patterns in the drift of embryonic and larval fishes and lamprey ammocoetes in two coastal rivers. Environmental Biology of Fishes 67:369–378.

Wolf, C. M., Griffith, C. Reed, and S. A. Temple. 1996. Avian and mammalian translocations: update and reanalysis of 1987 survey data. Conservation Biology 10:1142–1154.

Yun, S. S., A. P. Scott, J. M. Bayer, J. G. Seelye, D.A. Close, and W. Li. 2003. HPLC and ELISA analyses of larval bile acids from Pacific and western brook lampreys. Steroids 68:515–523.

American Fisheries Society Symposium 73:255–268, 2009

# Considering Pacific Lampreys When Implementing Instream Activities

BIANCA STREIF*
*U.S. Fish and Wildlife Service, Oregon Fish and Wildlife Office*
*2600 SE 98th Avenue, Suite 100, Portland, Oregon 97266, USA*

*Abstract.*—Efforts to minimize negative effects to fish species during instream activities generally do not consider the life history characteristics of native nongame fish species, particularly lamprey. While there is still much to be learned about Pacific lamprey *Entosphenus tridentatus* (formerly *Lampetra tridentata*) distribution, abundance, and status, the need to actively conserve lampreys is evident. Populations of the Pacific lamprey have been reduced in many river drainages along the West Coast of the United States. The purposes of this paper are to raise awareness of the need to consider Pacific lampreys in project implementation and to stimulate development of methods to address instream project impacts.

Projects that alter passage, change flow hydraulics, alter stream substrates, and/or decrease habitat complexity can negatively affect lampreys. Of particular importance to lampreys are low velocity areas where larvae (ammocoetes) live in the substrate as filter feeders for 3–7 years. Since several age-classes of ammocoetes may congregate in high densities, a single dewatering event during construction activities may have a significant effect on a local lamprey population. Methods to reduce effects to the various life history stages of lampreys are just beginning to be investigated and developed. However, basic measures for lamprey conservation will also benefit other aquatic species by providing for a diversity of habitats and stream structure, and complex velocity distributions, while also appropriately modifying the duration and timing of instream actions.

## Introduction

Pacific lampreys *Entosphenus tridentatus* (formerly *Lampetra tridentata*) are the most widely distributed lamprey species on the West Coast of the United States and have been found in streams from Hokkaido Island, Japan and around the Pacific Rim to the Revillagigedo Archipelago, Mexico (Renaud 2008). Historically, Pacific lampreys are thought to have been distributed wherever anadromous salmonids occurred (Simpson and Wallace 1982; Close et al. 1995, 2002). Similar to populations trends in salmonids, recent data indicate that the distribution of the Pacific lamprey has been reduced in many river drainages. Pacific lampreys are extirpated above dams and other impassable barriers in many larger rivers throughout coastal Washington, Oregon, and California (Close et al. 2002; Kostow 2002; U.S. Fish and Wildlife Service 2004). In addition to extirpations, available data also indicate that Pacific lampreys have declined in abundance in many systems, including the Columbia River basin (Close et al. 2002; Kostow 2002; U.S. Fish and Wildlife Service

* Corresponding author: bianca_streif@fws.gov

2004; Hyatt et al. 2007). Thus, the need for immediate conservation of lampreys is evident.

Stream restoration projects in the Pacific Northwest region have typically focused on anadromous salmonids because of their listing under the Endangered Species Act. Hence, restoration efforts that provide benefits to salmonids and minimize negative effects during instream activities may not adequately consider the unique life history characteristics and needs of other fish species, in particular lamprey. The life history and habitat requirements of lampreys must be understood by the project designer, manager, and construction crew in order to avoid unintended negative impacts and conserve lamprey populations.

The purposes of this paper are to raise awareness of the need to consider Pacific lampreys in project implementation and to stimulate development of methods to address instream project impacts. While the focus of this paper is the Pacific lamprey, many of the recommendations apply to all lamprey species in the Pacific Northwest, including western brook lamprey *Lampetra richardsoni* and river lamprey *L. ayresii*.

## Overview of Pacific Lamprey Life History

Pacific lampreys play an important and complex role in their native stream ecosystems (Figure 1). As adults in the marine environment, Pacific lampreys are parasitic and feed on a variety of marine and anadromous fish (Close et al. 2002). After the adult feeding phase, Pacific lampreys migrate to freshwater, cease feeding, and may hold over for almost a year. Most of their movement occurs at night (Ocker et al. 2001; Kostow 2002; Daigle et al. 2005). The degree of homing to natal streams is unknown, but adult lampreys probably cue in on the presence of ammocoetes, which release pheromones thought to aid adult lampreys in identifying successful spawning areas (Close 2001). They spawn in similar habitats to salmon, in gravel-bottomed streams, at the upstream end of riffle habitat. Both sexes construct the nest, often moving stones with their mouth. After the eggs are deposited and fertilized, adults typically die within 3–36 d of spawning (Pletcher 1963; Kan 1975).

In about 20 d, the eggs hatch (Scott and Crossman 1973). The ammocoetes drift downstream to areas of low velocity and fine substrates where they burrow, grow, and live as filter feeders for 3–7 years, feeding primarily on algae (Close et al. 2002). Suitable burrowing and filter-feeding habitats have flow regimes that are typically slow enough to promote fine-sediment deposition yet fast enough for influx of food (Torgerson and Close 2004).

The distribution and movement of ammocoetes in most systems is poorly understood. Ammocoetes are found throughout a river system, in tributary and main-stem reaches (Luzier et al. 2006). Torgerson and Close (2004) found that 80% of ammocoetes in their study area occurred along stream margins in the Middle Fork John Day River, Oregon. They also noted that ammocoete distribution within rivers is dependent upon a number of variables, including channel gradient, bedform morphology, water velocity, pools, and other factors that vary depending on spatial scale. Several age-classes of ammocoetes are typically present in a given site (Kostow 2002); thus, impacts to the ammocoete population will impact a broad series of recruitment cohorts.

Ammocoetes (Figure 2) often disperse downstream during peak flows (Stone et al. 2001, 2002; Pirtle et al. 2003; Lê et al. 2004; Luzier and Silver 2005). White and Harvey (2003) documented large numbers of ammocoetes moving almost exclusively at night in the Smith and Van Duzen rivers in California.

Metamorphosis to the juvenile phase (macropthalmia) occurs gradually over several

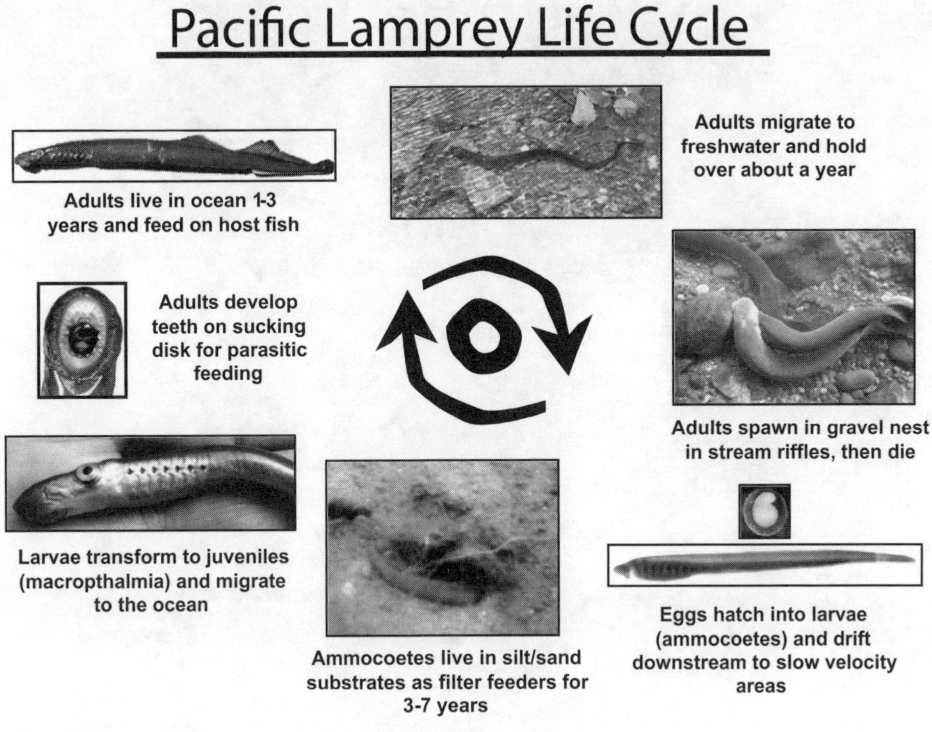


Figure 1. Diagram of Pacific lamprey life cycle.

months as developmental changes occur (including the appearance of eyes and teeth) and the macropthalmia leave the substrate to enter the water column. Transformation typically begins in July and is complete by December, although in inland populations it begins in late fall, coinciding with the beginning of emigration. The macropthalmia drift and swim downstream (mostly at night) between late fall and spring as they emigrate to the ocean where they mature into adults (Close et al. 2002; Moyle 2002).

## Considerations When Implementing Instream Activities

Instream activities of concern to lamprey conservation include a wide variety of channel maintenance and stream restoration project types. Channel maintenance actions can include dredging, bank stabilization, or structure modifications. Stream restoration activities can include projects that facilitate fish passage (e.g., culverts and fish ladders), various types of habitat structures (e.g., large wood, weirs, and boulders), or complete channel reconstruction.

The unique life history of the Pacific lamprey requires project modifications that consider long instream residency; use of substrate by ammocoetes; movements and migrations of ammocoetes, macropthalmia, and adults; and poor swimming abilities at all life stages. Pacific lampreys have not been studied to the same degree as salmon; thus, our knowledge of specific requirements is deficient. Nonetheless, many restoration projects that benefit salmon aimed at restoring natural stream functions and floodplain habitat will also benefit lampreys by providing diversity in habitats, substrates, off-channel habitats, cover, and good water quality.

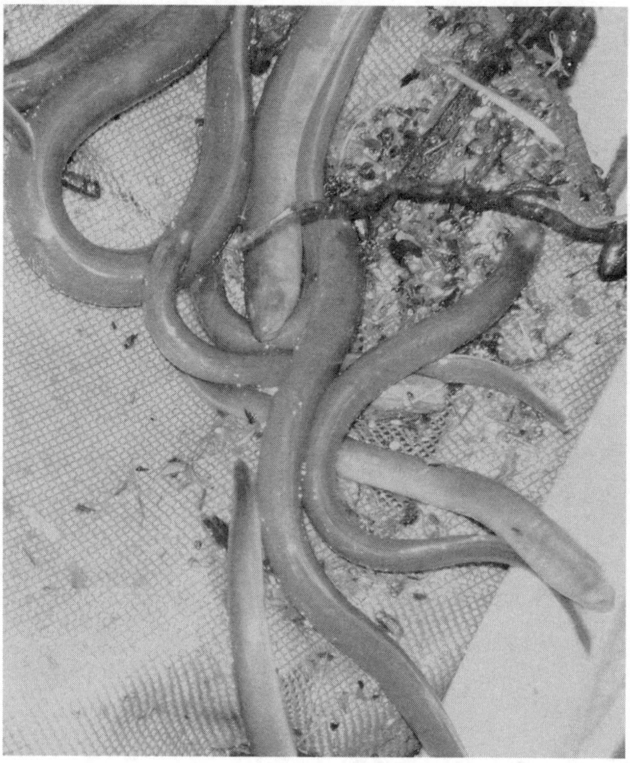

Figure 2. Ammocoetes.

Table 1 identifies potential short-term and long-term adverse effects of actions for various Pacific lamprey life stages when implementing instream activities based on current knowledge. Only potential adverse impacts are identified in the table. However, it is important to recognize that some projects may provide long-term benefit to lampreys or their habitat. Furthermore, adverse effects can be reduced or mitigated through modifications in project design and implementation, and with appropriate management plans.

## Specific Recommendations

Before beginning any instream project, it is important to identify if lampreys are present. The site can be surveyed or data can be used from local biologists to determine their potential use of the stream reach. If no data are available, then anadromous salmonid presence in the stream reach can be used as a surrogate to assume potential Pacific lamprey presence, with the understanding that site specific habitat use is different between the species. If lampreys are present, the instream work window may be adjusted to avoid spawning adults, ammocoetes can be moved outside of the project area, or work can avoid occupied reaches. Below is a general summary of survey methods; a detailed review of the capture and collection of lampreys can be found in Moser et al. 2007.

### *Survey and Salvage Techniques for Ammocoetes*

Ammocoetes are found in the stream substrates, usually in sands and silts, which are naturally deposited in slower velocity areas such as stream edges, pools, and backwater ar-

Table 1. Examples of potential adverse effects of actions for various Pacific lamprey life stages when implementing instream activities.

| | Potential short-term (ST) and long-term (LT) adverse effects of action on various life stages | | |
|---|---|---|---|
| Project type | Ammocoete | Macropthalmia | Adult |
| Temporary dewatering of channel typically conducted for project actions such as those listed below | ST – desiccation<br>LT – loss of several age-classes and generations | ST – desiccation, delay in out-migration<br>LT – loss of year-class | ST – desiccation of redd, disruption in migration<br>LT – loss of year-class |
| Dredging | ST – entrainment with dredge material<br>LT – loss of substrate | ST/LT – unknown | ST – loss of redds, disruption in migration<br>LT – loss of spawning habitat, loss of cues for migration |
| Road crossing modifications such as culvert replacements | ST – direct loss with substrate removal/ disturbance<br>LT – passage barrier | ST – unknown<br>LT – passage barrier | ST – disruption in migration<br>LT – passage barrier |
| Instream structures for habitat restoration or grade control | ST – direct loss with disturbance to substrate removal/disturbance<br>LT – passage barrier | ST – unknown<br>LT – passage barrier | ST – Loss of redds, disruption in migration<br>LT – passage barrier, loss of spawning habitat |
| Bank stabilization | ST – direct loss with disturbance<br>LT – loss of habitat | ST/LT - unknown | ST/LT – unknown |
| Irrigation diversions/ modifications | ST – direct loss with disturbance, desiccation<br>LT – mortality due to high water temperature, impingement/ entrainment, desiccation, passage barrier | ST – desiccation, delay in out-migration<br>LT – mortality due to high water temperature, impingement/ entrainment, desiccation, passage barrier | ST – desiccation of redd, disruption in migration<br>LT – mortality due to high water temperature, impingement/ entrainment, passage barrier |
| Projects in side channels | ST – direct loss with disturbance<br>LT – loss of habitat | ST/LT – loss of low energy refuges during high flows | ST/LT – loss of refuges during high flows |
| Estuarine projects | ST/LT – unknown | ST – effect on emigration<br>LT – depends on the project – tidal gates block passage | ST–disruption in migration<br>LT – depends on the project – tidal gates block passage |

eas. They may also be found in artificial habitats such as ditches and canals. The depth and extent of area utilized by ammocoetes within the stream substrates is largely unknown. Pacific lamprey ammocoetes within the substrate can be found in water depths from 1 cm (Claire 2004) to more than 15 m, as documented in the main-stem Willamette River (Luzier et al.

2009). If adult spawning is known to occur upstream of the site, it is likely that ammocoetes are present in the fine substrates downstream. Ammocoetes can be difficult to detect since they range in size from approximately 2 to 150 mm. The smaller (younger) ammocoetes are easily overlooked.

The primary method used in ammocoete surveys is electrofishing, although this technique is somewhat inefficient and the probability of detecting and capturing ammocoetes is low.

Because of the difficulty in capturing ammocoetes, multiple passes (at least five) are required to effectively sample (or salvage) juvenile lamprey. To improve sampling efficiency, Luzier et al. (2006) recommends using an electrofisher specifically designed to capture ammocoetes (the ABP-2 backpack electrofisher developed by the Engineering Technical Services, University of Wisconsin, Madison) with the following setting: 3 pulses/s (125 V DC) at 25% duty cycle, with a 3:1 burst pulse train (three pulses on, one pulse off) at 30 pulses/s to remove larvae from the substrate. Since ammocoetes can be difficult to capture with standard electrofishing techniques, an exhaustive search may be required before concluding that ammocoetes are not present in a stream reach. Smaller ammocoetes may be more sensitive to shocking and more easily captured than larger ammocoetes (Luzier et al. 2006). Seining, traps, and other techniques can also be used (Moser et al. 2007).

### *Survey and Salvage Techniques for Macropthalmia*

Transforming ammocoetes and macropthalmia are found in the water column, usually moving at night, as they emigrate to the ocean. They tend to swim low in the water column because they lack a swim bladder (Moursund et al. 2000). Their use within the estuary is poorly understood. Methods used to survey or conduct salvage operations include electrofishing, traps, and nets.

### *Survey and Salvage Techniques for Adult Lampreys*

Adults are the easiest to observe as they migrate upstream to their spawning areas. However, overwintering adults may be difficult to detect since they may hide under boulders, within bedrock crevices, or other structures (Luzier et al. 2009). Snorkeling and other visual techniques, traps, and electrofishing are methods used to survey adults lampreys (Moser et al. 2007). Salvage techniques (typically electrofishing) are similar to those used for other fish species.

Lampreys spawn in habitats similar to those utilized by salmonids, but lamprey nests may be difficult to identify (Moser et al. 2007). Since the eggs hatch is a short period of time (about 20 d), salvage would not be practical and avoidance of active spawning areas would be the simplest method to avoid impacts.

## Timing of Instream Work

Ammocoetes are found in the stream substrate year-round. Thus, there is no instream work window to avoid impacting this life stage.

Macropthalmia and transforming ammocoetes are generally found from July through December or even early spring (Luzier et al. 2009; Close et al. 2002). Regional differences are found in the timing, and triggers for metamorphosis are unknown. To avoid impacts to emigrating macropthalmia, instream actions should take place February through June.

Adult lampreys move to their spawning stream sometime between February and September. Differences in movement times vary by region and can be influenced by flow, temperature, and other factors. Eggs hatch in about 20 d depending on stream temperature. Adults can be present at any time of the year since they may overwinter before moving to their spawning habitat. Bayer et al. (2001) documented adult overwinter holding from September to mid-March in the John Day River, Oregon and noted that the timing coincided with decreasing temperature, day length, and discharge.

The instream work window for adult Pacific lampreys depends on the habitat and the kind of action occurring. Potential holding areas may need to be surveyed or avoided. Spawning areas can be avoided by working between late summer and February. Specific regional information can be used to adjust the instream work window for local conditions.

## Dewatering and Substrate Disturbance

An instream action that may dewater a stream includes culvert replacements, some instream habitat projects, and irrigation withdrawals and diversions. Substrate disturbance projects can include dredging, road-crossing modifications (such as culvert replacements), and instream structures for grade.

### *Ammocoetes (Figure 2)*

Since ammocoetes are found in the stream substrate year-round, taking into account the potential for them to utilize an area is essential to their conservation. A single dewatering event, physical disturbance, or contamination event may have a significant effect on multiple year-classes of a local lamprey population due to the inability of ammocoetes to move quickly from a disturbed area (Figure 3).

The goal of reducing the impact to ammocoetes during a dewatering event is to give them enough time to move out of the area on their own, leaving suitable wet areas near the project vicinity for them to move to, or salvage and remove them from the project site. Ideally, dewatering should be avoided. If dewatering is necessary, allow ammocoetes a chance to volitionally move out of a project area before construction by ramping down flows slowly over several days. Ammocoetes will emerge from the substrate as the subsurface flows are reduced. It is recommended that salvaging of ammocoetes occur as they emerge from the substrate and be continued until they stop emerging from the streambed.

The research and development of techniques to protect ammocoetes during channel dewatering events is necessary. An example

Figure 3. Stranded ammocoete from dewatering event.

that needs testing is a method used by some restoration practitioners: areas are identified adjacent to ammocoete habitat outside of the disturbance area but within the channel, and shallow holes are dug where ammocoetes may take refuge. Anecdotal information suggests that ammocoetes will move into areas that retain water and that covering these "refuge holes" will protect them from predators if left overnight.

Substrate disturbance, including dredging and chemical treatment of streams, should be avoided where ammocoetes are known to exist. When this is not possible, salvage efforts should take place prior to the activity.

### *Macropthalmia (Figure 4)*

Dewatering can hinder the emigration of transforming ammocoetes and macropthalmia to the ocean. Irrigation (water withdrawals), especially during the night, may have a greater impact on macropthalmia since they tend to move during that period.

Aspects of project design, construction timing, and implementation in regard to the alteration of timing (due to dewatering or channel disturbance) of emigrating lampreys should be considered. Project timing to avoid peak migration periods and diverting water during daylight can minimize impacts to macropthalmia.

### *Adult Lampreys*

Dewatering can alter the migration timing of adult lampreys, desiccate lamprey nests and eggs (Figure 5), and possibly affect overwintering adults, and should be considered in project design, construction timing, and implementation.

Many methods to reduce short-term effects to salmonids often protect spawning lampreys (Figure 6) since they utilize similar substrates and habitats within the stream channel. While little is known about spawning cues, it may be important to protect ammocoetes that provide pheromones, which are thought to aid adult lampreys in identifying successful spawning areas (Close 2001). Dewatering and channel disturbance actions need to avoid lamprey migration and spawning periods for reasons listed below:

- Passage to spawning areas must be maintained.
- Spawning lampreys and their nests should not be physically disturbed.
- Excessive sedimentation can decrease egg survival.
- Flows for egg incubation need to be maintained.
- Stream temperatures should not exceed 22°C (Meeuwig et al. 2005).

Overwintering adult lampreys may utilize substrates that consist of large cobbles, boulders, and bedrock. Adult lampreys will hide in these substrates when they overwinter, but are difficult to detect. No methods are known to salvage them from those locations.

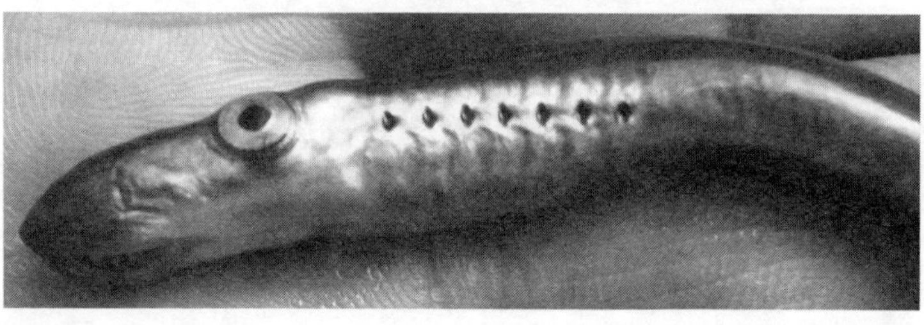

Figure 4. Macropthalmia.

Figure 5. Desiccated lamprey nest. (Photo courtesy of Abel Brumo.)

## Passage

### *Juvenile Lampreys (Ammocoetes and Macropthalmia)*

Little information is available regarding habitat requirements for juvenile lampreys, and their very nature makes gathering information on them difficult. Stream flow and passage are two primary considerations critical to juvenile movement when designing a project. When juvenile lampreys move downstream, they may be entrained in water diversions or turbine intakes. In many cases, these water diversions and hydroelectric projects have been screened to bypass juvenile salmonids. However, due to their size and weak swimming ability, juve-

Figure 6. Adult Pacific lamprey spawning.

niles are frequently impinged on the screens, resulting in injury or death (Hammond 1979; Jackson et al. 1996; Moursund et al. 2000). Moursund et al. (2003) observed that 70% of juvenile lampreys became impinged on 0.125-in (0.318 cm) bar screens within 1 min of exposure at water velocities of 1.5 ft/s (0.457 m/s). He also noted that because their tails are narrower than the rest of their body, they became wedged between screen bars as they attempted to push themselves off. Juvenile lampreys are also more likely to be impinged on horizontal bars than vertical ones (Moursund et al. 2003).

Passage impacts include inadequate flows for downstream movement, physical barriers that inhibit passage, or other disturbances during peak emigration periods. In addition, juvenile lamprey's poor swimming ability make them prone to predation, especially in systems with an altered predator–prey balance. General recommendations are listed below:

- Design intakes, bypass structures, and diversions with juvenile lamprey in mind.
- Avoid projects during peak emigration periods.
- Replace culverts with stream simulation types or bridges.
- Size diversion screens and adjust flows to avoid impingement on diversion structures. Newer irrigation ditch screen bypass systems may function to protect older (>age 2) juvenile lamprey. Younger ammocoetes sometimes move between drum screen seals and into ditches.
- Using screens with smaller mesh and bar spacing and lower approach velocities than those designed for salmonids may be more effective for protecting ammocoetes at water diversions (Gilmore 2005). Bar spacing less than 0.069-in (0.175 cm) and 0.125-in (0.318 cm) square mesh with velocities up to 4 ft/s impinged the fewest juvenile lampreys (Moursund et al. 2003).
- Avoid diverting water at night when movement may be greatest. Midwater intakes might reduce entrainment: ammocoetes appear to move downstream close to the surface at night (Moursund et al. 2000).
- Evaluate tide gates to ensure downstream macropthalmia passage in estuarine habitats.

### *Adults*

Most of the research and project design work for lampreys has been for the adult stage, particularly passage and swimming performance, though more study needs to occur. The greatest disparity between salmonids and adult lampreys is in the lamprey's poor swimming ability and inability to jump and navigate high velocities. Fish ladders designed to pass salmonids may not effectively pass lampreys (Close et al. 1995; Vella et al. 1999; Kostow 2002). Poor swimming ability (Mesa et al. 2003) combined with sharp angles and high water velocities effectively block or restrict lamprey passage (Ocker et al. 2001). A hanging culvert, even a couple of inches, is a barrier to lampreys. Adult lampreys move mostly at night with bursts of swimming, followed by attaching to a relatively smooth surface (CRBLTW 2004). Adult lampreys are able to move up waterfalls using the smooth, wet, vertical surfaces for attachment and the rock crevices to rest. Lampreys also travel deeper in the water column (as they have no air bladder) than salmonids; therefore, traditional spill gates may block passage. Some general considerations are as follows:

1. Maintain natural water flows and temperatures for migration and spawning. The preferred temperature for spawning is approximately 10–15°C (Kan 1975).
2. Design tide gates to allow upstream lamprey passage.

3. Provide passage by
   - Replacing hanging culverts using the stream simulation design or a bridge.
   - Piling rock at a culvert outlet in the water overflow or fitting outlet with attachment surface (e.g., flat plate) for temporary passage at hanging culverts.
   - Using removable spillway weirs during peak movement periods.
   - Creating smooth surfaces and round ed corners with no 90° bends in high velocity areas.
   - Using flat plates (e.g., steel, aluminum, and polycarbonate) to provide attachment points for passage ramps over flashboards or grating.
   - Designing and installing lamprey passage systems at barriers (Moser et al. 2006). These passage systems are relatively simple in design and provide alternate routes for adult lamprey passage where salmonid fish ladders or barriers from dams and diversions exist (Figure 7).

## Important Habitat Features and Conditions

Stream restoration actions that provide conditions to benefit Pacific lampreys are listed below.

### *For All Life Stages*

- Year-round flow and relatively stable flow conditions (sustained increases or decreases that take place over days and weeks) are essential to ammocoete rearing, juvenile emigration, adult holding, migration, and spawning.
- Habitat conditions that do not favor nonnative fish species can reduce predation on juvenile lampreys.
- Water temperatures below 22°C may decrease death or deformation of eggs or ammocoetes (Meeuwig et al. 2005) and possibly reduce predation and disease. It may also affect adult lamprey migration timing.
- Suitable screens or barriers to keep juvenile and adult lampreys out of undesirable habitats such as irrigation ditches and canals.
- Streams and rivers with unimpeded upstream and downstream passage for all life stages.

### *Ammocoetes*

- Depositional areas, including alcoves, backwater areas, pools, and low velocity stream and river margins that recruit sands and silts, downstream of spawning areas, provide ideal ammocoete rearing

Figure 7. Lamprey passage system at Bonneville Dam, Washington.

areas. Ammocoetes use a variety of substrate sizes based in part on availability. While they are often utilize silt and sand substrates, ammocoetes also use coarse sand and small gravel substrate (Almeida and Quintella 2002). They may also utilize low velocity pockets with sand behind boulders (Cochnauer et al. 2006).

- To protect ammocoete rearing habitat, maintain channel stability (gradients) by preventing channel incision.

*Adults*

- Stream reaches with a variety of large cobble, boulder, and bedrock substrates may provide habitat for adults to overwinter.
- Adult lampreys will often wedge them selves into natural or manmade crevices.
- Low gradient riffles and pool tailouts with substrate ranging in size from 27 to 88 mm with low embeddedness provide spawning habitats (Gunckel et al. 2006). Spawning habitats utilized by salmonids are generally compatible with the needs of lampreys.

## Conclusions

In the Pacific Northwest, most stream restoration projects focus on salmonids and generally do not consider the needs of lampreys. The unique life history of the Pacific lamprey requires project modifications that consider long instream residency; use of substrates by ammocoetes; movements and migrations of ammocoetes, macropthalmia, and adults; and poor swimming abilities at all life stages. General awareness by project designers, managers, and construction crews of the need to consider Pacific lampreys when implementing instream activities is currently limited and needs to be improved. In addition, few methods have been developed to address lamprey conservation needs, even though data indicate that their abundance and distribution has declined.

Implementation of actions for instream projects can be modified to not only minimize impacts, but also to help benefit and protect lampreys by modifying dewatering regimes, adjusting the project work area and seasonal timing, protecting preferred habitats, and designing structures to accommodate upstream and downstream lamprey passage requirements for all life stages. Torgerson and Close (2004) stated that, "Lamprey conservation and restoration efforts in rivers and streams need to focus on the importance of habitat heterogeneity at multiple scales to promote complexity in channel morphology and sediment composition."

The decline in Pacific lampreys and our lack of knowledge of them should result in cautious actions and diligent monitoring of conditions and lamprey responses to various types of projects. However, this lack of understanding should not prevent inclusion of measures that are likely to limit harm and promote survival and reproduction of lampreys while implementing day-to-day instream actions. We can protect and benefit lampreys through modifications in project design and implementation. Steps that will lead to lamprey protection and minimize project impacts are (1) raising awareness among stream management practitioners that Pacific lampreys may occur in project areas, (2) dissemination of lamprey life history information to anyone involved with stream related work, (3) identification of locations within streams where activities have the greatest potential to affect lampreys and developing maps that identify Pacific lamprey distribution, (4) advancement of techniques to restore or enhance lamprey habitats and minimize affects to the various life history stages during project design and implementation, (5) refinement of methods to address passage for all life stages, and (6) continued research, monitoring, and evaluation to refine steps 2 through 5.

## Acknowledgments

Special thanks to Janine Castro and Christina Luzier for their useful comments and editing of this paper. Thanks to Jennifer Thompson and Jody Brostrom for their review and suggestions. Thanks to Mary Moser, Abel Brumo Stewart Reid, Janet Oatney, and many others for their interest in discussing this topic and providing input.

## References

Almeida, P. R., and B. R. Quintella. 2002. Larval habitat of the sea lamprey (*Petromyzon marinus*) in the River Mondego (Portugal). Freshwater fish conservation: options for the future. Blackwell Science, Oxford, UK.

Bayer, J., T. Robinson, and J. Seelye. 2001. Upstream migration of Pacific lampreys in the John Day River: behavior, timing, and habitat use. 2000 Technical Report, Project No. 200005200 to Bonneville Power Administration, BPA Report DOE/BP-26080-1 2000, Portland, Oregon.

Claire, C. 2004. Pacific lamprey larvae life history, habitat utilization, and distribution in the South Fork Clearwater River drainage, Idaho. 2000–2002 Technical Report, Project No. 200002800 to Bonneville Power Administration, BPA Report DOE/BP-00004039-3, Portland, Oregon.

Close, D. A., M. S. Fitzpatrick, H. W. Li, B. Parker, D. Hatch, and G. James. 1995. Status report of the Pacific lamprey (*Lampetra tridentata*) in the Columbia River basin. Report to Bonneville Power Administration, Portland, Oregon.

Close, D. 2001. Pacific Lamprey Research and Restoration Project. Annual Report 2001, Project No. 1994–02600 to Bonneville Power Administration, BPA Report DOE/BP-00005455-3, Portland, Oregon.

Close, D. A., M. S. Fitzpatrick, and H. W. Li. 2002. The ecological and cultural importance of a species at risk of extinction, Pacific lamprey. Fisheries 27(7):19–25.

Cochnauer, T., C. Claire, and S. Putnam. 2006. Evaluate status of Pacific lamprey in the Clearwater River and Salmon River drainages, Idaho. 2005 Annual Report, Project No. 200002800 to Bonneville Power Administration, BPA Report DOE/BP-00020623-1, Portland, Oregon.

CRBLTW (Columbia River Basin Lamprey Technical Workgroup). 2004. Passage considerations for Pacific lamprey. Columbia Basin Fish and Wildlife Authority, Portland, Oregon.

Daigle, W. R., C. A. Peery, S. R. Lee, M. L. Moser. 2005. Evaluation of adult Pacific lamprey passage and behavior in an experimental fishway at Bonneville Dam. Technical Report 2005-1 for U.S. Army Corps of Engineers Portland District, Portland, Oregon and Bonneville Power Administration, Portland, Oregon.

Gilmore, T. 2005. Response of lamprey ammocoetes (*Lampetra* sp.) to four types of water diversion screens. Master's thesis. Washington State University, Pullman.

Gunckel, S. L., K. K. Jones, and S. E. Jacobs. 2006. Spawning distribution and habitat use of adult Pacific and western brook lamprey in Smith River, Oregon. Oregon Department of Fish and Wildlife, Information Report 2006–1, Corvallis.

Hammond, R. J. 1979. Larval biology of the Pacific lamprey, *Entosphenus tridentata* (Gairdner), of the Potlatch River, Idaho. Master's thesis. University of Idaho, Moscow.

Hyatt, M. W., C. Claire, and T. Cochnauer. 2007. Evaluate status of Pacific lamprey in the Clearwater and Salmon River drainages, Idaho. Annual Report 2006. Prepared for the Bonneville Power Administration, Portland, Oregon

Jackson, A. D., P. D. Kissner, D. R. Hatch, B. L. Parker, D. A. Close, M. S. Fitzpatrick, and H. Li. 1996. Pacific lamprey research and restoration. Annual Report 1996. Prepared for the Bonneville Power Administration, Portland, Oregon.

Kan, T. T. 1975. Systematics, variation, distribution, and biology of lampreys of the genus *Lampetra* in Oregon. Doctoral dissertation. Oregon State University, Corvallis.

Kostow, K. 2002. Oregon lampreys: natural history, status, and analysis of management issues. Oregon Department of Fish and Wildlife, Corvallis.

Lê, B., C. W. Luzier, and T. Collier. 2004. Evaluate habitat use and population dynamics of lamprey in Cedar Creek. 2003 Annual Report prepared for the Bonneville Power Administration, Portland, Oregon.

Luzier, C. W., G. Silver, and T. Whitesel, 2006. Evaluate habitat use and population dynamics of lampreys in Cedar Creek. 2005 Annual Report prepared for the Bonneville Power Administration, Portland, Oregon.

Luzier, C. W. and seven coauthors. 2009. Proceedings of the Pacific Lamprey Conservation Initiative Work Session - October 28–29, 2009. U.S. Fish and Wildlife Service, Regional Office, Portland, Oregon.

Luzier, C. W., and G. S. Silver. 2005. Evaluate habitat use and population dynamics of lamprey in Cedar Creek. 2004 Annual Report prepared for the Bonneville Power Administration, Portland, Oregon.

Meeuwig, M. H., J. M. Bayer, and J. G. Seelye. 2005. Effects of temperature on survival and development of early life stage Pacific and western brook lampreys. Transactions of the American Fisheries Society 134:19–27.

Mesa, M., J. Bayer, J. Seelye, and L. Weiland. 2003. Swimming performance and physiological responses to exhaustive exercise in radio-tagged and untagged Pacific lampreys. Transactions of the American Fisheries Society 132:483–492.

Moser, M. L., J. M. Butzerin, D. B. Dey. 2007. Capture and collection of lampreys: the state of the science. Reviews in Fish Biology and Fisheries 17(1):45–56.

Moser, M. L., D. A. Ogden, D. L. Cummings, and C. A. Peery. 2006. Development and evaluation of a lamprey passage structure in the Bradford Island Auxiliary Water Supply Channel, Bonneville Dam, 2004. Report of Research to the U.S. Army Corps of Engineers, Portland District, Portland, Oregon.

Moursund R. A., D. D. Dauble, and M. J. Langesley. 2003. Turbine intake diversion screens: investigating effects on Pacific lamprey. HydroReview 22(1):40–46.

Moursund, R. A., D. D. Dauble, and M. D. Bletch. 2000. Effects of John Day Dam bypass screens and project operations on the behavior and survival of juvenile Pacific lamprey (*Lampetra tridentata*). U.S. Army Corps of Engineers, Portland, Oregon.

Moyle, P. B. 2002. Inland fishes of California. University of California Press, Berkeley.

Ocker, P. A., L. C. Stuehrenberg, M. L. Moser, A. L. Matter, J. J. Vella, B. P. Sandford, T. C. Bjornn, and K. R. Tolotti. 2001. Monitoring adult Pacific lamprey (*Lampetra tridentata*) migration behavior in the lower Columbia River using radiotelemetry, 1998–1999. Report of Research to the U.S. Army Corps of Engineers, Portland District, Portland, Oregon.

Pirtle, J., J. Stone and S. Barndt. 2003. Evaluate habitat use and population dynamics of lamprey in Cedar Creek. 2002 Annual Report prepared for the Bonneville Power Administration, Portland, Oregon.

Pletcher, F. T. 1963. The life history and distribution of lampreys in the Salmon and certain other rivers in British Columbia, Canada. Master's thesis. University of British Columbia, Vancouver.

Renaud, C. B., 2008 Petromyzontidae, *Entosphenus tridentatus*: southern distribution record, Isla Clarión, Revillagigedo Archipelago, Mexico. Check List 4(1):82–85.

Scott, W. B., and E. J. Crossman. 1973. Freshwater fishes of Canada. Fisheries Research Board of Canada Bulletin 184.

Simpson, J., and R. Wallace. 1982. Fishes of Idaho. University of Idaho Press, Moscow.

Stone, J., T. Sundlov, S. Barndt, and T. Coley. 2001. Evaluate habitat use and population dynamics of lamprey in Cedar Creek. 2000 Annual Report prepared for the Bonneville Power Administration, Portland, Oregon.

Stone, J., J. Pirtle, and S. Barndt. 2002. Evaluate habitat use and population dynamics of lamprey in Cedar Creek. 2001 Annual Report prepared for the Bonneville Power Administration, Portland, Oregon.

Torgerson, C. E., and D. A. Close. 2004. Influence of habitat heterogeneity on the distribution of larval Pacific lamprey (*Lampetra tridentata*) at two spatial scales. Freshwater Biology 49:614–630.

U.S. Fish and Wildlife Service. 2004. Endangered and threatened wildlife and plants; 90-day finding on a petition to list three species of lampreys as threatened or endangered. Federal Register 69:247(27 December 2004):77158.

Vella, J., L. Stuehrenberg, and T. Bjornn. 1999. Radio telemetry of Pacific lamprey (*Lampetra tridentata*) in the lower Columbia River, 1996. Annual Report of Research to the U.S. Army Corps of Engineers, Portland, Oregon.

White, J. L., and B. C. Harvey. 2003. Basin-scale patterns in the drift of embryonic and larval fishes and lamprey ammocoetes in two coastal rivers. Environmental Biology of Fishes 67:369–378.

American Fisheries Society Symposium 72:269–278, 2009


# Current Status and Distribution of the Pacific Lamprey South of Point Conception, Coastal Southern California, USA

Camm C. Swift*
*ENTRIX, Inc., 2140 Eastman Avenue, Suite 200, Ventura, California 93003, USA*

Steve R. Howard
*United Water Conservation District, 106 North 8th Street, Santa Paula, California 93060, USA*

*Abstract.*—Since the 1970s, Pacific lamprey *Entosphenus tridentatus* (formerly *Lampetra tridentata*) have been documented in various drainages from coastal Southern California. These collections and observations document a relatively stable population especially in the Santa Clara River from the 1970s to 1990s, followed by a significant decline in numbers since the late 1990s, most abruptly after about 2000. This decline may be steeper than documented since collection efforts have increased in this same period. Pacific lamprey counts taken in downstream migratory steelhead *Oncorhynchus mykiss* traps were documented by the California Department of Fish and Game in Sespe Creek, a Santa Clara River tributary, in 1983 and 1984. The Highway 126 crossing of Sespe Creek, our sample site, is near this trapping site and was sampled repeatedly with seines in 1981–1988, 1995–1998, and 2005–2007. The most robust Pacific lamprey data set in Southern California is from intermittent annual counts at the Freeman Diversion on the Santa Clara River since 1991. Other localities south of Point Conception have also been sampled intermittently. Since the late 1990s, Pacific lamprey trapped during annual collections has dropped to a few individuals. The past 7 years (2002–2009) of efforts have encountered one ammocoete in lower Sespe Creek and 12 macrothalmia or ammocoetes in the main-stem Santa Clara River through both wet and dry years. Five ammocoetes collected in the lower Ventura River on March 8, 2005 are the only other individuals recorded south of Point Conception in the 2000s. This decline in Pacific lamprey south of Point Conception is not easily explained since many conditions have not changed appreciably in that time. Current and long-standing impacts such as dams and diversions that lack adequate Pacific lamprey passage, extended droughts, frequent high-intensity fires, poor water quality, fluctuating ocean conditions, and predation by nonnative piscivores may be sufficient to explain the decline.

## Introduction

The southern range limit for Pacific lamprey *Entosphenus tridentatus* (formerly *Lampetra tridentata*) is within coastal Southern California and northern Baja California, and both marine and freshwater records are sparse for this area. The Pacific lamprey is one of four north temperate coastal freshwater fish species that reach their southern limit in this area and have been strongly affected by human disturbances, especially south of Point Conception. The other three species include steelhead *Oncorhynchus mykiss*, threespine stickleback *Gasterosteus aculeatus*,

* Corresponding author: cswift@entrix.com or camswift@pacbell.net

and prickly sculpin *Cottus asper*. South of Point Conception, the abruptly elevated transverse range of mountains creates relatively cool and wet aquatic conditions in the coastal lowlands southward to and beyond the Los Angeles Basin (Swift et al. 1993; Gumprecht 1999; Boughton et al. 2006, 2007). However, the Mediterranean climate regime of seasonal rains led the first and subsequent European, Mexican, and American inhabitants to strongly affect the reliability, continuity, and quality of the existing natural watercourses from water diversions and urban and agricultural runoff (Van Wormer 1991). Thus, today, Pacific lamprey encounter intermittent streamflows and limited migration opportunities, artificially extended droughts, and poor water-quality conditions, including high water temperatures. Currently, increasing groundwater and surface water extractions, importation of water from northern California and the Colorado River, introductions of nonnative aquatic species, and increased frequency of natural and human-caused fires have collectively affected the habitat of this coolwater fauna.

Marine records for lamprey in the eastern North Pacific are rare in general. Freshwater habitat for Pacific lamprey in all coastal Southern California rivers deteriorated considerably before scientists or other investigators noted this species in the permanent record. Hubbs (1967) summarized the known marine and freshwater records for Southern California up to 1967 and noted that ammocoetes were common in several streams before 1920. Hubbs also summarized information on the seldom-recognized, and by then extinct, small species of exclusively freshwater, nonparasitic lamprey in the genus *Lampetra*. Swift et al. (1993) updated records for both the Pacific lamprey and the smaller freshwater nonparasitic lamprey, including the discovery of a few old specimens located at Whittier College verifying the early records of Ewy (1945). These specimens, currently located in the Los Angeles County Museum (LACM), completed the documentation of the nonparasitic form in all three drainages of the Los Angeles Basin: the Los Angeles, San Gabriel, and Santa Ana River systems. Chase (2001) reported on extensive catch data of Pacific lamprey from the Freeman Diversion in the lower Santa Clara River, Ventura County in the late 1990s. Ruiz-Campos and Gonzalez-Guzman (1996) and Ruiz-Campos et al. (2000) reported the first known Pacific lamprey occurrences in Mexican freshwaters. These records included evidence of reproduction in the Rio Santo Domingo drainage on the Pacific side of northern Baja California, about 330 km south of the United States–Mexican border. Moyle (2002) summarized these and some of the additional records documented below. We have continued to monitor and survey for Pacific lamprey in coastal Southern California at some of these same sites and have also collected information on catches of Pacific lamprey in other coastal Southern California streams from Point Conception south to the Mexican border. These efforts have resulted in few lamprey observations (reviewed below). The objective of this paper is to update the status and distribution of Pacific lamprey at the southern limit of its range.

## Methods

We have compiled all records known to us from the primary literature, gray literature, and museum records and have documented our own and other investigators' collections and observations. In a few cases, Pacific lampreys were documented by early observers or steelhead anglers (Lang et al. 1998). Three localities have been sampled repeatedly for many years that resulted in some quantitative measure of population abundance or status, including (1) lower Malibu Creek and lagoon, sampled by Swift and other investigators since the early 1970s, (2) the Freeman Diversion located 16.8 km upstream of the ocean in the Santa Clara River where anadromous migration monitoring began in 1991, and (3) in lower Sespe Creek, a tributary of the Santa Clara River, about 2 km upstream of its mouth in the river and 36 km from the ocean since the early 1970s. Malibu and Sespe creeks were sampled primarily by seine and direct observations. Upstream migration monitoring was conducted at the Freeman Diversion on the Santa Clara River with an upstream migrant trap that was designed to capture steelhead and by periodically turning water out of the fish ladder to conduct a thorough visual inspection throughout the ladder. The upstream migrant fish trap was decommissioned in 1998 following the federal listing of Southern California steelhead populations as endangered. Downstream migration monitoring was conducted at the Freeman Diversion from 1991 to 2009 with a downstream migrant fish trap that collected all fish species migrating downstream through the diversion. When flow is adequate to allow for volitional passage to the ocean, trapping at the Freeman Diversion does not occur. Additionally, an unknown fraction of downstream migrants does not enter the diversion facility and migrates over the dam during spill event in wet years.

Although Point Conception is often considered the dividing line between northern and Southern California fish faunas, we include the Santa Ynez River and Jalama Creek in our analysis. These drainages are located just north of Point Conception and may be important in assessing the southern limit of Pacific lamprey. The majority of the data collected for Pacific lamprey in Southern California are from the Santa Clara River watershed and Malibu Creek.

## Results

Since 1970, Pacific lamprey records from the Point Conception area south to Mexico have been primarily limited to the largest river drainages (Figure 1). The observations and catch data for Southern California stream systems start at the Santa Ynez River and continue southward to the currently known extent of the species range in California.

### *Santa Ynez River*

Hubbs (1967) documented a specimen taken in the lagoon at the mouth of the Santa Ynez River on May 4, 1941. This specimen is in the University of Michigan Museum of Zoology (UMMZ) collection (UMMZ 13374). The California Academy of Sciences (CAS) has one specimen: (CAS 15167), from the Santa Ynez River below the town of Santa Ynez, collected on August 15, 1948, field number Q48–56 taken by Clark Hubbs.

The Santa Ynez River Consensus Committee (1997) noted an adult lamprey taken at the San Lucas Bridge (Highway 154 Bridge) in 1993, an ammocoete just upstream of Alisal Road (near the town of Solvang) on May 26, 1994, and a spawning pair near the tail of the spill basin of Bradbury Dam on March 14, 1996.

Biologists at the Cachuma Operations and Maintenance Board (COMB) trap steelhead in Santa Ynez River tributaries and intermittently in the main stem. Two Pacific lamprey adults were captured in the main-stem Santa Ynez River near the town of Solvang on February 14, 1996 (490 and 480 mm) and another on March 20, 1996 (475 mm). The COMB biologists also observed a mating pair near the base of the Bradbury Dam in February of 1995 (S. Engblom, Cachuma Operations and Maintenance Board, personal communication). No Pacific lampreys have been observed or trapped by COMB biologists in the Santa Ynez River since 1996.

### *Jalama Creek*

An anecdotal Pacific lamprey record exists in Jalama Creek, a known steelhead stream historically that lies just north of Point Conception (Swift et al. 1993; Titus et al. 2000). On March 15, 2000, Brad Lundberg, foreman of Jalama Creek Ranch, remembered an "old timer" (87 years old in 2000) who fished for steelhead near the railroad trestle at the upper end of the lagoon on lower Jalama Creek and occasionally observed "eels" that were probably adult Pacific lampreys. These observations were most likely in or before the 1960s.

### *Ventura River*

No formal Pacific lamprey sampling efforts have been documented in the Ventura River.

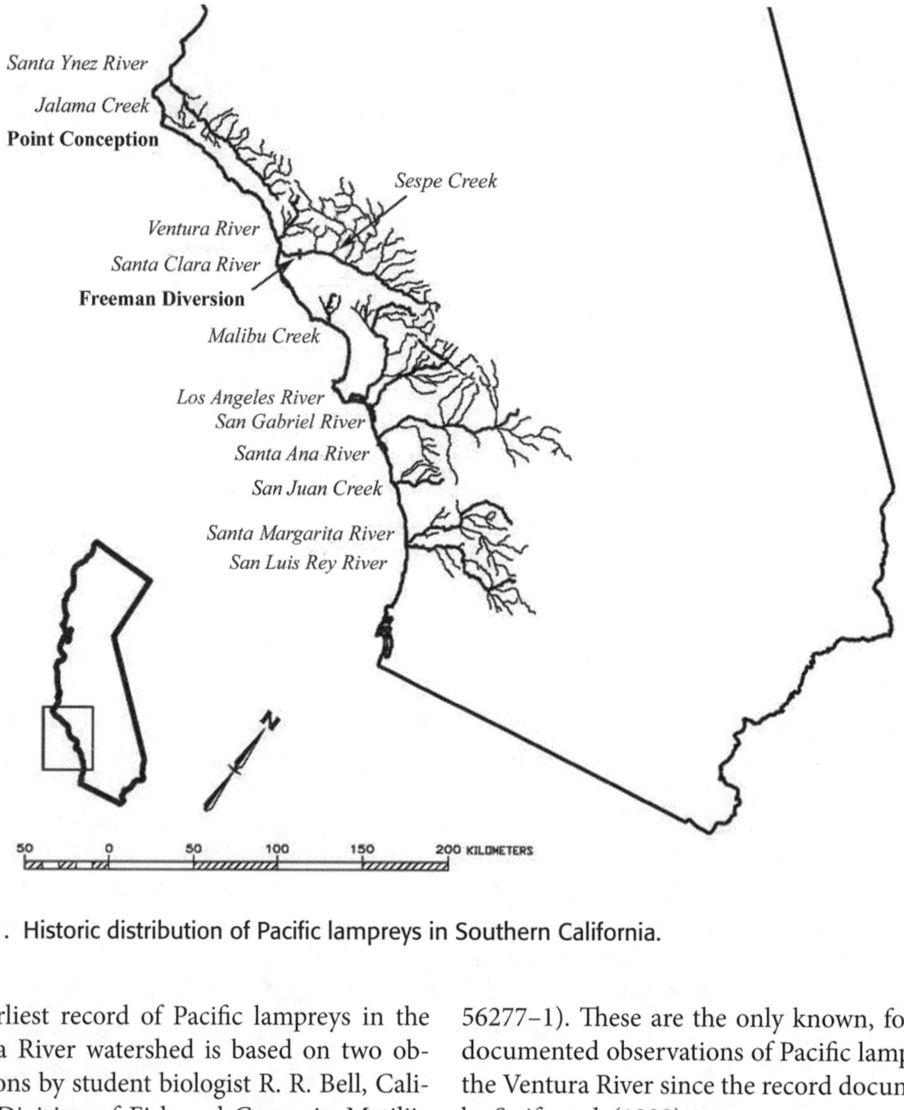


Figure 1. Historic distribution of Pacific lampreys in Southern California.

The earliest record of Pacific lampreys in the Ventura River watershed is based on two observations by student biologist R. R. Bell, California Division of Fish and Game, in Matilija Creek in 1951 after the construction of Matilija Dam in 1947. He identified these fish as brook lampreys, *Lampetra planeri*, but they were most likely recently landlocked Pacific lamprey.

A steelhead population survey was conducted in 2006 and 2007 in the Ventura River and no Pacific lampreys were observed or captured (M. Allen, Thomas R. Payne and Associates, personal communication). Goodman et al. (2008) sampled for Pacific lampreys in the lower Ventura River in 2005 and no lamprey were observed or collected. We collected five ammocoetes in the lower Ventura River approximately 3 km from the ocean on March 8, 2005 (LACM 56277–1). These are the only known, formally documented observations of Pacific lamprey in the Ventura River since the record documented by Swift et al. (1993).

## Santa Clara River and Sespe Creek

Sampling and observations at the Freeman Diversion have recorded numerous Pacific lampreys over the years. The total collections of Pacific lampreys at the Freeman Diversion from 1991 to 2009 include 2,179 adults, and 309 macrothalmia and ammocoetes. Since 2000, a total of 33 adults and 27 ammocoetes have been collected at the Freeman Diversion, which is 1.5% and 8.7%, respectively, of the total catches from 1991 to 2009 (Table 1). The Freeman Diversion upstream migrant fish trap was decom-

Table 1. Freeman diversion catches on the Santa Clara River from 1991 to 2009.

| Year | Adults | Macrothalmia and ammocoetes |
|---|---|---|
| 1991 | 74 | 5 |
| 1992 | No data | No data |
| 1993 | 465 | No data |
| 1994 | 908 | 70 |
| 1995 | 368 | 113 |
| 1996 | 309 | 54 |
| 1997 | 20 | 39 |
| 1998 | 0 | 1 |
| 1999 | 2 | 0 |
| 2000 | No data | No data |
| 2001 | 33 | 15 |
| 2002 | 0 | 0 |
| 2003 | 0 | 5 |
| 2004 | 0 | 6 |
| 2005 | 0 | 0 |
| 2006 | 0 | 1 |
| 2007 | 0 | 0 |
| 2008 | 0 | 0 |
| 2009 | 0 | 0 |
| Total | 2,179 | 309 |

missioned in 1998, but numerous ladder shutdowns occurred from 1991 to 2009. The entire fish ladder was surveyed visually for lamprey during shutdowns. These ladder shutdowns would result in observations of numerous Pacific lampreys prior to 2000. The downstream migrant trap has remained in operation from 1991 to 2009.

In Sespe Creek, a tributary of the Santa Clara River, Pacific lampreys have been observed as far upstream as 100 km from the ocean near Howard Creek at an elevation of more than 900 m. Sam Sweet and John Endler observed and collected Pacific lamprey ammoecotes in Sespe Creek below the confluence of Bear Creek during an Arroyo toad *Bufo californicus* survey conducted on May 8, 1981 (LACM 45700-1). Additionally, Sam Sweet and Steve Norton observed numerous Pacific lamprey ammocoetes and collected a small sample of them in Sespe Creek just upstream from Howard Creek on July 5, 1981. Currently, the July 5, 1981 sample is in a collection at the University of California Santa Barbara (S. Sweet, University of California Santa Barbara, personal communication).

The data from Sespe Creek (Table 2) show that until the late 1990s, sampling efforts generally resulted in catches of Pacific lampreys from December to June when flows were adequate for anadromous migration. Sampling efforts in Sespe Creek were most frequent in the late 1970s to early 1980s, in the late 1990s, and from 2005 to 2007. The winter and spring of 2006–2007 was so dry that no flow was observed at the Sespe Creek sampling site. The data for 1983 and 1984 are inflated due to extensive steelhead trapping by the California Department of Fish and Game in Sespe Creek that also took many Pacific lampreys of all three swimming life stages (adults, macrothalmia, and ammmocoetes). Until 1998, Pacific lampreys were taken virtually every time a reasonable seining effort was made. The only Pacific lamprey recorded in Sespe Creek after 1998 was of an observation of one ammocoete during an electrofishing survey in late June 2004. The extreme low rainfall year of 2006–2007 prevented Sespe Creek flow from extending downstream to the Highway 126 Bridge sampling site in the spring of 2007. The strong flows of 2005 were expected to be particularly good for the species, but only one was taken during sampling events in 2004, 2005, and 2006.

Table 2. Sespe Creek catches at the Highway 126 Bridge from 1975 to 2006.

| Year | Adults | Macrothalmia | Ammocoetes |
|---|---|---|---|
| 1975 | 1 | 0 | 2 |
| 1979 | 0 | 0 | 2 |
| 1981 | 0 | 0 | 26 |
| 1982 | 0 | 0 | 1 |
| 1983 | 2 | 0 | 10 |
| 1984 | 0 | 25 | 10 |
| 1987 | 1 | 0 | 0 |
| 1988 | 0 | 0 | 0 |
| 1995 | 0 | 0 | 3 |
| 1996 | 0 | 2 | 1 |
| 1997 | 0 | 0 | 5 |
| 1998 | 0 | 0 | 15 |
| 2004 | 0 | 0 | 1 |
| 2005 | 0 | 0 | 0 |
| 2006 | 0 | 0 | 0 |
| Total | 4 | 27 | 75 |

## Malibu Creek

In Malibu Creek, Pacific lampreys are limited to the lower 4.0 km below the impassable Rindge Dam that was constructed in 1929. Substantial collection efforts in Malibu Creek often failed to take Pacific lampreys, and presence records are limited to a few random events, namely sightings of dead adults and collections of ammocoetes (Table 3). The first record was of an ammocoete dip-netted during an invertebrate survey on July 30, 1981 (LACM 42596-1). Multiple freshwater fish surveys conducted between 1970 and 1981 resulted in no Pacific lampreys. On January 12, 1982, seven ammocoetes were collected in Malibu Creek (LACM 42694-1). On March 31, 1987 Shoken Sasaki, California Department of Fish and Game, collected one adult Pacific lamprey at the base of Rindge Dam. On May 20, 1991 Leo French and Lori Willis observed a dead adult in Malibu Creek Lagoon (specimen not retained). Finally, on March 30, 1993, approximately 10 Pacific lamprey ammocoetes were collected and released at the Malibu Creek Lagoon interface about 200 m upstream of the Highway 1 Bridge from a large sand bar at the mouth of a storm drain (C. C. Swift, unpublished data). Several subsequent sampling efforts for Pacific lampreys in Malibu Creek have resulted in negative results, including electroshocking efforts in August of 2005 (Goodman et al. 2008) in Malibu Creek and near the lagoon interface (Table 3). Based on multiple sampling efforts in Malibu Creek and the fact that those efforts before 1981 failed to detect Pacific lampreys, this species appears to be rare, difficult to detect, and/or only sporadically present in the system.

## Los Angeles River

The Los Angeles County Museum has one specimen that was collected in the Los Angeles River. The record is LACM 00005, which is of one specimen taken on May 2, 1925 by G. H. Reeve in the Los Angeles River, most likely near the City of Los Angeles between downtown and the current location of the Los Angeles Zoo.

Table 3. Malibu Creek catches from 1971 to 2007.

| Year | Adults | Macrothalmia | Ammocoetes |
|---|---|---|---|
| 1971 | 0 | 0 | 0 |
| 1972 | 0 | 0 | 0 |
| 1973 | 0 | 0 | 0 |
| 1974 | 0 | 0 | 0 |
| 1981 | 0 | 0 | 1 |
| 1982 | 0 | 0 | 7 |
| 1987 | 0 | 0 | 0 |
| 1991 | 1 | 0 | 0 |
| 1992 | 0 | 0 | 0 |
| 1993 | 0 | 0 | 10 |
| 2000 | 0 | 0 | 0 |
| 2006 | 0 | 0 | 0 |
| 2005 | 0 | 0 | 0 |
| 2007 | 0 | 0 | 0 |
| Total | 11 | 0 | 13 |

## San Gabriel River

The Los Angeles County Museum has three collections of Pacific lampreys from the San Gabriel River. The first collection (LACM 31614-1) is of four specimens (141–145) collected on October 21, 1945 by Chris Henne in the San Gabriel Canyon, Los Angeles County. The second collection (LACM 32593-1, 2[109–114]) is of Pacific lampreys collected near Pelliseer Road and the San Gabriel Dam south of Woodland Park (Whittier vicinity) in 1943 or 1944 by E. E. Ewy and L. James. The third collection (LACM 32594-1, 1[498]) was collected in the San Gabriel River approximately 3 mi from Cate Dam, also in the Whittier vicinity by Roy Newsom on April 20, 1944. The last specimen was cited by Hubbs (1967).

## Santa Ana River

The only definite records of Pacific lamprey in the Santa Ana River are of a single adult observed at the mouth of the river in March of 1991 (Swift et al. 1993) and from a note by Culver and Hubbs (1917) that migratory lamprey occur in this drainage. On February 23, 2009, Michael Shrosbee, bulldozer operator on the Santa Ana River at the River Road crossing north of the city of Norco, related catching two "eels" at this site in the 1960s. His description of these as "six to eight inches long and about the diameter of a pencil" strongly suggests lampreys. However, they could have been the smaller, freshwater, nonparasitic species of *Lampetra*, the last specimen of which was collected in that same vicinity in 1955 (Hubbs 1967; Swift et al. 1993).

### *Santa Margarita River*

The only known record of Pacific lampreys in the Santa Margarita River was of an adult lamprey approximately 68 cm long observed migrating upstream by Alan Brain in De Luz Creek, a tributary of the Santa Margarita River, sometime between the mid-1930s and early 1940s (Lang et al. 1998).

### *San Luis Rey River*

The only known record of Pacific lamprey in the San Luis Rey River was of a single ammocoete that was collected, photographed, and released in the lower San Luis Rey River at the Foussat (Foussant on some maps) Road crossing on June, 25 1997 by Linda Pardy, then of the California Department of Fish and Game.

## Discussion

A reduction in Pacific lamprey distribution and abundance in Southern California appears to be partially the result of dams and diversions, surface and groundwater extractions, flood control projects, environmental stochastic events, treated wastewater discharge, and potentially declining ocean conditions. Current stochastic events that occurred in Southern California include an extended dry period from 1999 to 2004 followed by one of the wettest years on record in 2005 that was then followed by the driest year on record in many parts of Southern California in 2007. These dynamic weather patterns were then coupled with large, high-intensity fires. Two major fires occurred within the Santa Clara River watershed in 2003 and 2006, as well as in Los Angeles and San Diego county areas where Pacific lampreys historically occurred. The fire that occurred in the Santa Clara River in 2003 burned approximately 11% of the watershed and the Day Fire in 2006 burned approximately 16% of the of the watershed area, including much of the Sespe Creek watershed. The affect of these events on Pacific lamprey habitats is exacerbated by the exceptionally steep, youthful, rapidly rising, and thus highly erosive terrain of Southern California (Norris and Webb 1990). In addition, the extensive urbanization of the lower elevation areas bordering the steep mountains also increases the magnitude of runoff discharge to natural habitats of anadromous and other fishes (Swift et al. 1993; Gumprecht 1999; National Marine Fisheries Service 2007). Many of these lowland stretches receive varying amounts of wastewater discharge that increases nutrient loading and contributes to anoxia in sediments of streams and lagoons that are preferred habitats for Pacific lamprey ammocoetes (Lee and Jones 1991). Possibly lagoons hold some importance for lamprey recruitment, and the combination of these factors in the past 10 years or so could have resulted in a lack of Pacific lamprey recruitment in Southern California drainages.

Our records confirm the observations of Shapovalov and Taft (1954) that Pacific lampreys are restricted to larger drainages as compared to steelhead, which migrate up both large- and small-order coastal drainages. Many smaller streams in this area that have steelhead or held them historically do not have records of Pacific lampreys. However, this could be partially or completely due to the lack of formal surveys for the presence of this species, difficulties in capturing early life histories of this species, and the alteration of habitat in smaller streams before any biological observations had been made. All of the records where we discovered historic or current Pacific lamprey presence are primarily from the largest drainages south of Point Conception that often correspond to those known, at least historically if not currently, as steelhead streams. However, we have no records from the many smaller steelhead streams, such as those between Point Conception and the Ventura River, and no records exist for several small streams between the larger Southern California rivers. Unlike steelhead, Pacific lampreys seldom maintain reproducing populations above natural or artificial barriers (Beamish and Northcote 1989; McPhail 2007), and no such populations have been detected in Southern California. Extensive sampling efforts above dams in Southern California streams have never revealed Pacific lampreys reproducing upstream. The ultimate survival of Pacific lampreys appears more dependent on the persistence of adequate migration corridors between the ocean and freshwater spawning and rearing areas than for steelhead, which can maintain resident freshwater populations.

The most recent Pacific lamprey records are limited to the largest rivers, including the Ventura River (2005), the Freeman Diversion in the Santa Clara River (2006), Sespe Creek (2005), Malibu Creek (1993), Los Angeles River (1925), San Gabriel River, (early 1940s), Santa Ana River (March 1991), Santa Margarita River (mid-1930s to early 1940s), and San Luis Rey River (1997). The recent records in the Ventura River and the Santa Clara River are only of a few ammocoetes. The records in drainages south of Malibu Creek are either very old or, if recent, appear to be stray adults since they are so rare. The majority of these records all date from the 1950s or earlier except for the single adult observed at the mouth of the Santa Ana River in March of 1991 (Swift et al. 1993). Hubbs (1967), Swift et al. (1993) and Miller et al. (2005) considered the Santa Ana River the southernmost freshwater record for the migratory or anadromous Pacific lamprey in the United States. However, a single ammocoete was collected, photographed, and released in the lower San Luis Rey River in the spring of 1997. Lang et al. (1998) cite an observation by Alan Brain of an adult Pacific lamprey approximately 68 cm long observed migrating upstream in De Luz Creek, a tributary of the Santa Margarita River, sometime between the mid-1930s and early 1940s. Brain was a preteen boy at the time and had several photographs of himself holding adult steelhead caught by hand in the same vicinity in those years. This observation of at least one adult Pacific lamprey along with steelhead in the same few years when steelhead appeared to be a regular occurrence in northern San Diego County suggests consistent runs in these years and possibly earlier. A thorough, multi-year survey of the Santa Margarita drainage from 1997 to 1999 failed to document the presence of Pacific lampreys or steelhead (Warburton et al. 2000) despite spanning a record rainfall year. Ruiz-Campos and Guzman (1996) have documented the potential for invasion of suitable streams much farther south, namely the Rio Santo Domingo in northern Baja California. Like steelhead, Pacific lampreys appear to still have the potential and capability to invade neighboring streams when conditions are favorable. Pacific lampreys are potentially affected by warming ocean temperatures associated with the El Niño Southern Oscillation, the Pacific Decadal Oscillation, and the effects of climate change (Horn and Stephens 2006). The collective effects of these patterns are predicted by Horn and Stephens (2006) to cause a northward shifting of warmwater marine faunas, with coldwater faunas displaced to the north. Under these scenarios, it may become increasingly difficult to restore Pacific lampreys in Southern California.

The presence in Southern California of nonnative piscivorous aquatic species and those that disrupt the soft benthic substrate likely have contributed to the decline of Pacific lamprey. These effects are intensified during drought periods when fishes and other aquatic organisms become concentrated resulting in increased vulnerability to predation (Swift et al. 1993). Largemouth bass *Micropterus salmoides*, redeye bass *M. coosae*, green sunfish *Lepomis cyanellus*, catfish *Ictalurus punctatus* and bullhead catfish *Ameiurus* sp., brown trout *Salmo trutta*, introduced strains of rainbow trout (nonanadromous *O. mykiss*), common carp *Cyprinus carpio*, bullfrog *Rana catesbeiana*, African clawed frog *Xaenopis* sp., and red swamp crayfish *Procambarus clarki* occur in some combination in all the drainages where Pacific lampreys have been recorded currently or historically. Most of these nonnative species directly prey on Pacific lampreys or disrupt the spawning substrate and areas where ammocoetes burrow. Before the influx of these nonnative species, Pacific lampreys and steelhead in Southern California streams outside the Los Angeles Basin were competing with only threespine stickleback *Gasterosteus aculeatus*, and from the Ventura River northward, threespine stickleback and prickly sculpin (Swift et al. 1993; Moyle 2002). In the Los Angeles Basin, a larger native fauna of an additional freshwater nonparasitic lamprey, two minnow species, a sucker, and a freshwater shrimp existed, somewhat analogous to a similar combination in lower San Francisco Bay streams (Swift et al. 1993; Moyle 2002).

The Pacific lamprey populations in Southern California became much reduced before genetic studies utilizing mitochondrial DNA could be used to assess their relationship to more northern populations (Goodman et al. 2008). Goodman et al. (2008) found very little variation between fish from central California to British Columbia, indicating considerable mixing of stocks and lack of genetically distinct populations. However, using amplified fragment length polymorphism fingerprint of genomic DNA, Lin et al. (2008) found statistically significant differences between all seven populations sampled from northern California to southern Alaska. Since considerable endemism is known already from the Los Angeles Basin in Southern California among other imperiled inland fishes (Swift et al. 1993; Moyle 2002; Jelks et al. 2008), it will be important for conservation purposes to establish whether Pacific lamprey populations in Southern California are genetically distinct as well.

## Acknowledgments

We appreciate the time Stewart Reid and Damon Goodman took in sharing the results of their collecting efforts. Dan Tormey gave valuable guidance to the manuscript. Justin Campbell and Dan Detmer provided essential assistance with maps. Scott Engblom shared valuable Santa Ynez River data. Sam Sweet shared valuable observations in Sespe Creek. Mark Allen shared important Ventura River survey data. Mike Gibson shared valuable Matilija Creek data. Since the earlier survey (Swift et al. 1993), many agency and other biologists continued to share their observations during field surveys, particularly California Department of Fish and Game, U.S. Forest Service, National Marine Fisheries Service, U.S. Geological Survey, Topanga-Las Virgenes Resource Conservation District, and the Resource Conservation District of the Santa Monica Mountains. In addition, many consultants who work on freshwater and anadromous fishes in Southern California have shared their experiences with the local fauna.

## References

Beamish, R. J., and T. G. Northcote. 1989. Extinction of a population of anadromous parasitic lamprey, *Lampetra tridentata*, upstream of an impassable dam Canadian Journal of Fish and Aquatic Sciences 46(3):420–425.

Boughton, D. A., P. Adams, E. Anderson, C. Fusaro, E. Keller, E. Kelley, L. Lenstch, J. Nielsen, K. Perry, H. Regan, J. Smith, C. Swift, L. Thompson, and F. Watson. 2006. Steelhead of the south-central/southern California coast: population characterization for recovery planning. NOAA Technical Memorandum NMFS-SWFSC-394.

Boughton, D. A., P. Adams, E. Anderson, C. Fusaro, E. Keller, E. Kelley, L. Lenstch, J. Nielsen, K. Perry, H. Regan, J. Smith, C. Swift, L. Thompson, and F. Watson. 2007. Viability criteria for steelhead of the south-central and southern California coast. NOAA Technical Memorandum NMFS-SWFSC-407.

Chase, S. 2001. Contributions to the life history of adult Pacific lamprey (*Lampetra tridentata*) in the Santa Clara River of southern California. Bulletin of the Southern California Academy of Sciences 100(2):74–85.

Culver, G. B., and C. L. Hubbs. 1917. The fishes of the Santa Ana system streams in southern California. Lorquinia 1:82–83.

Ewy, E. E. 1945. An ecological study of certain portions of the San Gabriel River basin. Master's thesis. Whittier College, Whittier, California.

Goodman, D. H., S. B. Reid, M. F. Docker, G. R. Haas, and A. P. Kinziger. 2008. Mitochondrial DNA evidence for high levels of gene flow among populations of a widely distributed anadromous lamprey *Entosphenus tridentatus*. Journal of Fish Biology 72:400–417.

Gumprecht, B. 1999. The Los Angeles River: its life, death, and possible rebirth. The Johns Hopkins University Press, Baltimore, Maryland.

Horn, M. H. and J. S. Stephens, Jr. 2006. Chapter 25. Climate change and overexploitation. Pages 621–635 *in* L. G. Allen, D. J. Pndella, II, and M. H. Horn, editors. The ecology of marine fishes. California and adjacent waters. University of California Press, Berkeley.

Hubbs, C. L. 1967. Occurrence of the Pacific lamprey, *Entosphenus tridentatus*, off Baja California and in streams of southern California, with remarks on its nomenclature. Transactions of the San Diego Society of Natural History 14:301–312.

Jelks, H. L., S. J. Walsh, N. M. Burkhead, S. Contreras-Balderas, E. Diaz-Pardo, D. A. Hendrickson, J. Lyons, N. E. Mandrak, F. McCormick, J. S. Nelson, S. P. Platania, B. A. Porter, C. B. Renaud, J. J. Schmiter-Soto, E. B. Taylor, and M. L. Warren. 2008. Conservation status of imperiled North American freshwater and diadromous fishes. Fisheries 33(8):372–407.

Lang, J. S., B. F. Oppenheim, and R. N. Knight. 1998. Southern steelhead, *Oncorhynchus mykiss*, habitat suitability survey of the Santa Margarita River, San Mateo and San Onofre creeks on Marine Corps Base Camp Pendleton, California. Coastal California Fish and Wildlife Office, Arcata, California for Assistant Chief of Staff, Environmental Security, Marine Corps Base Camp Pendleton.

Lee, G. F. and R. A. Jones. 1991. Effects of eutrophication on fisheries. Reviews in Aquatic Sciences 5:287–305. CRC Press, Boca Raton, Florida.

Lin, B., Z. Ahang, Y. Wang, K. P. Currens, A. Spidle, Y. Yamazaki, and D. A. Close. 2008. Amplified fragment length polymorphism assessment of genetic diversity in Pacific lampreys. North American Journal of Fisheries Management 28:1182–1193.

McPhail, J. D. 2007. The freshwater fishes of British Columbia. The University of Alberta Press, Edmonton, Alberta, Canada.

Miller, R. R., W. L. Minckley, and S. N. Norris. 2005. The freshwater fishes of Mexico. University of Chicago Press, Chicago.

Moyle, P. B. 2002. Inland fishes of California. Revised and expanded. University of California Press, Berkeley.

National Marine Fisheries Service. 2007. Federal recovery outline for the distinct population segment of southern California coast steelhead. National Marine Fisheries Service, Southwest Regional Office, Santa Barbara, California.

Norris, R. M., and R. W. Webb. 1990. The geology of California. Wiley, New York.

Ruiz-Campos, G., S. Contreras-Baldera, M. L. Lozano-Vilano, S. Gonzales-Guzman, and J. Alaniz-Garcia. 2000. Ecological and distributional status of the continental fishes of northwestern Baja California, Mexico. Bulletin of the Southern California Academy of Sciences 99:245–267.

Ruiz-Campos, G., and S. Gonzalez-Guzman. 1996. First record of Pacific lamprey, *Lampetra tridentata*, from Baja California, Mexico. California Fish and Game 82:144–146.

Shapovalov, L., and A. C. Taft. 1954. The life histories of the steelhead rainbow trout (*Salmo gairdneri*) and silver salmon (*Oncorhynchus kisutch*). California Department of Fish and Game, Fish Bulletin 98.

Santa Ynez River Consensus Committee 1997. Synthesis and analysis of information collected on the fishery resources and habitat conditions of the lower Santa Ynez River: 1993–1996. Prepared for the Santa Ynez River Consensus Committee, Santa Barbara, California.

Swift, C. C., T. R. Haglund, M. Ruiz, and R. Fisher. 1993. Status and distribution of the freshwater fishes of southern California. Bulletin of the Southern California Academy of Sciences 92(3):101–167.

Titus R. G., D. C. Erman, and W. M. Snider. 2000. History and status of steelhead in California coastal drainages south of San Francisco Bay. Draft. California Department of Fish and Game, Sacramento.

Van Wormer, S. R. 1991. A history of flood control in the Los Angeles County drainage area. Southern California Quarterly 73:55–94.

Warburton, M. L., C. C. Swift, and R. N. Fisher. 2000. Status and distribution of fishes in the Santa Margarita River drainage. Report to The Nature Conservancy, Newport Beach, California by U.S. Geological Survey, San Diego State University, San Diego, California.

American Fisheries Society Symposium 72:279–292, 2009


# Status and Conservation of Lampreys in California

Peter B. Moyle*

*Department of Wildlife, Fisheries, Conservation Biology, University of California One Shields Ave, Davis, California 95616, USA*

Larry R. Brown

*U.S. Geological Survey, California Water Science Center Placer Hall, 6000 J Street, Sacramento, California 95819, USA*

Shawn D. Chase

*Sonoma County Water Agency Post Office Box 11628, Santa Rosa, California 95406, USA*

Rebecca M. Quiñones

*Department of Wildlife, Fisheries, Conservation Biology, University of California One Shields Ave, Davis, California 95616, USA*

*Abstract.*—Lampreys are among the least studied group of fishes in California. At least seven species inhabit freshwater habitats within the state, including the Kern brook lamprey *Lampetra hubbsi*, a California endemic. Four species are micropredators on fish, Pacific lamprey *Entosphenus tridentatus* (formerly *L. tridentata*), river lamprey *L. ayresii*, Klamath lamprey *E. similis* (formerly *L. similis*) and Goose Lake lamprey *Entosphenus* sp. The remaining three species are nonfeeding as adults and are presumed to have many populations isolated from one another. Pacific lamprey and river lamprey are anadromous and may have increased diversity through multiple runs. A systematic analysis of the limited information available indicates that, with the possible exception of the Pit-Klamath brook lamprey *E. lethophagus* (formerly *L. lethophaga*), all species are either declining, in low numbers, or in isolated populations. Causes of the declines are multiple and species-specific, but in general, alteration of watersheds by humans, resulting in increased siltation, temperatures and pollution, as well as other habitat changes are the principal causes. Protecting lampreys has the benefit of protecting stream ecosystems throughout the state because of the wide historic presence of lampreys and because ammocoetes require clean, cool water and relatively complex habitat, including stable backwaters.

## Introduction

Lampreys are among the least studied groups of fishes in California despite their diversity (seven species) and presence in many California streams, including most coastal streams. There has been a general recognition that anadromous lampreys have declined and that most nonanadromous species are rare, making them of special concern (Moyle 2002). However systematic evaluation of their status is lacking. This lack of study was made explicit for California and other Pacific coast states when a recent petition to list four species of lamprey under the Federal Endangered Species Act (Klamath-Siskiyou Wildlands Center and 10 other conservation organizations 2003) was rejected because of a lack of quanti-

* Corresponding author: pbmoyle@ucdavis.edu

tative information (U.S. Fish and Wildlife Service 2004a, 2004b). Despite their importance in the diet and culture of Native American tribes (Kroeber and Barrett 1960), lampreys have never been important in commercial or recreational fisheries, perhaps explaining the lack of interest on the part of management agencies.

California is home to seven species of lamprey, although recent taxonomic work suggests that there may be more than seven. Two of the species are anadromous micropredators (i.e., they prey on fish larger than themselves), the Pacific lamprey *Entosphenus tridentatus* (formerly *Lampetra tridentata*) and the river lamprey *L. ayresii*. Three of the species are nonmigratory, nonpredatory forms, including Pit-Klamath brook lamprey *E. lethophagus* (formerly *L. lethophaga*), Kern brook lamprey *L. hubbsi*, and western brook lamprey *L. richardsoni*. Two other micropredator species are Klamath lamprey *E. similis* (formerly *L. similis*), which is endemic to the Klamath River system of California and Oregon, and Goose Lake lamprey *Entosphenus* sp., which is native to the Goose Lake basin.

Moyle (2002) and Moyle et al. (in press) review what little is known about California lampreys. In this article, we conduct an analysis of the status of the seven lampreys and of the factors responsible for their apparent decline, using methods derived from more general ones (for fishes) in Moyle et al. (in press). Although our approach is semi-quantitative, similar analyses have been shown to be useful (e.g., Moyle and Williams 1990; Molnar et al. 2008). The purpose of this article is to review the status of the lamprey species present in California. Because so little information is available on lampreys, this analysis relies on the professional judgment and unpublished observations of the authors. Our goal is to demonstrate the need for more studies on lamprey biology and status, as well as to provide the best available judgment of their status in California.

## Methods

We evaluated the status of California lampreys using information compiled in Moyle (2002) and in Moyle et al. (in press), as well as personal observations. Using this information, we conducted a limiting factors analysis to identify the major factors likely limiting lamprey populations (Table 1). Factors were rated as high (major limiting factor), moderate (potential to be a major factor limiting population viability in near future) and low (low or unknown effect on population viability). Because the information for all lampreys is limited, we developed a reliability score for our assessments, on a 1–4 scale (Table 1).

Following Moyle et al. (in press), the status of each lamprey species was evaluated using seven criteria, which included aspects of lamprey biology, vulnerability to environmental change, and limiting factors. All criteria were scored on a 1–5 scale where 1 was a low score and 5 was a high score (Table 2). The seven criteria were then averaged to produce an overall score for each species. A lamprey species scoring 1 or 2 was regarded as being in serious danger of extirpation in California or, for endemic species, extinction. A species scoring 4 or 5 was regarded as reasonably secure for the immediate future. A reliability score (above) was also calculated for the information for each species.

## Results and Discussion

### *Overall*

The limiting factors analysis (Table 3) indicates that multiple factors are reducing the populations of all seven lampreys in California but that the major factors affect each species differently. Reliability of information was low for all species, reflecting very limited quantitative data and little published research. Potential limiting factors common to four or more species included major dams, agriculture, grazing, and transportation. The only factors considered major limiting factors for more than one species were major dams and estuary modification. The limiting factors analysis (Table 3) indicates that the cumulative effects of many modifications to watersheds cause streams containing lampreys to be degraded as lamprey habitat, reducing

Table 1. Anthropogenic factors limiting, or potentially limiting, viability of populations of lamprey species in California. The factors are interactive and overlapping.

| Factor | Effects on lampreys |
|---|---|
| Major dams | Block passage, reduce flows or alter flow regimes, alter temperatures, increase embeddedness |
| Agriculture | Pollution from return water, increase temperatures and sediments, reduce flows, entrain ammocoetes |
| Grazing | Increase sedimentation, trample stream banks and rearing areas, reduce cover, pollution from wastes |
| Urban/suburban development | Pollution from runoff and sewage, reduced backwater rearing areas, reduced flows, block passage |
| Mining | Destruction of spawning and rearing habitat by instream mining (gravel, gold dredging), increased siltation, pollution from mine runoff |
| Transportation | Roads and railroads confine streams, reduce habitat, limit access to tributaries, increase pollution (e.g., spills) |
| Logging | Increase erosion rates and sediment loads, remove riparian cover, reduce logs in streams for cover |
| Fire | Direct effects on streams, indirect effects through increases in erosion, nutrients, water temperatures |
| Estuary modification | Estuaries important for anadromous forms, which are affected by reduced inflows, access, pollution, habitat loss, prey species reductions |
| Recreation | Some disturbance of ammocoetes and migration adults by in-river recreation |
| Harvest | Pacific lamprey harvest for bait and for food by Native Americans |
| Prey abundance | Decline in salmon and other prey fish |
| Alien species | Increased predation rates by alien centrarchids and salmonids |
| Reliability of information | Scored 1–4, where 1 was unreliable because little solid information was available; 2 was somewhat reliable, based on expert opinion and limited data; 3 was moderately reliable, based mainly on gray literature; and 4 was highly reliable, based on numerous accounts in the published and agency literature |

lamprey population viability. The ammocoetes (larvae) are particularly vulnerable because they spend multiple years in limited habitats and can be affected by transitory events such as rapid drops in river level or spills of pollutants.

Although our analysis of factors determining the status of lampreys in California (Table 4) reflects our limited knowledge of them, the information we have suggests that most species/populations are stressed by the classic problems of declining species. These include limited ranges of endemic species, reduced genetic diversity, small population size, isolation of populations, and alteration of habitats to conditions beyond physiological tolerances. Climate change is a growing problem, creating additional stress on individuals and their habitats. The average score of the seven metrics evaluated ranged from 2.7 to 4.3 (out of 5.0). Thus, as species in California, none of the seven lampreys seems to be in immediate danger of extinction. However, none of the species can be regarded as secure because populations are disappearing one at a time. The lampreys on the lower end of the status scale (Table 4) clearly need protection and management if they are going to persist.

## *Pacific lamprey*

Pacific lamprey is the species most often observed in California because of its size (>40 cm total length [TL]) and anadromous habits. Historically, it was extremely important as a tribal food source (Kroeber and Barrett 1960) because lamprey runs occurred somewhat earlier than the first salmon runs; thus, lampreys were available when other food was scarce. A small Yurok tribal fishery continues on the Klamath River, but the fishery is relatively small and not well monitored. Estimates of Yurok harvest in 2002, 2003, and 2004 were 1,400, 350, and 2,000

Table 2. Metrics used to evaluate the status of California lampreys, derived from Moyle et al. (in press). Evaluation applies only to California populations.

1. Distribution in California (only)[a]
   1. One watershed/stream system
   2. 2–3 watersheds/stream systems
   3. 3–5 watersheds/stream systems
   4. 6–10 watersheds/stream systems
   5. More than 10 watersheds/stream systems in California
2. Effective population size of largest population
   1. <50
   2. 50–100
   3. 100–1,000
   4. 1,000–10,000
   5. 10,000+
3. Dependence on human intervention (water management, habitat improvement) for persistence
   1. Captive broodstock program or similar extreme measures required to prevent extinction.
   2. Population persistence requires continuous active management (e.g., water addition or similar measures).
   3. Population persistence requires annual monitoring and intervention when needed (e.g., management of barriers, special flows, removal of alien species, establishment of refuge populations).
   4. Persistence requires periodic habitat improvements (e.g., gravel augmentation, habitat restoration).
   5. Self-sustaining populations require minimal intervention.
4. Environmental tolerance of adults and ammocoetes in freshwater (mainly physiological tolerances in relation to existing conditions)[b]
   1. Found naturally in extremely narrow range of conditions.
   2. Found in limited range of conditions, reduced by watershed alterations.
   3. Moderate tolerance to variable environmental conditions encountered.
   4. Broad tolerance to variable environmental conditions but some limitations. from pollution, etc.
   5. Tolerance of environmental conditions rarely an issue.
5. Genetic risks/problems
   1. Fragmentation, genetic drift, and isolation by distance are major forces shaping genetic diversity within and among extant California populations.
   2. As above, but limited gene flow among populations reduces risk, although hybridization can continue to be a threat.
   3. Moderately diverse genetically; genetic risks low but present.
   4. Genetically diverse but limited gene flow to other populations.
   5. Genetically diverse with gene flow to other populations (good metapopulation structure).
6. Vulnerability to extirpation from climate change
   1. Highly vulnerable in all watersheds inhabited.
   2. Highly vulnerable in most watersheds inhabited (possible refuges present).
   3. Vulnerable in portions of watersheds inhabited (e.g., headwaters, lowermost reaches of coastal streams).
   4. Low vulnerability due to location, coldwater sources, and/or active management.
   5. Not vulnerable to significant population loss.
7. Limiting factors analysis (Table 1)
   1. One or more threats could push species to extinction in <10 years.
   2. One or more threats could push species to extinction in 10–50 years.
   3. One or more threats could push species to extinction in 50–100 years.
   4. One or more threats may reduce populations, but extinction unlikely as a result.
   5. Known threats do not imperil the species.

[a] Moyle and Marchetti (2006) determined that there were 44 large, independent watersheds in California.
[b] A lamprey species may have apparent broad physiological tolerances based on historic distributions, but if it lives in a region (e.g., Southern California) where habitat conditions (e.g., temperature) naturally reach close to the limits of their likely tolerance, the environmental tolerance is scored lower. However, the physiological tolerances of lampreys are poorly understood.

Table 3. Anthropogenic factors limiting, or potentially limiting, viability of populations of lamprey species in California that were evaluated using expert opinion and available literature. H–factor has high effect, a major limiting factor; M–factor has a moderate effect, potential major limiting factor in future; L–factor currently has low or unknown effect.

| | Pacific lamprey | Klamath River lamprey | Goose Lake lamprey | Pit-Klamath brook lamprey | Kern brook lamprey | Western brook lamprey | River lamprey |
|---|---|---|---|---|---|---|---|
| Major dams | H | M | – | L | H | H | M |
| Agriculture | M | M | M | L | M | H | M |
| Grazing | L | M | H | M | L | M | L |
| Urban/suburban development | M | L | – | L | L | M | M |
| Mining | M | L | – | L | L | L | L |
| Transportation | M | M | M | M | L | H | M |
| Logging | L | M | L | M | – | M | L |
| Fire | L | L | H | M | L | H | – |
| Estuary modification | H | – | – | – | – | – | H |
| Recreation | L | – | – | – | L | L | – |
| Harvest | M | – | – | – | – | – | – |
| Prey abundance | H | M | L | – | – | – | H |
| Alien species | M | L | L | L | M | M | L |
| Reliability of information | 2 | 1 | 2 | 2 | 2 | 2 | 2 |

Table 4. Factors determining the status of lampreys in California, where 1 is a major negative factor contributing to status, 5 is a factor with no or positive effects on status, and 2–4 are intermediate values.

| Factor | Pacific lamprey | Klamath River lamprey | Goose Lake lamprey | Pit-Klamath brook lamprey | Kern brook lamprey | Western brook lamprey | River lamprey | Comments |
|---|---|---|---|---|---|---|---|---|
| Area occupied | 4 | 5 | 2 | 2 | 3 | 5 | 4 | All have multiple populations. |
| Effective population size | 4 | 4 | 2 | 4 | 3 | 2 | 3 | Most populations isolated. |
| Intervention dependence | 4 | 4 | 4 | 5 | 4 | 4 | 5 | Likely to be more important in future. |
| Tolerance | 3 | 3 | 3 | 3 | 3 | 3 | 3 | All moderately tolerant of poor water quality. |
| Genetic risk | 5 | 5 | 3 | 4 | 1 | 2 | 4 | Largely unknown, but isolated populations likely vulnerable to extirpation or inbreeding. |
| Climate change | 2 | 3 | 2 | 2 | 2 | 2 | 3 | Increased temperatures and reduced flows. |
| Limiting factors analysis | 3 | 5 | 3 | 4 | 3 | 3 | 4 | See Table 3. |
| Average | 3.6 | 4.3 | 2.7 | 3.4 | 2.7 | 3.0 | 3.7 | |
| Certainty (1–4) | 2 | 1 | 2 | 2 | 2 | 2 | 1 | Little published on status. |

lampreys (D. Williams, Yurok Tribal Fisheries, personal communication). Spawning Pacific lampreys are also harvested in California as sturgeon bait, but the extent of the fishery is unknown.

Pacific lampreys are found in Pacific coast streams from Japan (Morrow 1980) to Baja California (Ruiz-Campos and Gonzalez-Guzman 1996). In California, they were historically found along the entire coast and far inland along the major river systems. Pacific lampreys continue to occur in many northern and central coast streams of California. Historically, their distribution became more scattered and disjunct south of Point Conception (Swift et al. 1993), and they have apparently been extirpated from most Southern California streams in recent years (Swift and Howard 2009, this volume), with the exception of the Santa Clara River and possibly the Ventura River. The only remaining regular spawning runs have been in the Santa Clara River (Chase 2001), but even here, the numbers of migrants has declined severely in recent years (Swift and Howard 2009).

Pacific lamprey may be maintaining their wide distribution because of a high rate of "straying" from natal streams. In a study of Pacific lamprey genetics, Goodman (2006) examined populations from British Columbia, Canada to Point Conception on the California coast. Goodman (2006) found little differentiation among lampreys from the 81 sites sampled, indicating sufficient gene flow to homogenize the populations genetically. This suggests that Pacific lamprey have low fidelity to natal streams, resulting in significant gene flow through straying of adults returning to freshwaters to spawn.

In addition to anadromous populations, a number of landlocked populations with small body size have been identified in California. Some of these populations have been suggested as deserving species status (Moyle 2002). The actual relationships among these forms are complex and need resolution using techniques presented by Docker et al. (1999). A natural landlocked population has been identified from the upper Klamath River basin, including the Lost River in California (Vladykov and Kott 1979; Hubbs 1971). Populations associated with reservoirs above impassable dams include one in Clair Engle Reservoir on the Trinity River (Moyle 2002). Ammocoetes that were likely Pacific lamprey have been collected from Clear Creek upstream of Whiskeytown Reservoir (Brown and May 2007).

There is evidence that some rivers have at least two runs of Pacific lampreys (S. Chase, Sonoma County Water Agency, unpublished data). Pacific lampreys were counted during underwater video monitoring at an inflatable dam on the Russian River (at Mirabel, 37 km upstream from the mouth) between 2000 and 2008 (S. Chase, Sonoma County Water Agency, unpublished data). The video system is in place primarily from August 1 until the onset of the rainy season, usually between mid-November and early January. Video monitoring during the spring was also conducted in 2000 and 2007. During the spring, low but consistent numbers of Pacific lamprey (up to 46 per day) were observed migrating upstream between April 9 (earliest date monitored) and mid-June, with very few observed through early July. In 2000, the year with the most complete data set (May 12, 2000 through January 10, 2001), no lampreys were observed between July 3 and October 24. Partial springtime counts were 207 in 2000 and 570 in 2007. Low numbers of lamprey (<5 per day) were observed in the fall from late October through November, with one lamprey observed in early January (monitoring ended January 10, 2001). In all years sampled (2000–2008), low numbers (2–34, annually) of adult Pacific lamprey were counted, primarily in October and November. These data suggest a large spring run and smaller fall run.

Because there has been little interest in Pacific lamprey, it is difficult to assess its conservation status. The American Fisheries Society lists it as "vulnerable" throughout its range (Jelks et al. 2008), indicating that declines are not confined to California. Our analysis and Moyle (2002) suggest that Pacific lamprey populations are in decline throughout California, but the species is not in immediate danger of extirpation (Table 4), although local populations are probably being lost. Anecdotal evidence suggests that, similar to anadromous salmonids (Moyle et al. 2008), populations are much reduced from historical levels. The declines of salmon and lampreys may be interrelated, given the importance of salmon as prey for Pacific lamprey. The environmental causes of declines are multiple and interactive, including dams that cut off access to upstream spawning and rearing areas, modifications to estuaries that can effect passage of upstream and downstream migrants, and changes in land use (Table 3). Besides blocking migrations, dams have downstream effects such as disrupting upstream sediment inputs required to maintain habitat for ammocoetes. Ammocoetes are vulnerable to rapid decreases in stream flow. For example, on the Russian River, releases from Coyote Valley Dam are reduced every fall for a routine dam inspection and maintenance. Although the ramping rates are sufficient to protect salmonids, ammocoete and metamorphosing individuals are often stranded in large numbers and are vulnerable to desiccation and avian predation (Chase, unpublished data). Ammocoete habitat tends to be in low-gradient areas that are vulnerable to the effects of agriculture, urbanization, and channel maintenance (i.e., flood control projects, sediment removal). Also, culverts have been shown to block migration. Some indications of the status of some regional populations is as follows:

- Populations have been lost from almost all streams in Southern California, coincident with the urbanization of watersheds (Swift and Howard 2009).
- In the Sacramento-San Joaquin drainage, assuming that Pacific lamprey spawned in roughly the same areas as Chinook salmon *Oncorhynchus tshawytscha*, approximately 80% of their spawning habitat has been lost (Yoshiyama et al. 2001).
- In coastal watersheds south of San Francisco Bay, the flow and habitat requirements of Pacific lamprey are presumably similar to those of steelhead *O. mykiss*, which are listed as endangered (south-central coast and south coast distinct population segments). Limited data suggests that lampreys are increasingly scarce in these watersheds.
- In coastal watersheds north of San Francisco Bay, lampreys seem to still occur in most of their historic range, but surveys are lacking and numbers appear to be low compared to what they were in the past.
- Pacific lampreys are still abundant enough to be harvested in the Klamath River, but the population is likely vulnerable to the same problems associated with declines of anadromous salmonids (National Research Council 2004).

## *Klamath River lamprey*

The Klamath River lamprey is a predatory species with a mean adult length of about 21 cm TL. The species is distinct both morphologically and biochemically from Pacific lamprey (Docker et al. 1999; Lorion et al. 2000). It was originally described from the upper Klamath River and Upper Klamath Lake in Oregon, but it is widespread within the lower Klamath and Trinity River systems, including many tributary streams in California (Boyce 2002). A population of predatory lampreys in Copco Reservoir on the Klamath River (Coots 1955) presumably belongs to this species.

Although there is no quantitative information available on the abundance of Klamath River lamprey, it is likely they have declined from the same causes as the native resident and anadromous salmonids in the drainage (National Research Council 2004). These species are major prey for the lamprey. The potential important limiting factors for this species include major dams, agricultural activities, transportation, and logging (Table 3). These factors are not currently seen as having major negative effects on the species (Table 4). Moyle (2002) lists the species as declining but not yet in serious trouble, which fits our analysis as well. The American Fisheries Society indicates that it should have "vulnerable" status (Jelks et al. 2008).

### Goose Lake lamprey

The Goose Lake lamprey, first described by Carl Hubbs (1925), is a relatively small predatory lamprey confined to the Goose Lake basin in northeastern California and southeastern Oregon. Given its long isolation (ca. 1 million years) from other populations, the Goose Lake lamprey presumably deserves species status (Moyle 2002). The adults feed largely on Goose Lake tui chubs *Siphatales thallasinus* (also known as *Gila bicolor thallasina*), although the native redband trout *O. mykiss* subsp. (Moyle 2002) was probably a major prey (Hubbs 1925) before they became uncommon. The Goose Lake population is remarkable because this shallow lake in an arid region has gone dry several times in the past, most recently in 1992. The populations of lamprey and other endemic species are maintained in tributary streams with recolonization of the lake occurring when it refilled.

The Goose Lake lamprey is likely less abundant than it was historically and vulnerable to habitat loss. The species occupies a relatively small area of California and the populations may become quite small during times of drought when the species relies on small streams as refuges, resulting in one of the lowest average scores for factors affecting determining status (Table 4). Many of the tributary streams that likely serve as refuges during times of drought are dammed, diverted, and degraded by grazing, road building, and other human activities. The effects of grazing and fire are considered major potential limiting factors (Table 3). Fortunately, ongoing conservation efforts are reversing many of these problems and lampreys may persist by using small reservoirs as well.

The Goose Lake lamprey is listed in California as a species of special concern (Moyle et al., in press). Jelks et al. (2008) list it as a threatened species, with declining populations, a status similar to that given by NatureServe (www.natureserve.org). Oregon Department of Fish and Wildlife calls it a sensitive-critical species while the U.S. Forest Service lists it as a sensitive species due to its highly restricted range and vulnerability.

### Pit-Klamath brook lamprey

Genetic evidence indicates that this nonpredatory species is closely related to Pacific lamprey (Docker et al. 1999). It is found in the Pit River and Goose Lake drainages of northeastern California and southeastern Oregon. It is also found in the upper Klamath River of south-central Oregon, in the region upstream of Klamath Lakes (Hubbs 1971; Moyle and Daniels 1982). However, the taxonomy of this species is still unresolved and it is possible that the two populations may be independently derived from Pacific lamprey and are separate species (Moyle 2002).

This species is most commonly found in low-gradient reaches of clear streams with sand-silt bottoms or edges, where temperatures rarely exceed 25°C (Moyle and Daniels 1982). Such habitat is apparently still common in the Pit River drainage. Although Pit-Klamath brook lamprey populations have not been systematically surveyed since the studies of Moyle and Daniels (1982), Moyle (2002) indicates that the species is probably still widespread in the Pit River drainage and does not appear to be in decline. None of the limiting factors considered were scored as major limiting factors, although grazing, transportation, logging, and fire were considered moderate factors (Table 3). Our analysis reflects this judgment, but our status score was lowered by the species limited range and its potential to be affected by climate change (Table 4). The American Fisheries Society indicates that it should have "vulnerable" status (Jelks et al. 2008).

### Kern brook lamprey

This small (adults 8–14 cm TL), nonpredatory species is endemic to the east-side rivers of the San Joaquin Valley (Vladykov and Kott 1976; Brown and Moyle 1993). Based on tooth patterns, the species was thought to be derived from Pacific lamprey (Vladykov and Kott 1976); however, genetic evidence indicates derivation from river lamprey (Docker et al. 1999). Kern brook lampreys have been collected from the lower Merced, San Joaquin, Kings, and Kaweah rivers. Ammocoetes collected upstream of major storage reservoirs on the San Joaquin and Kings rivers are also likely this species. Kern brook lampreys were also collected from the siphons of the Friant-Kern Canal, which connects all these rivers, when the canal was poisoned to remove invasive white bass *Morone chrysops*.

Most of the known populations occur below large dams with limited connectivity to other populations through downstream reaches (Brown and Moyle 1993). Thus, major dams are considered a major limiting factor (Table 3) and limited connectivity of populations may represent a genetic risk (Table 4). Agriculture and introduced species are considered moderate limiting factors (Table 3). If a population is extirpated from a particular river, there is little likelihood of re-establishment without human intervention. Local extirpations seem possible given that the upstream reservoirs are operated without regard to protection of the species. A series of such extirpations could eventually result in extinction. These factors resulted in Kern brook lamprey sharing the lowest average status score (2.7) with Goose Lake lamprey (Table 4).

The Kern brook lamprey is listed as a species of special concern by the California Department of Fish and Game (Moyle et al., in press), a designation supported by our study. The American Fisheries Society considers the species to be threatened and declining (Jelks et al. 2008) while NatureServe (www.natureserve.org) considers its status to be somewhere between imperiled (G2) and critically imperiled (G1). In 2003, it was petitioned for listing as a threatened species under the Federal Endangered Species Act, but the petition was denied because "the petition did not provide sufficient information to warrant initiating a status review (U.S. Fish and Wildlife Service 2004a)." As an endemic species with a very limited range in regulated rivers, its populations should be monitored and managed.

## *Western brook lamprey*

Like Kern brook lamprey, western brook lampreys are nonpredatory and small, with a maximum size of about 18 cm TL. Western brook lampreys have been collected from coastal streams from California north to southeastern Alaska (Vladykov 1973; Morrow 1980). In California, most collections have been from the Sacramento River portion of the Sacramento-San Joaquin drainage system, including very isolated areas such as Kelsey Creek, a tributary to Clear Lake, Lake County, and St Helena Creek, a tributary to upper Putah Creek, Lake County (Moyle et al. in press). Hubbs (1925) collected them from Coyote Creek, a tributary to San Francisco Bay. Western brook lamprey have also been collected from a few California coastal streams, including the Eel River (Brown and Moyle 1997), the Navarro River (J. B. Feliciano, University of California Davis, personal communication), the Russian River and several of its tributaries (M. H. Fawcett, personal communication; Chase, unpublished data), and Prairie Creek (W. Duffy, U.S. Geological Survey, personal communication). This species may have also been present in the Los Angeles River basin of southern California, based on collections of ammocoetes (Culver and Hubbs 1917; C. L. Hubbs, personal communication); however, the population is now extirpated (Swift et al. 1993). Because adult brook lampreys are small and difficult to capture, while the ammocoetes are difficult to identify, it seems likely that western brook lampreys are (or were) more widely distributed in California streams and rivers than the limited number of collections would suggest.

Like Kern brook lamprey, the western brook lamprey is closely related to river lamprey based on biochemical evidence (Docker et al. 1999). It seems likely that the many coastal populations of western brook lamprey represent individual, independent evolutionary events similar to those described for threespine stickleback *Gasterosteus aculeatus* and resident populations of rainbow trout (Moyle 2002).

Major dams, transportation, agriculture practices, and wildfire are considered major limiting factors for this species because they all have strong effects on the small streams generally preferred by these lampreys (Table 3). Most collections have been from smaller streams located in the middle to upper reaches of watersheds that tend to located above dam sites, which isolate the populations. Such streams are also vulnerable to damage from such things as channelization, road crossings, and culverts associated with roads and railroads, which tend to run along streams. Agriculture can limit lampreys because water diversions and land reclamation in floodplains have similar effects to major dams while polluted return water can be too warm and toxic for brook lampreys. Fire can be important because most populations are primarily in upstream reaches within forests prone to wildfire (Moyle et al., in press). The conservation status of western brook lamprey in California is uncertain. For the species as a whole, Moyle (2002) and Moyle et al. (in press) consider it to be declining but not yet in trouble. It is listed as a species of special concern by the California Department of Fish and Game (Moyle et al., in press). However, if the separate populations of western brook lamprey have resulted from individual speciation events, then genetically distinct populations deserving of protection may exist in each major coastal watershed. Loss of such populations could be an important genetic risk for the species (Table 4). Evaluating this hypothesis would require evaluation of populations on a watershed basis.

### *River lamprey*

The river lamprey is a predatory species with a mean adult length of about 17 cm TL. The species has been collected from near Juneau, Alaska to the San Francisco Bay region (Vladykov and Follett 1958; Wydoski and Whitney 1979). The species has been reported from tributaries to San Francisco Bay, including Napa River, Sonoma Creek, and Alameda Creek (Leidy 2007). Migrating individuals are caught in small numbers in annual trawl surveys of the Sacramento-San Joaquin Delta and rotary screw traps monitoring Central Valley salmonid populations, so most reports are for the lower Sacramento and San Joaquin rivers and tributaries. In the Sacramento River system, river lampreys have been reported from the American and Feather rivers. In the San Joaquin River system, it has been reported from the Stanislaus and Tuolumne rivers. Spawning river lampreys have also been reported from Salmon Creek, a coastal stream, and tributaries to the Russian River (Sonoma County) (Fawcett, personal communication). A single adult female was collected from Cape Horn Dam on the Eel River, 250 km upstream of the ocean (Moyle 2002).

Quantitative data on river lamprey populations are lacking even though the Sacramento-San Joaquin watershed represents the southern limit of its range. It is likely that the species has declined as have anadromous salmonids at the southern end of their range (e.g., Chinook salmon, coho salmon *O. kisutch*, southern steelhead; Moyle et al. 2008) and likely for many of the same reasons. We identified estuary modification as a major limiting factor for this species because many of the collections have taken place in the highly modified Sacramento-San Joaquin Delta and most estuaries of coastal streams have been highly altered (Table 3). Major dams, agriculture, urbanization, and transportation were identified as moderate factors.

Moyle (2002) classified the species as declining, but there is little basis for making a status determination except the generally poor condition of most of the watersheds it inhabits combined with few recent collection localities. The American Fisheries Society recommends a "vulnerable" status (Jelks et al. 2008). Our analysis suggests that river lamprey are not in danger of extinction in California (Table 4), but given the lack of information on them and their apparent scattered distribution in larger rivers, this may not be true. Data are needed on all aspects of this species, including distribution, abundance, basic biology, and life history.

## Conclusions

With seven species, California has one of the most diverse lamprey assemblages in North America. It is likely that there are several other species that await formal description in the Klamath River and Goose Lake drainages. Regardless of taxonomy, the limited information available on California's lampreys indicates that all species, with the possible exception of the Pit-Klamath brook lamprey, appear to be less abundant and less widely distributed than they once were. Most are likely in steady decline, including the once-abundant Pacific lamprey. Especially worrisome are the four species of nonpredatory lampreys that are usually overlooked (or not identified) in stream surveys. These species are vulnerable to continuing habitat loss and fragmentation from human land-use practices. The apparent ongoing loss of lampreys from California streams may have implications for both ecology and conservation of stream systems.

Although the role of lampreys in California aquatic ecosystems has not been explicitly studied, they can be important in several ways. Ammocoetes are likely an important trophic pathway from organic detritus to predators. Ammocoetes are highly effective filter feeders and are very efficient at digesting organic detritus, although they ingest and process material relatively slowly (Mallatt 1982; Sutton and Bowen 1994; Mundahl et al. 2005). The slow-growing ammocoetes are a food source for some fishes. For example, Brown and Moyle (1997) documented consumption of lamprey ammocoetes by Sacramento pikeminnow *Ptychocheilus grandis* in the Eel River, California. In some California rivers, areas formerly or currently inhabited by ammocoetes now support large populations of filter-feeding introduced clams (Brown et al. 2007) that are not available to stream fishes as food and might even compete with ammocoetes for food or reduce space for their burrows. Adult Pacific lampreys serve as food for sea lions and seals (Roffe and Mate 1984; Browne et al. 2002; Orr et al. 2004) and likely other predators. Large runs of Pacific lampreys likely also represent, or did in the past, a large subsidy of marine-derived nutrients to streams and rivers, as has been demonstrated for salmon in California (Merz and Moyle 2006) and elsewhere (Willson and Halupka 1995; Wipfli et al. 1998, 2003).

Lampreys also represent fascinating evolutionary phenomena and a legacy of cultural use. They deserve protection for those reasons alone; however, protecting lampreys would have the added benefit of protecting stream ecosystems throughout the state because of the wide distribution of species and because of the habitat requirements of ammocoetes, especially clean, cool water and complex habitats that include relatively stable backwaters. The fact that all species of lampreys are most likely in decline in California suggests that many streams and rivers are in poor condition, as are the populations of fishes that serve as prey for predatory lamprey species. Thus, conserving lampreys has implications far beyond the protection of a few odd species that, at first glance, might seem to be of interest mainly to scientists. Lampreys are managed in Europe because some species have commercial value, with specific targets for abundance in specific river systems, so lessons can be learned from there (e.g., Nunn et al. 2008).

To accomplish lamprey conservation in California, the following actions should be considered:

1. Establish a systematic California lamprey survey for all species. Ideally, the survey would be conducted by a small team of biologists skilled in lamprey collection and identification.
2. Establish a lamprey data center, perhaps as part of the survey program. The data center would provide a forum for collection and integration of all lamprey information collected in California. The many rotary screw traps used to monitor outmigration of juvenile salmonids represent a potentially important and largely untapped source of data. Many trap operators record emigrating lamprey "smolt" and ammocoetes. Although specimens are rarely identified to species, the data can still

provide an indication of general lamprey status. Part of the program should be to provide materials to facilitate species identification.

3. Determine if conservation efforts for salmonids also benefit lampreys, especially in regulated streams. Do passage structures constructed for salmonids also pass lampreys? Do habitat restoration programs also create backwater habitat for lampreys? Are populations of Pacific lamprey tied to those of salmon and steelhead?

The underlying basis of the three suggestions is the need to pay more attention to lamprey conservation before species are lost or must be protected through listing under the Endangered Species Act. By the time species are listed, the emergency restoration efforts required to prevent extinction will be expensive and can limit other ecosystem services that humans previously obtained from lamprey habitats, such as clean water for diversion to cities.

## Acknowledgments

We thank the many people that provided unpublished data on lamprey populations. Comments from reviewers significantly improved the manuscript. Funding for the analysis of lamprey status was provided by California Department of Fish and Game, thanks to Glenn Yoshiyoka.

## References

Boyce, J. 2002. Genetic differentiation of lamprey populations within the genus *Lampetra* as determined by mtDNA control region sequence divergence and allozymes. Master's thesis. Humboldt State University, Arcata, California.

Brown, L. R., and J. T. May. 2007. Aquatic vertebrate assemblages of the upper Clear Creek Watershed, California. Western North American Naturalist 67:439–451.

Brown, L. R., and P. B. Moyle. 1993. Distribution, ecology, and status of the fishes of the San Joaquin River drainage, California. California Fish and Game 79:96–113.

Brown, L. R., and P. B. Moyle. 1997. Invading species in the Eel River, California: successes, failures, and relationships with resident species. Environmental Biology of Fishes 49:271–291.

Brown, L. R., J. Thompson, K. Higgins, and L. V. Lucas. 2007. Population density, biomass, and age-class structure of an invasive clam *Corbicula fluminea* in rivers of the lower San Joaquin River watershed, California. Western North American Naturalist 67:572–586.

Browne, P., J. L. Laake, and R. L. DeLong. 2002. Improving pinniped diet analyses through identification of multiple skeletal structures in fecal samples. Fishery Bulletin 100:423–433.

Chase, S. D. 2001. Contributions to the life history of adult Pacific lamprey (*Lampetra tridentata*) in the Santa Clara River of Southern California. Bulletin of the Southern California Academy of Sciences 100:74–85.

Coots, M. 1955. The Pacific lamprey, *Entosphenus tridentatus*, above Copco Dam, Siskiyou County, California. California Fish and Game 41:118–119.

Culver, G. B., and C. L. Hubbs. 1917. The fishes of the Santa Ana system streams in southern California. Lorquina 1:82–83.

Docker, M. F., J. H. Youson, R. J. Beamish, and R. H. Devlin. 1999. Phylogeny of the lamprey genus *Lampetra* inferred from mitochondrial cytochrome b and ND3 gene sequences. Canadian Journal of Fisheries and Aquatic Sciences 56:2340–2349.

Goodman, D.H. 2006. Evidence for high levels of gene flow among populations of a widely distributed anadromous lamprey, *Entosphenus tridentatus* (Petromyzontidae). Master's thesis. Humboldt State University, Arcata, California.

Hubbs, C. L. 1925. The life cycle and growth of lampreys. Papers of the Michigan Academy of Science Arts and Letters 4:587–603.

Hubbs, C. L. 1971. *Lampetra* (*Entosphenus*) *lethophaga*, new species, the nonparasitic derivative of the Pacific lamprey. Transactions of the San Diego Natural History Society 16:125–164.

Jelks, H. L., S. J. Walsh, N. M. Burkhead, S. Contreras-Balderas, E. Diaz-Pardo, D. A. Hendrickson, J. Lyons, N. E. Mandrak, F. McCormick, J. S. Nelson, S. P. Plantania, B. A. Porter, C. B. Renaud, J. J. Schmitter-Soto, E. B. Taylor, and M. J. Warren, Jr. 2008. Conservation status of imperiled North American freshwater and anadromous fishes. Fisheries 33:372–407.

Klamath-Siskiyou Wildlands Center and 10 other conservation organizations. 2003. A petition for rules to list: Pacific lamprey (*Lampetra tridentata*); river lamprey (*Lampetra ayresi*); western brook lamprey (*Lampetra richardsoni*); and Kern brook lamprey (*Lampetra hubbsi*) as threatened or endangered under the Endangered Species Act. Klamath-Siskiyou Wildlands Center, Ashland, Oregon.

Kroeber, A. L., and S. A. Barrett. 1960. Fishing among the Indians of northwestern California. Anthropological Records University of California 21.

Leidy, R. 2007. Ecology, assemblage structure, distribution, and status of fishes in streams tributary to the San Francisco Estuary, California. The San Francisco Estuary Institute, Contribution 530, Oakland, California.

Lorion, C. M., D. F. Markle, S. B. Reid, and M. F. Docker. 2000. Redescription of the presumed-extinct Miller Lake lamprey *Lampetra minima*. Copeia 2000:1019–1028.

Mallatt, J. 1982. Pumping rates and particle retention efficiencies of the larval lamprey, an unusual suspension feeder. Biological Bulletin 163:197–210.

Merz, J. E., and P. B. Moyle. 2006. Salmon, wildlife, and wine: marine-derived nutrients in the human-dominated ecosystems of central California. Ecological Applications 16:999–1009.

Molnar, J. L. R. L Gamboa, C. Revenga, and M. D. Spalding. 2008. Assessing the global threat of invasive species to marine biodiversity. Frontiers in Ecology and the Environment 6:485–492.

Morrow, J. E. 1980. The freshwater fishes of Alaska. Alaska Northwest Publishing, Anchorage.

Moyle, P. B. 2002. Inland fishes of California, revised and expanded. University of California Press, Berkeley.

Moyle, P. B., and R. A. Daniels. 1982. Fishes of the Pit River system, McCloud River system, and Surprise Valley Region. University of California Publications in Zoology 115:1–82.

Moyle, P. B., and J. E. Williams. 1990. Biodiversity loss in the temperate zone: decline of the native fish fauna of California. Conservation Biology 4:275–284.

Moyle, P. B., J. A. Israel, S. A. Purdy. 2008. Salmon, steelhead, and trout in California: status of an emblematic fauna. California Trout, Inc., San Francisco.

Moyle, P. B., R. M. Quiñones, and J. V. Katz. In press. Fish species of special concern in California, 3rd edition. California Department of Fish and Game, Sacramento.

Moyle, P. B., and M. P. Marchetti. 2006. Predicting invasion success: freshwater fishes in California as a model. Bioscience 56:515–524.

Mundahl, N. D., C. Erickson, M. R. Johnston, G. A. Sayeed, and S. Taubel. 2005. Diet, feeding rate, and assimilation efficiency of American brook lamprey larvae. Environmental Biology of Fishes 72:67–72.

National Research Council. 2004. Endangered and threatened fishes in the Klamath River basin: causes of decline and strategies for recovery. National Academies Press, Washington, D.C.

Nunn, A. D., J. P. Harvey, R. A. A. Noble, and I. G. Cowx. 2008. Condition assessment of lamprey populations in the Yorkshire Ouse catchment, north-east England, and the potential influence of physical migration barriers. Aquatic Conservation: Marine and Freshwater Ecosystems 18:175–189.

Orr, A. J., A. S. Banks, S. Mellman, H. R. Huber, R. L. DeLong, and R. F. Brown. 2004. Examination of the foraging habits of Pacific harbor seal (*Phoca vitulina richardsi*) to describe their use of the Umpqua River, Oregon, and their predation on salmonids. Fishery Bulletin 102:108–117.

Roffe, T. J., and B. R. Mate. 1984. Abundances and feeding habits of pinnipeds in the Rogue River, Oregon. The Journal of Wildlife Management 48:1262–1274.

Ruiz-Campos, G., and S. Gonzalez-Guzman. 1996. First freshwater record of Pacific lamprey, *Lampetra tridentata*, from Baja California, Mexico. California Fish and Game 82:144–146.

Sutton, T. M., and S. H. Bowen. 1994. Significance of organic detritus in the diet of larval lampreys in the Great Lakes basin. Canadian Journal of Fisheries and Aquatic Sciences 51:2380–2387.

Swift, C. C., and S. R. Howard. 2009. Current status and distribution of the Pacific lamprey south of Point Conception, coastal Southern California, USA. Pages 269–278 *in* L. R. Brown, S. D. Chase, M. G. Mesa, R. J. Beamish, and P. B. Moyle, editors. Biology, management, and conservation of lampreys in North America. American Fisheries Society, Symposium 72, Bethesda, Maryland.

Swift, C. C., T. R. Haglund, M. Ruiz, and R. N. Fisher. 1993. The status and distribution of the freshwater fishes of southern California. Bulletin of the Southern California Academy of Sciences 92:101–167.

U.S. Fish and Wildlife Service. 2004a. Endangered and threatened wildlife and plants; 90-day finding on a petition to list the Kern brook lamprey as threatened or endangered. Federal Register 69:247(27 December 2004):77152–77158.

U.S. Fish and Wildlife Service. 2004b. Endangered and threatened wildlife and plants; 90-Day finding on a petition to list three species of lampreys as threatened or endangered. Federal Register 69:247(27 December 2004):77158–77167.

Vladykov, V. D. 1973. *Lampetra pacifica*, a new nonparasitic species of lamprey (Petromyzontidae) from Oregon and California. Journal of the Fisheries Research Board of Canada 30:205–213.

Vladykov, V. D., and W. I. Follett. 1958. Redescription of *Lampetra ayresii* (Gunther) of western North America, a species of lamprey (Petromyzontidae) distinct from *Lampetra fluviatilis* (Linnaeus) of Europe. Journal of the Fisheries Research Board of Canada 15:47–77.

Vladykov, V. D., and E. Kott. 1976. A new nonparasitic species of lamprey from the genus *Entosphenus* Gill 1862, (Petromyzonidae) from south central California. Bulletin of the Southern California Academy of Sciences 75:60–67.

Vladykov, V. D., and E. Kott. 1979. A new parasitic species of the holarctic lamprey genus *Entosphenus* Gill, 1862 (Petromyzonidae) from Klamath River, in Oregon and California. Canadian Journal of Zoology 57:808–823.

Wydoski, R. S., and R. R. Whitney. 1979. Inland fishes of Washington. University of Washington Press, Seattle.

Willson, M. F., and K. C. Halupka. 1995. Anadromous fish as keystone species in vertebrate communities. Conservation Biology 9:489–497.

Wipfli, M. S., J. P. Hudson, and J. P. Caouette. 1998. Influence of salmon carcasses on stream productivity: response of biofilm and benthic macroinvertebrates in southeastern Alaska, U.S.A. Canadian Journal of Fisheries and Aquatic Sciences 55:1503–1511.

Wipfli, M. S., J. P. Hudson, J. P. Caouette, and D. T. Chaloner. 2003. Marine subsidies in freshwater ecosystems: salmon carcasses increase the growth rates of stream-resident salmonids. Transactions of American Fisheries Society 132:371–381.

Yoshiyama, R. M., E. R. Gerstung, F. W. Fisher, and P. B. Moyle. 2001. Historical and present distribution of chinook salmon in the Central Valley drainage of California. Pages 71–76 *in* R. L. Brown, editor. Fish Bulletin 179: Contributions to the biology of Central Valley salmonids. Volume 1. California Department of Fish and Game, Sacramento.

American Fisheries Society Symposium 72:293–309, 2009


# Taxonomy, Distribution, and Conservation of Lampreys in Canada

Claude B. Renaud*
*Canadian Museum of Nature*
*Post Office Box 3443, Station D, Ottawa, Ontario K1P 6P4, Canada*

Margaret F. Docker
*Department of Biological Sciences, University of Manitoba*
*Winnipeg, Manitoba R3T 2N2, Canada*

Nicholas E. Mandrak
*Fisheries and Oceans Canada, Burlington, Ontario L7R 4A6, Canada*

*Abstract.*—There are 11 lamprey species in Canada: Vancouver lamprey *Entosphenus macrostomus* (formerly *Lampetra macrostoma*), Pacific lamprey *Entosphenus tridentatus* (formerly *Lampetra tridentata*), chestnut lamprey *Ichthyomyzon castaneus*, northern brook lamprey *Ichthyomyzon fossor*, silver lamprey *Ichthtyomyzon unicuspis*, river lamprey *Lampetra ayresii*, western brook lamprey *Lampetra richardsoni*, Alaskan brook lamprey *Lethenteron alaskense*, American brook lamprey *Lethenteron appendix* (formerly *Lampetra appendix*), Arctic lamprey *Lethenteron camtschaticum* (formerly *Lampetra camtschatica*), and sea lamprey *Petromyzon marinus*. *Entosphenus* and *Lethenteron* were previously synonymized with *Lampetra*, but Nelson (2006) recognized these as three distinct genera. Conservation status has been assessed in only five species and in two of these (western brook lamprey and chestnut lamprey), only for portions of their Canadian population. The 2007 Committee on the Status of Endangered Wildlife in Canada list indicates that the enigmatic population of western brook lamprey in Morrison Creek, British Columbia, is endangered; the Vancouver lamprey in British Columbia is threatened; the chestnut lamprey in Saskatchewan and Manitoba is special concern; the northern brook lamprey in Manitoba, Ontario, and Quebec is of special concern in the latter two provinces and data deficient in the former; and the Alaskan brook lamprey in the Northwest Territories is data deficient. The threats to the four species at risk were collectively related to habitat degradation and loss, sensitivity to a catastrophic event, and sensitivity to lampricide used to control the invasive sea lamprey in the Great Lakes basin. Despite much lamprey work being conducted in the past decade, there still remain a number of knowledge gaps. These gaps include unequivocal evidence as to whether parasitic and nonparasitic members of a paired species should be considered distinct species and information on the distribution and population sizes and trends of the native lamprey species.

* Corresponding author: crenaud@mus-nature.ca

## Introduction

The freshwater fish fauna of Canada is depauperate relative to the North American fauna overall (199 of 1224 North American species are found in Canada; Nelson et al. 2004), and lampreys are no exception, although they are considerably better represented (11 of 20 North American lampreys are found in Canada; Nelson et al. 2004). This general pattern is the result of the relatively recent deglaciation of Canada following the Wisconsinan glacial period (ca. 10,000 years before present), limiting colonization and speciation events in Canada, and the relatively cool climate of much of Canada limiting overall richness (Crossman and McAllister 1986; McAllister et al. 1986). Like many freshwater fishes in North America, several lamprey species in Canada are in need of conservation. Four of the 11 species have been assessed as requiring some level of protection by the Committee on the Status of Endangered Wildlife in Canada, and one species has been assessed as data deficient.

Lamprey taxonomy is presently in a state of flux with advocates for both lumping or splitting of taxa based on morphological, physiological, and molecular evidence (e.g., Beamish 1982, 1987; Beamish and Withler 1986; McPhail 2007). This taxonomic debate is particularly relevant to concerns regarding the conservation of lampreys and the efforts to prevent the loss of biodiversity. The purpose of this paper is to review the current taxonomy and highlight uncertainty surrounding it, to provide an update on the distribution and conservation status of lampreys in Canada, and to identify key knowledge gaps, including the parasitic–nonparasitic conundrum of the "paired" (Zanandrea 1959) or "satellite" species concept (Vladykov and Kott 1979b).

The paired species concept applies to lamprey species that are closely related and morphologically very similar to each other but differ in characters associated with their different mode of life as adults; one species assumes a feeding or parasitic phase following metamorphosis while the other species remains nonfeeding or nonparasitic (Zanandrea 1959). The satellite species concept is simply an extension of the former concept in that, in some cases, there are two or more nonparasitic species (i.e., satellites) believed to have evolved from a single parasitic (i.e., stem) species (Vladykov and Kott 1979b). The course of evolution in lampreys is thought to have progressed from a wide-ranging anadromous parasitic species to a more restricted freshwater-resident parasitic species to a freshwater nonparasitic species (Hardisty and Potter 1971c; Beamish 1985). The recognition of the parasitic and nonparasitic counterparts of a paired species as comprising valid species has been the prevailing practice, at least since the seminal work of Hubbs and Trautman (1937), but this situation is being challenged with increased frequency in recent years, and a resolution to this question is required.

## Taxonomic Issues

The taxonomy of lampreys is based primarily on the dentition characteristics of adults (Vladykov and Follett 1967). Other adult morphological characters useful for taxonomic purposes are the number of trunk myomeres (Vladykov and Kott 1979a) and the number and arrangement of velar tentacles (Vladykov and Kott 1976). Ammocoetes are notoriously difficult to identify to species, but the number of trunk myomeres and the pigmentation pattern of various body regions have been used with success in several species (Vladykov 1950; Richards et al. 1982).

Eleven lamprey species are currently recognized in Canada (Table 1), the nine species that were treated by Scott and Crossman (1973) plus Vancouver lamprey *Entosphenus macrostomus* (formerly known as *Lampetra macrostoma*) and Alaskan brook lamprey *Lethenteron alaskense* (described in 1982 and 1978). In addition, R. J. Beamish (Pacific Biological Station, Fisheries and Oceans Canada, personal communication) believes that there are several undescribed species of lampreys in British Columbia, notably in the Nass River

Table 1. Scientific and common names and life history characteristics for the 11 lamprey species in Canada. Life history types are abbreviated as follows: P = parasitic, NP = nonparasitic, FW = freshwater-resident, A = anadromous.

| Scientific name | Common name | Life history |
|---|---|---|
| *Entosphenus macrostomus* (Beamish 1982) | Vancouver lamprey | P, FW |
| *E. tridentatus* (Gairdner in Richardson 1836) | Pacific lamprey | P, A |
| *Ichthyomyzon castaneus* Girard 1858 | chestnut lamprey | P, FW |
| *I. fossor* Reighard and Cummins 1916 | northern brook lamprey | NP, FW |
| *I. unicuspis* Hubbs and Trautman 1937 | silver lamprey | P, FW |
| *Lampetra ayresii* (Günther 1870) | western river lamprey[a] | P, A |
| *L. richardsoni* Vladykov and Follett 1965 | western brook lamprey | NP, FW |
| *Lethenteron alaskense* Vladykov and Kott 1978 | Alaskan brook lamprey | NP, FW |
| *L. appendix* (DeKay 1842) | American brook lamprey | NP, FW |
| *L. camtschaticum* (Tilesius von Tilenau 1811) | Arctic lamprey | P, A; P, FW |
| *Petromyzon marinus* Linnaeus 1758 | sea lamprey | P, A; P, FW |

[a] Nelson et al. (2004) give the common name as river lamprey. However, to avoid confusion with the European river lamprey *Lampetra fluviatilis* (Linnaeus 1758), we suggest calling it western river lamprey.

basin on the mainland and Copper (or Zymoetz) River basin on Haida Gwaii.

The taxonomic recognition of lamprey species has primarily relied on the biological species concept in which reproductive isolation is the central tenet (Mayr 1942). Because reproductive isolation is difficult to test, especially in allopatric populations, any character, morphological or molecular, that is diagnostic in identifying two putative species is taken to indicate reproductive isolation and, therefore, specific distinction of those two entities. Hence, the morphological species concept is often used to infer reproductive isolation. With the advent of cladistics (Hennig 1966), the phylogenetic species concept has gained wide acceptance. By this concept, species are defined as the smallest identifiable units that are diagnosable and monophyletic (i.e., share a common ancestor). The phylogenetic species concept does not conflict with the biological or morphological species concepts except in cases of parallel evolution (see Schluter and Nagel 1995; Taylor 1999). In this paper, where species status has been debated, we evaluate the applicability of these different species concepts.

Comments on selected Canadian lampreys for which the taxonomy is uncertain or that have been the object of recent nomenclatural change are presented below in alphabetical order by genus name (as they appear in Table 1).

## Genus Names Entosphenus and Lethenteron

Some authors (Hubbs and Potter 1971; Bailey 1980) considered *Entosphenus* and *Lethenteron* as subgenera within the genus *Lampetra*, whereas others (Vladykov and Follett 1967; Vladykov and Kott 1979a, 1979b) classified them as distinct genera. Recent morphological (Gill et al. 2003) and genetic (Docker et al. 1999) evidence suggests that *Entosphenus* is sufficiently distinct to be considered a separate genus, although *Lethenteron* and *Lampetra* were found to be less distinct (Docker et al. 1999; Gill et al. 2003). The American Fisheries Society Committee on Names of Fishes supported synonymizing *Entosphenus* and *Lethenteron* with *Lampetra* (Nelson et al. 2004), but Nelson (2006) recognized *Entosphenus* and *Lethenteron* as valid genera based on the cladistic interpretations of Gill et al. (2003). In this context, the American Fisheries Society Endangered Species Committee has treated *Entosphenus* as a genus in their recently published list of imperiled North American freshwater and diadromous fishes (Jelks et al. 2008).

*Vancouver lamprey.*—The Vancouver lamprey was initially believed to be a dwarf race of the Pacific lamprey *Entosphenus tridentatus* (formerly known as *Lampetra tridentata*), which either spent 1 year in freshwater prior to going to sea or was landlocked (see Beamish 1985). However, it was described as a distinct species on the basis of morphological and physiological differences (Beamish 1982), as well as differences in spawning time and spawning location (lake versus stream) that would likely lead to reproductive isolation (Beamish 1982, 2001). It can be morphologically distinguished from Pacific lamprey predominantly by its distinctive larger oral disk relative to total length. Other morphological differences include smaller total length, larger relative eye diameter, and longer relative prebranchial length (Beamish 1982). Internally, the velar tentacles in Vancouver lamprey are very weakly pigmented relative to the darkly pigmented base and lower portion of the velar tentacles of Pacific lamprey (Beamish 1982). Physiologically, Vancouver lamprey differs from Pacific lamprey in its ability to osmoregulate in freshwater throughout its entire life cycles, whereas feeding-phase Pacific lamprey is unable to survive in freshwater for more than 6–10 months (Beamish 1982; Clarke and Beamish 1988). However, the taxonomic status of Vancouver lamprey remains controversial (McPhail 2007). Although it appears to conform to the morphological and biological species concepts, Vancouver lamprey is probably one of several independently evolved postglacial freshwater derivatives of Pacific lamprey in British Columbia, including others on Quadra and Nelson islands (Beamish 2001). The lack of observed genetic differences between Vancouver lamprey and Pacific lamprey (Docker et al. 1999) suggest that the two species have diverged recently. The phylogenetic identity of such recently and independently derived forms is complex (e.g., whether each independently derived freshwater form merits species status). Despite morphological similarity, independently evolved freshwater derivatives would not be monophyletic and would not conform to the phylogenetic species concept (see Schluter and Nagel 1995; Taylor 1999). As these other purported freshwater derivatives of Pacific lamprey and their relationship to Vancouver lamprey are studied in more detail, the taxonomic status of Vancouver lamprey will need to be re-examined.

*Morrison Creek lamprey.*—The Morrison Creek lamprey *Lampetra richardsoni* var. *marifuga* has been shown under laboratory conditions to feed parasitically as an adult and has been interpreted to be a parasitic variety of the western brook lamprey *L. richardsoni* (Beamish and Withler 1986; Beamish 1987). All larvae in this creek are morphologically identified as western brook lampreys, and ammocoetes of river lamprey *L. ayresii*—which is the parasitic counterpart of river lamprey and which can be morphologically distinguished from them as larvae (Richards et al. 1982)—have not been collected in Morrison Creek (Beamish 1985; Beamish and Withler 1986). However, following metamorphosis, some of the lampreys in this creek retain sharp teeth for at least 8 months (while the teeth of nonparasitic western brook lampreys taken at the same time are obsolete), take on the silver coloration typical of downstream migrating river lampreys, are capable of feeding in the laboratory, and spawn and die 1 year later than the sympatric nonparasitic western brook lampreys (Beamish 1987). The parasitic morphotype, however, is unlike river lampreys in that it is unable to survive in seawater (hence the name *marifuga*, which means literally to run away from the sea), and anadromous river lampreys have never been reported in Morrison Creek (Beamish 1985; Beamish and Withler 1986). It may be possible that the Morrison Creek lamprey is a freshwater-resident river lamprey population as was proposed by Renaud (1997), especially in light of the experimental study of Beamish and Youson (1987), who were able to keep a few river lampreys in freshwater through their entire postmetamorphic period. Although nonanadromous river lampreys have not been reported to date, freshwater-resident populations of other anadromous lamprey species (e.g., sea lamprey *Petromyzon marinus* in the Laurentian Great Lakes and the more closely related European river lamprey *L. fluviatilis*) are common (Hardisty 2006).

Since a "variety" published after 1960 is an excluded taxonomic rank in zoological nomenclature (Article 15.2, International Commission on Zoological Nomenclature 1999), the Morrison Creek lamprey has not been given formal taxonomic status. The two varieties could not be separated electrophoretically (Beamish and Withler 1986), and there is no evidence that the parasitic variety forms a self-sustaining population. Although the parasitic variety is capable of feeding in the laboratory, with nearly sexually mature individuals having an intestine full of ingested muscle (Youson and Beamish 1991), it may not be feeding in the stream; no scars have been reported on fishes, no increase in lamprey body length after metamorphosis has been noted, and lamprey digestive tracts were either empty or contained only small amounts of mucus-like material (Beamish 1985). Similarly, although the parasitic variety has spawned in the laboratory and produced viable eggs (Beamish 1985), there is no evidence that it spawns in the wild; development of the gonads and secondary sex characters are delayed and the sex ratios are highly male-biased (Youson and Beamish 1991). The situation in Morrison Creek, where one population apparently gives rises to two morphotypes, while highly unusual, is not unique. Kucheryavyi et al. (2007) reported a single population of Arctic lamprey *Lethenteron camtschaticum* (formerly known as *Lampetra camtschatica*) in the Utkholok River basin in Kamchatka that produces three different morphs: a nonparasitic form, a resident parasitic form, and an anadromous parasitic form. According to these authors, there was no temporal reproductive isolation among the forms with different adult feeding and migratory habits, which they propose may be linked to different diets at the ammocoete stage (see Docker 2009, this volume, for a review of feeding type plasticity in lampreys).

*Alaskan brook lamprey*.—The Alaskan brook lamprey was described in 1978, but its taxonomic status remains unresolved. It was recognized as a distinct species by Mecklenburg et al. (2002), but the most recent *Common and Scientific Names of Fishes* (Nelson et al. 2004) still considers Alaskan brook lamprey to be a synonym of American brook lamprey *Lethenteron appendix* (formerly *Lampetra appendix*). Alaskan brook lamprey and American brook lamprey have similar life history and morphology, suggesting that they are a single species under the morphological species concept. However, they occur 2,400 km apart and may therefore represent independent derivatives of the parasitic Arctic lamprey. As such, they would not be recognized as a single species under the phylogenetic species concept. Further debate has involved the relationship of Alaskan brook lamprey to Arctic lamprey (McPhail and Lindsey 1970). Nonparasitic Alaskan brook lamprey was originally considered a dwarf freshwater-resident form of Arctic lamprey by Heard (1966). However, adult life history type is generally considered a species-specific character, in agreement with the biological species concept, given the observation that size-assortative mating results in reduced mating success between parasitic and nonparasitic lampreys (Beamish and Neville 1992; see Knowledge Gaps below). Further analysis of these three taxa using samples from various localities, including both mitochondrial and nuclear markers, will be required to confirm reciprocal monophyly.

The Committee on the Status of Endangered Wildlife in Canada used "darktail lamprey" as the common name for *Lethenteron alaskense* (COSEWIC 2007). However, we use "Alaskan brook lamprey" because it is the one proposed by Vladykov and Kott (1979a) and used in Mecklenburg et al. (2002), and additionally, the former name is not particularly diagnostic as many species have dark tails. Also, we use the specific epithet *alaskense* as proposed in the original description instead of *alaskensis*, as proposed by Nelson et al. (2004:187), to agree with *Lethenteron*, which is neuter, whereas *Lampetra* is feminine.

*The Arctic lamprey*.—Although largely known as *Lethenteron japonicum* (or *Lampetra japonica*), Kottelat (1997) showed *L. japonicum* (Martens 1868) to be a junior synonym of

*L. camtschaticum*. Although the International Commission on Zoological Nomenclature could have been petitioned to conserve the long-standing species name *japonicum*, Nelson et al. (2004) felt that with the Kottelat publication and the use of *camtschaticum* in recent publications, it is now too late to do so.

## Distribution and Postglacial Origins

Lampreys occur in all provinces and territories except Nunavut (Table 2; Figure 1). There are two centers of lamprey diversity in Canada—the Pacific coast and the Great Lakes-St. Lawrence basin. Four species (Vancouver lamprey, Figure 1a; Pacific lamprey, Figure 1a; river lamprey, Figure 1e; and western brook lamprey, Figure 1f) and one distinct population of western brook lamprey (Morrison Creek lamprey, Figure 1f) are found along the Pacific coast of Canada in the Province of British Columbia. These include both marine and freshwater species that likely survived Wisconsinan glaciation in a Pacific refugium south of the ice sheet and colonized the northern Pacific coast and, for some species, the adjacent freshwaters following the retreat of the ice sheet ca. 10,000 years before present (McPhail and Lindsey 1970).

Four freshwater species (chestnut lamprey *Ichthyomyzon castaneus*, Figure 1b; northern brook lamprey *I. fossor*, Figure 1c; silver lamprey *I. unicuspis*, Figure 1d; and American brook lamprey, Figure 1g) and one marine species (sea lamprey, Figure 1i) of lampreys are present in the Great Lakes basin in the Province of Ontario, as well as the St. Lawrence River basin in the Province of Quebec. The four freshwater species all likely survived in the Mississippian refugium during the Wisconsinan glaciation and colonized the Great Lakes early in postglacial times (Mandrak and Crossman 1992). The marine sea

Table 2. Provincial and territorial (subnational) NatureServe and General Status ranks for lampreys found in Canada. AB = Alberta; BC = British Columbia, MB = Manitoba; NB = New Brunswick; NL = Newfoundland and Labrador; NS = Nova Scotia; NT = Northwest Territories; NU = Nunavut; ON = Ontario; PE = Prince Edward Island; QC = Quebec; SK = Saskatchewan; and YT = Yukon.

| Scientific name | AB | BC | MB | NB | NL | NS | NT | NU | ON | PE | QC | SK | YT |
|---|---|---|---|---|---|---|---|---|---|---|---|---|---|
| *Entosphenus macrostomus* | | S1; 1 | | | | | | | | | | | |
| *E. tridentatus* | | S4; 4 | | | | | | | | | | | |
| *Ichthyomyzon castaneus* | | | S3S4; 3 | | | | | | S1?; 2 | | SNR; 5 | S3S4;1 | |
| *I. fossor* | | | S2; 2 | | | | | | S3; 3 | | S2S3; 2 | | |
| *I. unicuspis* | | | S3; 3 | | | | | | S3; 3 | | S3S4; 3 | | |
| *Lampetra ayresii* | | S4; 4 | | | | | | | | | | | |
| *L. richardsoni* | | S4; 4 | | | | | | | | | | | |
| *L. richardsoni* (Morrison Creek population) | | S1 | | | | | | | | | | | |
| *Lethenteron alaskense* | | | | | | | SNR; 5[a] | | | | | | |
| *L. appendix* | | | | | | | | | S3; 3 | | S3S4; 3 | | |
| *L. camtschaticum* | S1; 4 | | | | | | SNR; 5 | | | | | | 5 |
| *Petromyzon marinus* | | | | | S5; 4 | S1; 3 | S5; 4 | | | SNA; 7 | 8 | S4; 4 | |

[a] Identified as *Lampetra appendix* in CESCC (2006).

*Sx*; *y*: *Sx* = NatureServe subnational rank; *y* = General Status rank. *Sx* = subnational rank (1 = Critically Imperiled; 2 = Imperiled; 3 = Vulnerable; 4 = Apparently Secure; 5 = Secure; NA = not applicable because the species is not a suitable target for conservation activities; and NR = not ranked); General Status: (1 = At risk; 2 = May Be at Risk; 3 = Sensitive; 4 = Secure; 5 = Undetermined; 7 = Exotic; and 8 = Accidental).

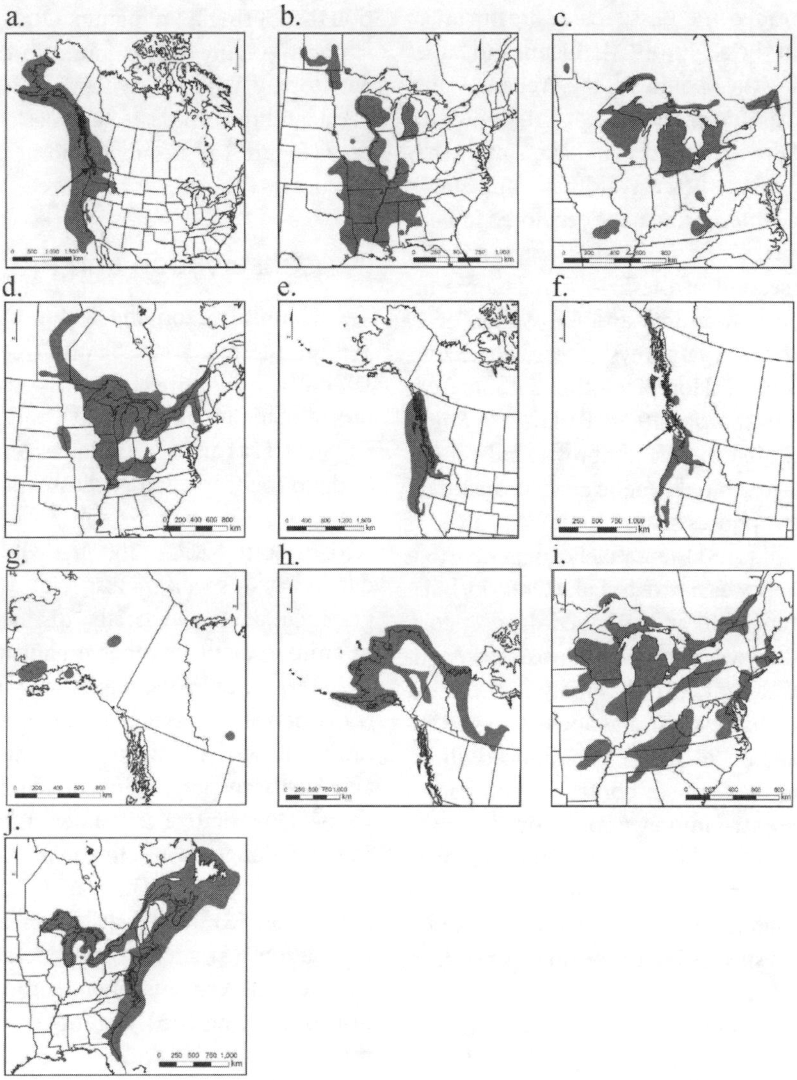

Figure 1. (a) Canadian distribution of Vancouver lamprey (indicated by an arrow) based on Beamish (1982) and Canadian/U.S. distribution of Pacific lamprey (shaded) based on Vladykov and Kott (1979b), Mecklenburg et al. (2002), and Page and Burr (in press). (b) Canadian/U.S. distribution of chestnut lamprey based on Vladykov and Kott (1979b), Atton and Merkowsky (1983), Page and Burr (1991), Renaud et al. (1996), Renaud and de Ville (2000), Bailey et al. (2004), and Stewart and Watkinson (2004). (c) Canadian/U.S. distribution of northern brook lamprey based on Vladykov and Kott (1979b), Bailey et al. (2004), Stewart and Watkinson (2004), SARA (2007), and Page and Burr (in press). (d) Canadian/U.S. distribution of silver lamprey based on Vladykov and Kott (1979b), Bailey et al. (2004), Stewart and Watkinson (2004), and Page and Burr (in press). (e) Canadian/U.S. distribution of river lamprey based on Vladykov and Kott (1979b), McPhail (2007), and Page and Burr (in press). (f) Canadian/U.S. distribution of Morrison Creek lamprey (shaded) based on Vladykov and Kott (1979b), Mecklenburg et al. (2002), and McPhail (2007). Canadian distribution of the Morrison Creek population (indicated by an arrow) based on Beamish et al. (2001). (g) Canadian/U.S. distribution of Alaskan brook lamprey based on Vladykov and Kott (1978), Page and Burr (1991), and Mecklenburg et al. (2002). (h) Canadian/U.S. distribution of Arctic lamprey based on Vladykov and Kott (1979b), Nelson and Paetz (1992), Mecklenburg et al. (2002), and Sawatzky et al. (2007). (i) Canadian/U.S. distribution of American brook lamprey based on Vladykov and Kott (1979b; as *Lampetra lamottenii*), Mandrak and Crossman (1992), and Page and Burr (in press). (j) Canadian/U.S. distribution of sea lamprey based on Vladykov and Kott (1979b), Scott and Scott (1988), Halliday (1991), Dempson and Porter (1993), and Page and Burr (in press).

lampreys invaded the Great Lakes upstream of Lake Ontario (Erie, Huron, Michigan, and Superior) following the opening of the Welland Canal between Lake Ontario and Lake Erie; however, the origin of the sea lampreys in Lake Ontario has been widely debated (see Waldman et al. 2004). Since the sea lamprey had not been noted in Lake Ontario until 1835, 7 years after completion of the eastern section of the Erie Canal, some investigators (e.g., Mandrak and Crossman 1992) concluded that sea lampreys entered Lake Ontario via this route. However, other authors (e.g., Radforth 1944; Lawrie 1970; Bailey and Smith 1981) argued that the sea lamprey is indigenous to Lake Ontario. Initial genetic analyses based on protein electrophoresis were equally equivocal. Brussard et al. (1981) tentatively suggested that the sea lamprey was native to Lake Ontario, but a follow-up study with an expanded data set concluded that they were probably invasive (Wright et al. 1985). Recent mtDNA analysis indicated that the sea lamprey is indigenous to Lake Ontario (Waldman et al. 2004); pronounced differences in the frequency of control region haplotypes between sea lamprey populations from the Atlantic coast and Lake Ontario support the idea of post-Pleistocene natural colonization from an Atlantic refugium. However, for conservation purposes, this species is still considered exotic in Ontario (Table 2).

The anadromous sea lamprey is native to all provinces east of Ontario (Table 2; Figure 1j). Three of the freshwater species present in the Great Lakes and St. Lawrence River basins (chestnut lamprey, Figure 1b; northern brook lamprey, Figure 1c; and silver lamprey, Figure 1d) dispersed naturally early in postglacial times westward into Manitoba and one of these species (chestnut lamprey, Figure 1b) further west into Saskatchewan. The same three freshwater species dispersed eastward into Quebec while American brook lamprey (Figure 1i) either also used this dispersal route or entered Quebec from an Atlantic coastal refugium (Mandrak and Crossman 1992; Table 2).

Two species have a western Arctic distribution and likely survived the Wisconsinan glaciation in a Beringian refugium. One species, Arctic lamprey (Figure 1h), is found in the Northwest Territories, Yukon Territory, and northern Alberta, and the other species, Alaskan brook lamprey (Figure 1g), is only found in the Northwest Territories (Table 2).

## Conservation and Protection

The Committee on the Status of Endangered Wildlife in Canada (COSEWIC) has existed since 1977. The federal Species at Risk Act (SARA) was proclaimed on June 5, 2003 and it establishes COSEWIC as an independent scientific advisory body to the Canadian government in matters of species assessments for conservation purposes. According to SARA (2007), a wildlife species is defined as "a species, subspecies, variety or geographically or genetically distinct population of animal, plant or other organism, other than a bacterium or virus, that is wild by nature and (a) is native to Canada; or (b) has extended its range into Canada without human intervention and has been present in Canada for at least 50 years." Quantitative criteria exist for the categories of endangered and threatened and fall under five broad headings. These are (1) declining total population, (2) small distribution, and decline or fluctuation, (3) small total population size and decline, (4) very small population or restricted distribution, and (5) quantitative analysis. The category Special Concern is reserved for species that are particularly sensitive to human activities or natural events but do not meet the thresholds for either endangered or threatened. More information on these criteria is available on the COSEWIC Web site (www.cosewic.gc.ca). COSEWIC status reports, lists of species protected by SARA, and recovery strategies and action plans are available from the SARA (2007) public registry (www.sararegistry.gc.ca). A determination of the potential for rescue effect by conspecifics from the contiguous United States can be made by consulting NatureServe Explorer (www.natureserve.org/explorer).

The conservation status of 5 of the 11 Canadian lampreys has been assessed by COSEWIC

(COSEWIC 2007), but for two, chestnut lamprey and western brook lamprey, the assessment is based on only a part of their Canadian range (Table 3). A status report for a sixth species, silver lamprey, is underway. Currently, only four species are listed under SARA (2007). Species are considered for listing under SARA after their conservation status has been formally assessed by COSEWIC. The federal Minister of the Environment makes the final decision of whether to list COSEWIC-assessed fish species under the Species at Risk Act. This decision is based on the COSEWIC report and additional biological and socioeconomic analyses conducted by the federal Department of Fisheries and Oceans. The Morrison Creek population of western brook lamprey is listed as endangered, Vancouver lamprey as threatened, and northern brook lamprey (Great Lakes – Upper St. Lawrence biogeographic zone) as special concern under Schedule 1 of SARA (2007) and are afforded full protection under the act. Chestnut lamprey is listed as special concern under Schedule 3 of SARA (2007) and requires an updated COSEWIC report before it can be potentially listed under Schedule 1 and be afforded full protection.

The species, in the sense of SARA (2007), most at risk of extinction is the endangered Morrison Creek population of western brook lamprey occurring on Vancouver Island. COSEWIC determined that this population constituted a designatable unit and, therefore, was eligible for assessment based on its apparent uniqueness in producing both parasitic and nonparasitic lampreys from a single population (Beamish and Withler 1986). The threats to this species are residential development and road construction, both of which would result in loss of habitat. The extremely circumscribed area occupied by this lamprey, at less than 50 $km^2$, also means that a single stochastic catastrophic event could put the entire population in jeopardy. The apparent uniqueness of the population means that there is no possible rescue effect from the United States

Table 3. Global and national NatureServe, COSEWIC, and General Status ranks for lampreys found in Canada.

| Scientific name | NatureServe[a] | COSEWIC[b] | General Status[c] |
|---|---|---|---|
| *Entosphenus macrostomus* | G1; N1 | T (2000) | 1 |
| *E. tridentatus* | G5; NNR | | 4 |
| *Ichthyomyzon castaneus* | G4; N3N4 | SC (1991) | 3 |
| *I. fossor* | G4; N3 | SC; DD (2007) | 3 |
| *I. fossor* (Great Lakes – Upper St. Lawrence Freshwater Biogeographic Zone) | G4; N4 | SC (2007) | |
| *I. fossor* (Saskatchewan – Nelson Freshwater Biogeographic Zone) | G4; NNR | DD (2007) | |
| *I. unicuspis* | G5; N4 | | 3 |
| *Lampetra ayresii* | G4; N4 | | 4 |
| *L. richardsoni* | G4G5; N4N5 | | 4 |
| *L. richardsoni* (Morrison Creek population) | G4G5T1Q; N1 | END (2000) | |
| *Lethenteron alaskense*[d] | GNR; NNR | DD (2006) | |
| *L. appendix* | G4; N4 | | 3 |
| *L. camtschaticum* | G4; N3N4 | | 4 |
| *Petromyzon marinus* | G5; N5 | | 4 |

[a] G*x* = global rank (1 = Critically Imperiled; 2 = Imperiled; 3 = Vulnerable; 4 = Apparently Secure; 5 = Secure; NR = Not Ranked; and T1Q = infraspecific taxon, critically imperiled, questionable taxonomy); N*x* = Canadian national rank (1 = Critically Imperiled; 2 = Imperiled; 3 = Vulnerable; 4 = Apparently Secure; 5 = Secure; and NR = Not Ranked).
[b] COSEWIC (most recent assessment year in parentheses): DD = Data Deficient; END = Endangered; SC = Special Concern; and T = Threatened.
[c] General Status: 1 = At Risk; 3 = Sensitive; and 4 = Secure.
[d] Grouped with *Lampetra* (=*Lethenteron*) appendix in General Status list (CESCC 2006).

or immigration from neighboring Canadian populations of nonparasitic western brook lamprey. Additionally, marine waters surrounding Vancouver Island constitute a barrier to dispersal. A final recovery strategy is available for the Morrison Creek lamprey (National Recovery Team for Morrison Creek Lamprey 2007), but no action plan has been developed yet.

The next species most at risk of extinction is another lamprey endemic to Vancouver Island, Vancouver lamprey. This parasitic species is restricted to two freshwater lakes, Cowichan and Mesachie, with a total area of 63 km$^2$. It has been assessed as threatened because of its endemic nature and very restricted distribution. A single stochastic catastrophic event could have dire consequences, and there is no possible rescue effect from the United States. A proposed recovery strategy is available for the Vancouver lamprey (Vancouver Lamprey Recovery Team 2007), but no action plan has been developed yet.

Two other species were assessed as being of special concern: the parasitic chestnut lamprey and the nonparasitic northern brook lamprey occurring in the Great Lakes—Upper St. Lawrence freshwater biogeographic zone (Ontario–Quebec). Northern brook lamprey occurring in the Saskatchewan–Nelson freshwater biogeographic zone (Manitoba) was assessed as data deficient. For chestnut lamprey, only the populations of Saskatchewan and Manitoba were considered because at the time of the assessment, this was the known Canadian range of the species. The range now includes Ontario and Quebec (Renaud et al. 1996; Renaud and de Ville 2000), and a reassessment will have to encompass the entire Canadian range. Species assessed in the categories Extirpated, Endangered, Threatened, or Special Concern are termed "at risk" and these are normally reassessed every 10 years based on an update status report. The threat to chestnut lamprey is its sensitivity to habitat degradation, whereas the threat to northern brook lamprey in the Great Lakes–Upper St. Lawrence freshwater biogeographic zone is its sensitivity to lampricide treatments aimed at controlling the sea lamprey in the Great Lakes portion of its Ontario–Quebec range (Davis 1970; Schuldt and Goold 1980). The rescue effect for chestnut lamprey is determined to be negligible. Although conservation ranks for the contiguous states of North Dakota and Minnesota are given as not ranked and those for Wisconsin and Michigan are apparently secure, there are large discontinuities between the occurrences in those states and those in Canada (NatureServe 2007). Additionally, there is no record of chestnut lamprey in the states bordering Canada east of Michigan (NatureServe 2007). The rescue effect for northern brook lamprey in the Great Lakes–Upper St. Lawrence freshwater biogeographic zone is determined to be either negligible or nonexistent. Although the conservation status for northern brook lamprey in the contiguous states of Wisconsin and Michigan is apparently secure, and that of Minnesota is vulnerable (equivalent to COSEWIC's special concern), it is critically imperiled (equivalent to COSEWIC's endangered) in Ohio, Pennsylvania, New York, and Vermont (NatureServe 2007). Furthermore, one of the characteristics of a nonparasitic lamprey such as northern brook lamprey is its low dispersal ability, as inferred by their much shorter spawning migration distances relative to those of parasitic species (Hardisty and Potter 1971b). Hence, one would not expect a nonparasitic species to be a good colonizer in the short term. Northern brook lamprey in the Saskatchewan–Nelson freshwater biogeographic zone in Manitoba was assessed as data deficient because it has not been the subject of any targeted and comprehensive search since its discovery in the late 1970s, and therefore, its distribution and population abundance and trends are not well known.

Alaskan brook lamprey was assessed as data deficient by COSEWIC because of taxonomic uncertainty and the lack of information regarding its biology, especially in terms of population trends and abundance, as well as its distribution.

In addition to the above species assessed by COSEWIC, we draw attention to a localized extinction event of Pacific lampreys in Elsie Lake, British Columbia, brought about by the construction in the late 1950s of a series of dams (Beamish and Northcote 1989), evidence that dams may have a deleterious effect on lamprey survival.

NatureServe has ranked all fish species in Canada at global, national, and subnational (province and territory) levels (Tables 2 and 3). Vancouver lamprey is ranked as critically imperiled at the global, national (Canada), and subnational (British Columbia) levels. Alaskan brook lamprey is not ranked at any level. At the global level, all other lampreys are ranked as apparently secure or secure, with the Morrison Creek population of western brook lamprey being identified as critically imperiled but of questionable taxonomy. At a national level, Pacific lamprey and Alaskan brook lamprey have not been ranked, Vancouver lamprey and the Morrison Creek population of western brook lamprey are ranked as critically imperiled, and all other lampreys are ranked as vulnerable (N3) to secure (N5).

The "General Status of Wildlife in Canada" (General Status) has also ranked all major taxonomic groups in Canada, including fishes (CESCC 2006). The national, provincial, and territorial rankings are generally consistent with the NatureServe rankings (Tables 2 and 3). However, the General Status ranks chestnut lamprey as at risk (1) in Saskatchewan compared to vulnerable/apparently secure (S3S4), Arctic lamprey as secure (4) in Alberta compared to critically imperiled (S1), and sea lamprey as sensitive (3) in Newfoundland compared to critically imperiled (S1) (Table 2). The General Status also identifies Arctic lamprey as undetermined (5) in the Yukon and sea lamprey as accidental (8) in Prince Edward Island, whereas NatureServe does not provide a rank for these species in those jurisdictions (Table 2). The Morrison Creek population of western brook lamprey was not ranked by the General Status. The difference in the rankings of these two methods may be attributed to the different approaches used to develop the ranks; NatureServe is based on number of occurrences, and the General Status is based on regional expert opinion.

From a protection perspective, species at risk are legally protected in Canada only if they are listed under the federal Species at Risk Act. Once listed under SARA, a species is protected as it is illegal to "kill, harm, harass, capture or take an individual of a wildlife species that is listed as an extirpated species, an endangered species or a threatened species" (SARA Section 32(1)). The habitat of a species at risk is also protected as "no person shall destroy any part of the critical habitat of any listed endangered species or of any listed threatened species—or of any listed extirpated species if a recovery strategy has recommended the reintroduction of the species into the wild in Canada —if … (b) the listed species is an aquatic species" (SARA Section 58(1)). Protection is also given to the residence of the species (habitat constructed and maintained by a species, e.g., ammocoete burrows). Recovery strategies must be developed within a year of a species being listed as endangered, and within 2 years for threatened and extirpated species. Once a recovery strategy is completed, a recovery action plan to implement the strategy must be developed. Management plans must be completed within 3 years after listing.

Additionally, all fish species and their habitat are also protected by the federal Fisheries Act. Further protection may be afforded to species under various provincial and territorial legislations and in national and provincial parks.

## Knowledge Gaps

The main knowledge gap related to lampreys in Canada is whether it is valid to recognize as distinct species the parasitic and nonparasitic members of paired and satellite species (see review by Docker 2009). While northern brook lamprey and silver lamprey have been treated as separate species for more than 70 years (Hubbs and Trautman 1937), recent unpublished molecular studies question their specific distinctiveness. The two species appear to be morphologically indistinguishable as larvae (Neave et al. 2007), but the adults of northern brook lamprey are nonparasitic and have unpigmented lateral line neuromasts while those of silver lamprey are parasitic and have pigmented lateral line neuromasts. However, no species-specific differences were detected in more than 10,000 base pairs of the mitochondrial genome, and microsatellite allele frequencies were not significantly different between sympatric northern brook lamprey and silver lamprey (M. F. Docker, University of Manitoba, unpublished data). The mitochondrial DNA data showed two distinct lineages (i.e., monophyletic groups), but both occurred within each species, suggesting that nonparasitism may have evolved independently in each lineage (i.e., that northern brook lamprey is polyphyletic). Consequently, Docker et al. (2005) suggest that northern brook lamprey and silver lamprey should be considered a single species according to the phylogenetic species concept. This would result in the silver and northern brook lamprey being identified as *Ichthyomyzon fossor*, in compliance with the principle of priority (Article 23.1) in the International Code of Zoological Nomenclature (International Commission on Zoological Nomenclature 1999). Silver lamprey and northern brook lamprey may represent two trophic forms of a single species, one that feeds in the adult stage and one that does not. This issue is not yet resolved.

The taxonomic status of some other paired lamprey species similarly may be questionable. Mitochondrial DNA sequences for nonparasitic western brook lamprey and parasitic river lamprey were identical (Docker et al. 1999), and a population of western brook lamprey that apparently also produces some individuals capable of feeding parasitically (at least under laboratory conditions) has been discovered on Vancouver Island (see Morrison Creek lamprey above). The species status of nonparasitic Alaskan brook lamprey and parasitic Arctic lamprey has likewise been questioned (McPhail and Lindsey 1970; see Alaskan brook lamprey, above). However, Alaskan populations of Alaskan brook lamprey and Arctic lamprey sampled to date show no shared haplotypes (N. J. Lang, Field Museum of Natural History, Chicago, personal communication), and genetic differences were observed between populations of nonparasitic American brook lampreys from Ontario and Arctic lampreys from Russia (Docker et al. 1999). Given that the evolution of nonparasitism has occurred independently and repeatedly in multiple lamprey taxa, at different times and in different places (Hubbs and Potter 1971), morphological and molecular divergence and reproductive isolation (i.e., speciation by all of the species concepts defined above) may be complete in some pairs but not in others. The taxonomic status of each pair should be determined individually.

These taxonomic re-examinations are not restricted to the North American continent. Recently, the evolutionary history of the paired European species, the parasitic European river lamprey and the nonparasitic European brook lamprey *Lampetra planeri*, was examined by comparing sequences of three mtDNA genes from populations of these two lampreys from across their ranges (Espanhol et al. 2007). Their results were not sufficiently conclusive to distinguish between the alternative single species or very recently diverged but distinct species hypotheses, but their results did suggest that European brook lamprey was polyphyletic. They suggested that further testing was required.

The debate over whether it is valid to recognize as distinct species the parasitic and nonparasitic members of a paired species has been ongoing for over a century with proponents for trophic plasticity within a species (Wajgel 1884; Enequist 1937; McPhail and Lindsey 1970; Youson and Beamish 1991; Kucheryavyi et al. 2007) on the one side and, for specific distinctiveness (Hubbs and Trautman 1937; Vladykov and Follett 1965), on the other. The key to resolving this issue may be common garden experiments (i.e., rearing two putative species under the same conditions from the zygote to the adult stage) that seek to elucidate the triggering mechanism for trophic interactions, or the lack thereof, in the adult stage. These experiments will be particularly challenging as lamprey larval life is long (usually 3–5 years; Hardisty and Potter 1971a), and it is difficult to keep larvae in aquaria free of fungal infections (that often lead to death) for such prolonged periods. This issue has conservation implications, such as whether paired nonparasitic and parasitic lampreys should be treated as separate or single conservation units and whether it might be possible to rehabilitate or rescue endangered populations of one feeding type with the other. Until strong evidence rejects the hypothesis that parasitic and nonparasitic members of a paired species represent distinct species, however, we continue to follow conventional taxonomy recognizing them as distinct.

Another significant knowledge gap involves the distribution and abundance of many of the lamprey species in Canada. Except for the Great Lakes region where intensive surveys targeting sea lamprey have been carried out for several decades, the distribution of lampreys in Canada is generally poorly known. This is largely the result of the lack of concern and sampling for nongame fishes, and where fish sampling does occur, it rarely specifically targets lampreys or uses the most appropriate sampling methods for lampreys (i.e., pulsed DC electrofishing). Knowledge of the distribution of those species that appear to occur in relatively low numbers or in more remote regions (e.g., *Ichthyomyzon* species in Manitoba or Alaskan brook lamprey) is particularly lacking, and the distribution and population trends of the paired northern brook lamprey and silver lamprey is difficult to assess given our inability to distinguish between the larvae of these two species. Furthermore, an important piece of information that is often lacking, but useful when assigning conservation status, is the amount of effort that was expended in order to collect specimens (i.e., catch per unit effort); documentation of sampling sites that did not result in any specimens (absence data) is also lacking. More targeted and ongoing sampling using the most appropriate methods is required to get a better understanding of the distribution and abundance, as well as the population trends, of lampreys in Canada. A taxonomic key to all Canadian lamprey larvae would greatly help in achieving this objective. While keys to ammocoetes exist for a certain number of species (e.g., Vladykov 1950; Richards et al. 1982), the ammocoetes of a few species have not yet been described (Alaskan brook lamprey and Arctic lamprey) and, therefore, are not included in any keys. Additionally, some keys conflict with each other and lead to divergent identifications for the *Ichthyomyzon* species (Vladykov and Kott 1980; Lanteigne 1988).

More research should be done to determine the threats that affect lamprey survival as these are seldom well known. Threats, such as declining prey, fluctuating water levels and pollutants, have been hypothesized, but the causal relationships between these threats and the distribution and abundance of lampreys have generally not been well tested. A better understanding of these factors and threats would lead to more accurate assessments of conservation status and more directed recovery actions.

## Conclusion

Two parasitic lamprey species, one population that produces both parasitic and nonparasitic adults, and one nonparasitic species are currently listed as at risk in Canada, either in the endangered, threatened, or special concern categories. Despite their different modes of life, they all share two common traits, namely all are confined to freshwater and, while the threats acting upon them are varied, these are generally linked to habitat degradation. Additionally, they are similar in that the potential for rescue for each of them is either nonexistent or negligible at best. Recovery strategies are available for only two of the species, the Morrison Creek lamprey and the Vancouver lamprey, but no action plans are yet available to implement either of these strategies. Three facts appear clear: (1) more work is required to resolve taxonomic uncertainties; (2) more needs to be done in terms of surveys and population monitoring; and (3) it is not sufficient for species to be assessed at a certain level of risk; action plans need to be developed and implemented if we are to conserve our biodiversity. Finally, it is alarming to note that the two lampreys most at risk in Canada, the endangered Morrison Creek lamprey and the threatened Vancouver lamprey, were both uplisted to higher risk of extinction categories following reassessments made 1 year and 2.5 years after the initial assessments. Therefore, not only has the situation for these species become worse, it has done so in a very short time span.

## Acknowledgments

We thank Ruben Boles, COSEWIC Secretariat, Ottawa, for making available some status reports on lampreys that were not yet posted on the Species at Risk Act Public Registry.

## References

Atton, F. M., and J. J. Merkowsky. 1983. Atlas of Saskatchewan fish. Saskatchewan Parks and Renewable Resources, Fisheries Technical Report 83–2, Regina, Saskatchewan, Canada.

Bailey, R. M. 1980. Comments on the classification and nomenclature of lampreys—an alternative view. Canadian Journal of Fisheries and Aquatic Sciences 37:1626–1629.

Bailey, R. M., W. C. Latta, and G. R. Smith. 2004. An atlas of Michigan fishes with keys and illustrations for their identification. Miscellaneous Publications Museum of Zoology University of Michigan 192.

Bailey, R. M., and G. R. Smith. 1981. Origin and geography of the fish fauna of the Laurentian Great Lakes basin. Canadian Journal of Fisheries and Aquatic Sciences 38:1539–1561.

Beamish, R. J. 1982. *Lampetra macrostoma*, a new species of freshwater parasitic lamprey from the west coast of Canada. Canadian Journal of Fisheries and Aquatic Sciences 39:736–747.

Beamish, R. J. 1985. Freshwater parasitic lamprey on Vancouver Island and a theory of the evolution of the freshwater parasitic and nonparasitic life history types. Pages 123–140 *in* R. E. Foreman, A. Gorbman, J. M. Dodd, and R. Olsson, editors. Evolutionary biology of primitive fishes. Plenum, New York.

Beamish, R. J. 1987. Evidence that parasitic and nonparasitic life history types are produced by one population of lamprey. Canadian Journal of Fisheries and Aquatic Sciences 44:1779–1782.

Beamish, R. J. 2001. Updated status of the Vancouver Island Lake Lamprey, *Lampetra macrostoma*, in Canada. Canadian Field-Naturalist 115:127–130.

Beamish, R. J., and C.-E. M. Neville. 1992. The importance of size as an isolating mechanism in lampreys. Copeia 1992:191–196.

Beamish, R. J., and T. G. Northcote. 1989. Extinction of a population of anadromous parasitic lamprey, *Lampetra tridentata*, upstream of an impassable dam. Canadian Journal of Fisheries and Aquatic Sciences 46:420–425.

Beamish, R. J., and R. E. Withler. 1986. A polymorphic population of lampreys that may produce parasitic and a nonparasitic varieties. Pages 31–49 *in* T. Uyeno, R. Arai, T. Taniuchi, and K. Matsuura, editors. Indo-Pacific fish biology: Proceedings of the Second International Conference on Indo-Pacific Fishes. Ichthyological Society of Japan, Tokyo.

Beamish, R. J., and J. H. Youson. 1987. Life history and abundance of young adult *Lampetra ayresi* in the Fraser River and their possible impact on salmon and herring stocks in the Strait of Georgia. Canadian Journal of Fisheries and Aquatic Sciences 44:525–537.

Beamish, R. J., J. H. Youson, and L. A. Chapman. 2001. Status of the Morrison Creek western brook lamprey, *Lampetra richardsoni*, in Canada. Canadian Field-Naturalist 115:573–578.

Brussard, P. F., M. Collings Hall, and J. Wright. 1981. Structure and affinities of freshwater sea lamprey (*Petromyzon marinus*) populations. Canadian Journal of Fisheries and Aquatic Sciences 38:1708–1714.

CESCC (Canadian Endangered Species Conservation Council). 2006. Wild species 2005: the general status of species in Canada. Environment Canada, Ottawa, Ontario. Available: www.wildspecies.ca (August 2007).

Clarke, W. C., and R. J. Beamish. 1988. Response of recently metamorphosed anadromous parasitic lamprey (*Lampetra tridentata*) to confinement in fresh water. Canadian Journal of Fisheries and Aquatic Sciences 45:42–47.

COSEWIC (Committee on the Status of Endangered Wildlife in Canada). 2007. Canadian species at risk. Available: www.cosewic.gc.ca (July 2007).

Crossman, E. J., and D. E. McAllister. 1986. Zoogeography of freshwater fishes of the Hudson Bay drainage, Ungava Bay and the Arctic Archipelago. Pages 53–104 *in* C. H. Hocutt and E. O. Wiley, editors. The zoogeography of North American freshwater fishes. Wiley, New York.

Davis, W. A. 1970. The comparative susceptibility of three genera of larval lampreys to a lampricide. Master's thesis. The University of Guelph, Guelph, Ontario.

DeKay, J. E. 1842. Zoology of New-York, or the New-York fauna; comprising detailed descriptions of all the animals hitherto observed within the state of New-York, with brief notices of those occasionally found near its borders, and accompanied by appropriate illustrations. Part IV. Fishes. W. and A. White and J. Visscher, Albany, New York.

Dempson, J. B., and T. R. Porter. 1993. Occurrence of sea lamprey, *Petromyzon marinus*, in a Newfoundland river, with additional records from the northwest Atlantic. Canadian Journal of Fisheries and Aquatic Sciences 50:1265–1269.

Docker, M. F. 2009. A review of the evolution of nonparasitism in lampreys and an update of the paired species concept. Pages 71–114 *in* L. R. Brown, S. D. Chase, M. G. Mesa, R. J. Beamish, and P. B. Moyle, editors. Biology, management, and conservation of lampreys in North America. American Fisheries Society, Symposium 72, Bethesda, Maryland.

Docker, M. F., N. E. Mandrak, D. D. Heath, and K. T. Scribner. 2005. Genetic markers to distinguish and quantify the level of gene flow between northern brook and silver lampreys. Great Lakes Fishery Commission. Available: www.glfc.org/research/scr.php (October 2008).

Docker, M. F., J. H. Youson, R. J. Beamish, and R. H. Devlin. 1999. Phylogeny of the lamprey genus *Lampetra* inferred from mitochondrial cytochrome *b* and ND3 gene sequences. Canadian Journal of Fisheries and Aquatic Sciences 56:2340–2349.

Enequist, P. 1937. Das Bachneunauge als Ökologische Modifikation des Flussneunauges – Über die Fluss- und Bachneunaugen Schwedens Vorläufige Mitteilung. Arkiv för Zoologi 29A(20):1–22.

Espanhol, R., P. R. Almeida, and M. J. Alves. 2007. Evolutionary history of lamprey paired species *Lampetra fluviatilis* (L.) and *Lampetra planeri* (Bloch) as inferred from mitochondrial DNA variation. Molecular Ecology 16:1–16.

Gill, H. S., C. B. Renaud, F. Chapleau, R. L. Mayden, and I. C. Potter. 2003. Phylogeny of living parasitic lampreys (Petromyzontiformes) based on morphological data. Copeia 2003:687–703.

Girard, C. F. 1858. Fishes. In General report on the zoology of the several Pacific railroad routes, volume 10. Beverly Tucker, Washington, D.C.

Günther, A. 1870. Catalogue of the fishes in the British Museum. Catalogue of the Physostomi and of the Lophobranchii, Plectognathi, Dipnoi, Ganoidei, Chondropterygii, Cyclostomata, Leptocardii in the British Museum. Taylor and Francis, London.

Halliday, R. G. 1991. Marine distribution of the sea lamprey (*Petromyzon marinus*) in the northwest Atlantic. Canadian Journal of Fisheries and Aquatic Sciences 48:832–842.

Hardisty, M. W. 2006. Lampreys: life without jaws. Forrest Text, Ceredigion, UK.

Hardisty, M. W., and I. C. Potter. 1971a. The behaviour, ecology and growth of larval lampreys. Pages 85–125 *in* M. W. Hardisty and I. C. Potter, editors. The biology of lampreys, volume 1, Academic Press, New York.

Hardisty, M. W., and I. C. Potter. 1971b. The general biology of adult lampreys. Pages 127–206 *in* M. W. Hardisty and I. C. Potter, editors. The biology of lampreys, volume 1. Academic Press, New York.

Hardisty, M. W., and I. C. Potter. 1971c. Paired species. Pages 249–277 *in* M. W. Hardisty and I. C. Potter, editors. The biology of lampreys, volume 1. Academic Press, New York.

Heard, W. R. 1966. Observations on lampreys in the Naknek River system of southwest Alaska. Copeia 1966:322–339.

Hennig, W. 1966. Phylogenetic systematics. University of Illinois Press, Urbana.

Hubbs, C. L., and I. C. Potter. 1971. Distribution, phylogeny and taxonomy. Pages 1–65 *in* M. W. Hardisty and I. C. Potter, editors. The biology of lampreys, volume 1. Academic Press, New York.

Hubbs, C. L., and M. B. Trautman. 1937. A revision of the lamprey genus *Ichthyomyzon*. Miscellaneous Publications of the Museum of Zoology University of Michigan 35.

International Commission on Zoological Nomenclature. 1999. International code of zoological nomenclature, 4th edition. The International Trust for Zoological Nomenclature c/o The Natural History Museum, London.

Jelks, H. L., S. J. Walsh, N. M. Burkhead, S. Contreras-Balderas, E. Díaz-Pardo, D. A. Hendrickson, J. Lyons, N. E. Mandrak, F. McCormick, J. S. Nelson, S. P. Platania, B. A. Porter, C. B. Renaud, J. J. Schmitter-Soto, E. B. Taylor, and M. L. Warren, Jr. 2008. Conservation status of imperiled North American freshwater and diadromous fishes. Fisheries 33:372–407.

Kottelat, M. 1997. European freshwater fishes. Biologia 52 (Supplement 5):1–271.

Kucheryavyi, A. V., K. A. Savvaitova, D. S. Pavlov, M. A. Gruzdeva, K. V. Kuzishchin, and J. A. Stanford. 2007. Variations of life history strategy of the Arctic lamprey *Lethenteron camtschaticum* from the Utkholok River (western Kamchatka). Journal of Ichthyology 47:37–52.

Lanteigne, J. 1988. Identification of lamprey larvae of the genus *Ichthyomyzon* (Petromyzontidae). Environmental Biology of Fishes 23:55–63.

Lawrie, A. H. 1970. The sea lamprey in the Great Lakes. Transactions of the American Fisheries Society 99:766–775.

Linnaeus, C. 1758. Systema Naturae per regna tria naturae, secundum classes, ordines, genera, species, cum characteribus, differentiis, synonymis, locis. 10th edition. Volume 1. Laurentii Salvii, Stockholm, Sweden.

Mandrak, N. E., and E. J. Crossman. 1992. Postglacial dispersal of freshwater fishes into Ontario. Canadian Journal of Zoology 70:2247–2259.

Martens, E. von. 1868. Ueber einige ostasiatische Süsswasserthiere. Archiv für Naturgeschichte 34:1–64.

Mayr, E. 1942. Systematics and the origin of species, from the viewpoint of a zoologist. Harvard University Press, Cambridge, Massachusetts.

McAllister, D. E., S. P. Platania, F. W. Schueler, M. E. Baldwin and D. S. Lee. 1986. Ichthyofaunal patterns on a geographic grid. Pages 17–51 *in* C. H. Hocutt and E. O. Wiley, editors. The zoogeography of North American freshwater fishes. Wiley, New York.

McPhail, J. D. 2007. The freshwater fishes of British Columbia. The University of Alberta Press, Edmonton, Alberta, Canada.

McPhail, J. D., and C. C. Lindsey. 1970. Freshwater fishes of Northwestern Canada and Alaska. Fisheries Research Board of Canada Bulletin 173.

Mecklenburg, C. W., T. A. Mecklenburg, and L. K. Thorsteinson. 2002. Fishes of Alaska. American Fisheries Society, Bethesda, Maryland.

National Recovery Team for Morrison Creek Lamprey. 2007. Recovery strategy for the Morrison Creek lamprey (*Lampetra richardsoni* var. *marifuga*) in Canada. Species at Risk Act Recovery Strategy Series, Fisheries and Oceans Canada, Ottawa.

NatureServe. 2007. NatureServe Explorer: an online encyclopedia of life [Web application]. Version 6.2. NatureServe, Arlington, Virginia. Available: www.natureserve.org/explorer (July 2007).

Neave, F. B., N. E. Mandrak, M. F. Docker, and D. L. Noakes. 2007. An attempt to differentiate sympatric *Ichthyomyzon* ammocoetes using meristic, morphological, pigmentation, and gonad analyses. Canadian Journal of Zoology 85:549–560.

Nelson, J. S. 2006. Fishes of the world, 4th edition. Wiley, Hoboken, New Jersey.

Nelson, J. S., E. J. Crossman, H. Espinosa-Pérez, L. T. Findley, C. R. Gilbert, R. N. Lea, and J. D. Williams. 2004. Common and scientific names of fishes from the United States, Canada, and Mexico, 6th edition. American Fisheries Society, Special Publication 29, Bethesda, Maryland.

Nelson, J. S., and M. J. Paetz. 1992. The fishes of Alberta, 2nd edition. The University of Alberta Press, Edmonton, Alberta, Canada.

Page, L. M., and B. M. Burr. In press. A field guide to freshwater fishes of North America north of Mexico, 2nd edition. Houghton Mifflin Company, Boston.

Radforth, I. 1944. Some considerations on the distribution of fishes in Ontario. Contributions of the Royal Ontario Museum of Zoology 25:1–116.

Reighard, J., and H. Cummins. 1916. Description of a new species of lamprey of the genus *Ichthyomyzon*. Occasional Papers of the Museum of Zoology University of Michigan 31.

Renaud, C. B. 1997. Conservation status of Northern Hemisphere lampreys (Petromyzontidae). Journal of Applied Ichthyology 13:143–148.

Renaud, C. B., and N. de Ville. 2000. Three records of the chestnut lamprey, *Ichthyomyzon castaneus*, new to Québec. Canadian Field-Naturalist 114:333–335.

Renaud, C. B., S. C. Ribey, and F. Chapleau. 1996. Four records of the chestnut lamprey, *Ichthyomyzon castaneus*, new to Ontario. Canadian Field-Naturalist 110:450–453.

Richards, J. E., R. J. Beamish, and F. W. H. Beamish. 1982. Descriptions and keys for ammocoetes of lampreys from British Columbia, Canada. Canadian Journal of Fisheries and Aquatic Sciences 39:1484–1495.

Richardson, J. 1836. Fauna Boreali-Americana; or the zoology of the northern parts of British America: containing descriptions of the objects of natural history collected on the late northern land expeditions under command of Captain Sir John Franklin, R.N. Part 3. The fish. J. Murray, London.

SARA 2007 (Species at Risk Act) Public Registry. Available: www.sararegistry.gc.ca. (July 2007).

Sawatzky, C. D., D. Michalat, J. D. Reist, T. J. Carmichael, N. E. Mandrak, and L. G. Heuring. 2007. Distributions of freshwater and anadromous fishes from the mainland Northwest Territories, Canada. Canadian Manuscript Report of Fisheries and Aquatic Sciences 2793, Ottawa.

Schluter, D., and L. M. Nagel. 1995. Parallel speciation by natural selection. American Naturalist 146:292–301.

Schuldt, R. J., and R. Goold. 1980. Changes in the distribution of native lampreys in Lake Superior tributaries in response to sea lamprey (*Petromyzon marinus*) control, 1953–77. Canadian Journal of Fisheries and Aquatic Sciences 37:1872–1885.

Scott, W. B., and E. J. Crossman. 1973. Freshwater fishes of Canada. Fisheries Research Board of Canada Bulletin 184.

Scott, W. B., and M. G. Scott. 1988. Atlantic fishes of Canada. Canadian Bulletin of Fisheries and Aquatic Sciences No. 219:1–731.

Stewart, K. W., and D. A. Watkinson. 2004. The freshwater fishes of Manitoba. University of Manitoba Press, Winnipeg.

Taylor, E. B. 1999. Species pairs of north temperate freshwater fishes: evolution, taxonomy, and conservation. Reviews in Fish Biology and Fisheries 9:299–324.

Tilesius von Tilenau, W. G. 1811. Piscium camtschaticorum descriptions et icones. Mémoires de l'Académie Impériale des Sciences de St. Pétersbourg 3:225–285 + pls. 8–13.

Vancouver Lamprey Recovery Team. 2007. Recovery strategy for the Vancouver lamprey (*Lampetra macrostoma*) in Canada [Proposed]. Species at Risk Act Recovery Strategy Series, Fisheries and Oceans Canada, Ottawa.

Vladykov, V. D. 1950. Larvae of eastern American lampreys (Petromyzonidae). I. Species with two dorsal fins. Le Naturaliste Canadien 77:73–95.

Vladykov, V. D., and W. I. Follett. 1965. *Lampetra richardsoni*, a new nonparasitic species of lamprey (Petromyzonidae) from western North America. Journal of the Fisheries Research Board of Canada 22:139–158.

Vladykov, V. D., and W. I. Follett. 1967. The teeth of lampreys (Petromyzonidae): their terminology and use in a key to holarctic genera. Journal of the Fisheries Research Board of Canada 24:1067–1075.

Vladykov, V. D., and E. Kott. 1976. The taxonomic significance of velar tentacles in holarctic lampreys (Petromyzonidae). Revue des Travaux de l'Institut des Pêches Maritimes 40:787–789.

Vladykov, V. D., and E. Kott. 1978. A new nonparasitic species of the holarctic lamprey genus *Lethenteron* Creaser and Hubbs, 1922, (Petromyzontidae) from northwestern North America with notes on other species of the same genus. Biological Papers of the University of Alaska 19:1–74.

Vladykov, V. D. and E. Kott. 1979a. List of Northern Hemisphere lampreys (Petromyzonidae) and their distribution. Department of Fisheries and Oceans Miscellaneous Special Publication 42, Ottawa.

Vladykov, V. D., and E. Kott. 1979b. Satellite species among the holarctic lampreys (Petromyzonidae). Canadian Journal of Zoology 57:860–867.

Vladykov, V. D., and E. Kott. 1980. Description and key to metamorphosed specimens and ammocoetes of Petromyzonidae found in the Great Lakes region. Canadian Journal of Fisheries and Aquatic Sciences 37:1616–1625.

Wajgel, L. 1884. Die Zusammenziehung der Zwei Arten von *Petromyzon* (*P. Planeri* und *P. fluviatilis*) in Eine. Verhandlungen der kaiserlich-königlichen zoolgisch-botanischen Gesellschaft in Wien 33:311–319.

Waldman, J. R., C. Grunwald, N. K. Roy, and I. I. Wirgin. 2004. Mitochondrial DNA analysis indicates sea lampreys are indigenous to Lake Ontario. Transactions of the American Fisheries Society 133:950–960.

Wright, J., C. C. Krueger, P. F. Brussard, and M. Collings Hall. 1985. Sea lamprey (*Petromyzon marinus*) populations in northeastern North America: genetic differentiation and affinities. Canadian Journal of Fisheries and Aquatic Sciences 42:776–784.

Youson, J. H., and R. J. Beamish. 1991. Comparison of the internal morphology of adults of a population of lampreys that contains a nonparasitic life-history type, *Lampetra richardsoni*, and a potentially parasitic form, *L. richardsoni* var. *marifuga*. Canadian Journal of Zoology 69:628–637.

Zanandrea, G. 1959. Speciation among lampreys. Nature 184:380.

American Fisheries Society Symposium 72:311–321, 2009

# Critical Uncertainties and Research Needs for the Restoration and Conservation of Native Lampreys in North America

MATTHEW G. MESA* AND ELIZABETH S. COPELAND

*U.S. Geological Survey, Western Fisheries Research Center*
*Columbia River Research Laboratory*
*5501 Cook Underwood Road, Cook, Washington 98605, USA*

*Abstract*.—We briefly reviewed the literature, queried selected researchers, and drew upon our own experience to describe some critical uncertainties and research needs for the conservation and restoration of native lampreys in North America. We parsed the uncertainties and research needs into five general categories: (1) population status; (2) systematics; (3) passage at dams, screens, and other structures; (4) species identification in the field; and (5) general biology and ecology. For each topic, we describe why the subject is important for lampreys, briefly summarize our current state of knowledge, and discuss the key data or information gaps.

## Introduction

Lampreys are distinctive fishes that serve diverse functions in aquatic ecosystems. As members of the superclass *Agnatha*, they share with the hagfishes a special place in evolutionary history as the only living representatives of the primitive jawless fishes. In nature, lampreys are an important and nutritious food source for a variety of birds, fish, and mammals (Close et al. 2002). In some cases, they may serve as predation buffers for more commercially valuable species such as Pacific salmon (Close et al. 2002). In rivers and streams, anadromous lampreys are probably a significant source of marine-derived nutrients after spawning and dying (Wipfli et al. 1998; Fisher Wold and Hershey 1999). The ammocoetes, which are filter feeders and can remain buried in stream substrates for several years, may help maintain stream water quality by processing, storing, and cycling nutrients (Close et al. 2002). Tribal peoples in the Pacific Northwest have long pointed out the importance of lampreys to their cultural, spiritual, and subsistence living (Close et al. 2002). The evolutionary history of lampreys and their colloquial reference as "living fossils" has made them common subjects in studies of comparative anatomy and physiology.

Clearly, lampreys are important to the structure and function of some marine and freshwater ecosystems and have scientific and cultural value to humans. However, many people have a negative perception of lampreys because of the sea lamprey *Petromyzon marinus* problem in the Great Lakes, their allegedly parasitic nature, eel-like morphology, and their large, rasping, sucker-like mouth. As such, many would not consider these fish "charismatic" fauna and they do not garner the research or management attention they deserve, despite populations of most species being in decline. There are currently about 20 recognized species of lamprey in the United States, Canada, and Mexico, although the taxonomic status of many is unresolved (Table 1; Nelson et al. 2004). Currently, no lampreys are listed under the federal Endangered Species Act (ESA) in the United States, but more than half are listed as threatened, en-

* Corresponding author: mmesa@usgs.gov

Table 1. List of recognized lamprey species in North America, their distribution, feeding strategy (P = parasitic, NP = not parasitic), and migratory life history (A = anadromous, R = resident), current threats to population status, and state and provincial listings under various statutes. Distribution and threat information were collated from the NatureServe Web site (www.natureserve.org/explorer) unless otherwise noted. Species listings by government entities were compiled by examining state, provincial, and federal Web sites.

| Species | Distribution | Feeding strategy | Migratory life history | Threats | State, provincial, or federal listings |
|---|---|---|---|---|---|
| Ohio lamprey *Ichthyomyzon bdellium* | Ohio River basin, New York west to Indiana and Illinois, south to Alabama. | P | R | Dredging, siltation, and hydrological alteration. | Endangered: Ohio<br>Threatened: Pennsylvania |
| Chestnut lamprey *I. castaneus* | Great Lakes, St. Lawrence River and Mississippi River basins, Ontario, Quebec, Michigan, Wisconsin, and Minnesota south to Louisiana, Tennessee west to Kansas and Oklahoma; Manitoba; Saskatchewan, North Dakota; and Gulf Coast drainages. | P | R | Loss of spawning areas. | Threatened: Kansas, Iowa<br>Special concern: Saskatchewan |
| Northern brook lamprey *I. fossor* | Mississippi and Great Lakes drainages from New York, Quebec, Ontario, Michigan, and Wisconsin; north shore of Lake Superior to Indiana and Ohio. Disjunct population in Vermont in a tributary to Lake Champlain. | NP | R | Siltation, pollution of spawning areas; lampricides where range overlaps with invasive populations of sea lamprey. | Endangered: Illinois, Indiana, Ohio, Minnesota, Ontario, Vermont |
| Southern brook lamprey *I. gagei* | Gulf Slope drainages from Texas to Florida; Mississippi River basin of southern Missouri, Oklahoma, Arkansas, and Louisiana; Tennessee River drainage, Kentucky and Alabama; disjunct population in Wisconsin and Minnesota. | NP | R | | Special concern: Minnesota |
| Mountain brook lamprey *I. greeleyi* | Fragmented range in the Mississippi–Ohio basin:,New York, Pennsylvania, Ohio, Tennessee, Alabama, Kentucky, and Virginia. | NP | R | Siltation, agricultural runoff. | Endangered: Ohio<br>Threatened: Pennsylvania<br>Special concern: New York |

Table 1. Continued.

| Species | Distribution | Feeding strategy | Migratory life history | Threats | State, provincial or federal listings |
|---|---|---|---|---|---|
| Silver lamprey *I. unicuspis* | St. Lawrence-Great Lakes basins, south through upper Mississippi and Ohio river basins to Tennessee; recorded also from Nelson River (Manitoba), Missouri River (Nebraska), and Mississippi River (Mississippi). | P | R | Habitat deterioriation; lampricides where its range overlaps with the invasive sea lamprey. | Special concern: Vermont |
| Least brook lamprey *L.ampetra aepyptera* | Atlantic Slope drainages, Mississippi River basin from Pennsylvania to Missouri, Arkansas, Georgia, Alabama, and Mississippi. | NP | R | Siltation; acid-mine damage in Ohio. | Threatened: Illinois |
| American brook lamprey *L. appendix* | Atlantic Slope drainages, Great Lakes, and Mississippi River basins, Missouri and Arkansas; Martin River (Arctic basin), Northwest Territories, Alaska. | NP | R | Habitat loss, sedimentation, extreme water level changes. | Endangered: Connecticut, New Hampshire<br>Threatened: Iowa, Maryland, Maine, Vermont, Massachusetts, North Carolina<br>Special concern: Rhode Island |
| River lamprey *L. ayresii* | Alaska–California. | P | A | Dredging; dams. | Species of concern: Washington, Oregon<br>State candidate endangered: California |
| Arctic lamprey *L. camtschatica* | Arctic and Pacific drainages of Northwest Territories, Yukon, and Alaska (Circumpolar). | P | A/R | | |
| Mexican brook lamprey *L. geminis* | Endemic to highlands of central Mexico in the upper Rio Duero and Rio Grande de Morelia drainages (Lyons et al. 1994). | NP | R | Water diversions, increased turbidity, and total dissolved solids. | Endangered: Mexico |
| Kern brook lamprey *L. hubbsi* | Limited to San Joachin basin in California. | NP | R | Habitat loss and fragmentation due to dams and diversions. | State candidate endangered: California |

Table 1. Continued.

| Species | Distribution | Feeding strategy | Migratory life history | Threats | State, provincial, or federal listings |
|---|---|---|---|---|---|
| Pit-Klamath lamprey *L. lethophaga* | Pit River drainage of northeastern California and Klamath River. drainage in south-central Oregon | NP | R | Habitat loss and fragmentation due to dams and diversions. | |
| Vancouvery lamprey *L. macrostoma* | Cowichan and Mesachie lakes on southern Vancouver Island, British Columbia. | P | R | Narrow endemic population. | Threatened: British Columbia |
| Miller Lake lamprey *L. minima* | Formerly thought to be endemic to Miller Lake, last collected there in the early 1950s before the lake was treated with ichthyocides. Recently documented in Miller Creek, Jack Creek, and the upper sections of the Williamson and Sycan rivers, Oregon (Lorion et al. 2000). | P | R | Historical use of ichthyocides extirpated Miller Lake lamprey from Miller Lake. | |
| Western brook lamprey *L. richardsoni* | Southeastern Alaska–California. | NP | R | Dams, diversions, and habitat fragmentation. | Endangered: Canada (Morrison Creek lamprey only) |
| Klamath lamprey *L. similis* | Klamath River and Upper Klamath Lake, northern California and southern Oregon. | P | R | Dams, diversions, and pollution. | State candidate endangered: California |
| Mexican lamprey *L. spadicea* | Endemic to highlands of central Mexico (Lyons et al. 1994). | P | R | Water diversions, increased turbidity and total dissolved solids, and reduction of host species. | Endangered: Mexico |
| Pacific lamprey *L. tridentata* | Pacific coast from Alaska to the Rio Santo Domingo in Baja California. | P | A/R | Dams, diversions, and pollution. | Species of concern: Oregon, Washington |
| Sea lamprey *Petromyzon marinus* | Native from Newfoundland to Florida, Lake Ontario; invasive in the upper Great Lakes. | P | A/R | | |

dangered, or species of concern by individual states. In January 2003, 11 conservation groups petitioned the U.S. Fish and Wildlife Service (FWS) to list four species of West Coast lampreys (the Pacific lamprey, western brook lamprey, river lamprey, and Kern brook lamprey) under the ESA. However, the FWS denied the petition because more information was needed on their basic biology and ecology, distribution and abundance, habitat requirements, threats, and ammocoete identification to document the status of these populations. Similar information needs exist for other species of lampreys. The notion that populations of lampreys in North America face many threats is real and there is still much to be learned about their biology, ecology, and conservation (Jelks et al. 2008).

For this paper, we present a brief review of critical uncertainties and research needs for the restoration and conservation of lampreys in North America. We drew heavily from our own work and that of others in the western United States and queried researchers in other locations for their needs and concerns. After reviewing some key literature and collating responses from other researchers, we parsed the uncertainties and research needs into five categories: (1) population status; (2) systematics; (3) passage at dams, screens, and other diversions; (4) species identification in the field; and (5) general biology and ecology. For each of these topics, we describe why the subject is important for lampreys, briefly summarize our current state of knowledge, and discuss key information gaps. We hope our review will facilitate the efforts of researchers and managers trying to conserve, restore, and understand lampreys not only in North America, but throughout the world.

## Population Status

Little is known about the current abundance and distribution of native lampreys in key watersheds throughout North America. Without knowledge of the relative abundance of adult and juvenile (including both larvae or ammocoetes and macrophthalmia or recently metamorphosed individuals with large eyes) lampreys in a river basin, it is difficult to evaluate the importance or efficacy of management actions designed to conserve or restore lamprey populations. Perhaps the best example of the impact of our poor understanding of lamprey population status comes from the recent decision by the FWS to deny listing of four species of lampreys under the ESA. Sparse information on the abundance and distribution of these fish was a key reason why the FWS denied the petition. In the eastern United States, the impacts of lampricide treatments in streams of the Great Lakes on native lampreys are not well understood. For example, although native populations of lampreys in downstream areas of tributaries that receive lampricide treatment may be depleted, the persistence of lampreys higher up in the basin may replenish downstream populations (P. Cochran, St. Mary's University, personal communication). In contrast, sea lamprey control efforts in Vermont do kill native American brook lamprey and silver lampreys, but annual surveys suggest that populations of these fish are stable (B. Young, U.S. Fish and Wildlife Service, personal communication). These examples illustrate the need for detailed and recurring information on the population status of native lampreys in many areas.

To obtain information on the population status of lampreys requires routine, standardized surveys of fish in key watersheds, such as those recommended by Moser and Close (2003) for Pacific lampreys in the Columbia River basin. Such surveys could be quite effective if methods were developed to differentiate among species at all life stages in the field (discussed below). By implementing standardized procedures in the field, information collected within a region could be validly compared between different watersheds. More information could also be gained by reviewing historic databases, improving or developing new methods to monitor lamprey populations at dams and other structures, and recording lamprey catch data from other sampling efforts (e.g., smolt traps for monitoring juvenile salmonids). Without question, the listing of species and criteria for recovery of species under the ESA are in large part based on population status information. Thus, given the widespread concern for many lamprey populations in North America, assessing the abundance and distribution of lampreys will form the foundation for their conservation and restoration in the future.

## Systematics

Despite many recent studies (Docker et al. 1999; Lorion et al. 2000; Neave 2004; Neave et al. 2007), there are still some unresolved questions about species identification in the so-called paired species of lampreys (Zanandrea 1959). The problem with these paired species is that the ammocoetes are morphologically similar (e.g., Docker et al. 1999; Meeuwig et al. 2006) but the adults differ in life history type (e.g., parasitic forms that feed as adults and nonparasitic forms that do not). Further, in some of these groups, the presumed species cannot be separated genetically (Docker et al. 1999, 2005; Neave 2004). Thus, for some paired species of lampreys, the question is whether there are two distinct species or different life history types of the same species. Resolving the taxonomic status of paired lamprey species is important because conservation efforts rely on documenting species-specific declines in abundance and distribution. Lacking clear species delineations complicates conservation efforts by possibly over or under estimating current and historical abundance estimates and distributions. Docker (2009, this volume) reviews the concept of paired species in lampreys and Renaud et al. (2009, this volume) discuss this issue in detail for lampreys in Canada.

For example, in the Great Lakes region, the native northern brook lampreys and silver lamprey are of conservation concern because their ranges overlap that of the invasive sea lamprey and populations of the native lampreys appear to be declining from sea lamprey control measures. As we stated above, sometimes morphological and genetic analyses cannot differentiate between ammocoetes of recognized paired species (Docker et al. 1999; Lorion et al. 2000; Mandrak et al. 2004), and that seems to be the case with this group. While genetic markers can differentiate the ammocoetes of sea lampreys from northern brook lamprey and silver lamprey, Neave (2004) could not identify any genetic, morphological, or meristic characters that could unequivocally separate northern brook lamprey and silver lamprey ammocoetes. In contrast, Filcek et al. (2005) found that ammocoetes of these two species could be separated with high probability using multilocus microsatellite analyses. However, in this study, the fish were collected from different lakes in the region. When Docker et al. (2005) tested the same genetic markers on populations from Lake Huron, there was no significant genetic difference between the species throughout their range. Thus, perhaps northern brook lamprey and southern lamprey are not distinct species but rather the nonparasitic and parasitic forms of a single species.

In the west, molecular phylogenies produced by Docker et al. (1999) showed that most members of paired and satellite species—including European brook lampreys *Lampetra planeri* and European river lampreys *L. fluviatilis*, western brook lampreys, river lampreys, and Pacific lampreys and its presumed derivatives—were genetically indistinguishable. In addition, Docker et al. (1999) also concluded that the Kern brook lamprey was more closely related to river lamprey and western brook lamprey than to any other *Entosphenus* species and should be placed in the subgenus *Lampetra*. These examples illustrate some of the confusion surrounding taxonomic assignments in lampreys and suggest that many lamprey species designations should be re-evaluated (Docker 2006), especially as genetic markers become more refined. Microsatellite markers and other genetic tools should be developed to investigate the quantity of gene flow among putative lamprey species, especially in areas with sympatric populations like the Klamath basin or the Great Lakes.

In addition to resolving issues surrounding the paired species concept in lampreys and lamprey taxonomy in general, there is a need to better define the population structure of Pacific lampreys, particularly in the western United States. Currently, there is contrasting information about whether Pacific lampreys in this region show a distinct genetic structure. On the one hand, Goodman et al. (2008) analyzed mitochondrial DNA variation in two gene fragments among individuals from British Columbia to central California and concluded that there was a high level of gene flow among populations and little evidence of genetic differentiation. This is similar to findings on native sea lampreys, which show a panmictic genetic structure on both sides of the Atlantic Ocean (Bryan et al. 2005; Almada et al. 2007). Goodman et al. (2008) suggest that the lack of mtDNA differentiation among widely distributed populations of Pacific lampreys could result from a variable oceanic dispersal and lack of natal homing. On the other hand, Lin et al. (2008) used an analysis of amplified fragment length polymorphisms of genomic DNA to assess the genetic structure of Pacific lampreys throughout their range. Although their analysis indicated significant levels of historical gene flow, they did discern genetic differences in populations of fish over large geographic regions (i.e., Japan, Alaska, and the Pacific Northwest) and among Pacific Northwest samples. The contrasting results of Lin et al. (2008) and Goodman et al. (2008) could be due in part to the precise mechanism facilitating gene flow among anadromous lamprey populations, and perhaps other things. For example, as we mentioned earlier, Goodman et al. (2008) suggest that passive ocean dispersal would facilitate genetic mixing of populations, especially when coupled with little or no homing behavior. In contrast, Lin et al. (2008) propose that lampreys from different river systems might interbreed on a relatively local scale but that interbreeding might not occur across their range. Details on the oceanic life phase of anadromous lampreys, including prey selection, migratory habits, and distribution and duration, should receive further study. Resolution of life history discrepancies and a thorough assessment of the genetic structure of Pacific lampreys are important for their conservation and management in the Pacific Northwest.

Finally, another important aspect of genetic structure of anadromous lampreys is the possibility of multiple runs or races of fish, similar to spring and fall Chinook salmon *Oncorhynchus tshawytscha*. Ongoing studies of Pacific lamprey migrations in the Willamette River, Oregon, examining variation in morphology and physiology, may provide more evidence of distinct runs (B. Clemens, Oregon State University, personal communication). Moyle et al. (2009, this volume) suggest that multiple runs of lampreys may occur in some California streams.

## Passage at Dams, Culverts, Screens, and Other Structures

The paper by Moser and Mesa (2009, this volume) outlines the importance of, issues with, and research needs for passage of lampreys at dams and other structures. However, their paper focuses on issues with adult lampreys and large hydropower facilities because more information is available on this topic. Passage at dams, culverts, screens, and other structures is a serious limiting factor for lampreys because most systems were not designed to facilitate the passage of lampreys. These structures can lead to fragmented habitat and may prevent lampreys from spawning in areas of historical distribution. One area in dire need of more research is passage problems with juvenile lampreys (ammocoetes and macrophthalmia), which could be a serious bottleneck to the maintenance of lamprey populations.

The performance of juvenile lampreys, particularly anadromous species, at passage obstacles is an area of research where almost nothing is known. This is unfortunate because any benefits gained by facilitating adult lamprey passage at dams and other structures could be negated by large losses of juveniles as they encounter the same or different structures during their downstream migration. Juvenile lampreys are poor swimmers (Dauble et al. 2006), may migrate passively downstream, and could be especially vulnerable to entrainment and impingement at dams and associated structures. Indeed, Moursund et al. (2003) showed that Pacific lamprey macrophthalmia were routinely impinged on diversion screens located in front of turbine intakes on Columba River dams. The magnitude, severity, and consequences of these impingement events have yet to be evaluated seriously. Juvenile lampreys may be especially vulnerable to irrigation diversions, many of which are unscreened or poorly screened (Rose et al. 2008). We are unaware of any studies quantifying entrainment levels of juvenile lampreys at water diversions, even though the effect could be tremendous given the thousands of diversions in North America. Even if water diversions were properly screened, the criteria for safe passage of fish at screens have been developed for teleosts only, particularly salmonids. We do not know the effects of encountering a vertical drum screen or passing over a horizontal flat-plate screen on juvenile lampreys. In contrast to the well-studied juvenile salmonids, juvenile lampreys do not lend themselves well to movement and passage studies using tools such as telemetry or passive integrated transponder (PIT) tagging because of their small size, fragile nature, and difficulty in catching enough specimens for statistical and scientific rigor (Schreck et al. 1999; Mueller et al. 2006; Moser et al. 2007). Nevertheless, new techniques must be developed and implemented to gain more understanding on a topic that has been largely overlooked and could present serious challenges to the conservation and preservation of lampreys in North America.

## Species Identification in the Field

A major impediment to understanding lampreys is the difficulty of identifying species at all life stages, particularly ammocoetes. Species identification in lampreys is largely based on adult dentition (even for nonparasitic species), physical proportions, size at spawning, and pigmentation (Hubbs and Potter 1971; Richards et al. 1982). The protracted ammocoete life stage of lampreys complicates species identification because ammocoetes lack dentition, different species often look the same, and often multiple age- and size-classes are present. For example, the Pacific lamprey, river lamprey, and western brook lamprey all occur in the Columbia River basin. In a status report of Oregon lampreys, Kostow (2002) noted that state agencies were unable to identify ammocoetes within the basin and migrating adult Pacific lamprey were the only adults regularly observed. In fact, river lamprey has not been documented in the Columbia River basin or anywhere in Oregon since 1980, but whether the species is now absent from the area or simply unobserved or misidentified is unknown (Kostow 2002). Richards et al. (1982) developed a descriptive key for identifying juveniles of these three species in Canada based on subtle differences in pigmentation. However, the key had limited use because ammocoetes lack diagnostic pigmentation patterns in their first year (age 0–1) and geographic variability in pigmentation is suspected within and among species in the Columbia River basin (Meeuwig et al. 2006). Recently, however, Goodman et al. (2009, this volume) used pigmentation of the caudal fin and caudal ridge to provide valid genus assignments of lamprey ammocoetes within *Entosphenus* and *Lampetra*. Using meristics, Meeuwig et al. (2006) were able to reliably differentiate young Pacific lampreys and western brook lampreys but admit that the accuracy of counts required was not suitable for use in the field. The difficulty in positively identifying ammocoetes in the field makes documenting the distribution and abundance of lamprey species difficult, especially in areas where species are sympatric. Ammocoete identification is important because they are the most frequently observed life stage in streams and could provide information on habitat preference, abundance, distribution, and recruitment. The robustness of published taxonomic keys for ammocoetes should be validated with individuals varying in size and from different geographic areas. Because rapid and precise techniques for identifying ammocoetes in the field may not be possible, specimens should be collected for further analysis, genetic or otherwise, when encountered.

## General Biology and Ecology

In our queries to other researchers, there was a common call for more research on the general biology and ecology of lampreys. Common themes included the timing and location of spawning, reproductive success, basic habitat requirements, parasite–host relations, growth, and assessments of regional biological diversity. In particular, study of parasite–host relations that have evolved for long periods would be ecologically interesting and, when combined with techniques such as PIT tagging, could reveal individual growth rates, movements, and host choice. Throughout North America, there is a need to examine taxonomic and life history diversity among populations of currently recognized species with widespread or disjunct distributions. Other needs included a greater understanding of some life history variants such as the neotenic least brook lamprey and the giant American brook lamprey, including information on their origins, early development, growth, and feeding. For native parasitic *Ichthyomyzon* in the Midwest, there is a need to understand the size of watershed necessary to support viable populations. This would include knowledge of host assemblages and densities and may be critical in areas receiving lampricide treatments where native lamprey populations are being impacted.

In the western United States, and particularly the Columbia River basin, how far advanced would our knowledge of Pacific lamprey movements, distribution, homing, and growth be if a large-scale program of PIT-tagging macrophthalmia had been implemented a decade ago? Such a program is now the foundation of anadromous salmonid management in the Pacific Northwest and has yielded a wealth of useful, unprecedented information. We encourage the development of new and innovative techniques for tagging macrophthalmia that would allow researchers to embark on studies of movements, homing, growth, and survival. Finally, a clearer understanding of the ecological role of lampreys in aquatic ecosystems is needed to help justify efforts aimed at conservation and protection. Topics of particular interest include the role of ammocoetes as filter feeders, any effects they may have on water quality in streams, the importance of juvenile lampreys as prey for a variety of animals, the nutrient dynamics of semelparous lampreys in streams and watersheds, and the importance of lampreys in predator–prey interactions. Research on these topics would not only increase our scientific understanding of lamprey ecology and life history, but would also inform the general public about the importance of lampreys to aquatic ecosystems. Highlighting the importance of lampreys in aquatic ecosystems could help encourage public support for lamprey preservation and recovery, especially when the condition of such ecosystems is important for more commercially or socially relevant species.

## Conclusion

Lampreys are fascinating animals and iconic when it comes to the study of evolutionary biology, comparative anatomy, and physiology. Despite an impressive amount of study on lamprey biology over the years, there is still much to be learned about these fish from both basic and applied perspectives. Because so many populations of lampreys in North America seem to be declining and are of special concern, the time is right to embark on research designed to answer questions relevant to their conservation and preservation. Because federal listing of some species under the ESA seems inevitable, we encourage researchers and managers to start collecting information on population status, systematics, limiting factors, and basic biology and ecology. Educating the general public and state, Federal, and other administrators about the links between the health of lamprey populations, watersheds, and other more commercially important species would facilitate the release of more funding directed towards lamprey research and a greater understanding of these special fishes.

## Acknowledgments

We thank John Lyons, Phil Cochran, Bradley Young, and Matt White for sharing their insights on lamprey research needs. Peter Moyle and an anonymous reviewer provided useful editorial comments and improved the manuscript.

## References

Almada, V. C., J. I. Pereira, J. P. Fonseca, A. Levy, C. Maia, and A. Valente. 2007. Mitochondrial DNA fails to reveal genetic structure in sea-lampreys along European shores. Molecular Phylogenetics and Evolution 46:391–396.

Bryan, M. B., D. Zalinski, K. B. Fileck, S. Libants, W. Li, and K. T. Scribner. 2005. Patterns of invasion and colonization of the sea lamprey (*Petromyzon marinus*) in North America as revealed by microsatellite genotypes. Molecular Ecology 14:3757–3773.

Close, D. A., M. S. Fitzpatrick, and H. W. Li. 2002. The ecological and cultural importance of a species at risk of extinction, Pacific lamprey. Fisheries 27:19–25.

Dauble, D. D., R. A. Moursund, and M. D. Bleich. 2006. Swimming behavior of juvenile Pacific lamprey, *Lampetra tridentata*. Environmental Biology of Fishes 75:167–171.

Docker, M. F. 2006. Bill Beamish's contributions to lamprey research and recent advances in the field. Guelph Ichthyology Reviews 7.

Docker, M. F. 2009. A review of the evolution of nonparasitism in lampreys and an update of the paired species concept. Pages 71–114 *in* L. R. Brown, S. D. Chase, M. G. Mesa, R. J. Beamish, and P. B. Moyle, editors. Biology, management, and conservation of lampreys in North America. American Fisheries Society, Symposium 72, Bethesda, Maryland.

Docker, M. F., N. E. Mandrak, D. D. Heath, and K. T. Scribner. 2005. Genetic markers to distinguish and quantify the level of gene flow between northern brook and silver lampreys. Great Lakes Fishery Commission Project Completion Report, Ann Arbor, Michigan.

Docker, M. F., J. H. Youson, R. J. Beamish, and R. H. Devlin. 1999. Phylogeny of the lamprey genus *Lampetra* inferred from mitochondrial cytochrome *b* and ND3 gene sequences. Canadian Journal of Fisheries and Aquatic Sciences 56:2340–2349.

Filcek, K. B., S. A. Gilmore, K. T. Scribner, and M. L. Jones. 2005. Discriminating lamprey species using multilocus microsatellite genotypes. North American Journal of Fisheries Management 25:502–509.

Fisher Wold, K. A., and A. E. Hershey. 1999. Effects of salmon carcass decomposition on biofilm growth and wood decomposition. Canadian Journal of Fisheries and Aquatic Sciences 56:767–773.

Goodman, D. H., A. P Kinziger, S. B. Reid, and M. F. Docker. 2009. Morphological diagnosis of *Entosphenus* and *Lampetra* ammocoetes (Petromyzontidae) in Washington, Oregon, and California. Pages 223–232 *in* L. R. Brown, S. D. Chase, M. G. Mesa, R. J. Beamish, and P. B. Moyle, editors. Biology, management, and conservation of lampreys in North America. American Fisheries Society, Symposium 72, Bethesda, Maryland.

Goodman, D. H., S. B. Reid, M. F. Docker, G. R. Haas, and A. P. Kinziger. 2008. Mitochondrial DNA evidence for high levels of gene flow among populations of a widely distributed anadromous lamprey *Entosphenus tridentatus* (Petromyzontidae). Journal of Fish Biology 72:400–417.

Hubbs, C. L., and I. C. Potter. 1971. Distribution, phylogeny and taxonomy. Pages 1–65 *in* M. W. Hardisty and I. C. Potter, editors. The biology of lampreys, volume 1. Academic Press, London.

Jelks, H. L., S. J. Walsh, N. M. Burkhead, S. Contreras-Balderas, E. Díaz-Pardo, D. A. Hendrickson, J. Lyons, N. E. Mandrak, F. McCormick, J. S. Nelson, S. P. Platania, B. A. Porter, C. B. Renaud, J. J. Schmitter-Soto, E. B. Taylor, and M. L. Warren, Jr. 2008. Conservation status of imperiled North American freshwater and diadromous fishes. Fisheries 33(8):372–407.

Kostow, K. 2002. Oregon lampreys: natural history, status, and analysis of management issues. Oregon Department of Fish and Wildlife, Fish Division, Portland.

Lin, B., Z. Zhang, Y. Wang, K. P. Currens, A. Spidle, Y. Yamazaki, and D. A. Close. 2008. Amplified fragment length polymorphism assessment of genetic diversity in Pacific lampreys. North

America Journal of Fisheries Management 28:1182–1193.

Lorion, C. M., D. F. Markle, S. B. Reid, and M. F. Docker. 2000. Redescription of the presumed-extinct Miller lake lamprey, *Lampetra minima*. Copeia 2000(4):1019–1028.

Lyons, J., P. A. Cochran, O. J. Polaco, and E. Merino-Nambo. 1994. Distribution and abundance of the Mexican lampreys (Petromyzontidae: *Lampetra*: subgenus *Tetrapleurodon*). Southwestern Naturalist 39:105–113.

Mandrak, N. E., M. F. Docker, and D. D. Heath. 2004. Native Ichthyomyzon lampreys of the Great Lakes: development of genetic markers and a morphological key to ammocoetes. Great Lakes Fishery Commission Project Completion Report, Ann Arbor, Michigan.

Meeuwig, M. H., J. M. Bayer, and R. A. Reiche. 2006. Morphometric discrimination of early life stage *Lampetra tridentata* and *L. richardsoni* (Petromyzonidae) from the Columbia River basin. Journal of Morphology 267:623–633.

Moser, M. L., J. M. Butzerin, and D. B. Dey. 2007. Capture and collection of lampreys: the state of science. Reviews in Fish Biology and Fisheries 17:45–56.

Moser, M. L., and D. A. Close. 2003. Assessing Pacific lamprey status in the Columbia River basin. Northwest Science 77:116–125.

Moser, M. L., and M. G. Mesa. 2009. Passage considerations for anadromous lampreys. Pages 115–124 *in* L. R. Brown, S. D. Chase, M. G. Mesa, R. J. Beamish, and P. B. Moyle, editors. Biology, management, and conservation of lampreys in North America. American Fisheries Society, Symposium 72, Bethesda, Maryland.

Moursund, R. A., D. D. Dauble, and M. Langeslay. 2003. Turbine intake diversion screens: investigating effects on Pacific lamprey. Hydro Review 22(1):40–46.

Moyle, P. S., L. R. Brown, S. D. Chase, and R. M. Quiñones. 2009. Status and conservation of lampreys in California. Pages 279–292 *in* L. R. Brown, S. D. Chase, M. G. Mesa, R. J. Beamish, and P. B. Moyle, editors. Biology, management, and conservation of lampreys in North America. American Fisheries Society, Symposium 72, Bethesda, Maryland.

Mueller, R. P., R. A. Moursund, and M. D. Bleich. 2006. Tagging juvenile Pacific lamprey with passive integrated transponders: methodology, short-term mortality, and influence on swimming performance. North American Journal of Fisheries Management 26:361–366.

Neave, F. B. 2004. The utility of meristic, morphometric, pigmentation, and gonad analysis in the identification of *Ichthyomyzon* lamprey larvae. Master's thesis. University of Guelph, Guelph, Ontario, Canada.

Neave, F. B., N. E. Mandrak, M. F. Docker, and D. L. Noakes. 2007. An attempt to differentiate sympatric *Ichthyomyzon* ammocoetes using meristic, morphological, pigmentation, and gonad analyses. Canadian Journal of Zoology 85:549–560.

Nelson, J. S., E. J. Crossman, H. Espinosa-Perez, L. T. Findley, C. R. Gilbert, R. N. Lea, and J. D. Williams. 2004. Common and scientific names of fishes from the United States, Canada, and Mexico, 6th edition. American Fisheries Society, Special Publication 29, Bethesda, Maryland.

Renaud, C. B., M. F. Docker, and N. E. Mandrak. 2009. Taxonomy, distribution, and conservation of lampreys in Canada. Pages 293–309 *in* L. R. Brown, S. D. Chase, M. G. Mesa, R. J. Beamish, and P. B. Moyle, editors. Biology, management, and conservation of lampreys in North America. American Fisheries Society, Symposium 72, Bethesda, Maryland.

Richards, J. E., R. J. Beamish, and F. W. H. Beamish. 1982. Descriptions and keys for ammocoetes of lampreys from British Columbia, Canada. Canadian Journal of Fisheries and Aquatic Sciences 39:1484–1495.

Rose, B. P., M. G. Mesa, and G. Barbin-Zydlewski. 2008. Field-based evaluations of horizontal flat-plate fish screens. North American Journal of Fisheries Management 28:1702–1713.

Schreck, C. B., M. S. Fitzpatrick, and D. L. Lerner. 1999. Determination of passage of juvenile lamprey: development of a tagging protocol. Oregon State University, Oregon Cooperative Fish and Wildlife Research Unit, Corvalis.

Wipfli, M. S., J. Hudson, and J. Caouette. 1998. Influence of salmon carcasses on stream productivity: response of biofilm and benthic macroinvertebrates in southeastern Alaska, U.S.A. Canadian Journal of Fisheries and Aquatic Sciences 55:1503–1511.

Zanandrea, G. 1959. Speciation among lampreys. Nature (London) 184.